Lecture Notes in Computer Science 3519

Commenced Publication in 1973
Founding and Former Series Editors:
Gerhard Goos, Juris Hartmanis, and Jan van Leeuwen

T0223579

Hongbo Li   Peter J. Olver
Gerald Sommer (Eds.)

Computer Algebra and Geometric Algebra with Applications

6th International Workshop, IWMM 2004
Shanghai, China, May 19-21, 2004
and International Workshop, GIAE 2004
Xian, China, May 24-28, 2004
Revised Selected Papers

 Springer

Volume Editors

Hongbo Li
Academy of Mathematics and Systems Science
Chinese Academy of Sciences, Beijing 100080, China
E-mail: hli@mmrc.iss.ac.cn

Peter J. Olver
University of Minnesota
School of Mathematics, Minneapolis, MN 55455, USA
E-mail: olver@math.umn.edu

Gerald Sommer
Christian-Albrechts-Universität zu Kiel
Institut für Informatik, Olshausenstr.40, 24098 Kiel, Germany
E-mail: gs@ks.informatik.uni-kiel.de

Library of Congress Control Number: 2005927771

CR Subject Classification (1998): F.2.1-2, G.1, I.3.5, I.4, I.2, I.1

ISSN   0302-9743
ISBN-10   3-540-26296-2 Springer Berlin Heidelberg New York
ISBN-13   978-3-540-26296-1 Springer Berlin Heidelberg New York

Springer is a part of Springer Science+Business Media

springeronline.com

© Springer-Verlag Berlin Heidelberg 2005
Printed in Germany

Typesetting: Camera-ready by author, data conversion by Scientific Publishing Services, Chennai, India
Printed on acid-free paper   SPIN: 11499251   06/3142   5 4 3 2 1 0

Preface

Mathematics Mechanization consists of theory, software and application of computerized mathematical activities such as computing, reasoning and discovering. Its unique feature can be succinctly described as AAA (Algebraization, Algorithmization, Application). The name *"Mathematics Mechanization"* has its origin in the work of Hao Wang (1960s), one of the pioneers in using computers to do research in mathematics, particularly in automated theorem proving. Since the 1970s, this research direction has been actively pursued and extensively developed by Prof. Wen-tsun Wu and his followers. It differs from the closely related disciplines like Computer Mathematics, Symbolic Computation and Automated Reasoning in that its goal is to make algorithmic studies and applications of mathematics the major trend of mathematics development in the information age.

The International Workshop on Mathematics Mechanization (IWMM) was initiated by Prof. Wu in 1992, and has ever since been held by the Key Laboratory of Mathematics Mechanization (KLMM) of the Chinese Academy of Sciences. There have been seven workshops of the series up to now. At each workshop, several experts are invited to deliver plenary lectures on cutting-edge methods and algorithms of the selected theme. The workshop is also a forum for people working on related subjects to meet, collaborate and exchange ideas.

There were two major themes for the IWMM workshop in 2004. The first was "Constructive and Invariant Methods in Algebraic and Differential Equations," or, in short, "Computer Algebra with Applications." The second was "Geometric Invariance and Applications in Engineering"(GIAE), or, in short, "Geometric Algebra with Applications." The two themes are closely related to each other. On the one hand, essentially due to the efforts of D. Hestenes and his followers, recent years have witnessed a dramatic resurgence of the venerable subject Geometric Algebra (which dates back to the 1870s), with dramatic new content and applications, ranging from mathematics and physics, to geometric reasoning, neural networks, robotics, computer vision and graphics. On the other hand, the rise of computer algebra systems and algorithms has brought previously infeasible computations, in particular those in geometric algebra and geometric invariance, within our grasp. As a result, the two intertwined subjects hold a particular fascination, not only for students and practitioners, but also for mathematicians, physicists and computer scientists working on effective geometric computing.

Since it is very difficult to put the two major themes into a single conference without parallel sessions, the organizers decided to split this year's *IWMM* workshop into two conferences, one for each theme. The first workshop was held in a beautiful quiet riverside town near Shanghai, called ZhuJiaJiao, from May 19 to 21. The second workshop was held in a glorious conference hall of the Xi'an

Hotel, Xi'an, from May 24 to 28. Altogether 169 scholars from China, USA, UK, Germany, Italy, Japan, Spain, Canada, Mexico and Singapore were attracted to the conferences and presented 65 talks. The following invited speakers presented the plenary talks:

Wen-tsun Wu (China)	Gerald Sommer (Germany)
Peter J. Olver (USA)	Alyn Rockwood (USA)
Anthony Lasenby (UK)	Joan Lasenby (UK)
Quan Long (Hong Kong, China)	Jingzhong Zhang (China)
Neil White (USA)	Timothy Havel (USA)
Jose Cano (Spain)	Greg Reid (Canada)
Andrea Brini (Italy)	Dongming Wang (China & France)
Xiaoshan Gao (China)	William Chen (China)
Hongqing Zhang (China)	Ke Wu (China)

The two conferences were very successful, and the participants agreed on the desirability of publication of the postproceedings by a prestigious international publishing house, these proceedings to include the selected papers of original and unpublished content. This is the background to the current volume.

Each paper included in the volume was strictly refereed. The authors and editors thank all the anonymous referees for their hard work. The copyediting of the electronic manuscript was done by Ms. Ronghua Xu of KLMM. The editors express their sincere appreciation for her dedication.

It should be emphasized that this is the first volume to feature the combination and interaction of the two closely related themes of Computer Algebra and Geometric Algebra. It is the belief of the editors that the volume will prove to be valuable for those interested in understanding the state of the art and for further combining and developing these two powerful tools in geometric computing and mathematics mechanization.

Beijing, Minneapolis, Kiel

March 2005

Hongbo Li

Peter J. Olver

Gerald Sommer

Table of Contents

Part I : Computer Algebra and Applications

Part II: Geometric Algebra and Applications

On Wintner's Conjecture About Central Configurations

Wen-tsun Wu

Mathematics Mechanization Key Laboratory,
Academy of Mathematics and Systems Science,
Chinese Academy of Sciences,
Beijing 100080, P. R. China

Abstract. According to Wintner, the study of central configurations in celestial mechanics may be reduced to an extremality problem. Wintner's Conjecture amounts to saying that the corresponding extremal zeroes for each fixed number n of different masses is finite. By the author's Finite Kernel Theorem it follows that the corresponding number of extremal values is finite for each fixed n. Thus, Wintner's Conjecture will be true or false according to whether there will be only a finite number of extremal zeroes or not. This gives thus a new way of attacking Wintner's Conjecture.

Keywords: Celestial Mechanics, Central Configurations, Wintner's Conjecture, Extremal Values, Extremal Zeroes, Finite Kernel Theorem.

The determination of central configurations in astromechanics is a fascinating problem since the time of Euler and Lagrange. In fact, Euler had shown that for three arbitrarily given distinct masses, there exist exactly three distinct collinear central configurations, and Lagrange had shown that the equilateral triangle and only the equilateral triangle is a non-collinear central configuration for three arbitrary masses. In the last century Moulton had proved that for arbitrary n distinct masses the number of collinear central configurations is $n!$ or $n!/2$, cf. [1]. In more recent time, A. Wintner had shown that, for four masses the regular tetrahedron and only the regular tetrahedron is a non-flat central configuration. Wintner had even announced a conjecture to the effect that for arbitrarily fixed number n of masses, the number of possible central configurations is always **finite**. This conjecture had resisted all attacks by quite a number of specialists, including e.g. Smale and his followers by some topological method created *ad hoc*, cf. [5, 6] and [2].

The actual determination of central configurations for a fixed n is not simple. Thus, in the literature there are only sporadic results of not great significance. In the nineties of the last century the present author had formulated a method of the determination of central configurations in reducing it to the solving of a system of polynomial equations under restrictions also in the form of polynomial equations. It is applied to give an alternative proof of theorems of Euler and Lagrange that the central configurations found by them are the only possible

H. Li, P. J. Olver and G. Sommer (Eds.): IWMM-GIAE 2004, LNCS 3519, pp. 1–4, 2005.

ones for $n = 3$, cf. [9]. In recent years H. Shi and others had applied this method to find various central configurations of special types for $n \geq 4$, cf. their papers, [3, 4, 11].

What is of great significance is this: Wintner in his classic on celestial mechanics, viz. [8], had shown that the central configurations are in correspondence with the extremal zeroes of some extremalization problem of rational function type and hence also of polynomial type. In applying the *Finite Kernel Theorem*, (cf. [10], Chap. 5, §5) on such extremality problems, we know that the extremal values of such problems are necessarily *finite* in number. If it can be shown that for each extremal value there can only be associated a *finite* number of extremal zeroes, then the total number of extremal zeroes of the problem will be *finite*, which is just the Wintner Conjecture in question. In any way the above gives an alternative method to attack the interesting and difficult conjecture of Wintner.

To begin with, let us first recall the definition of *Central Configuration*. Thus, let t be the time which is supposed to be fixed, and $m_i, (i = 1, \cdots, n)$ be n masses in question. Let us take a barycentric coordinate system with the center of mass of the system $\{m_1, \cdots, m_n\}$ at the origin (see §322 of [8], similar for references below). With respect to such a barycentric coordinate system let $\xi_i = (x_i, y_i, z_i)$ be the barycentric position of $m_i, (i = 1, \cdots, n)$. Set

$$\rho_{jk} = |\xi_j - \xi_k| = [(x_j - x_k)^2 + (y_j - y_k)^2 + (z_j - z_k)^2]^{\frac{1}{2}}. \tag{1}$$

Then the potential energy of the system is given by

$$U = \frac{\Sigma_{1 \leq j < k \leq n} m_j m_k}{\rho_{jk}}. \tag{2}$$

The Newtonian force acting on the mass m_i is then given by

$$U_{\xi_i} = (U_{x_i}, U_{y_i}, U_{z_i}), \tag{3}$$

in which $U_{x_i} = \frac{\partial U}{\partial x_i}$, etc.

By Wintner's definition, the system $\{m_i, \xi_i | i = 1, \cdots, n\}$ is said to form a **Central Configuration** if the force of gravitation acting on each m_i, ξ_i is proportional to both the mass m_i and the barycentric position vector ξ_i, i.e.,

$$U_{\xi_i} = \sigma m_i \xi_i, \quad \text{for} \quad i = 1, \cdots, n, \tag{4}$$

where σ is some scalar independent of i.

It is proved by Wintner that

$$\sigma = -\frac{U}{J}, \tag{5}$$

in which (§322 bis)

$$J = \frac{1}{\mu} \Sigma_{1 \leq j < k \leq n} m_j m_k \rho_{jk}^2, \qquad \mu = \Sigma_{i=1,\cdots,n} m_i. \tag{6}$$

Wintner proved further (§355) that (4), (5), (6), are equivalent to the following equation:

$$JU_{\xi_i} = -\frac{1}{2}UJ_{\xi_i}, \quad i.e. \ (JU^2)_{\xi_i} = 0, \quad \text{for} \ i = 1, \cdots, n. \tag{7}$$

He further proved (§355, 357) that the system is a central configuration if and only if

$$(JU^2)_{\rho_{ik}} + \Sigma_s \chi_s (R_s)_{\rho_{ik}} = 0, \quad \text{for} \ 1 \le i < k \le n, \tag{8}$$

in which the χ_s's are the Lagrangian multipliers and

$$R_s(\rho_{12}, \cdots, \rho_{n-1,n}) = 0, \quad \text{for} \ s = 1, \cdots, p \tag{9}$$

are the necessary geometrical relations among the distances ρ_{ik} for $n \ge 4$ in planar case and $n \ge 5$ in spatial case.

In other words, *the system $\{m_i, \xi\}$ forms a central configuration if and only if*

$$JU^2 = Extremum \tag{10}$$

under the restrictions (9).

The Wintner's conjecture is thus reduced to the *extremum* problem (10) under the restricted conditions (9) which are equations in rational functions of various variables involved. We hope that our Finite Kernel Theorem of the present author concerning extremum problems of such type may be useful to arrive at a final proof of **Wintner Conjecture**, as indicated in the beginning of the present paper.

References

1. F. R. Moulton, The Straight Line Solutions for the Problem of n-Bodies, *Annals of Math.*, 12 (1910) 1–17.
2. J. I. Palmore, Measure of Degenerate Relative Equilibria, I, *Annals of Math.*, 104 (1976) 421–429.
3. H. Shi, F. Zou, The Flat Central Configurations of Four Planet Motions, *MM Research Preprints*, No. 15, (1997) 113–121.
4. H. Shi, F. Zou, Square and Rhombus Central Configurations, *System Science and Mathematical Sciences*, 2000, v.13, No.1, 74–84.
5. S. Smale, Topology and Mechanics, I,II. *Invent. Math*, 10 (1970) 305–331; 11 (1970) 45–64.
6. S. Smale, Problems on the Nature of Relative Equilibria in Celestial Mechanics, in Manifolds-Amsterdam 1970, 194–198.
7. J. Waldvogel, Note Concerning a Conjecture by A. Wintner, *Celestial Mechanics*, 5 (1972) 37–40.
8. A. Wintner, *Analytic Foundations of Celestial Mechanics*, 1941.
9. Wen-tsun Wu, Central Configurations in Planet Motions and Vortex Motions, *MM Research Preprints*, No.13, (1995) 1–14.

10. Wen-tsun Wu, Mathematics Mechanization: Mechanical Geometry Theorem-Proving, Mechanical Geometry Problem-Solving and Polynomial Equations-Solving, Science Press/Kluwer Academic Publishers, (2000).
11. Y. Wu & H. Shi, A Special Central Configuration, *MM Research Preprints*, No. 20, (2001) 221–228.

Polynomial General Solutions for First Order Autonomous ODEs*

Ruyong Feng and Xiao-Shan Gao

Key Laboratory of Mathematics Mechanization,
Institute of Systems Science,AMSS, Academia Sinica,
Beijing 100080, P. R. China

Abstract. For a first order autonomous ODE, we give a polynomial time algorithm to decide whether it has a polynomial general solution and to compute one if it exists. Experiments show that this algorithm is quite effective in solving ODEs with high degrees and a large number of terms.

1   Introduction

To find elementary function solutions for differential equations could be traced back to the work of Liouville. As a consequence, such solutions of differential equations are called *Liouvillian solutions*. In [16], Risch gave an algorithm for finding Liouvillian solutions for the simplest differential equation $y' = f(x)$, that is, to find elementary function solutions to the integration $\int f(x)\mathrm{d}x$. Kovacic presented a method for solving second order linear homogeneous differential equations [13]. Singer established the general framework to find Liouvillian solutions of general homogeneous linear differential equations [18]. Many interesting results on finding Liouvillian solutions of linear ODEs are given in [1, 2, 3, 6, 11, 20, 19]. In [14], Li and Schwarz gave the first method to find rational solutions for a class of partial differential equations.

All these results are limited to linear cases. There seems no general methods to find Liouvillian solutions of nonlinear differential equations. With respect to ODEs of the form $y' = R(x, y)$ where $R(x, y)$ is a rational function, Poincaré made important contributions [15]. More recently, Carnicer also made important progresses in solving the Poincaré problem [5], which is equivalent to finding the degree bound for the algebraic solutions of $y' = R(x, y)$. For ODEs of this form, other work includes: Cano proposed an algorithm to find polynomial solutions [4]; Singer studied the Liouvillian first integrals [18]. On the other hand, Hubert gave a method to compute a basis of the general solutions of first order ODEs and applied it to study the local behavior of the solutions[10]. Bronstein gave an effective method to compute rational solutions of Ricatti equations [2]. In [9], we propose an algorithm to find rational solutions for first order autonomous ODEs. But this algorithm has exponential complexity.

---

* Partially supported by NKBRP of China and by a USA NSF grant CCR-0201253.

H. Li, P. J. Olver and G. Sommer (Eds.): IWMM-GIAE 2004, LNCS 3519, pp. 5–17, 2005.
© Springer-Verlag Berlin Heidelberg 2005

In this paper, we will give a polynomial time algorithm to find polynomial solutions of first order autonomous ODEs. Instead of finding arbitrary polynomial solutions, we will find the general solutions for ODEs of polynomial type. For example, the general solution for $(\frac{dy}{dx})^2 - 4y = 0$ is: $y = (x + c)^2$, where c is an arbitrary constant. Three main results are given in this paper. First, we give a sufficient and necessary condition for an ODE to have polynomial general solutions. Second, we give a detailed analysis of the structure of the first order autonomous ODEs which have polynomial general solutions. This leads to an almost *explicit formula* for the polynomial solutions of the first order autonomous ODE. Third, by introducing a novel method of substituting a polynomial solution into a first order ODE, we get a polynomial time algorithm to find polynomial general solutions of first order autonomous ODEs. Our experiments show that this algorithm is quite effective in solving ODEs with high degree and a large number of terms.

The paper is organized as follows. In section 2, a criterion for an ODE to have polynomial general solutions is given. In section 3, we give the degree bound of polynomial solutions of first order autonomous ODEs. In section 4, we analyze the structure of the first order autonomous ODEs which have polynomial solutions. In section 5, we present a polynomial time algorithm to find polynomial general solutions of first order autonomous ODEs. In section 6, we present the conclusion.

2   Polynomial General Solution to ODEs

Let $\mathbf{K} = \mathbf{Q}(x)$ be the differential field of rational functions in x with differential operator $\frac{d}{dx}$ and y an indeterminate over $\mathbf{K}$. We denote by y_i the i-th derivative of y. We use $\mathbf{K}\{y\}$ to denote the ring of differential polynomials over the differential field $\mathbf{K}$, which consists of the polynomials in the y_i with coefficients in $\mathbf{K}$. All differential polynomials in this paper are in $\mathbf{K}\{y\}$, if there is no other statement. Let Σ be a system of differential polynomials in $\mathbf{K}\{y\}$. A *zero* of Σ is an element in a universal extension field of $\mathbf{K}$ [17], which vanishes every differential polynomial in Σ. The totality of the zeros in $\mathbf{K}$ is denoted by $\text{Zero}(\Sigma)$. In this paper, we will use $\mathcal{C}$ to denote the constant field of the universal extension of $\mathbf{K}$.

Let $P \in \mathbf{K}\{y\}/\mathbf{K}$. We denote $\text{ord}(P)$ the highest derivative of y in P, called the *order* of P. Let $o = \text{ord}(P) > 0$ be the order of P. We may write P as follows

$$P = a_d y_o^d + a_{d-1} y_o^{d-1} + \ldots + a_0$$

where a_i are polynomials in $y_1, \ldots, y_{o-1}$ for $i = 0, \ldots, d$ and $a_d \neq 0$. a_d is called the *initial* of P and $S = \frac{\partial P}{\partial y_o}$ is called the *separant* of P. The k-th derivative of P is denoted by $P^{(k)}$. Let S be the separant of P, $o = \text{ord}(P)$ and $k > 0$. Then we have

$$P^{(k)} = S y_{o+k} - R_k \tag{1}$$

where R_k is of lower order than $o + k$.

Let P be a differential polynomial of order o. A differential polynomial Q is said to be *reduced* with respect to P if $\text{ord}(Q) < o$ or $\text{ord}(Q) = o$ and $\deg(Q, y_o) < \deg(P, y_o)$. For two differential polynomials P and Q, let $R = \text{prem}(P, Q)$ be the differential pseudo-remainder of P with respect to Q. We have the following *differential remainder formula* for R (see [12, 17])

$$JP = \sum_i B_i Q^{(i)} + R$$

where J is a product of certain powers of the initial and separant of Q and B_i are differential polynomials. For a differential polynomial P with order o, we say that P is *irreducible* if P is irreducible when P is treated as a polynomial in $\mathbf{K}[y, y_1, \ldots, y_o]$.

Let $P \in \mathbf{K}\{y\}/\mathbf{K}$ be an irreducible differential polynomial and

$$\Sigma_P = \{A \in \mathbf{K}\{y\} | \text{prem}(A, P) = 0\}. \tag{2}$$

In [17], Ritt proved that

Lemma 1. Σ_P *is a prime differential ideal.*

Let Σ be a non-trivial prime ideal in $\mathbf{K}\{y\}$. A zero η of Σ is called *a generic zero* of Σ if for any differential polynomial P, $P(\eta) = 0$ implies that $P \in \Sigma$. It is well known that an ideal Σ is prime iff it has a generic zero [17].

A *universal constant extension* of $\mathbf{Q}$ is obtained by first adding an infinite number of arbitrary constants to $\mathbf{Q}$ and then taking the algebraic closure. We further assume that the universal field in this paper contains a universal constant extension of $\mathbf{Q}$.

Definition 1. *Let* $F \in \mathbf{K}\{y\}/\mathbf{K}$ *be an irreducible differential polynomial. A general solution of* $F = 0$ *is defined as a generic zero of* Σ_F. *A polynomial general solution of* $F = 0$ *is defined as a general solution of* $F = 0$ *of the form*

$$\hat{y} = \sum_{i=0}^{n} a_i x^i, \ (a_n \neq 0) \tag{3}$$

where a_i *are in a universal constant extension of* $\mathbf{Q}$.

Example 1. In this example, we give three ODEs $E_i = 0$ which have polynomial general solutions $S_i = 0$ respectively.

$$E_1 = y_1^2 - 4y \qquad\qquad S_1 = (x - c)^2$$
$$E_2 = xy_1 - ny \qquad\qquad S_2 = cx^n$$
$$E_3 = y_1(y_1 - 1)(y_1 - 2) - (xy_1 - y)^2 \quad S_3 = cx + \sqrt{c(c-1)(c-2)}$$

where c is an arbitrary constant and n is a fixed positive integer.

In the literature in general, a *general solution* of $F(y) = 0$ is defined as a family of solutions with o independent parameters in a loose sense where $o = \text{ord}(F(y))$. From Theorem 6 in [12] (Chapter 2, section 12), we can see that the above definition of general solutions are essentially the same to the definition in the literature. But, the definition given by Ritt is more precise.

Theorem 1. *Let $F(y)$ be an irreducible differential polynomial. Then $F(y) = 0$ has a polynomial general solution of degree n iff n is the least integer such that $\text{prem}(y_{n+1}, F(y)) = 0$.*

Proof. ($\Longrightarrow$) Suppose that $F(y) = 0$ has a polynomial general solution $\hat{y}$ with degree n. Since $y_{n+1}(\hat{y}) = 0$, $y_{n+1} \in \Sigma_F$ which means that $\text{prem}(y_{n+1}, F(y)) = 0$ by Lemma 1.

($\Longleftarrow$) Assume that there exists an n such that $\text{prem}(y_{n+1}, F(y)) = 0$ and n is the least. If $n = -1$, then $F(y) = y$. It is obvious. Now we suppose that $n \geq 0$. From Lemma 1, $y_{n+1} \in \Sigma_F$. Hence, all the elements in the zero set of Σ_F must have the form: $\bar{y} = \sum_{i=0}^{n} \bar{a}_i x^i$. In particular, the generic zero $\hat{y}$ has the form: $\hat{y} = \sum_{i=0}^{n} a_i x^i$. If $a_n = 0$, then $y_n(\hat{y}) = 0$ which implies that $y_n \in \Sigma_F$. Hence $\text{prem}(y_n, F(y)) = 0$, a contradiction.   ∎

3   A Criterion for First Order Autonomous ODEs

In what follows, if there is no other statement, $F(y)$ will always be a non-zero first order irreducible differential polynomial with coefficients in $\mathcal{C}$ which are not arbitrary constants.

Lemma 2. *Let $\bar{y} = \sum_{i=0}^{n} \bar{a}_i x^i$ be a solution of $F(y) = 0$, where $\bar{a}_i \in \mathcal{C}$, $n > 0$ and $\bar{a}_n \neq 0$. Then for an arbitrary constant c,*

$$\hat{y} = \sum_{i=0}^{n} \bar{a}_i (x + c)^i \tag{4}$$

is a polynomial general solution for $F(y) = 0$.

Proof. It is easy to show that $\hat{y}$ is still a zero of Σ_F. For any $G(y) \in \mathbf{K}\{y\}$ satisfying $G(\hat{y}) = 0$, let $R(y) = \text{prem}(G(y), F(y))$. Then $R(\hat{y}) = 0$. Suppose that $R(y) \neq 0$. Since $F(y)$ is irreducible and $\deg(R(y), y_1) < \deg(F(y), y_1)$, there are two differential polynomials $P(y), Q(y) \in \mathbf{K}(c_{k,l})\{y\}$ such that $P(y)F(y) + Q(y)R(y) \in \mathbf{K}(c_{k,l})[y]$ and $P(y)F(y) + Q(y)R(y) \neq 0$ where $c_{k,l}$ are the coefficients of F as a polynomial in y, y_1. Thus $(PF + QR)(\hat{y}) = 0$. Because c is an arbitrary constant which is transcendental over $\mathbf{K}(c_{k,l})$ and $n > 0$, we have $P(y)F(y) + Q(y)R(y) = 0$, a contradiction. Hence $R(y) = 0$ which means that $G(y) \in \Sigma_F$. So $\hat{y}$ is a generic zero of Σ_F.   ∎

The above theorem reduces the problem of finding a polynomial general solution to the problem of finding a polynomial solution. In what below, we will show how to find such a solution.

Lemma 3. *Suppose that $\deg(F(y), y_1) = m > 0$. If $\bar{y} = \sum_{i=0}^{n} \bar{a}_i x^i$ ($\bar{a}_i \in \mathcal{C}, \bar{a}_n \neq 0$) is a solution of $F(y) = 0$, then $n \leq m$.*

Proof. Assume that $F(y) = \sum_{i=0}^{l} c_{\alpha_i \beta_i} y^{\alpha_i} y_1^{\beta_i}$, where $c_{\alpha_i \beta_i} \neq 0$ and $(\alpha_i, \beta_i) \neq (\alpha_j, \beta_j)$ if $i \neq j$. Substituting y in $F(y)$ by $\bar{y} = \sum_{i=0}^{n} \bar{a}_i x^i$, we get a polynomial $F(\bar{y})$ in x. Assume that $n > m > 0$. Then $n \geq 2$. We consider the highest degree of x in $F(\bar{y})$ which is the largest number in $\{n\alpha_i + (n-1)\beta_i \quad \text{for } i = 0, \cdots, l\}$. If $\bar{y}$ is a solution of $F(y) = 0$, all the coefficients of $F(\bar{y})$ are zero. Hence the number of the terms $y^{\alpha_i} y_1^{\beta_i}$ such that $n\alpha_i + (n-1)\beta_i$ is the largest is at least two. Without loss of generality, we suppose that two of them are $n\alpha_1 + (n-1)\beta_1$ and $n\alpha_2 + (n-1)\beta_2$. Then we have $n(\alpha_1 - \alpha_2) = (n-1)(\beta_2 - \beta_1)$. Assume that $\beta_2 \geq \beta_1$. Since $(n, n-1) = 1$, we have $n|(\beta_2 - \beta_1)$. But $0 \leq \beta_2 - \beta_1 \leq m < n$, which implies that $\beta_1 = \beta_2$. Hence $\alpha_1 = \alpha_2$. This contradicts $(\alpha_1, \beta_1) \neq (\alpha_2, \beta_2)$. Hence $n \leq m$. ∎

The following theorem gives a criterion for $F(y) = 0$ to have polynomial general solutions.

Theorem 2. *Let $F(y)$ be a first order autonomous and irreducible differential polynomial and $n = \deg(F(y), y_1)$. Then $F(y) = 0$ has a polynomial general solution iff $\operatorname{prem}(y_{n+1}, F(y)) = 0$.*

Proof. From Lemma 2, we need only to consider polynomial solutions of $F(y) = 0$ with coefficients in $\mathcal{C}$. Now the result is a direct consequence of Lemma 3 and Theorem 1. ∎

Algorithm 1. *The input is a first order autonomous ODE $F(y) = 0$. The output is a polynomial general solution of $F(y) = 0$ if it exists.*

1. Let n be the degree of $F(y)$ in y_1. If $\operatorname{prem}(y_{n+1}, F(y)) \neq 0$ then $F(y) = 0$ has no polynomial solutions and the algorithm exists; otherwise goto the next step.
2. Let d be the smallest number such that $\operatorname{prem}(y_{d+1}, F(y)) = 0$. By Theorem 1, the polynomial solution of $F(y) = 0$ is of degree d.
3. Substitute $z = a_d x^d + a_{d-1} x^{d-1} + \cdots + a_1 x + a_0$ into $F(y) = 0$ and let **PS** be the set of the coefficients of $F(z)$ as polynomials in x.
4. Solve equations **PS** $= 0$ with Wu's method [22]. Any solution with $a_d \neq 0$ of **PS** $= 0$ will provide a polynomial general solution of $F(y) = 0$. If **PS** $= 0$ has no solutions, then $F(y) = 0$ has no polynomial general solutions.

Now we give a simple example to show how the algorithm works.

Example 2. Consider the equation

$$F(y) = 31 - 54y + 27y^2 - 3y_1^2 - y_1^3.$$

1. $n := 3$. Since four is the smallest number k such that $\operatorname{prem}(y_k, F(y)) = 0$, $F(y) = 0$ has a polynomial general solution with degree three.

2. $y := a_0 + a_1 x + a_2 x^2 + a_3 x^3$. Substituting y into $F(y)$, we get $PS =$

$$\{ \; 27a_3^2 - 27a_3^3, 54a_2a_3 - 54a_2a_3^2,$$
$$54a_1a_3 + 27a_2^2 - 27a_3^2 - 9a_1a_3^2 - 24a_2^2a_3 - 3a_3(6a_1a_3 + 4a_2^2),$$
$$-54a_3 + 54a_0a_3 + 54a_1a_2 - 36a_2a_3 - 24a_1a_2a_3 - 2a_2(6a_1a_3 + 4a_2^2),$$
$$-54a_2 + 54a_0a_2 + 27a_1^2 - 18a_1a_3 - 12a_2^2 - a_1(6a_1a_3 + 4a_2^2) - 8a_2^2a_1 - 3a_3a_1^2,$$
$$-54a_1 + 54a_0a_1 - 12a_1a_2 - 6a_1^2a_2, 31 - 54a_0 + 27a_0^2 - 3a_1^2 - a_1^3 \; \}.$$

3. Solve $PS = 0$. We have the solutions

$$\{a_3 = 1, a_1 = \frac{a_2^2}{3} + 1, a_0 = 1 + \frac{a_2}{3} + \frac{a_2^3}{27}, a_2 = a_2\}.$$

Let $a_2 = 0$. Then $F(y) = 0$ has a polynomial general solution

$$\hat{y} = (x + c)^3 + (x + c) + 1$$

by Lemma 2.

It is known that the general methods of equation solving are exponential algorithms. Therefore, the above algorithm might be ineffective. In the next section, we will analyze the structure of the first order autonomous ODEs which have polynomial solutions. After doing so, we can obtain a polynomial time algorithm.

4   Structure of the First Order Autonomous ODEs with Polynomial General Solutions

If $\bar{y} = \sum_{i=0}^{n} \bar{a}_i x^i$ is a polynomial solution of $F(y) = 0$, we regard x, y, y_1 as independent indeterminants and eliminate x in the polynomial set $\{\sum_{i=0}^{n} \bar{a}_i x^i - y, \sum_{i=1}^{n} i\bar{a}_i x^{i-1} - y_1\}$. Then we will obtain a new differential polynomial $R(y)$. Theorem 3 below will give the relation between $R(y)$ and $F(y)$.

Lemma 4. Let $f_1(y) = \sum_{i=0}^{n} \bar{a}_i x^i - y$, $f_2(y) = \sum_{i=1}^{n} i\bar{a}_i x^{i-1} - y_1$ $(n \geq 1, \bar{a}_n \neq 0, \bar{a}_i \in \mathcal{C})$. If $n \geq 2$, let $R(y)$ be the Sylvester-resultant of $f_1(y)$ and $f_2(y)$ with respect to x and if $n = 1$, let $R(y) = f_2(y)$. Then $R(y)$ is an irreducible polynomial in $\mathcal{C}[y, y_1]$ and has the form

$$R(y) = (-1)^n \bar{a}_n^{n-1} y_1^n + (-1)^{n-1} n^n \bar{a}_n^n y^{n-1} + G(y, y_1) \qquad (5)$$

where $tdeg(G)$ (the total degree of G) $\leq n - 1$ and G does not contain the term y^{n-1}.

Proof. When $n = 1$, it is clear. Assume that $n \geq 2$. We know that $R(y)$ is the following determinant which has 2n-1 columns and rows:

$$
\begin{vmatrix}
\bar{a}_n & \bar{a}_{n-1} & \bar{a}_{n-2} & \cdots & \bar{a}_1 & \bar{a}_0 - y \\
 & \bar{a}_n & \bar{a}_{n-1} & \cdots & \bar{a}_2 & \bar{a}_1 & \bar{a}_0 - y \\
 & & \cdots & \cdots & & & \\
 & & \bar{a}_n & \bar{a}_{n-1} & \bar{a}_{n-2} & \cdots & \bar{a}_1 & \bar{a}_0 - y \\
n\bar{a}_n & (n-1)\bar{a}_{n-1} & (n-2)\bar{a}_{n-2} & \cdots & \bar{a}_1 - y_1 & & & \\
 & n\bar{a}_n & (n-1)\bar{a}_{n-1} & \cdots & \bar{a}_2 & \bar{a}_1 - y_1 & & \\
 & & \cdots & \cdots & & & \\
 & & n\bar{a}_n & (n-1)\bar{a}_{n-1} & (n-2)\bar{a}_{n-2} & \cdots & \bar{a}_2 & \bar{a}_1 - y_1
\end{vmatrix}.
$$

Regard $\bar{a}_i$ in the above determinant as indeterminants. Let $R = \sum c_{\alpha_i \beta_i} y^{\alpha_i} y_1^{\beta_i}$, where $(\alpha_i, \beta_i) \neq (\alpha_j, \beta_j)$ if $i \neq j$ and $c_{\alpha_i \beta_i}$ are non-zero polynomials in $\bar{a}_0, \cdots, \bar{a}_n$. We define a weight

$$w : \mathcal{C}[x, y, y_1, \bar{a}_i] \to \mathcal{Z}$$

which satisfies $w(x) = 1$, $w(\bar{a}_i) = n - i$, $w(y) = n$, $w(y_1) = n - 1$, $w(st) = w(s) + w(t)$ and $w(k) = 0$ for $k \in \mathcal{C}$. Then f_1 and f_2 are the isobaric polynomials with the weight n and $n - 1$. From [7], we know that the resultant of two homogeneous polynomials is still homogeneous. By the same way, we can show that the resultant of two isobaric polynomials with the weight n and $n - 1$ is still an isobaric polynomial with weight $n(n - 1)$. Hence $R(y)$ is an isobaric polynomial with the weight $n(n - 1)$. We have $w(y^{\alpha_i} y_1^{\beta_i}) = n\alpha_i + (n - 1)\beta_i = \alpha_i + (n - 1)(\alpha_i + \beta_i) \leq n(n - 1)$, which implies if $\alpha_i > 0$ then we have $\alpha_i + \beta_i < n$. By the computation of the above determinant, the coefficients of y^n and y^{n-1} in $R(y)$ are $(-1)^n \bar{a}_n^{n-1}$ and $(-1)^{n-1} n^n \bar{a}_n^n$. Then the form of $R(y)$ is as (5).

In the following, we take $\bar{a}_i$ as complex numbers. If $R(y)$ is reducible, we assume that $R(y) = F_1(y)F_2(y)$, where $0 < \mathrm{tdeg}(F_1(y)), \mathrm{tdeg}(F_2(y)) < n$. Since $R(y) = P(y)f_1(y) + Q(y)f_2(y)$, where $P(y)$, $Q(y)$ are two differential polynomials, we have $R(\bar{y}) \equiv 0$ which implies that $F_1(\bar{y}) = 0$ or $F_2(\bar{y}) = 0$. But we know that it is impossible by Lemma 3, because $\deg(F_1(y), y_1)$, $\deg(F_2(y), y_1) < n$, a contradiction. ∎

Theorem 3. *Use the same notations as in Lemma 4. If $\bar{y} = \sum_{i=0}^n \bar{a}_i x^i$ is a polynomial solution of $F(y) = 0$, then $R(y)|F(y)$. Since $F(y)$ is irreducible, $F(y) = \lambda R(y)$, where $\lambda \in \mathcal{C}$ and $\lambda \neq 0$.*

Proof. From Lemma 3 and Lemma 4, we know $\deg(F(y), y_1) \geq n = \deg(R(y), y_1)$. Let $T(y) = \mathrm{prem}(F(y), R(y))$. Then we have the remainder formula $J(y)^k F(y) = Q(y)R(y) + T(y)$, where $Q(y)$, $T(y) \in \mathcal{C}[y, y_1]$, $J(y)$ is the initial of $R(y)$ and $\deg(T(y), y_1) < \deg(R(y), y_1)$. Since $F(\bar{y}) = 0$ and $R(\bar{y}) = 0$, we have $T(\bar{y}) = 0$. By Lemma 3, $T(y) = 0$. That is $J(y)^k F(y) = Q(y)R(y)$ which implies that $R(y)|F(y)$ because $R(y)$ is irreducible. Since $F(y)$ is irreducible, it is clear that $F(y) = \lambda R(y)$ where $\lambda \in \mathcal{C}$ and $\lambda \neq 0$. ∎

From Lemma 4 and Theorem 3, if $F(y)$ has a polynomial solution $\bar{y} = \sum_{i=0}^n \bar{a}_i x^i$, it must be of the following form

$$F(y) = ay_1^n + by^{n-1} + G(y, y_1) \tag{6}$$

where $a, b \in \mathcal{C}$ are not zero, $\mathrm{tdeg}(G) \le n-1$ and G does not contain the term y^{n-1}.

As a consequence of Theorem 3 and Lemma 4, we have

Corollary 1. *Let $F(y)$ be of the form (6) and have a polynomial solution of the form (4). Then*

$$\bar{a}_n = -\frac{b}{n^n a}. \tag{7}$$

Lemma 5. *Let $F(y)$ be of the form (6) and have a polynomial general solution of the form (4). Then we may construct a new general solution of the following form for $F(y) = 0$*

$$\hat{y} = \bar{a}_n(x+c)^n + \sum_{i=0}^{n-2} \tilde{a}_i(x+c)^i. \tag{8}$$

In other words, we may assume that $\bar{a}_{n-1} = 0$ in the general solution of $F(y) = 0$.

Proof. It is clear that

$$\hat{y} = \bar{a}_n(x + c - \frac{\bar{a}_{n-1}}{n\bar{a}_n})^n + \sum_{i=0}^{n-2} \tilde{a}_i(x + c - \frac{\bar{a}_{n-1}}{n\bar{a}_n})^i.$$

Since $c - \frac{\bar{a}_{n-1}}{n\bar{a}_n}$ is still an arbitrary constant, replacing $c - \frac{\bar{a}_{n-1}}{n\bar{a}_n}$ by c in the above equation, we get the form (8) and it is still a general solution of $F(y) = 0$. ∎

The following theorem tells us that the value of $\bar{a}_k$ only depends on the values of $\bar{a}_i$ for $i \ge k$ if $F(y) = 0$ has polynomial general solutions.

Lemma 6. *Let $F(y)$ be of the form (6) and $z = (-b/n^n a)x^n + a_{n-1}x^{n-1} + a_{n-2}x^{n-2} + \cdots + a_0$ where a_i are indeterminants. Substituting y by z in $F(y)$, the coefficients of $x^{(n-1)^2+i-1}$ in $F(z)$ are of the following form*

$$(\frac{-b}{n^n a})^{n-2}(n-1-i)ba_i + h_i(a_{n-1}, \cdots, a_{i+1}), \quad \text{for } i = n-2, \cdots, 0 \tag{9}$$

where $h_i(a_{n-1}, \cdots, a_{i+1})$ are the polynomials in $a_{n-1}, \cdots, a_{i+1}$.

Proof. Let $\mathcal{C}_i$ be the coefficient of $x^{(n-1)^2+i-1}$ in $F(z)$ for $i = 0, \cdots, n-2$ where

$$F(z) = az_1^n + bz^{n-1} + G(z, z_1).$$

As in the proof of Lemma 4, we define a weight w. Then $z = (-b/n^n a)x^n + a_{n-1}x^{n-1} + \cdots + a_0$ is an isobaric polynomial with the weight n. Hence $z^{\alpha_j}z_1^{\beta_j}$ is still an isobaric polynomial with the weight $n\alpha_j + (n-1)\beta_j$. Now we consider $\mathcal{C}_i$. By computation, we know that the highest weight of the terms in $F(z)$ is $n(n-1)$. Hence the highest weight in $\mathcal{C}_i$ is not greater than $n-i$. So a_k can not appear in $\mathcal{C}_i$ for $k \le i-1$ and if a_i appears in $\mathcal{C}_i$, then it must be linear and its coefficient must be constant. In the coefficients of $x^{(n-1)^2+i-1}$ in $az_1^n + bz^{n-1}$, the term in which a_i appears are $(\frac{-b}{n^n a})^{n-1}in^n aa_i + (\frac{-b}{n^n a})^{n-2}(n-1)ba_i$. In the coefficients of $x^{(n-1)^2+i-1}$ in $G(z, z_1)$, since the weight of each term is less than $n-i$ (for $\mathrm{tdeg}(G) < n$), a_i can not appear. Therefore $\mathcal{C}_i$ has the form (9). ∎

5   A Polynomial-Time Algorithm

From the results in section 4, we have the following algorithm.

Algorithm 2. *The input is $F(y)$. The output is a polynomial general solution of $F(y) = 0$ if it exists.*

1. If $F(y)$ can be written as the form (6), then goto step 2. Otherwise, by Theorem 3, $F(y) = 0$ has no polynomial general solutions and the algorithm terminates.
2. Let $F(y)$ be of degree n in y_1. Let $\bar{a}_n = -\frac{b}{n^n a}$, $\bar{a}_{n-1} = 0$,
$$\bar{a}_i = -\frac{h_i(\bar{a}_{n-1}, \cdots, \bar{a}_{i+1})}{(-b/n^n a)^{n-2}(n-1-i)b}, i = n-2, \cdots, 0, \text{ where } h_i \text{ are from Lemma 6. We}$$
have $\bar{a}_i \in \mathcal{C}$.
3. Let $\bar{y} = \sum_{i=0}^{n} \bar{a}_i x^i$. If $F(\bar{y}) \equiv 0$ then $\hat{y} = \sum_{i=0}^{n} \bar{a}_i (x + c)^i$ is a polynomial general solution of $F(y) = 0$. Otherwise, $F(y) = 0$ has no polynomial general solutions.

   The correctness of Step 3 is due to the following facts. By Corollary 1, Lemmas 5 and 6, if $F(y) = 0$ has polynomial general solutions, then $\hat{y} = \sum_{i=0}^{n} \bar{a}_i (x + c)^i$ must be such a solution. By Lemma 2, to check wether $\hat{y}$ is a polynomial general solution we need only to check whether $\bar{y}$ is a solution of $F(y) = 0$.
   Now we give some examples.

Example 3. Consider the differential polynomial:

$$F(y) = y_1{}^4 - 8\,y_1{}^3 + (6 + 24\,y)\,y_1{}^2 + 257 + 528\,y^2 - 256\,y^3 - 552\,y.$$

1. $F(y)$ can be written as the form:

$$F(y) = y_1{}^4 - 256\,y^3 + G(y, y_1)$$

   where

$$G(y, y_1) = -8\,y_1{}^3 + (6 + 24\,y)\,y_1{}^2 + 257 + 528\,y^2 - 552\,y,$$

   $tdeg(G) \leq 3$.
2. If $F(y) = 0$ has a polynomial general solution, then its degree is four and the coefficient of x^4 must be $a_4 = \frac{256}{4^4} = 1$.
3. Let $z = x^4 + a_2 x^2 + a_1 x + a_0$. Replacing y by z in $F(y)$ and collecting the coefficients of x^8, x^9, x^{10}, we obtain the following equations:

$$768\,a_2 + 528 - 384\,a_2{}^2 - 768\,a_0 = 0,$$
$$-512 - 512\,a_1 = 0,$$
$$384 - 256\,a_2 = 0.$$

Solving the above equations, we have $a_0 = \frac{17}{16}, a_1 = -1, a_2 = \frac{3}{2}$.

4. Let $\bar{y} = x^4 + \frac{3}{2}x^2 - x + \frac{17}{16}$. Substituting y by $\bar{y}$ in $F(y)$, $F(y)$ becomes zero. Hence a polynomial general solution of $F(y) = 0$ is

$$\bar{y} = (x+c)^4 + \frac{3}{2}(x+c)^2 - (x+c) + \frac{17}{16}.$$

In Step 2 of Algorithm 2, we need to compute $F(z)$ where $z = -\frac{b}{n^n a}x^n + a_{n-1}x^{n-1} + \cdots + a_0$ and a_i are indeterminants. A naive method of doing this evaluation is very costly. In order to give a polynomial-time algorithm, we need to give an efficient algorithm for Step 2. From Theorem 6, to compute the value of $\bar{a}_k$, we need only to compute $h_k(\bar{a}_{n-1}, \cdots, \bar{a}_{k+1})$. In other words, we need only to compute $F(\bar{z})$ where $\bar{z} = \sum_{k+1}^{n} \bar{a}_i x^i$ and $\bar{a}_i \in C$. Moreover, we need only to compute the coefficient of $x^{(n-1)^2+k-1}$ in $F(\bar{z})$. To compute $F(\bar{z})$, we need to compute multiplication of two univariate polynomials which can be computed by the classical Karatsuba method ([21]).

Algorithm 3. *The inputs are $F(y)$ as the form (6) and $\bar{z} = \bar{a}_n x^n + \cdots + \bar{a}_0$ where $\bar{a}_i \in C$. The output is the coefficient of $x^{(n-1)^2+k-1}$ in $F(\bar{z})$ for some k.*

1. Compute $\bar{z}^n$ and $\bar{z}_1^{n-1}$ where $\bar{z}_1$ is the derivative of $\bar{z}$ wrt x. We compute $\bar{z}^n$ step by step. That is, we compute $\bar{z}^2$ first, then compute the multiplication of $\bar{z}^2$ and $\bar{z}$, and so on. Note that, after we computed $\bar{z}^n$, we have also obtained $\bar{z}^i$ for $i < n$. For $\bar{z}_1^{n-1}$, we compute it in the same way.
2. Write $F(y)$ as the form: $F(y) = \sum_{i=0}^{n}(d_{0,i}+d_{1,i}y+\cdots+d_{n-i,i}y^{n-i})y_1^i$ where $d_{i,j} \in C$.
3. For i from 0 to n, compute $p_i = d_{0,i} + d_{1,i}\bar{z} + \cdots + d_{n-i,i}\bar{z}^{n-i}$.
4. result:=0.
   For j from 0 to n
   For i from 0 to $(n-1)^2 + k - 1$
   result:=result+coeff(p_j, x, i)*coeff$(y_1^j, x, (n-1)^2 + k - 1 - i)$
   where coeff(p, x, k) means the coefficient of x^k in p.
5. return(result).

Example 4. Let $F(y)$ be as in Example 3 and $\bar{z} = x^4$. Now we compute the coefficient of x^{10} in $F(\bar{z})$.

1. $\bar{z} := x^4, \bar{z}^2 := x^8, \bar{z}^3 := x^{12}$.
   $\bar{z}_1 := 4x^3, \bar{z}_1^2 := 16x^6, \bar{z}_1^3 := 64x^9, \bar{z}_1^4 := 256x^{12}$.
2.

$$p_0 := 257 + 528\,\bar{z}^2 - 256\,\bar{z}^3 - 552\,\bar{z} = 257 - 552x^4 + 528x^8 - 256x^{12},$$
$$p_1 := 0, \quad p_2 := 6 + 24\bar{z} = 6 + 24x^4, \quad p_3 := -8, \quad p_4 := 1.$$

3. result:=coeff$(p_2, x, 4)$*coeff$(\bar{z}_1^2, x, 6)$=24*16=384. Because for other i, j, coeff(p_j, x, i)*coeff$(\bar{z}_1^j, x, 10 - i)$= 0.

Since multiplication is the dominant factor for the running time of the algorithm, we will use the number of multiplications of rational numbers to measure the complexity of the algorithm. In Algorithm 3, the complexity of Step 1 is $O(n^4)$. The complexity of Step 2 is $O(n^2)$. The complexity of Step 3 is $O(n^4)$. The complexity of Step 4 is $O(n^3)$. Hence the complexity of Algorithm 3 is $O(n^4)$.

Algorithm 4. *The inputs are $F(y)$ as the form (6) and $z = a_n x^n + a_{n-1} x^{n-1} + \cdots + a_0$ where a_i are interminates. The output is $\bar{a}_i$ for $i = n, \cdots, 0$ in Step 2 of Algorithm 2.*

1. Let $\bar{a}_n = -\frac{b}{n^n a}$, $\bar{a}_{n-1} = 0$.
2. Let $i = n - 2$.
   while $i \geq 0$ do
   (a) $\bar{y} := \bar{a}_n x^n + \cdots + \bar{a}_{i+1} x^{i+1}$.
   (b) $C_i :=$ the coefficient of $x^{(n-1)^2+i-1}$ in $F(\bar{y})$ by Algorithm 3.
   (c) $\bar{a}_i := -\frac{C_i}{(-b/n^n a)^{n-2}(n-1-i)b}$.
   (d) $i := i - 1$.

Example 5. (Example 3 continued)

1. $n := 4, a := 1, b := -256$.
2. $\bar{a}_4 := 1, \bar{a}_3 := 0$.
3. $\bar{y} := x^4$. Substitute $\bar{y}$ into $F(y)$.
4. Then by Algorithm 3, the coefficient of x^{10} in $F(x^4)$ equals to 384. That is, $C_2 = 384$.
5. $\bar{a}_2 := -\frac{384}{-256} = \frac{3}{2}$ which equals to a_2 in Example 3. The other coefficients can be computed similarly.

It is easy to know that the complexity of Algorithm 4 is $O(n^5)$. In Step 3 of Algorithm 2, we verify whether $\bar{y}$ is a polynomial solution of $F(y) = 0$. If $\bar{y}$ is not a polynomial solution of $F(y) = 0$, that is $F(\bar{y}) \neq 0$, then $F(\bar{y})$ will be a polynomial in x with degree not greater than $n(n-1)$. Hence, if $F(\bar{y}) \neq 0$, then $F(\bar{y}) = 0$ as an equation in x has $n(n-1)$ roots at most. So we can verify it by numerical computation. If $F(\bar{y})(k) = 0$ for $k = -\frac{n(n-1)}{2}, -\frac{n(n-1)}{2}+1, \cdots, \frac{n(n-1)}{2}$, then $F(\bar{y}) = 0$. Otherwise, $F(\bar{y}) \neq 0$.

Algorithm 5. *The inputs are $F(y)$ as the form (6) and $\bar{y} = \sum_{i=0}^{n} \bar{a}_i x^i$ as in Algorithm 2. The output is "Yes" or "No" where "Yes" means $\bar{y}$ is a solution of $F(y) = 0$ and "No" means $\bar{y}$ is not a solution of $F(y) = 0$.*

1. Write $F(y)$ as the form: $F(y) = \sum_{j=0}^{n}(d_{0,j} + d_{1,j}y + \cdots + d_{n-j,j}y^{n-j})y_1^j$ where $d_{i,j} \in \mathcal{C}$.
2. For k from $-n(n-1)/2$ to $n(n-1)/2$
   (a) Substitute x^i by k^i in $\bar{y}$ and $\bar{y}_1$. Then we get the values of $\bar{y}$ and $\bar{y}_1$ at k, which are denoted by $\bar{y}(k)$ and $\bar{y}_1(k)$. Compute $\bar{y}_1(k)^i$ and $\bar{y}(k)^i$ for $i = 1 \cdots n$ in the same way as Step 1 of Algorithm 3.

(b) result:=0.

For j from 0 to n, compute $p_j(k) = d_{0,j} + d_{1,j}\bar{y}(k) + \cdots + d_{n-j,j}\bar{y}(k)^j$,
result:=result+$p_j(k)\bar{y}_1(k)^j$.

(c) If result$\neq 0$ then return(No).

3. return(Yes).

It is easy to check that the complexity of Algorithm 5 is $O(n^3)$. So we have the following theorem.

Theorem 4. *We can decide whether $F(y) = 0$ has a polynomial general solution and compute one if it exists with $O(n^5)$ multiplications of rational numbers.*

6   Conclusion

We have implemented the algorithms in Maple. The software is available at http://www.mmrc .iss.ac.cn/˜xgao/software.html. In Table 1, we present the statistic results of running our algorithm for twenty differential equations, which could be found in [8]. We only give the total degrees and numbers of terms in these differential equations. In Table 1, F_i and G_i denote the differential equations. Here, the coefficients of F_i and G_j are integers. The coefficients of F_i are less than 10^6 but that of G_j may be very large. The running time is in seconds. The column of "solution" means whether they have a polynomial general solution or not. The running time is collected on a computer with Pentium 4, 2.66GHzCPU and 256M memory.

From the experimental results, we could conclude that the algorithm can be used to find polynomial solutions for very large ODEs efficiently. Recently, we have extended the method proposed in this paper to find rational and algebraic solutions of first order autonomous ODEs. It is interesting to see whether the result can be extended to the case when the coefficients of first order ODEs are not constant.

Table 1. Statistics on Algorithm2

	tdegree	term	time(s)	solution		tdegree	term	time(s)	solution
G_6	6	17	0.077	Y	F_6	6	22	0.063	N
G_7	7	22	0.125	Y	F_7	7	28	0.266	N
G_8	8	30	0.312	Y	F_8	8	37	1.141	N
G_9	9	38	1.468	Y	F_9	9	45	3.500	N
G_{10}	10	47	8.108	Y	F_{10}	10	55	10.656	N
G_{11}	11	57	16.062	Y	F_{11}	11	60	31.345	N
G_{12}	12	68	34.250	Y	F_{12}	12	74	70.438	N
G_{13}	13	80	78.203	Y	F_{13}	13	80	148.984	N
G_{14}	14	93	178.469	Y	F_{14}	14	90	273.065	N
G_{15}	15	106	306.250	Y	F_{15}	15	110	434.514	N

References

1. Boucher, D., About the Polynomial Solutions of Homogeneous Linear Differential Equations Depending on Parameters, *Proc. ISSAC1999*, 261-268, ACM Press, 1999.
2. Bronstein, M., On Solutions of Linear Ordinary Differential Equations in their Coefficient Field, *J. Symb. Comput.*, 13(4), 1992.
3. Barkatou, M. A., Pflgel, E., An Algorithm Computing the Regular Formal Solutions of a System of Linear Differential Equations, *J. Symb. Comput.*, 28(4-5), 569-587, 1999.
4. Cano, J., An Algorithm to Find Polynomial Solutions of $y' = R(x, y)$, private communication, 2003.
5. Carnicer, M.M., The Poincaré Problem in the Nondicritical Case, *Ann. of Math.*, 140, 289-294, 1994.
6. Cormier, O., On Liouvillian Solutions of Linear Differential Equations of Order 4 and 5, *Proc. ISSAC2001*, 93-100, ACM Press, 2001.
7. Cox, D., Little, J. and O'Shea, D., *Ideals, Varieties, and Algorithms*, Springer, New York, 1991.
8. Feng, R.Y. and Gao, X.S., Polynomial General Solution for First Order ODEs with Constant Coefficients, *MM Research Preprints*, No. 22, 24-29, 2003. (Available at http://www.mmrc.iss.ac.cn/pub/mm-pre.html)
9. Feng, R.Y. and Gao, X.S., Rational General Solutions of Algebraic Ordinary Differential Equations, *Proc. ISSAC2004*, 155-162, ACM Press, 2004.
10. Hubert, E., The General Solution of an Ordinary Differential Equation, *Proc. ISSAC1996*, 189-195, ACM Press, 1996.
11. Kean, C., Taylor Polynomial Solutions of Linear Differential Equations, *Appl. Math. Comput.*, V.142, Issue 1, Sep.20, 155-165, 2003.
12. Kolchin, E.R., *Differential Algebra and Algebraic Groups*, Academic Press, New York, 1950.
13. Kovacic, J.J., An Algorithm for Solving Second Order Linear Homogeneous Differential Equations, *J. Symb. Comput.*, 2(1), 3-43, 1986.
14. Li, Z.M. and Schwarz, F., Rational Solutions of Riccati-like Partial Eifferential Equations, *J. Symb. Comput.*, 31(6), 691-716, 2001.
15. Poincaré, H., Sur L'intégration Algébrique des Équations Différentielles du Premier Ordre et du Premier Degré, *Rend. Circ. Mat. Palermo*, 11, 193-239, 1897.
16. Risch, R.H., The Problem of Integration in Finite Terms, *Trans. Amer. Math. Soc*, 139, 167-189, 1969.
17. Ritt, J.F., *Differential Algebra*, Amer. Math. Sco. Colloquium, New York, 1950.
18. Singer, M.F., Liouillian First Integrals of Differential Equations, *Trans. Amer. Math. Soc.*, 333(2), 673-688, 1992.
19. Ulmer, F. and Calmet, J., On Liouvillian Solutions of Homogeneous Linear Differential Equations, *Proc. ISSAC1990*, 236-243, ACM Press, 1990.
20. Van der Put, M. and Singer, M., *Galois Theory of Linear Differential Equations*, Springer, Berlin, 2003.
21. Winkler, F., *Polynomial Algorithm in Computer Algebra*, Spring Verlag, 1996.
22. Wu, W.T., *Mathematics Mechanization*, Science Press/Kluwer, Beijing, 2000.

The Newton Polygon Method for Differential Equations

José Cano

Dpto. Algebra, Geometría y Topología,
Fac. de Ciencias, Universidad de Valladolid, Spain
jcano@agt.uva.es

Abstract. We prove that a first order ordinary differential equation (ODE) with a dicritical singularity at the origin has a one-parameter family of convergent fractional power series solutions. The notion of a dicritical singularity is extended from the class of first order and first degree ODE's to the class of first order ODE's. An analogous result for series with real exponents is given.

The main tool used in this paper is the Newton polygon method for ODE. We give a description of this method and some elementary applications such as an algorithm for finding polynomial solutions.

1 Introduction

In this paper we give a sufficient condition for a first order ordinary differential equation (ODE) to have a one-parameter family of fractional power series (resp. generalized formal power series) solutions. If the family of solutions have rational exponents it turns out that each one is convergent, hence its sum is the parametrization of an analytical branch curve. The sufficient condition is a generalization of the notion of dicritical singularity of holomorphic foliations in dimension 2. The foliation defined by $a(x, y) \, dx + b(x, y) \, dy = 0$ is dicritical if there exists a dicritical blowup in the reduction of singularities process. It is well known that it is dicritical if and only if there exists a one-parameter family of analytical invariant curves passing through the origin. The above foliation corresponds to the first order and first degree ODE $a(x, y) + b(x, y) \, y' = 0$. We generalize this result for first order and arbitrary degree ODE. The dicritical property is described in terms of the Newton polygon process and in case of first order and first degree agrees with that of foliations.

Briot and Bouquet [1] in 1856 used the Newton polygon method for studying the singularities of first order and first degree ODE's and Fine [2] gives a complete description of the method for general ODE in 1889.

D.Y. Grigoriev and M. Singer [3] use it to give an enumeration of the set of formal power series solution with real exponents of an ODE $F(y) = 0$. They give restrictions (expressed as a sequence of quantifier-free formulas) over parameter C_i and μ_i for $\sum_{i=1}^{\infty} C_i \, x^{\mu_i}$ to be a solution of $F = 0$. The one-parameter families of solutions that we obtain in this paper are simpler than those obtained in [3]

H. Li, P. J. Olver and G. Sommer (Eds.): IWMM-GIAE 2004, LNCS 3519, pp. 18–30, 2005.

for the general case. (See Theorem 2 for a precise description.) In particular the parameter does not appear in the exponents μ_i. When the Newton polygon process to a first order ODE $F(y) = 0$ is applied, it is easy to see that the "first" parameter that we need to introduce is a coefficient and not an exponent (see Proposition 1). The problem arises in subsequent steps, because now we are dealing with a differential equation which has parameters as coefficients. For instance, $xy' - Cy = 0$ has $y(x) = C_1 x^C$ as solution and parameter C has "jumped" to the exponent. Here we proof that this phenomenon does not happen.

In a forthcoming paper we will give a complete description of the set of generalized power series solutions of a first order ODE.

In Section 2, we give a detailed description of the Newton polygon process in order to state some basic definitions and notations. This section should be read in connection with our main result, stated and proved in section 3, as a technical guide for the proof. All results in section 2 are not original and can be found in [3, 4, 5, 6, 7, 8]. As an elementary application of the Newton polygon method we have included an algorithm which gives a bound for degree of polynomial solutions of a first order ODE.

2   Description of the Classical Newton Polygon Method

Let K be a field, $\mathbb{C} \subseteq K$. We will denote $K((x))^*$ the field of Puiseux series over K. Hence, the elements of $K((x))^*$ are formal power series $\sum_{i \geq i_0} c_i x^{i/q}$, $i \in \mathbb{Z}$, $c_i \in K$ and $q \in \mathbb{Z}_+$. A well-ordered series with real exponents with coefficients in the field K is a series $\phi(x) = \sum_{\alpha \in A} c_\alpha x^\alpha$, where $c_\alpha \in K$, and A is a well ordered subset of $\mathbb{R}$. If there exists a finitely generated semi-group Γ of $\mathbb{R}_{\geq 0}$ and $\gamma \in \mathbb{R}$, such that, $A \subseteq \gamma + \Gamma$, then we say that $\phi(x)$ is a *grid-based* series. (This terminology comes from [7].) Let $K((x))^w$ and $K((x))^g$ be the set of well-ordered series and that of grid-based series respectively. Both are differential rings with the usual inner operations and the differential operator

$$\frac{d}{dx}(c) = 0, \ \forall c \in K, \quad \text{and} \quad \frac{d}{dx}\left(\sum c_\alpha x^\alpha\right) = \sum \alpha c_\alpha x^{\alpha-1} .$$

In fact, both rings are fields by virtue of Theorem 1.

Let $F(y_0, \ldots, y_n)$ be a polynomial on the variables $y_0, \ldots, y_n$ with coefficients in $K((x))^g$. The differential equation $F(y, \frac{dy}{dx}, \ldots, \frac{d^n y}{dx^n}) = 0$ will be denoted by $F(y) = 0$. We are going to describe the classical Newton polygon method for searching solutions of the differential equation $F(y) = 0$ in the field $K((x))^w$. We write F in a unique way as

$$F = \sum a_{\alpha, \rho_0, \ldots, \rho_n} x^\alpha y_0^{\rho_0} \cdots y_n^{\rho_n}, \quad a_{\alpha, \rho} \in K,$$

where $\alpha \in A$ and ρ belongs to a finite subset of $\mathbb{N}^{n+1}$. We define the cloud of points of F to be the set $\mathcal{P}(F) = \{P_{\alpha, \rho} \mid a_{\alpha, \rho} \neq 0\}$, where we denote

$$P_{\alpha, \rho} = (\alpha - \rho_1 - 2\rho_2 - \cdots - n\rho_n, \ \rho_0 + \rho_1 + \cdots + \rho_n) \in \mathbb{R} \times \mathbb{N} . \tag{1}$$

The Newton polygon $\mathcal{N}(F)$ of F is the convex hull of the set

$$\bigcup_{P \in \mathcal{P}(F)} (P + \{(a,0) \mid a \geq 0\}) \ .$$

We remark that $\mathcal{N}(F)$ has a finite number of vertices and all of them has as ordinate a non-negative integer.

Given a line $L \subseteq \mathbb{R}^2$ with slope $-1/\mu$, we say that μ is the *inclination* of L. Let $\mu \in \mathbb{R}$, we denote $L(F;\mu)$ to be the line with inclination μ such that $\mathcal{N}(F)$ is contained in the right closed half-plane defined by $L(F;\mu)$ and $L(F;\mu) \cap \mathcal{N}(F) \neq \emptyset$. We define the polynomial

$$\Phi_{(F;\mu)}(C) = \sum_{P_{\alpha,\underline{\rho}} \in L(F;\mu)} a_{\alpha,\underline{\rho}} C^{\rho_0 + \cdots + \rho_n} (\mu)_1^{\rho_1} \cdots (\mu)_n^{\rho_n}, \tag{2}$$

where $(\mu)_k = \mu(\mu-1)\cdots(\mu-k+1)$ and $P_{\alpha,\underline{\rho}}$ is as in (1).

2.1   Necessary Initial Conditions

Lemma 1. *Let* $y(x) = c\,x^\mu + \cdots$ *higher order terms* $\cdots \in K((x))^w$ *be a solution of the differential equation* $F(y) = 0$. *Then we have that* $\Phi_{(F;\mu)}(c) = 0$.

Proof. We have that $F(c\,x^\mu + \cdots) =$

$$\sum_{\alpha,\underline{\rho}} a_{\alpha,\underline{\rho}}\, x^\alpha\, (c\,x^\mu + \cdots)^{\rho_0} (\mu c\, x^{\mu-1} + \cdots)^{\rho_1} \cdots ((\mu)_n c\, x^{\mu-n} + \cdots)^{\rho_n} =$$

$$\sum_{\alpha,\underline{\rho}} \left\{ A_{\alpha,\underline{\rho}}\, c^{\rho_0 + \cdots + \rho_n} (\mu)_1^{\rho_1} \cdots (\mu)_n^{\rho_n}\, x^{\alpha - \rho_1 - 2\rho_2 - \cdots - n\rho_n + \mu(\rho_0 + \cdots + \rho_n)} + \cdots \right\} =$$

$$\left\{ \sum_{\nu(\alpha,\underline{\rho};\mu) = \nu(F;\mu)} a_{\alpha,\underline{\rho}}\, c^{\rho_0 + \cdots + \rho_n} (\mu)_1^{\rho_1} \cdots (\mu)_n^{\rho_n} \right\} x^{\nu(F;\mu)} + \cdots \text{ higher order terms,}$$

where $\nu(\alpha,\underline{\rho};\mu) = \alpha - \rho_1 - 2\rho_2 - \cdots - n\rho_n + \mu(\rho_0 + \cdots + \rho_n)$, and $\nu(F;\mu) = \min\{\nu(\alpha,\underline{\rho};\mu) \mid a_{\alpha,\underline{\rho}} \neq 0\}$.   □

The set $\mathrm{NIC}(F) = \{(c,\mu) \mid c \in \bar{K}, c \neq 0,\ \mu \in \mathbb{R},\ \Phi_{(F;\mu)}(c) = 0\}$, where $\bar{K}$ is the algebraic closure of K, is called the set of necessary initial conditions for F. We give a precise description of this set using the Newton polygon of F.

For each $\mu \in \mathbb{R}$, $L(F;\mu) \cap \mathcal{N}(F)$ is either a side or a vertex. Assume that $L(F;\mu) \cap \mathcal{N}(F) = S$ is a side. We call $\Phi_{(F;\mu)}(C)$ the *characteristic* polynomial of F associated to the side S. Let $A_S = \{c \in \bar{K} \mid c \neq 0,\ \Phi_{(F;\mu)}(c) = 0\}$. We say that the side S is of type (0) if $A_S = \emptyset$, of type (I) is A_S is a finite set and of type (II) if $A_S = \bar{K}$. We have that $A_S \times \{\mu\} = \mathrm{NIC}(F) \cap (\bar{K} \times \{\mu\})$.

Let $p = (a,h)$ be a vertex of $\mathcal{N}(F)$, and let $\mu_1 < \mu_2$ be the inclinations of the adjacent sides at p. For any μ such that $\mu_1 < \mu < \mu_2$, we have that $L(F;\mu) \cap \mathcal{N}(F) = \{p\}$. Then $\Phi_{(F;\mu)}(C) = C^h \Psi_{(F;p)}(\mu)$, where

$$\Psi_{(F;p)}(\mu) = \sum_{P_{\alpha,\underline{\rho}} = p} a_{\alpha,\underline{\rho}} (\mu)_1^{\rho_1} \cdots (\mu)_n^{\rho_n} \ , \tag{3}$$

Table 1. Necessary initial conditions for the equation $F(y) = 0$ of example 1

vertices

μ	$\Psi_{(F;p)}(\mu)$	type
$-\infty < \mu < -1/2$	$\mu^3(\mu-1)^2$	0
$-1/2 < \mu < 0$	0	IV
$0 < \mu < 2$	$\mu^3(\mu-1)$	III
$2 < \mu < 4$	$\mu - 4$	0
$4 < \mu < \infty$	-1	0

sides

μ	$\Phi_{(F;\mu)}(C)$	type
$\mu = -1/2$	$-9/32\,C^5$	0
$\mu = 0$	0	II
$\mu = 2$	$2\,C(4C+1)$	I
$\mu = 4$	-1	0

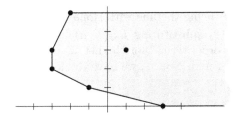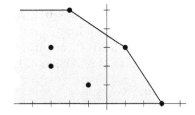

Fig. 1. The Newtonpolygon of F at the origin (left) and at infinity (right)

where $P_{\alpha,\rho}$ is as in (1). The polynomial $\Psi_{(F;p)}(\mu)$ is called the *indicial* polynomial associated to p. Let $A_p = \{\mu \in (\mu_1, \mu_2) \mid \Psi_{(F;p)}(\mu) = 0\}$. Then $\mathrm{NIC}(F) \cap \bar{K} \times (\mu_1, \mu_2) = \bar{K} \times A_p$. There are again three possibilities. Either $A_p = \emptyset$, in this case we say that the vertex p is of type (0); or A_p is a finite set, then p is of type (III); or $A_p = (\mu_1, \mu_2)$, and we call p of type (IV).

Let $(c, \mu) \in \mathrm{NIC}(F)$. Then μ is associated with either a side or a vertex of $\mathcal{N}(F)$. We say that (c, μ) is of type (I)–(IV) depending on the type of the vertex or of the side associated with μ.

Example 1. Let $F(y) = x^3\,y_0^2\,y_1\,y_2^2 - y_1^3 + y_0\,y_1\,y_2 + x^{-1}\,y_0\,y_1^2 + x\,y_2^2 - y_1\,y_2 - y_1 + 4x^{-1}\,y_0 - x^3\,y_0\,y_1\,y_2 - x^3$.

The necessary initial conditions for $F(y)$ are described in Table 1 and its Newton polygon is drawn on the left hand of Fig. 1.

Proposition 1. *Let $F(x, y, y') \in K((x))^w[y, y']$ be a polynomial first order ordinary differential equation. Let p be a vertex of $\mathcal{N}(F)$. Then p is not of type (IV).*

Proof. Let $p = (\alpha, h)$. Let F_p be the sum of all monomials of F whose corresponding point is p. We have that $F_p = \sum_{i=0}^h a_i\,x^\alpha\,y^i\,(xy')^{h-i}$. Then the indicial polynomial associated with the vertex p is $\Psi_{(F;p)}(\mu) = \sum_{i=0}^h a_i\,\mu^{h-i}$. If this polynomial has an infinite number of solutions, then $a_i = 0$ for all $0 \le i \le h$. Then F_p is identically null, which is an absurd because p belongs to the cloud of points of F. $\square$

2.2   The Newton Polygon at Infinity and Polynomial Solutions

Let us assume that the coefficients of $F(y)$ are rational functions. For any point x_0, in order to find formal solutions in the variable $x - x_0$ we only need to perform the change of variable $\bar{x} = x - x_0$ in the differential equation and work as before.

At infinity, we have two possibilities. Either we perform the change of variable $z = 1/x$ in the differential equation and we work as above, or we just look for series of the form $y(x) = \sum_{i=0}^{\infty} c_i x^{\mu_i}$, $\mu_i > \mu_{i+1}$, for all i. In this case we do not need to do any change to the original equation, but the monomials involved in the necessary initial condition are those which give the greatest order. Hence, the Newton polygon of $F(y)$ at infinity $\mathcal{N}_\infty(F)$ is defined as the convex hull of $\bigcup_{P \in \mathcal{P}(F)}(P + \{(a, 0) \mid a \leq 0\})$, $L_\infty(F; \mu)$ being the line with slope $-1/\mu$ which supports $\mathcal{N}_\infty(F)$, and $\Phi^\infty_{(F;\mu)}(C)$ as in (2) substituting $L_\infty(F; \mu)$ for $L(F; \mu)$. Then Lemma 1 holds substituting "lower order terms" for "higher ..." and $\Phi^\infty_{(F;\mu)}$ for $\Phi_{(F;\mu)}$. Using this version of Lemma 1 and Prop. 1, we have the following

Algorithm
Input: A first order ODE $F(y) = 0$ with rational coefficients.
Output: A bound for the degree of the polynomial solutions of $F(y) = 0$.
Let p the top vertex of $\mathcal{N}_\infty(F)$. Return N, where N is greater than any real root of $\Psi_{(F;p)}(\mu)$, and such that $N \geq \mu$, where μ is the inclination of the non-horizontal side of $\mathcal{N}_\infty(F)$ adjacent to p.

We may apply the above algorithm for any ODE provided the top vertex of $\mathcal{N}_\infty(F)$ is not of type (IV). For instance, the equation in Example 1 satisfies this hypothesis, so that $3/2$ as a bound for the degree of its polynomial solutions.

2.3   The Newton Polygon Process

Given a differential polynomial $F(y) \in K((x))^g[y_0, \ldots, y_n]$, the Newton polygon process constructs a tree $\mathcal{T}$. The root of $\mathcal{T}$ is τ_0. For each node τ of the tree there are three associated elements: a differential polynomial $F_\tau(y)$ with coefficients in $\bar{K}((x))^g$, an element $c_\tau \in \bar{K}$, and $\mu_\tau \in \mathbb{R} \cup \{-\infty, \infty\}$. For the root τ_0, we have $F_{\tau_0}(y) = F(y)$, $c_{\tau_0} = 0$ and $\mu_{\tau_0} = -\infty$.

Let τ be a node of $\mathcal{T}$ which is not a leaf. We are going to describe all its descendant nodes. First, if $y = 0$ is a solution of $F_\tau(y) = 0$, then there is a descendant σ of τ, which is a leaf and for which $F_\sigma = F_\tau$, $\mu_\sigma = \infty$ and $c_\sigma = 0$. The other descendant nodes of τ are in a bijective correspondence with the set $D_\tau = \{(c, \mu) \in \mathrm{NIC}(F_\tau); \mu > \mu_\tau\}$. For each $(c, \mu) \in D_\tau$, there is a descendant node σ for which $F_\sigma(y) = F_\tau(cx^\mu + y)$, $c_\sigma = c$ and $\mu_\sigma = \mu$.

The above tree depends on F and on the field K. If necessary, we shall write $\mathcal{T}(F)$ or even $\mathcal{T}(F; K)$ to clearly state which tree we are referring to. For instance, let $F = y'$, we may consider $K = \mathbb{C}$ or $L = \mathbb{C}(C)$. The tree $\mathcal{T}(F; L)$, has for each rational function $R(C)$ a node σ with $c_\sigma = R(C)$ and $\mu_\sigma = 0$.

As usual, the level of a node σ is k if the path of $\mathcal{T}$ from τ_0 to σ is $\tau_0, \tau_1, \ldots, \tau_k = \sigma$. We say that the tree $\mathcal{T}$ is *discrete* if for each $k \geq 0$, the number of nodes of level k is finite.

Example 2. Let $F(y)$ be as in example 1. We will describe here some nodes of the tree $\mathcal{T} = \mathcal{T}(F; \mathbb{C})$. The tree $\mathcal{T}$ has only one node of level zero: the root node τ_0. We have $F_{\tau_0}(y) = F(y)$, $c_{\tau_0} = 0$ and $\mu_{\tau_0} = -\infty$. From Table 1, we have that $\mathrm{NIC}(F_{\tau_0})$ is the set

$$D_{\tau_0} = \{(c,\mu)\,;\; -1/2 < \mu < 0,\, c \in \mathbb{C}\} \cup \{(c,\mu)\,;\, c \in \mathbb{C},\, \mu \in \{0,1\}\} \cup \{(-1/4, 2)\}$$

Since $y = 0$ is not a solutions of $F(y) = 0$, then the descendant nodes of τ_0 are in bijective correspondence with set D_{τ_0}. In particular, the tree $\mathcal{T}$ is not discrete. Let us consider σ_1 and σ_2 the descendant nodes of τ_0 corresponding to the elements $(2,0)$ and $(-1/3,0)$ of D_{τ_0} respectively. Thus, $F_{\sigma_1}(y) = F(2 + y)$, $F_{\sigma_2}(y) = F(-1/3 + y)$ and $\mu_{\sigma_1} = \mu_{\sigma_2} = 0$. Using the Newton polygon of F_{σ_1} and that of F_{σ_2}, we see that $D_{\sigma_1} = D_{\sigma_2} = \{(2\sqrt{-1}, 1), (-2\sqrt{-1}, 1)\}$, and $y = 0$ is not a solution of neither $F_{\sigma_1}(y) = 0$ nor $F_{\sigma_2}(y) = 0$. Hence, each σ_i, for $i = 1$ or $i = 2$, has exactly two descendant nodes $\sigma_{i,j}$, with $j = 1$ or $j = 2$. Now we have that $D_{\sigma_{2,j}} = \emptyset$ and $y = 0$ is not a solution of $F_{\sigma_{2,j}}(y) = 0$. Thus, $\sigma_{2,j}$ is a leaf of $\mathcal{T}$ because it has no descendant nodes. Lemma 1 implies that there are not solutions of $F(y) = 0$ of type $-1/3 + \sqrt{-1}\, x + \cdots$, where dots stands for higher order terms. We remark that $\mu_{\sigma_{2,j}} = 1 \neq \infty$. In following subsection we will see that branches of $\mathcal{T}$ with a leaf σ satisfying $\mu_\sigma \neq \infty$ does not correspond to solutions of $F(y) = 0$. On the other hand, we have that $D_{\sigma_{1,j}} = \{(3/14, 2)\}$ and $y = 0$ is not a solution $F_{\sigma_{1,j}}(y) = 0$, hence $\sigma_{1,j}$ has exactly one descendant node $\sigma_{2,j}$, where $c_{\sigma_{2,j}} = 3/14$ and $\mu_{\sigma_{2,j}} = 2$. Moreover, after the next subsection, we will be able to prove that there are two branches $(\sigma_{k,j})_{k \geq 0}$ of $\mathcal{T}$, where $j = 1, 2$ and we put $\sigma_{0,j} = \tau_0$, each one corresponding to a solution of $F(y) = 0$ of the form $2 + (-1)^j \sqrt{-1}\, x + 3/14\, x^2 + \cdots$.

2.3.1 Relation Between Branches of $\mathcal{T}$ and Solutions of $F(y) = 0$

The tree $\mathcal{T}$ has three types of branches: (a) infinite ones; (b) finite branches whose leaf σ has $\mu_\sigma = \infty$; and (c) finite branches whose leaf σ has $\mu_\sigma < \infty$. In this paragraph we will see that there exists a one to one correspondence between formal power series solution of $F(y) = 0$ and the set of branches of $\mathcal{T}$ of types (a) or (b).

Let $\tau_0, \tau_1, \ldots, \tau_k, \tau_{k+1}$ be a finite branch of $\mathcal{T}$ such that $\mu_{\tau_{k+1}} = \infty$. Let $\phi(x) = \sum_{i=1}^{k} c_{\tau_i} x^{\mu_{\tau_i}}$. We have that $F_{\tau_k}(y) = F(\phi(x) + y)$. Since $\mu_{\tau_{k+1}} = \infty$ we have that $y = 0$ is a solution of $F_{\tau_k}(y) = 0$. Hence, $\phi(x)$ is a solution of $F(y) = 0$.

Reciprocally, let $\phi(x) = \sum_{i=1}^{k} c_i x^{\mu_i}$ be a solution of $F(y) = 0$. By Lemma 1, there exists a finite branch $\tau_0, \tau_1, \ldots, \tau_k, \tau_{k+1}$ of $\mathcal{T}$ with $c_{\tau_i} = c_i$ and $\mu_i = \mu_{\tau_i}$ for $1 \leq i \leq k$.

Let $\tau_0, \tau_1, \ldots, \tau_k$ be a finite branch of $\mathcal{T}$ such that $\mu_{\tau_{k+1}} < \infty$. Let $\phi(x) = \sum_{i=1}^{k} c_{\tau_i} x^{\mu_{\tau_i}}$. We have that $F_{\tau_k}(y) = F(\phi(x) + y)$. Since τ_k is a leaf, one has $\mathrm{NIC}(F_{\tau_k}) \cap \bar{K} \times (\mu_{\tau_k}, \infty) = \emptyset$, and $y = 0$ is not a solution. By Lemma 1, this means that there is no solution $y(x)$ of $F(y) = 0$ such that $y(x) = \phi(x) + \psi(x)$ and with the terms of $\psi(x)$ of order greater than μ_{τ_k}. Hence, the above branch does not correspond to any solution of $F(y) = 0$.

In view of Lemma 1, it is obvious that if $y(x) = \sum_{i=1}^{\infty} c_i\, x^{\mu_i}$, where $c_i \neq 0$ for all i, is a solution of $F(y) = 0$, then there is an infinite branch $(\tau_i)_{i \geq 0}$ of $\mathcal{T}$. The reciprocal result is the following

Theorem 1. *Let $F(y) \in K((x))^g[y_0, \ldots, y_n]$. Let $B = (\tau_i)_{i \geq 0}$ be an infinite branch of the tree $\mathcal{T}$. Let $\phi(x) = \sum c_{\tau_i}\, x^{\mu_{\tau_i}}$. Then $\phi(x) \in \bar{K}((x))^g$ and is a solution of $F(y) = 0$. In particular, if all the exponents of $\phi(x)$ are rational, then they have a greatest common denominator and $\phi(x)$ is a Puiseux series.*

See [3, 4, 7, 8] for proofs of this theorem in different settings. They are based on the stabilization of the Newton polygon process, which also gives a recurrent formula for the coefficients of $\phi(x)$ which we will need later.

2.3.2 Stabilization of the Newton Polygon Process

In this paragraph we will see that given an infinite branch $(\tau_i)_{i \geq 0}$ of $\mathcal{T}$, there exists i_1, such that, for $i \geq i_1$, the node τ_i has only one descendant, in other words, the coefficient $c_{\tau_{i+1}}$ and the exponent $\mu_{\tau_{i+1}}$ are completely determined. We will give a recurrent formula for the coefficients $c_{\tau_{i+1}}$ and prove Theorem 1.

Lemma 2. *Let $G(y)$ be a differential polynomial and $H(y) = G(c\,x^{\mu} + y)$. Let $P = (t, h)$ be the point of highest ordinate in $L(G; \mu) \cap \mathcal{N}(G)$. Let $A = \{(t', h') \in \mathbb{R}^2 \mid h' \geq h\}$. Then $\mathcal{N}(G) \cap A = \mathcal{N}(H) \cap A$. Moreover, $\mathcal{N}(H)$ is contained in the right half-plane defined by $L(G; \mu)$. Finally, if $\Phi_{(G, \mu)}(c) = 0$, then the intersection point of $L(G, \mu)$ with the x-axis is not a vertex of $\mathcal{N}(H)$.*

Proof. Let $M(y) = a\,x^{\alpha}\,y_0^{\rho_0} \cdots y_n^{\rho_n}$ be a differential monomial. Let $P_{\alpha, \rho} = (t, h)$ its corresponding point. By simple computation, $M(c\,x^{\mu} + y) = M(y) + V(y)$, where $V(y)$ is a sum of differential monomials whose corresponding points have ordinate less than h and lie in the line passing through $P_{\alpha, \rho}$ with inclination μ. This implies the first two statements. In order to prove the last one, we remark that if $T(y)$ is the sum of monomials of $G(y)$ whose corresponding points lie in $L(G; \mu)$, then $T(c\,x^{\mu})$ is the coefficient of the monomial whose corresponding point is the intersection of $L(G, \mu)$ with the x-axis. Since $T(c\,x^{\mu}) = \Phi_{(G; \mu)}(c)$, we are done. $\square$

Lemma 3. *Let τ be a node of $\mathcal{T}$ with $\mu_{\tau} \neq \infty$. Then either $y = 0$ is a solution of $F_{\tau}(y) = 0$ or $\mathcal{N}(F_{\tau})$ has a side of inclination $\mu > \mu_{\tau}$.*

Proof. It is a consequence of Lemma 2 and the construction of $\mathcal{T}$.

Definition 1 (The Pivot Point). *For any non leaf node τ of $\mathcal{T}$, we will denote p_{τ} to be the point of least ordinate in $L(F_{\tau}; \mu_{\tau}) \cap \mathcal{N}(F_{\tau})$.*

Let $(\tau_i)_{i \geq 0}$ be an infinite branch of $\mathcal{T}$. Let $p_{\tau_i} = (\alpha_i, \beta_i)$. We have that $\beta_i \in \mathbb{N}$. By Lemma 2, $\beta_i \geq \beta_{i+1} \geq 1$ for $i \geq 1$. Hence, there exists i_0 such that, if $i \geq i_0$, then $\beta_{i_0} = \beta_i$. We have also that $p_{\tau_i} = p_{\tau_{i_0}}$ for $i \geq i_0$. We call point $p = p_{\tau_{i_0}}$ the pivot point of the branch.

We have that p is a vertex of $\mathcal{N}(F_{\tau_i})$ for $i \geq i_0$. Hence, the monomials of $F_{\tau_{i+1}}(y)$ and those of $F_{\tau_i}(y)$ corresponding to p are exactly the same for $i \geq i_0$.

In particular, the indicial polynomial $\Psi_{(F_{\tau_i};p)}(\mu)$ is the same for $i \geq i_0$; denote it by $\Psi(\mu)$. We say that the branch stabilizes at step $i_1 \geq i_0$ if μ_{τ_i} is not a root of $\Psi(\mu)$ for $i \geq i_1$.

Example 3. Following with the notation of example 2, let $(\tau_i)_{i \geq 0}$ be a branch of $\mathcal{T}$ such that $\tau_1 = \sigma_{1,1}$. We have that $p_{\tau_2} = (-2, 1)$. Since its ordinate is already equal to 1, the pivot point p of the branch is p_{τ_2}. The indicial polynomial $\Psi_{(F_{\tau_2};p)}(\mu) = \mu(1 + 3\mu)$. For $j \geq 2$ we have that $\mu_{\tau_j} \geq \mu_{\tau_2} = 2$, and so μ_{τ_j} is not a root of $\Psi(\mu)$. Hence, the branch $(\tau_i)_{i \geq 0}$ stabilizes at step 2.

By a simple use of the chain rule, we can prove (see Sect. 4 in [4]) the following

Lemma 4. *Let h be the ordinate of the pivot point p of the branch $(\tau_i)_{i \geq 0}$. Assume that $h \geq 2$ and that the differential variable $y^{(k)}$ actually appears in at least one of the monomials of $F_{\tau_{i_0}}(y)$ corresponding to the pivot point p. Then $\phi(x) = \sum c_{\tau_i} x^{\tau_i}$ is also a solution of $\frac{\partial F}{\partial y^{(k)}}(y) = 0$, and the pivot point of $\phi(x)$ with respect to $\frac{\partial F}{\partial y^{(k)}}(y)$ has ordinate $h - 1$.*

Remark 1. Let $(\tau_i)_{i \geq 0}$ be a branch of the tree $\mathcal{T}(F)$ associated with $F = 0$. By a successive application of the above lemma, there exists $G(y) = \frac{\partial^{|a|} F}{\partial y_0^{a_0} \cdots \partial y_n^{a_n}}(y) \neq 0$ such that the tree $\mathcal{T}(G)$ associated with $G(y) = 0$ has a branch $(\tau_i')_{i \geq 0}$ with $c_{\tau_i'} = c_{\tau_i}$ and $\mu_{\tau_i'} = \mu_{\tau_i}$ for $i \geq 0$, and whose corresponding pivot point p has ordinate equal to 1. Since p has ordinate equal to 1, we have that $\Psi_{(G_{\tau_i'};p)}(\mu)$ is a nonzero polynomial, hence it has a finite number of real roots. This implies that the branch $(\tau_i')_{i \geq 0}$ stabilizes at some step i_1.

Definition 2. *Let $F(y)$ be a differential polynomial. We say that $F = 0$ has quasi-linear solved form if $p = (0, 1)$ is a vertex of $\mathcal{N}(F)$, $\mathcal{N}(F) \subseteq \mathbb{R}_{\geq 0} \times \mathbb{R}$, and the indicial polynomial $\Psi_{(F;p)}(\mu)$ has no positive real roots.*

Proposition 2. *Let $F(y) \in R((x))^g[y_0, \ldots, y_n]$, where R is a ring, $\mathbb{C} \subseteq R \subseteq K$, and K is a field. Assume that $F = 0$ has quasi-linear solved form. Let us write*

$$F(y) = \sum_{\alpha, \rho} a_{\alpha, \rho} x^{\alpha + \rho_1 + 2\rho_2 + \cdots + n\rho_n} y_0^{\rho_0} y_1^{\rho_1} \cdots y_n^{\rho_n} \ ,$$

where the exponents α lie in a finitely generated semi-group Γ of $\mathbb{R}_{\geq 0}$. Then there exists a unique series solutions $\phi(x) \in K((x))^g$ of $F = 0$ with positive order. Moreover, let us write $\Gamma = \{\gamma_i\}_{i \geq 0}$, where $\gamma_i < \gamma_{i+1}$, for all i. Then, for each $i \geq 1$, there exists a polynomial $Q_i(\{A_{\alpha, \rho}\}, T_1, \ldots, T_{i-1})$, which only depends on Γ, and with coefficients in R, such that, if we write $\phi(x) = \sum_{i=1}^{\infty} d_i x_i^{\gamma}$, we have that

$$d_i = -\frac{Q_i(\{a_{\alpha, \rho}\}, d_1, \ldots, d_{i-1})}{\Psi_{(F;(0,1))}(\gamma_i)}, \quad i \geq 1 \ . \tag{4}$$

Proof. Consider $F(y) = M(y) + G(y)$, where $M(y)$ is the sum of those monomials of $F(y)$ whose corresponding point is $(0, 1)$. For any series $\phi(x) = \sum_{i \geq 1} d_i \, x^{\gamma_i}$, we have that $M(\phi(x)) = \sum_{i \geq 1} d_i \Psi_{(F;(0,1))}(\gamma_i) \, x^{\gamma_i}$ and $G(\phi(x)) = \sum_{i \geq 1} g_i \, x^{\gamma_i}$, where each g_i is a polynomial expression on $a_{\alpha,\rho}$ and $d_1, \ldots, d_{i-1}$. Since $\bar\Psi_{(F;(0,1))}(\gamma_i) \neq 0$, the relations $d_i \Psi_{(F;(0,1))}(\gamma_i) + g_i = 0$, $i \geq 1$, determine uniquely $\phi(x)$. □

Lemma 5. *Let $F(y) \in K((x))^g[y_0, \ldots, y_n]$ and $(\tau_i)_{i \geq 0}$ be an infinite branch of $\mathcal{T}$ with pivot point $p = (\alpha, 1)$ and which stabilizes at step i_1. Let ξ be a rational, $\mu_{\tau_i} < \xi < \mu_{\tau_{i+1}}$. Consider $H(y) = x^{-\alpha} F_{\tau_{i_1}}(x^\xi y)$. Then $H(y) \in K((x))^g[y, y']$ and has quasi-linear solved form.*

Proof. Let $G(y) = F_{\tau_{i_1}}(y)$. The affine map $(a, t) \mapsto (a - \alpha_0 + \xi t, t)$ is a bijection between the clouds of points of $G(y)$ and $H(y)$; hence it is a bijection between their Newton polygons, sending a side of inclination μ to a side of inclination $\mu - \xi$. Moreover, we have that $\Psi_{(H;(0,1))}(\mu) = \Psi_{(G;q)}(\mu + \xi)$. This proves that H has quasi-linear solved form. The fact that the series coefficients of $H(y)$ are grid-based is a consequence of the following fact: let Γ be a finitely generated semi-group and $\gamma \in \mathbb{R}$, then $A = (\gamma + \Gamma) \cap \mathbb{R}_{\geq 0}$ is contained in a finitely generated semi-group. To see this, let Γ be generated by $s_1, \ldots, s_k$. Let Σ be the set of $(n_1, \ldots, n_k) \in \mathbb{N}^k$ such that $\gamma + \sum n_i s_i > 0$ and for any $(n'_1, \ldots, n'_k) \neq (n_1, \ldots, n_k)$, with $n'_i \leq n_i$ for $1 \leq i \leq k$, one has that $\gamma + \sum n'_i s_i < 0$. Then Σ is finite, and A is generated by $1, s_1, \ldots, s_k$ and $\gamma + \sum n_i s_i$, where $(n_1, \ldots, n_k) \in \Sigma$.

Proof of theorem 1. By Remark 1, we may assume that the pivot point p of $(\tau_i)_{i \geq 0}$ has ordinate 1. Now apply Lemma 5 and Proposition 2. □

3   Formal Power Series Solutions of First Order ODE

Definition 3. *Let $F(y) \in \mathbb{C}((x))^g[y, y']$ and $\mathcal{T}$ be the tree constructed in the previous section. Let σ be a descendant node of τ which is not a leaf. We say that the node σ is of type (I)–(IV) if (c_σ, μ_σ) is a necessary condition of F_τ of the corresponding type (I)–(IV).*

Let τ be a node of $\mathcal{T}$. We call τ irrationally dicritical if it is of type (II) or (III). We say that τ is dicritical if the path $\tau_0, \ldots, \tau_{k+1} = \tau$ from τ_0 to τ satisfies that each τ_i is of type (I) for $1 \leq i \leq k$, τ is of type (II) or (III), and $\mu_\tau \in \mathbb{Q}$.

A branch of $\mathcal{T}$ is called dicritical (resp. irrationally dicritical) if it contains a dicritical (resp. irrationally dicritical) node. We will say that $F = 0$ is dicritical (resp. irrationally dicritical) at the origin if its tree has at least a dicritical (resp. irrationally dicritical) branch $(\tau_i)_{i \geq 0}$ with $\mu_{\tau_1} > 0$.

We remark that the tree $\mathcal{T}$ is discrete if and only if $F = 0$ has no irrationally dicritical branches.

This section is devoted to proving the following

Theorem 2. Let $F(y) \in \mathbb{C}((x))^g[y, y']$ be irrationally dicritical. Then there exists a one-parameter family of grid-bases series solutions of $F(y) = 0$ as follows

$$y_c(x) = \left(\sum_{i=1}^{k-1} b_i\, x^{\mu_i} \right) + c\, x^{\mu_k} + \left(\sum_{i=k+1}^{s} c_i\, x^{\mu_i} \right) + \left(\sum_{i=s+1}^{\infty} d_i\, x^{\mu_i} \right) \tag{5}$$

where

- each b_i is a fixed constant,
- the parameter $c \in \mathbb{C} \setminus E$, where E is a finite set,
- each c_i satisfies a polynomial equation $Q_i(c, c_{k+1}, \ldots, c_i) = 0$, and
- each $d_i = R_i(c, c_{k+1}, \ldots, c_s, d_{s+1}, \ldots, d_{i-1})/\mu_i\, g(c, c_{k+1}, \ldots, c_s)$, where R_i and g are polynomials with coefficients in $\mathbb{C}$.
- The exponents $\{\mu_i\}_{i \geq 1}$ do not depend on the parameter c (we allow zero coefficients in (5)).

Moreover, if $F(y) \in \mathbb{C}((x))^*[y, y']$ and $F = 0$ is dicritical, then there exists a one-parameter family as (5) with exponents $\{\mu_i\}_{i \geq 1} \subseteq \frac{1}{q}\mathbb{Z}$. If coefficients of $F(y)$ are convergent Puiseux series, then each $y_c(x)$ is also a convergent Puiseux series, hence it corresponds to the parametrization of an analytic branch curve.

Lemma 6. Let $F(x, y, y')$ be a differential polynomial with coefficients in $\mathbb{C}((x))^g$. Let C be an indeterminate over $\mathbb{C}$, and let L be an extension field of $\mathbb{C}(C)$. Let $C_1 \ldots, C_t \in L$. Consider the differential polynomial

$$G(x, y, y') = F(x,\ \phi + y,\ \phi_x + y') \in L((x))^g[y, y'],$$

where $\phi(x) = C + C_1\, x^{\mu_1} + \cdots + C_t\, x^{\mu_t}$, $0 < \mu_1 < \cdots < \mu_t$, and $\phi_x = \frac{d\phi}{dx}$. Let $p = (a, h)$ be any vertex of the bottom part of the Newton polygon of $G(x, y, y')$. (The bottom part of $\mathcal{N}$ is constituted by the sides of $\mathcal{N}$ with positive slope.)

Then there is only one monomial of $G(x, y, y')$ whose corresponding point is p and this monomial has the form $g\, x^a\, (x\, y')^h$, where g is a non-zero element of L. In particular, the characteristic polynomial associated to any side of the bottom part of $\mathcal{N}(G)$ has nonzero roots.

Proof. Multiplying F by a convenient x^α, we may assume that

$$F = \sum_{r \in \Gamma}\sum_{s \in \mathbb{N}} \varphi_{r,s}(y)\, x^r\, (xy')^s, \quad \varphi_{r,s}(y) \in \mathbb{C}[y]\ ,$$

where Γ is a finitely generated semi-group of $\mathbb{R}_{\geq 0}$ containing $1, \mu_1, \ldots, \mu_t$. We have that

$$G = \sum_{r \in \Gamma, s, l \geq 0, k \geq 0} \binom{s}{k} \frac{1}{l!}\, \varphi_{r,s}^{(l)}(\phi)\, x^r\, (x\phi_x)^{s-k}\, y^l\, (xy')^k\ . \tag{6}$$

Let $\frac{d}{dC}$ be the derivative with respect to C in $\mathbb{C}(C)$. We choose an extension of this derivation operator to L, and we extend it trivially to $L((x))^*[y, y']$. Hence,

$$\frac{d}{dC}\text{Coeff}_{x^a y^l (xy')^k}(G) = \text{Coeff}_{x^a}\left(\sum_{r,s}\binom{s}{k}\frac{1}{l!}\frac{d}{dC}\left\{\varphi_{r,s}^{(l)}(\phi)\,x^r\,(x\phi_x)^{s-k}\right\}\right) = A+B,$$

where

$$A = \text{Coeff}_{x^a}\left(\frac{d\phi}{dC}\sum_{r,s}\binom{s}{k}\frac{1}{l!}\varphi_{r,s}^{(l+1)}(\phi)\,x^r\,(x\phi_x)^{s-k}\right)$$

$$= \sum_{0\leq t\leq a}\text{Coeff}_{x^{a-t}}\left(\frac{d\phi}{dC}\right)\text{Coeff}_{x^t}\left(\sum_{r,s}\binom{s}{k}\frac{1}{l!}\varphi_{r,s}^{(l+1)}(\phi)\,x^r\,(x\phi_x)^{s-k}\right)$$

$$= \sum_{0\leq t\leq a}\text{Coeff}_{x^{a-t}}\left(\frac{d\phi}{dC}\right)(l+1)\text{Coeff}_{x^t y^{l+1}(xy')^k}(G),$$

$$B = \sum_{0\leq t\leq a}\text{Coeff}_{x^{a-t}}\left(x\frac{\partial\phi}{\partial x\partial C}\right)\text{Coeff}_{x^t}\left(\sum_{r,s}\binom{s}{k}\frac{s-k}{l!}\varphi_{r,s}^{(l)}(\phi)\,x^r\,(x\phi_x)^{s-k-1}\right)$$

$$= (k+1)\sum_{0\leq t<a}\text{Coeff}_{x^{a-t}}\left(x\frac{\partial\phi}{\partial x\partial C}\right)\text{Coeff}_{x^t y^l(xy')^{k+1}}(G)\ .$$

We remark that $t \in \Gamma$ and all the above sums are finite. The above equalities hold by (6) and because $\text{Coeff}_{x^0}\left(x\frac{\partial\phi}{\partial x\partial C}\right) = 0$.

Let $p = (a, h)$ be a vertex of the bottom part of $\mathcal{N}(G)$. Let $l'+k = h$. Assume that $k < h$, so that $l' \geq 1$. Let $l = l' - 1$. Then

$$\text{Coeff}_{x^t y^{l+1}(xy')^k}(G) = \text{Coeff}_{x^t y^l(xy')^{k+1}}(G) = 0, \quad \text{for } t < a\ .$$

From this, in the above computation, $A = \text{Coeff}_{x^a y^{l'}(xy')^h}(G)$ and $B = 0$. Hence

$$\text{Coeff}_{x^a y^{l'}(xy')^h}(G) = \frac{d}{dC}\text{Coeff}_{x^a y^l(xy')^k}(G) = 0\ .$$

The last equality holds because the point $(a, h - 1)$ does not belong to $\mathcal{N}(N)$. So, the only coefficient different from zero is that of $x^a y^0 (xy')^h$. Now let S be a side of the bottom part of $\mathcal{N}(G)$ with inclination $\mu_0 > 0$. Let (a, h) and $(a + k\mu_0, h - k)$ be the vertices of S. We have that

$$\Phi_{(G;\mu_0)}(T) = g_h\,(\mu_0\,T)^h + \cdots + g_{h-k}\,(\mu_0\,T)^{h-k}\ .$$

Since g_h and g_{h-k} are different from zero, $\Phi_{(G;\mu_0)}(T)$ has non-zero roots. $\quad\square$

Theorem 3. *Let $F(x, y, y') \in \mathbb{C}((x))^g[y, y']$. Assume that $F(y) = 0$ has a necessary initial condition (c, μ_0) of type (II) or (III). Let C be an indeterminate over $\mathbb{C}$ and $L = \mathbb{C}(C)$. Then $F = 0$ has a solution as follows*

$$y(x) = C\,x^{\mu_0} + \sum_{i=1}^{\infty} C_i\,x^{\mu_i} \in \bar{L}((x))^g\ .$$

Moreover, there exist i_1, a polynomial $g(T, T_1, \ldots, T_{i_1})$ with coefficients in $\mathbb{C}$ such that $\bar{g} = g(C, C_1, \ldots, C_{i_1}) \neq 0 \in \bar{L}$, and

$$C_i \in \mathbb{C}[C, C_1, \ldots, C_{i_1}, \frac{1}{\bar{g}}], \quad \text{for all } i \geq 1 .$$

Proof. Consider $G(y) = F(x^{\mu_0} y)$. Then $(c, 0)$ is a necessary initial condition for $G = 0$ of type (II) or (III). It suffices to prove the statement for $G(y)$ and $\mu_0 = 0$. We have that $\Phi_{(G;0)}(C) \equiv 0$. We consider $G(y) \in L((x))^g[y, y']$, hence $(C, 0)$ is a necessary initial condition for $G = 0$. Let $\mathcal{T} = T(G; L)$ be the tree of G constructed in the previous section. Let τ_1 be the node of $\mathcal{T}$ such that $c_{\tau_1} = C$ and $\mu_{\tau_1} = 0$. Let us prove that there exists a branch $(\tau_i)_{i \geq 0}$ of $\mathcal{T}$, passing through τ_1, which corresponds to a solution of $G(y) = 0$. Let $(\tau_i)_{i=0}^k$ be a path of $\mathcal{T}$. If $y = 0$ is a solution of $F_{\tau_k}(y) = 0$, we are done. If $y = 0$ is not a solution, by lemma 3 there exists a side S of $\mathcal{N}(F_{\tau_k})$ with inclination $\mu_{k+1} > \mu_k$. In particular, S is a side of the bottom part of $\mathcal{N}(F_{\tau_k})$. By Lemma 6, the characteristic polynomial $\Phi_{(F_{\tau_k}; \mu_{k+1})}(T)$ has a non zero root C_{k+1} in $\bar{L}$. Hence, the path $(\tau_i)_{i=0}^k$ can be continued to a branch $(\tau_i)_{i \geq 0}$ which corresponds to a solution $\phi(x) = C + \sum_{i \geq 1} C_i x^{\mu_i} \in \bar{L}((x))^g$ of $G(y) = 0$, where $C_i = c_{\tau_i}$ and $\mu_i = \mu_{\tau_i}$, for all $i \geq 1$.

It remains to prove the second part. If $\phi(x)$ is a finite sum of monomials we are done. Let $p = (\alpha, h)$ be the pivot point of $\phi(x)$ with respect to $G(y)$. If $h \geq 2$ then, by Lemmas 4 and 6, $\phi(x)$ is a solution of $\frac{\partial^{h-1} G}{\partial y'^{h-1}}$. Hence, we may assume that the pivot point is $p = (\alpha, 1)$ and it is reached at step i_1. By Lemma 6, we have that $\Psi_{(G;p)}(\mu) = g\mu$, where $g x^\alpha (xy')$ is the only monomial of $G_{\tau_{i_1}}(y)$ whose corresponding point is p. The element g is a polynomial expression on $C, C_1, \ldots, C_{i_1}$ with coefficients in $\mathbb{C}$. Let ξ be a rational number such that $\mu_{i_1} < \xi < \mu_{i_1+1}$. Consider $H(y) = x^{-\alpha} G_{\tau_1}(x^x i y)$. By Lemma 5 $H(y)$ has quasi-linear solved form. ¿From proposition 2, $\psi(x) = \sum_{i > i_1} C_i x^{\mu_i - \xi}$ is the only solution of $H(y) = 0$ with positive order and the coefficients C_i satisfy the recurrent equations (4). Let $R = \mathbb{C}[C, C_1, \ldots, C_{i_1}] \subseteq \bar{L}$, so that $H(y) \in R((x))^g[y, y']$. We have $\Psi_{(H;(0,1))}(\mu) = (\mu - \xi) g$. Using the recurrent equations (4), one sees that $C_i \in R[\frac{1}{g}] \subseteq \bar{L}$. $\square$

Corollary 1. *Let $\eta : \mathbb{C}[C, C_1, \ldots, C_{i_1}, \frac{1}{\bar{g}}] \to \mathbb{C}$ be a ring homomorphism. Let $c = \eta(C)$ and $c_i = \eta(C_i)$, for $i \geq 1$. Then $\eta(y(x)) = c x^{\mu_0} + \sum_{i=1}^\infty c_i x^{\mu_i}$ is a solution of $F(y) = 0$.*

Proof. Set $C_0 = C$ and $c_0 = c$. Let $F_k = F(\sum_{i=0}^k C_i X^{\mu_i})$ and $G_k = F(\sum_{i=0}^k c_i X^{\mu_i})$. We have that $\text{ord}(G_k) \geq \text{ord}(F_k)$, for all $k \geq 1$. $\square$

Remark 2. Since $C_1, \ldots, C_{i_1}$ are algebraic over $\mathbb{C}(C)$, there exist polynomials $Q_i \in \mathbb{C}[T, T_1, \ldots, T_i]$, $1 \leq i \leq i_1$, and $g \in \mathbb{C}[T, \ldots, T_{i_1}]$ such that if $\mathcal{C} = \{\underline{c} = (c, c_1, \ldots, c_{i_1}) \in \mathbb{C}^{i_1+1} \mid Q_i(\underline{c}) = 0, 1 \leq i \leq i_1, g(\underline{c}) \neq 0\}$ then there exists a homomorphism $\eta : \mathbb{C}[C, C_1, \ldots, C_{i_1}, \frac{1}{\bar{g}}] \to \mathbb{C}$ with $\eta(C_i) = c_i$ for $0 \leq i \leq i_1$ if and only if $\underline{c} \in \mathcal{C}$. Moreover, there exists a finite set $E \subseteq \mathbb{C}$ such that the projection $\pi : \mathcal{C} \to \mathbb{C} \setminus E$ over the first coordinate is onto. Then we obtain a one-parameter family of solutions of $F(y) = 0$ as described in Theorem 2.

Proof of Theorem 2. Let τ be an irrationally dicritical node of $\mathcal{T}(F; \mathbb{C})$. Let $(\tau_i)_{i=0}^{k+1}$ be the path from τ_0 to τ. Then $(c_{\tau_{k+1}}, \mu_{\tau_{k+1}})$ is a necessary initial condition of $F_{\tau_k}(y) = 0$ of type (II) or (III). Apply Theorem 3 and the above remark to $F_{\tau_k}(y) = 0$. If $F(y) \in \mathbb{C}((x))^*[y, y']$, consider a dicritical node τ and the path $(\tau_i)_{i=0}^{k+1}$, such that τ_i are of type (I) for $1 \leq 1 \leq \tau_k$. Hence $F_{\tau_{k+1}}$ has rational exponents. The solution constructed in Theorem 3 has also rational exponents because the necessary initial conditions used there correspond to sides. Hence, the family of solutions have rational exponents. Finally, by Lemma 6 the pivot point of all them has a corresponding monomial of type $\bar{g} \, x^\alpha \, (xy')$, with $\bar{g} \neq 0$. This guarantee the convergence of the solutions by a direct application of Theorem 2 of [4] or by the main theorem of [9]. □

References

1. Briot, C., Bouquet, J., Propriétés des fonctions définies par des équations différentielles. Journal de l'Ecole Polytechnique **36** (1856) 133–198
2. Fine, H., On the functions defined by differential equations, with an extension of the Puiseux Polygon construction to these equations. Amer. Jour. of Math. **XI** (1889) 317–328
3. Grigoriev, D.Y., Singer, M., Solving Ordinary Differential Equations in Terms of Series with Real Exponents. Trans A.M.S. **327** (1991) 329–351
4. Cano, J., On the series defined by differential equations, with an extension of the Puiseux Polygon construction to these equations. Analysis, Inter. Jour. Anal. its Appli. **13** (1993) 103–119
5. Cano, J., An extension of the Newton-Puiseux Polygon construction to give solutions of pfaffian forms. Ann. Inst. Fourier **43** (1993) 125–142
6. Della Dora, J., Jung, F., About the Newton Polygon Algorithm for Non Linear Ordinary Differential Equations. Proceedings of ISSAC'97 (1997)
7. van der Hoeven, J., Asymptotique automatique. PhD thesis, École Polytechnique (1997)
8. Cano, F., Moussu, R., Rolin, J.P., Non-oscillating integral curves and valuations. Crelle's journal (to appear)
9. Malgrange, B., Sur le théorème de Maillet. Asymptotic Anal. **2** (1989) 1–4

Implicit Reduced Involutive Forms and Their Application to Engineering Multibody Systems

Wenqin Zhou[1], David J. Jeffrey[1], Gregory J. Reid[1], Chad Schmitke[2], and John McPhee[2]

[1] Department of Applied Mathematics, The University of Western Ontario, London, Ontario, Canada N6A 5B7
[2] Systems Design Engineering, University of Waterloo, Waterloo, Ontario, Canada N2L 3G1

Abstract. The RIFSIMP package in Maple transforms a set of differential equations to Reduced Involutive Form. This paper describes the application of RIFSIMP to challenging real-world problems found in engineering design and modelling. RIFSIMP was applied to sets of equations arising in the dynamical studies of multibody systems. The equations were generated by the Maple package DYNAFLEX, which takes as input a graph-like description of a multibody mechanical system and generates a set of differential equations with algebraic constraints. Application of the standard RIFSIMP procedure to such Differential Algebraic Equations can require large amounts of computer memory and time, and can fail to finish its computations on larger problems.

We discuss the origin of these difficulties and propose an Implicit Reduced Involutive Form to assist in alleviating such problems. This form is related to RIFSIMP form by the symbolic inversion of a matrix. For many applications such as numerically integrating the multibody dynamical equations, the extra cost of symbolically inverting the matrix to obtain explicit RIFSIMP form can be impractical while Implicit Reduced Involutive Form is sufficient.

An approach to alleviating expression swell involving a hybrid analytic polynomial computation is discussed. This can avoid the excessive expression swell due to the usual method of transforming the entire input analytic differential system to polynomial form, by only applying this method in intermediate computations when it is required.

1   Introduction

A principal goal of multibody dynamics is the automatic generation of the equations of motion for a complex mechanical system, given a description of the system as input [14]. After generation, the set of equations must be analyzed or solved. Commercial programs exist that can generate and integrate such systems of equations. ADAMS, DADS and WORKING MODEL are examples of such products, and they are in widespread use in the automotive, aerospace and robotics industries [15]. These programs have many strengths, in particular they can handle systems containing many bodies (up to 100), but they have drawbacks.

H. Li, P. J. Olver and G. Sommer (Eds.): IWMM-GIAE 2004, LNCS 3519, pp. 31–43, 2005.
© Springer-Verlag Berlin Heidelberg 2005

For multibody systems, the general form of the dynamic equations is

$$M(t, q, \dot{q})\ddot{q} + \Phi_q^T \lambda = F(t, q, \dot{q}) , \tag{1}$$

$$\Phi(t, q) = 0 . \tag{2}$$

Here, q is a vector of generalized co-ordinates, $M(t, q, \dot{q})$ is the mass matrix, Φ is a vector of the constraint equations and λ is a vector of Lagrange multipliers [17, 18]:

$$\Phi = \begin{bmatrix} \Phi^1 \\ \vdots \\ \Phi^\ell \end{bmatrix} , \qquad \Phi_q = \begin{bmatrix} \Phi_{q_1}^1 & \cdots & \Phi_{q_n}^1 \\ \vdots & \vdots & \vdots \\ \Phi_{q_1}^\ell & \cdots & \Phi_{q_n}^\ell \end{bmatrix} , \qquad \lambda = \begin{bmatrix} \lambda^1 \\ \vdots \\ \lambda^\ell \end{bmatrix} .$$

Because the programs are purely numerical, it is difficult to check or comprehend the basic equations they have generated, and it is difficult to obtain analytic insight into the equations' properties. Also, when used for simulation, the programs are inefficient because the equations are effectively re-assembled at each time step, and the numerical assembly may include many multiplications in which one of the terms is 0 or 1. As a consequence, these programs are not well suited to real-time simulations and virtual reality, and, because of their large size, they cannot be downloaded onto the microprocessors that are typically used in real-time controllers [15, 3].

In contrast to numerically based programs, packages such as DYNAFLEX use symbolic programming to generate the equations of motion in a completely analytical form [16]. This approach offers several advantages [11]. The structure of the equations is easily obtained and manipulated, allowing one to gain a physical insight into the system; the equations can be easily exchanged with other researchers or engineers, something crucial to communication between different design groups; and real-time simulations are facilitated.

However, symbolic packages also have drawbacks. The equations generated by DYNAFLEX are usually too complicated to be solved symbolically. Even numerical solution is often difficult, inefficient or even impossible, because the equations are Differential Algebraic Equations (DAE), which typically are of second order but of high differential index. Also, the number of bodies that these programs can handle is not as large as for the numerically based programs.

Therefore, it is natural to develop methods for symbolically pre-processing the output of programs such as DYNAFLEX so that the output has desirable features such as being in simplified canonical form and including all constraints. Since any consistent initial value must satisfy all constraints, the inclusion of all constraints is a necessary condition for stating an existence and uniqueness theorem for such systems. The statement of such a theorem is another desirable feature for the output of such methods. Such features enable the consistent initialization of numerical solution procedures and can facilitate the identification of analytical solutions. In this paper we discuss how the RIFSIMP package can be used for the symbolic simplification of ODE and PDE systems and return canonical differential forms. It has the following features [9, 21]:

- Computation with polynomial nonlinearities.
- Advanced case splitting capabilities for the discovery of particular solution branches with desired properties.
- A visualization tool for the examination of the binary tree that results from multiple cases.
- Automatic generation of an existence and uniqueness theorem for the system.
- Algorithms for working with formal power series solutions of the system.

Applying RIFSIMP to the equations output by DYNAFLEX has the benefit of symbolically and automatically generating all special cases, through the RIF-SIMP case-split options [9, 21]. In a full case analysis, some cases can be very complicated while others can be simple enough to be solved analytically. The canonical form generated by RIFSIMP is of low (0 or 1) differential index [1, 10] which is suitable for the application of numerical solvers. An important option with RIFSIMP is the possibility of excluding special cases that are known to be not of interest. Thus if we know that a special case, say $m = 0$, is of no physical interest, then we can append the *inequation* $m \neq 0$ to the input system [9, 21].

Application of the RIFSIMP package to multibody systems revealed that it has difficulty handling large systems generated by DYNAFLEX. The symptoms are excessive time and memory requirements. It is a well-known effect in computer algebra that these are linked, in that a computation that overflows physical memory will cause the operating system to swap memory to disk. The swapping, however, essentially brings the system to a halt. There are a number of contributing factors to the growth in time and memory, as will be described below. Therefore, if one wants to handle industrial-scale problems, one must modify the RIFSIMP approach. The modification we introduce here is the possibility of relaxing the requirements on the canonical form.

We remark that these are common difficulties encountered during the application of computer algebra methods to obtain canonical forms (e.g. such as Gröbner Bases for systems of multi-variate polynomials). An underlying idea in this paper is that many application may not require the full potency of canonical simplification (canonicity can be very expensive); and it is important to explore weaker non-canonical forms when they may achieve the objective in the given application.

2   Two Examples of Mechanical Systems

Two examples will be used to illustrate the application of computer algebra methods to multi-body systems.

The three dimensional top example considered in the paper, is an example of a small open-loop system. Other examples of open loop systems include robot arms or similar devices with a free end and a fixed end. The slider-crank mechanism considered in the paper, is an example of a small closed-loop system. Another typical example of a closed loop is a piston turning a crank through connecting rods, so that there are constraints on both the crank and the piston. Simple textbook problems in dynamics use *ad hoc* choices of co-ordinate systems to produce simple systems of equations without constraints modelling the problems. But this method is usually not possible with complicated systems, which are automatically generated by packages such as DYNAFLEX and constraint equations can not be eliminated. In addition, the constraints introduce additional variables (essentially Lagrange multipliers) into the equations, representing the forces they exert.

2.1   Open Loop: Three-Dimensional Top

In this classic problem, the top is an axisymmetric body that can precess about a vertical (Z) axis, nutate about a rotated X axis, and spin about a body-fixed symmetry axis, see figure 1. The top is assumed to rotate without slipping on the ground; this is modelled by placing a spherical (ball-and-socket) joint at O. DYNAFLEX automatically generates co-ordinates using standard 3-1-3 Euler angles (ζ, η, ξ), which correspond to precession, nutation, and spin, respectively.

The top has a moment of inertia J about the symmetry axis, and A about an axis at O perpendicular to the symmetry axis. Two angles η (the angle of the axis of symmetry to the z axis), and ζ specify the orientation of the axis of symmetry of the top, while ξ specifies how a point on the top is moving relative to its axis of symmetry. There is a coordinate singularity when η is equal to 0 or π, which RIFSIMP will automatically detect as part of its case analysis. Coordinate singularities are ubiquitous in automatically generated mechanical systems, and their automatic detection is an important problem.

The dynamic equations (1,2) generated by DYNAFLEX are, after changing from DYNAFLEX-generated symbols to more conventional ones, as follows. For details, we refer to the DYNAFLEX Users Guide [16].

$$M = \begin{bmatrix} (A\sin^2\eta + J\cos^2\eta) & 0 & J\cos\eta \\ 0 & A & 0 \\ J\cos\eta & 0 & J \end{bmatrix}, \quad q = \begin{bmatrix} \zeta \\ \eta \\ \xi \end{bmatrix} \tag{3}$$

and

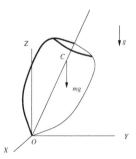

Fig. 1. The three-dimensional top. The centre of mass is at C and OC= l. Gravity acts in the $-Z$ direction

$$F = \begin{bmatrix} \sin(\eta) \left[2(J - A)\cos(\eta)\dot{\zeta} + J\dot{\xi} \right] \dot{\eta} \\ \sin(\eta) \left[(A - J)\cos(\eta)\dot{\zeta}^2 - J\dot{\xi}\dot{\zeta} + mgl \right] \\ J\sin(\eta)\dot{\eta}\,\dot{\zeta} \end{bmatrix} \qquad (4)$$

The fact that the top is an open-loop system is reflected in the fact that DYNAFLEX has generated 3 equations for 3 unknowns.

2.2   Two-Dimensional Slider Crank

The slider crank is a simple example of a closed-loop system with $q^T = (\theta_1, \theta_2)^T$ where $\theta_1 = \theta_1(t)$ and $\theta_2 = \theta_2(t)$ are the angles shown in figure 2. The system is given by (1)–(2) where:

$$M = \begin{bmatrix} l_1^2(\frac{m_1}{4} + m_2 + m_3) + J_1 & -l_1 l_2 \cos(\theta_1 + \theta_2)(\frac{m_2}{2} + m_3) \\ -l_1 l_2 \cos(\theta_1 + \theta_2)(\frac{m_2}{2} + m_3) & l_2^2(\frac{m_2}{4} + m_3) + J_2 \end{bmatrix} \qquad (5)$$

$$F = \begin{bmatrix} -l_1 g(\frac{m_1}{2} + m_2 + m_3)\cos\theta_1 - l_1 l_2 \dot{\theta}_2^2 \sin(\theta_1 + \theta_2)(\frac{m_2}{2} + m_3) \\ l_2 g(\frac{m_2}{2} + m_3)\cos\theta_2 - l_1 l_2 \dot{\theta}_1^2 \sin(\theta_1 + \theta_2)(\frac{m_2}{2} + m_3) \end{bmatrix} \qquad (6)$$

and there is a single constraint equation between the angles:

$$\Phi = l_1 \sin\theta_1 - l_2 \sin\theta_2 = 0 \qquad (7)$$

Therefore, $\Phi_q^T \lambda$ in (1) is given by $\Phi_q^T \lambda = \begin{pmatrix} l_1 \cos\theta_1 \\ -l_2 \cos\theta_2 \end{pmatrix} \lambda$. Note that in this example, λ is a scalar representing the normal reaction force of the constraint on the slider. In other words, in addition to generating the constraint equation (7), DYNAFLEX automatically generated the constraint force $\lambda(t)$. For both of these examples, the challenge now is to analyze the equations with computer algebraic methods such as RIFSIMP.

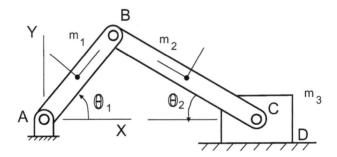

Fig. 2. The two-dimensional slider crank. The arm of length l_1 and mass m_1 rotates and causes the mass m_3 attached to the end of the arm of length l_2 to move left and right. Each arm has mass m_i and moment of inertia J_i, for $i = 1, 2$

3   Simplification Using RifSimp with Case Split

Given that symbolic algorithms such as DYNAFLEX exist for automatically producing the equations modelling multi-body systems, it is natural to exploit the further simplification and transformation of such systems using symbolic methods. In this section we discuss the simplification of such systems using the Maple algorithm `casesplit` which is part of RIFSIMP.

The theory underlying the Reduced Involutive Form given in [12] applies to systems of analytic nonlinear PDE in dependent variables $u_1, u_2, \ldots u_n$, which can be functions of several independent variables. While the general method applies to analytic systems, like most methods in computer algebra, the implemented algorithms apply to systems which are polynomial functions of their unknowns. This is to avoid well-known undecidability issues for classes of functions wider than polynomial or rational functions (e.g. there is no finite algorithm that can decide whether an analytic function is zero or non-zero at a point).

For the present application the only independent variable is time. In general the systems produced by DYNAFLEX are polynomial functions of sines and cosines of the angles $\theta_1(t), \theta_2(t), \ldots, \theta_n(t)$ between different components (arms and rotors etc).

One common method to convert such systems to rational form is the transformation

$$\cos \theta_j = x_j(t), \quad \sin \theta_j = y_j(t), \quad x_j^2 + y_j^2 = 1 \tag{8}$$

and another is the well-known Weierstrass transformation [5]:

$$\cos \theta_j = (1 - u_j(t)^2)/(1 + u_j(t)^2), \quad \sin \theta_j = 2u_j(t)/(1 + u_j(t)^2) \tag{9}$$

where $where u_j = tan(theta/2)$. If one solves for u_j, then the usual problems regarding choice of appropriate branch for the inverse arise. The transformation (9) has the advantage that the number of variables remains the same, and no new constraints are introduced. The transformation (8) has the disadvantage that additional constraints are introduced.

We will later discuss an alternative hybrid analytic-polynomial strategy. In that approach, the equations are manipulated in their original analytic form and conversions to polynomials by transformations such as those above are only used at intermediate computations and only for parts of the system which require the full algorithmic power of rational polynomial algebra. After resolving an analytic obstacle in this manner the inverse transformation yields the analytic form and the computation continues until the next algorithmic analytic obstacle is encountered.

RIFSIMP takes as its input a system of differential equations and a ranking of dependent variables and derivatives. Using its default ranking, RIFSIMP orders the dependent variables lexicographically and the derivatives primarily by total derivative order [9, 21]. For example for systems of ODE this default ranking is:

$$u_1 \prec u_2 \prec \ldots \prec u_n \prec u_1' \prec u_2' \prec \ldots \prec u_n' \prec u_1'' \prec \ldots \tag{10}$$

Each equation is then classified as being either leading linear or nonlinear, meaning linear or nonlinear in its highest derivative with respect to the ranking $\prec$.

RIFSIMP solves the leading linear equations for their highest derivatives until it can no longer find any such equations.

While solving explicitly for the highest derivatives, RIFSIMP splits cases based on the pivots (the coefficients of the leading derivatives) with which it divides. This yields a binary tree of cases.

Each leading-nonlinear equation (a so-called constraint) is differentiated and then reduced with respect to the current set of leading linear equations and then with respect to the leading nonlinear equations. A nonzero result means that this equation is a new constraint which should be appended to the system [9, 21].

For the current application if the solutions are 1 dimensional curves then each case output by the RIFSIMP algorithm has form:

$$v = f(t, w) \, , \tag{11}$$
$$g(t, w) = 0 \, , \tag{12}$$
$$h(t, w) \neq 0 \, . \tag{13}$$

Here v is the list of (highest-order) derivatives; w is a list of all derivatives, including dependent variables, lower in the ranking than v; g is a list of constraint equations; h is a list of inequations. From this form, it is possible to prove an existence and uniqueness theorem.

In particular, in our application, where there is a single independent variable t, the initial condition is $w(t^0) = w^0$ where the initial condition must satisfy the constraint $g(t^0, w^0) = 0$ and any inequations $h(t^0, w^0) \neq 0$ and this leads to a local analytic solution to the original system with this initial condition. Then the Existence and Uniqueness Theorem [12] states that there exists a local analytic solution satisfying the above initial conditions and inequations.

Application to Spinning Top

In order to apply RIFSIMP to equations (1,2) with M and F given by (3) – (4), we first convert the trigonometric functions to polynomials using the Weierstrass transformation, which is $\cos \eta = (1 - u(t)^2)/(1 + u(t)^2)$, $\sin \eta = 2u(t)/(1 + u(t)^2)$. This yields a rational polynomial differential system. The resulting case tree produced by `casesplit` is surprisingly large, containing 24 cases, see figure 3.

It is important to understand the reasons for the many cases discovered by RIFSIMP, because for more complicated systems, the case analysis can become overwhelming. We first note that rigid-body mechanics is inherently complicated, and flexible body mechanics more so. The motion is mostly rotational, meaning that linear and angular momentum must be considered (introducing mass and moment-of-inertia parameters), and that the equations contain trigonometric functions. Beyond this, however, we note that one of the advantages of symbolic analysis for the engineer is the use of *symbolic* parameters for the masses, moments of inertia, lengths, etc. This is useful for their design studies, but it is well known in symbolic computing that the introduction of large numbers of symbols causes expression swell, slowdowns in the computation, and the occurrence of many special cases. Finally, for RIFSIMP, there is the problem that the program

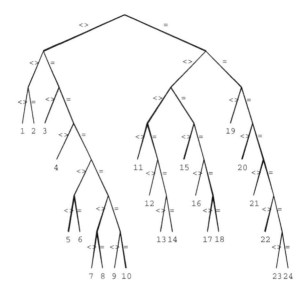

Fig. 3. Complete Case Tree for the 3D Top

assumes computation in the complex domain, whereas the engineering application only requires real variables. Thus the special cases identified in the complex plane are not relevant.

RIFSIMP has an input option that allows inequations to be appended to the system. These allow us to record physical facts such as $m \neq 0$, $g \neq 0$, etc. By recording as much information as possible in the list of inequations, the case tree can be significantly reduced. We find, for example, that the 24 cases in figure 3 can be reduced to 9. Amongst these special cases, RIFSIMP can identify dynamically degenerate cases. Examples are the cases of the top being oriented exactly vertically or exactly horizontally. In each case, one part of the precessional motion is not present. However, this strategy only delays the arrival of overwhelming expression swell, and we have therefore looked for an alternative to the standard RIFSIMP process.

4   Implicit Reduced Involutive Form Method

Two origins of expression swell for RIFSIMP are the following. If there are many equations containing many symbols, then inverting the matrix M to obtain explicit expressions for the $\ddot{q}$ results in division by many pivots that might be zero. After this, the reduction of equations modulo the existing equations is also difficult, because of the size of the equation set and the number of unknowns. From these observations we are led to seek a form weaker than a canonical solved form which is computationally feasible, while retaining as much of the power of RIFSIMP as possible. To this end, we introduce an *implicit* reduced involutive form.

Definition 4.1 [Implicit Reduced Involutive Form]: *Let $\prec$ be a ranking. A system $L = 0, N = 0$ is said to be in implicit reduced involutive form if there exist derivatives $r_1, ..., r_k$ such that L is leading linear in $r_1, ..., r_k$ with respect to $\prec$ (i.e. $L = A[r_1, \cdots, r_k]^T - b = 0$) and*

$$[r_1, \cdots, r_k]^T = A^{-1}b, \quad N = 0, \quad \det(A) \neq 0 \qquad (14)$$

is in reduced involutive form.

This form is of interest, since computing A^{-1} on examples can be very expensive. Sometimes implicit rif-form can be obtained very cheaply, just by appropriate differentiation of the constraints.

Example 4.1 [Spinning Top]: In this case the system has form $M\ddot{q} = F$ and this is in implicit reduced involutive form provided $\det(M) \neq 0$. In general implicit reduced involutive form can be regarded as a cheap way of obtaining some but not all of the cases resulting from a system (e.g. the cases in this example with $\det(M) = 0$ are not covered).

Example 4.2 [General Multi-Body Systems]: To convert a system of general form (1,2) with non-trivial constraints to implicit reduced involutive form one would have to at least differentiate the constraints twice. Carrying this out we obtain

$$M\ddot{q} + \Phi_q^T \lambda = F(t, q, \dot{q}) \qquad (15)$$
$$D_t^2\Phi = \Phi_q\ddot{q} + H\dot{q} + 2\Phi_{tq}\dot{q} + \Phi_{tt} = 0 \qquad (16)$$
$$D_t\Phi = \Phi_q\dot{q} + \Phi_t = 0 \qquad (17)$$
$$\Phi(t, q) = 0 \qquad (18)$$

where $\Phi_{tq} = \frac{\partial \Phi_q}{\partial t}$, $\Phi_{tt} = \frac{\partial^2 \Phi}{\partial t^2}$ and

$$H = \begin{bmatrix} \sum_i \Phi^1_{q_1 q_i}\dot{q}_i & \cdots & \sum_i \Phi^1_{q_n q_i}\dot{q}_i \\ \vdots & \vdots & \vdots \\ \sum_i \Phi^\ell_{q_1 q_i}\dot{q}_i & \cdots & \sum_i \Phi^\ell_{q_n q_i}\dot{q}_i \end{bmatrix}$$

We now show:

Theorem 4.1 *Consider the ranking $\prec$ defined by $q \prec \dot{q} \prec \lambda \prec \ddot{q} \prec \dot{\lambda} \prec \dddot{q} \cdots$ where the dependent variables q, λ are ordered lexicographically $q_1 \prec q_2 \prec ...$ and $\lambda_1 \prec \lambda_2 \prec ....$ The systems (15, 16, 17, 18) are in implicit rif-form with A, b, $[r_1, ..., r_k]^T$ in Definition 4.1 given by:*

$$A = \begin{bmatrix} M & \Phi_q^T \\ \Phi_q & 0 \end{bmatrix}, \quad b = \begin{bmatrix} F(t, q, \dot{q}) \\ -H\dot{q} - 2\Phi_{tq}\dot{q} - \Phi_{tt} \end{bmatrix}, \quad [r_1, ..., r_k]^T = \begin{pmatrix} \ddot{q} \\ \lambda \end{pmatrix} \qquad (19)$$

and $N = \{\Phi = 0, \Phi_q\dot{q} + \Phi_t = 0\}$, $\det(A) \neq 0$.

Proof: Set $N = \{\Phi = 0, D_t\Phi\} = \{\Phi = 0, \Phi_q\dot{q} + \Phi_t = 0\}$ in Theorem 4.1. Differentiating again yields $D_t^2\Phi$ given by (16).

Rewriting the system (15,16) in matrix form yields A, b, $[r_1, ..., r_k]^T$ in (19).

It remains to verify that $D_t \psi$ when reduced first with respect to L and then with respect to N yields zero for each $\psi \in N$. First $D_t \Phi = \Phi_q \dot{q} + \Phi_t$ is unaltered by reduction with respect to L and then reduces to zero with respect to N (since it is already in N). Next $D_t(\Phi_q \dot{q} + \Phi_t)$ is given in (16) and reduces to zero on simplification with respect to L (since it is a member of L). Note that we are working with analytic systems. Here as described in Rust [12] reduction to zero means detection as member of the analytic ring of functions (with coefficients again analytic functions) generated by the members of N. In general this is not algorithmic, but for multi-body systems by using one of the transformations to polynomial form, it can be converted into a polynomial ideal membership question which can be answered algorithmically, then transformed back to the analytic form. Here however the detection is trivial and does not require such techniques.

Again, we can note that implicit reduced involutive form easily obtained non-degenerate cases corresponding to $\det(A) \neq 0$. To determine whether a case is empty or not requires the analysis of whether there are any common solutions satisfying $N = 0$ and $\det(A) \neq 0$. This is a purely algebraic problem, which can be resolved in the worse case by applying one of the transformations to rational polynomial form. In that form one of the standard methods of commutative algebra (e.g. triangular sets) can be applied. Alternatively one can use some of the techniques of the new area of numerical algebraic geometry such as the methods of Sommese, Verschelde and Wampler [19]. That method determines so-called *generic* points on components of the variety determined by $N = 0$. Substitution of these points into A and application of some technique from numerical linear algebra (e.g. the singular value decomposition) can determine if $\det(A) \neq 0$.

A full analysis would have to consider the more difficult cases with $\det(A) = 0$. Indeed higher index problems (index > 2) yield $\det(A) = 0$ and further differentiations of the constraints need to be carried out to obtain implicit reduced involutive form.

Example 4.3 [Slider-Crank]: The example of the two-dimensional slider crank described above exactly fits into the class being discussed. Notice that even this simple case generates 5 independent parameters: each arm has a mass and a moment of inertia and the slider has a mass. The computation of this example is much harder to achieve in reduced involutive form.

5   Conclusion and Future Work

The mechanical systems generated by programs such as DYNAFLEX mean that they are ideal for testing new algorithms for dealing with DAE. The underlying idea is that such directly physical systems should lead to insights and new techniques for such DAE.

The implicit forms obtained can be useful in the numerical solution of such systems. For example, the matrix A above yields a system of DAE which can be solved using an implicit numerical method (i.e. along a solution curve, the constant matrix A evaluated at a certain time step is a constant matrix, which is inverted using stable numerical methods). Thus a very expensive symbolic (exact) inversion of a matrix has to be compared to the solution using LU decomposition at each step along the path. In many applications we stress that the repeated solution of these systems along the path, are *much cheaper than symbolically inverting the matrix once and then evaluating the solution along the path.* Finding a balance between paying the cost of symbolic simplification, on the one hand, and, on the other, finding ways of working with implicit representations is a subject of our ongoing research. This is important for example in being able to carry out real time simulations.

In some respects the method that we eventually are approaching is quite similar to that appearing in the literature (e.g. see Visconti [20]). The purpose of the article is to try to draw rigorous differential elimination approaches closer together with such methods. In addition we suggest the use of analytic systems to assist in efficiency (avoiding a full polynomialization of the problem, since this can increase the complexity of the problem). In our calculations full polynomialization led to many extra equations compared to the analytic approach. The total degree (which is a measure of the complexity of the system) was often dramatically increased by the transformations, and this was reflected in our experience with calculations.

Indeed it is quite surprising that some of the techniques in DAE have not produced analogous strategies in general differential elimination packages for ODE and PDE such as `diffalg` or the RifSimp package. Our article is an effort to try to bridge this gap. Indeed an interesting aspect of the article and the work was the interaction between the authors from mechanical engineering (McPhee and Schmitke) and those from computer algebra (Jeffrey, Reid and Zhou). It forced the computer algebraists to examine some of the underlying techniques and assumptions routinely made in computer algebra. For example the conversion of analytic systems to rational polynomial form, is almost automatic and unquestioned as desirable in computer algebra approaches. The restriction to polynomial or rational polynomial functions also arose historically in the largely algebraic earlier era of symbolic computation. But as indicated in this article such a conversion can lead to unnecessarily large expressions.

We briefly discuss and compare reduced involutive form [12, 21] with the coordinate independent involutive form of geometric PDE [8, 13]. (Geometric) involutive form, provided certain regularity conditions are satisfied, does not depend on the explicit form of the PDE, but instead on their locus in Jet Space. Reduced involutive form, although closely related to involutive form, is not always involutive but can simply be prolonged without eliminations to involutive form [7]. Implicit reduced involutive form is closer to involutive form than the coordinate dependent regular differential chains of differential algebra [6] and coordinate dependent reduced involutive form. Both these coordinate dependent forms depend on having systems triangularized or solved with respect to their leading derivatives in the given ranking. Roughly, the solved-form requirement is dropped in the introduction of implicit reduced involutive form.

Our planned work includes other strategies for controlling the generation of large expressions, since there will always be a desire on the part of design engineers to increase the number of bodies that can be modelled. One strategy for large expression management (LEM) has been described in [4]. The key idea is that large expressions are not arbitrary collections of terms, but contain a structure. An analogy can be drawn with the situation in the study of matrices arising in engineering applications: they almost always have a 'structure' to them. For example, they are banded, or otherwise sparse. By recognizing structure, we can solve larger problems. Returning to symbolic manipulation, we can note that simplification routines in computer algebra can cause a loss of structure, usually with the result that larger expressions are generated. A very simple example is the apparent simplification of $(1 + x)^9 - 1$, where a computer system will expand the bracket in order to cancel the 1 from the expansion with the 1 outside the bracket. Using the tools developed in [4] and now incorporated into Maple, we can preserve the structure inherent in engineering equations, such as those described here.

References

1. U. Ascher and L. Petzold. *Computer Methods for Ordinary Differential Equations and Differential-Algebraic Equations.* SIAM (1998).
2. B. Buchberger, G. E. Collins. *Computer Algebra Symbolic and Algebraic Computation.* Springer-Verlag (1983).
3. P. Rideau. *Computer Algegbra and Mechanics, The James Software.* Computer Algebra in Industry I. John Wiley (1993).
4. R. M. Corless, D. J. Jeffrey, M. B. Monagan, Pratibha. *Two Perturbation Calculations in Fluid Mechanics Using Large-Expression Management*, J. Symbolic Computation **11**, 1–17, (1996).
5. David A. Cox, John B. Little, Donal O'Shea. *Ideals, Varieties, And Algorithms.* Springer-Verlag (1997).
6. E. Hubert. *Factorization free decomposition algorithms in differential algebra.* J. Symbolic Computation 29: 641–662, (2000).
7. E. Mansfield. *A Simple Criterion for Involutivity.* Journal of the London Mathematical Society 54: 323–345, 1996.
8. J.F. Pommaret. *Systems of Partial Differential Equations and Lie Pseudogroups.* Gordon and Breach Science Publishers, Inc. (1978).
9. A.D. Wittkopf, G.J. Reid. *The Reduced Involutive Form Package.* Maple Software Package. First distributed as part of Maple 7. (2001).
10. G.J. Reid, P. Lin and A.D. Wittkopf. *Differential-Elimination Completion Algorithms for DAE and PDAE.* Studies in Applied Mathematics, **106**, 1–45, (2001).
11. Christian Rudolf. *Road Vehicle Modeling Using Symbolic Multibody System Dynamics.* Diploma Thesis, University of Waterloo in cooperation with University of Karlsruhe (2003).
12. C.J. Rust. *Rankings of Partial Derivatives for Elimination Algorithms and Formal Solvability of Analytic Partial Differential Equations.* Ph.D. Thesis, University of Chicago (1998).
13. W.M. Seiler. *Analysis and application of the formal theory of partial differential equations.* Ph.D. thesis, Lancaster University, (1994).

14. W. Schiehlen. *Multibody Systems Handbook.* Springer-Verlag (1990).
15. Pengfei Shi, John McPhee. *Symbolic Programming of a Graph-Theoretic Approach to Flexible Multibody Dynamics*; Mechanics of Structures and Machines, 30(1), 123-154 (2002).
16. Pengfei Shi, John McPhee. *DynaFlex User's Guide*, Systems Design Engineering, University of Waterloo (2002).
17. P. Shi, J. McPhee. *Dynamics of flexible multibody systems using virtual work and linear graph theory.* Multibody System Dynamics, 4(4), 355-381 (2000).
18. P. Shi, J. McPhee, G. Heppler. *A deformation field for Euler-Bernouli beams with application to flexible multibody dynamics.* Multibody System Dynamics, 4, 79-104 (2001).
19. A.J. Sommese, J. Verschelde, and C.W. Wampler. *Numerical decomposition of the solution sets of polynomial systems into irreducible components.* SIAM J. Numer. Anal. 38(6), 2022–2046 (2001).
20. J. Visconti. *Numerical Solution of Differential Algebraic Equations Global Error Estimation and Symbolic Index Reduction.* Ph.D. Thesis. Laboratoire de Modélisation et Calcul. Grenoble (1999).
21. A.D. Wittkopf. *Algorithms and Implementations for Differential Elimination.* Ph.D. Thesis. Simon Fraser University, Burnaby (2004).

Hybrid Method for Solving New Pose Estimation Equation System*

Greg Reid**, Jianliang Tang†, Jianping Yu, and Lihong Zhi‡

**Dept. of Applied Mathematics, University of Western Ontario,
London, Canada N6A 5B7
† Department of Mathematics, Shenzhen University,
Shenzhen 518060, P.R. China
‡ Key Laboratory of Mathematics Mechanization,
AMSS, Academia Sinica, Beijing 100080, P.R. China

Abstract. Camera pose estimation is the problem of determining the position and orientation of an internally calibrated camera from known 3D reference points and their images. We introduce a new polynomial equation system for 4-point pose estimation and apply our symbolic-numeric method to solve it stably and efficiently. In particular, our algorithm can also recognize the points near critical configurations and deal with these near critical cases carefully. Numerical experiments are given to show the performance of the hybrid algorithm.

1 Introduction

Given a set of correspondences between 3D reference points and their images, *4-point pose estimation* consists of determining the position and orientation of the camera with respect to four known reference points. It is a classical and common problem in computer vision and photogrammetry and has been studied in the past [1, 2, 6, 8, 11, 20, 23].

The well-known polynomial system (1) corresponding to the *4-point pose estimation* generically has a unique positive solution. It can be found successfully by linear algorithms proposed in [20, 2, 23]. But there are certain degenerate cases for which no unique solution is possible. These *critical configurations* are known precisely and include the following notable degenerate case: a 3D line and a circle in an orthogonal plane touching the line. In [2] an algorithm is presented that solves the problem including the critical configurations, but the relative error and failure rate (backward error) are significantly higher than one would like. In [23], the authors present a new linear algorithm which works well even in the degenerate cases. However, the matrices are much larger 70×90 compared with 24×24 matrices used in [2].

---

* Supported by NKBRPC 2004CB318000 and Chinese National Science Foundation under Grant 10401035 and Reids Canadian NSERC Grant.

H. Li, P. J. Olver and G. Sommer (Eds.): IWMM-GIAE 2004, LNCS 3519, pp. 44–55, 2005.

In this paper, we introduce a new variable and transform the polynomial system for *4-point pose estimation* to a new system with only five equations and three variables. Our symbolic-numeric method can also be applied to the new system and find solutions in general or critical cases. The matrices we used in the general or critical cases are of size 20×20; while in near critical cases, we are using a matrix of size 35×50 in order to recover the sensitive unique root.

The rest of the paper is organized as follows. In Section 2, we introduce the basic geometry of *the 4-point pose estimation problem*. A new system of equations is introduced. In Section 3, we briefly review the symbolic-numeric method for polynomial system solving. Then, we illustrate how to apply this method to solve the polynomial system corresponding to the critical or near critical cases. In Section 4, the simulated experimental results are given. Some conclusions are given in Section 5.

2   Geometry of Camera Pose from Four Points

In the following, we briefly introduce the geometry of camera pose from four points. Let C be the calibrated camera center, and P_1, P_2, P_3, P_4 be the reference points (see Fig. 1). Let $c_{12} = 2\cos \angle (P_1 C P_2), c_{13} = 2\cos \angle (P_1 C P_3), c_{14} = 2\cos \angle (P_1 C P_4),$ $c_{23} = 2\cos \angle (P_2 C P_3), c_{24} = 2\cos \angle (P_2 C P_4), c_{34} = 2\cos \angle (P_3 C P_4).$

From triangles $C P_1 P_2$, $C P_1 P_3$, $C P_2 P_3$, $C P_1 P_4$, $C P_2 P_4$ and $C P_3 P_4$, we obtain the $4-point$ *pose estimation equation system*:

$$
\begin{cases}
X_1^2 + X_2^2 - c_{12}X_1X_2 - |P_1P_2|^2 = 0, \\
X_1^2 + X_3^2 - c_{13}X_1X_3 - |P_1P_3|^2 = 0, \\
X_2^2 + X_3^2 - c_{23}X_2X_3 - |P_2P_3|^2 = 0, \\
X_2^2 + X_4^2 - c_{24}X_2X_4 - |P_2P_4|^2 = 0, \\
X_3^2 + X_4^2 - c_{34}X_3X_4 - |P_3P_4|^2 = 0, \\
X_1^2 + X_4^2 - c_{14}X_1X_4 - |P_1P_4|^2 = 0.
\end{cases}
\tag{1}
$$

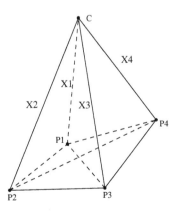

Fig. 1. The 4-point pose estimation problem

We are only interested in finding the positive solutions for X_1, X_2, X_3, X_4. Since $X_4 = |P_4C|$ is positive, we may make the following variable changes. Let

$$X_1 = x_1 X_4, \ X_2 = x_2 X_4, \ X_3 = x_3 X_4,$$
$$|P_1 P_4| = \sqrt{w} X_4, \ |P_1 P_2| = \sqrt{aw} X_4, \ |P_1 P_3| = \sqrt{bw} X_4,$$
$$|P_2 P_3| = \sqrt{cw} X_4, \ |P_2 P_4| = \sqrt{dw} X_4, \ |P_3 P_4| = \sqrt{ew} X_4.$$

Equation system (1) become the following equivalent equation system:

$$\begin{cases} x_1^2 + x_2^2 - c_{12} x_1 x_2 - aw = 0, \\ x_1^2 + x_3^2 - c_{13} x_1 x_3 - bw = 0, \\ x_2^2 + x_3^2 - c_{23} x_2 x_3 - cw = 0, \\ x_2^2 + 1 - c_{24} x_2 - dw = 0, \\ x_3^2 + 1 - c_{34} x_3 - ew = 0, \\ x_1^2 + 1 - c_{14} x_1 - w = 0. \end{cases} \tag{2}$$

From $x_1^2 + 1 - c_{14} x_1 - w = 0$ and $|c_{14}| < 2$ because $c_{14} = 2 \cos \angle(P_1 C P_4)$, we have

$$w = (|P_1 P_4|/X_4)^2 = x_1^2 + 1 - c_{14} x_1 = (x_1 - c_{14}/2)^2 + 1 - c_{14}^2/4 > 0.$$

X_4 can be uniquely determined by $X_4 = |P_1 P_4|/\sqrt{w}$ and the equivalent correspondence is:

$$(x_1, x_2, x_3, \sqrt{w}) \xleftarrow[\substack{X_1 = x_1 X_4, X_2 = x_2 X_4, X_3 = x_3 X_4}]{\substack{|P_1 P_4| = \sqrt{w} X_4}} \rightarrow (x_1, x_2, x_3, X_4)$$
$$\xleftarrow{} \rightarrow (X_1, X_2, X_3, X_4) \tag{3}$$

Substituting w into above equation system, we have the following equivalent equation system:

$$\begin{cases} (1-a)x_1^2 + x_2^2 - c_{12} x_1 x_2 - a(1 - c_{14} x_1) = 0, \\ (1-b)x_1^2 + x_3^2 - c_{13} x_1 x_3 - b(1 - c_{14} x_1) = 0, \\ (1-c)x_2^2 + x_3^2 - c_{23} x_2 x_3 - c(1 - c_{14} x_1) = 0, \\ (1-d)x_2^2 + 1 - c_{24} x_2 - d(1 - c_{14} x_1) = 0, \\ (1-e)x_3^2 + 1 - c_{34} x_3 - e(1 - c_{14} x_1) = 0. \end{cases} \tag{4}$$

The equation system (4) is simpler than the original system (1), and from the positive solution x_i we can get the coordinates X_i according to the equivalent correspondence. The recovered camera-point distances X_i are used to estimate the coordinates of the 3D reference points in a camera-centered 3D frame: $\bar{P}_i = X_i K^{-1} U_i$ (see [20]). The final step is the absolute orientation determination [21]. The determination of the translation and the scale follow immediately from the estimation of the rotation.

The system (4) is still an overdetermined polynomial system of five equations in 3 variables. The parameters c_{ij} ($1 \le i, j \le 4$) and a, b, c, d, e are data of limited accuracy. It is still very difficult to use Gröbner basis algorithms [4] or Ritt-Wu's characteristic algorithms [29, 31] to solve such approximate overdetermined polynomial systems. In the following, we briefly introduce our new developed complete linear method [23] for solving such system stably.

3 Linear Methods for Pose Determination from 4 Points

Consider a general polynomial system S in $x_1, \ldots, x_n$ of degree q and its corresponding vector of monomials of degree less than or equal to q. The system can be written as

$$M_0 \cdot [x_1^q, x_1^{q-1}x_2, \ldots, x_n^2, x_1, \ldots, x_n, 1]^T = [0, 0, \ldots, 0, 0, \ldots, 0, 0]^T \quad (5)$$

in terms of its coefficient matrix M_0. Here and hereafter, $[\ldots]^T$ means the transposition. Further, $[\xi_1, \xi_2, \ldots, \xi_n]$ is one of the solutions of the polynomial system, if and only if

$$[\xi_1^q, \xi_1^{q-1}\xi_2, \ldots, \xi_n^2, \xi_1, \ldots, \xi_n, 1]^T \quad (6)$$

is a null vector of the coefficient matrix M_0.

Since the number of monomials is usually bigger than the number of polynomials, the dimension of the null space can be big. The aim of completion methods, such as ours and those based on Gröbner bases and others [15, 12, 5, 14, 17, 18, 16, 25, 28], is to include additional polynomials belonging to the ideal generated by S, to reduce the dimension to its minima.

The bijection

$$\phi : x_i \leftrightarrow \frac{\partial}{\partial x_i}, \quad 1 \leq i \leq n, \quad (7)$$

maps the system S to an equivalent system of linear homogeneous PDEs denoted by R. Jet space approaches are concerned with the study of the jet variety

$$V(R) = \left\{ \left(\underset{q}{u}, \underset{q-1}{u}, \ldots, \underset{1}{u}, u \right) \in J^q : R \left(\underset{q}{u}, \underset{q-1}{u}, \ldots, \underset{1}{u}, u \right) = 0 \right\}, \quad (8)$$

where $\underset{j}{u}$ denotes the formal jet coordinates corresponding to derivatives of order exactly j.

A single prolongation of a system R of order q consists of augmenting the system with all possible derivatives of its equations, so that the resulting augmented systems, denoted by DR, has order $q + 1$. Under the bijection ϕ, the equivalent operation for polynomial systems is to multiply by monomials, so that the resulting augmented system has degree $q + 1$.

A single geometric projection is defined as

$$E(R) := \left\{ \left(\underset{q-1}{u}, \ldots, \underset{1}{u}, u \right) \in J^{q-1} : \exists \underset{q}{u}, R \left(\underset{q}{u}, \underset{q-1}{u}, \ldots, \underset{1}{u}, u \right) = 0 \right\}. \quad (9)$$

The projection operator E maps a point in J^q to one in J^{q-1} by simply removing the jet variables of order q (i.e. eliminating $\underset{q}{u}$). For polynomial systems of degree q, by the bijection ϕ, the projection is equivalent to eliminating the monomials of the highest degree q. To numerically implement an approximate involutive form method, we proposed in [30, 23] a numeric projection operator $\hat{E}$ based on singular value decomposition.

By the famous Cartan-Kuranishi Theorem [10, 19, 27], after application of a finite number of prolongations and projections, the algorithm above terminates with an involutive or an inconsistent system.

Suppose that R is involutive at prolonged order k and projected order l, and by the bijection ϕ has corresponding system of polynomials S with a finite number of solutions. Then the dimension of $\hat{E}^l(D^k R)$ allows us to determine the number of approximate solutions of S up to multiplicity. In particular these solutions approximately generate the null space of $\hat{E}^l(D^k R)$. We can compute eigenvalues and eigenvectors to find these solutions.

The following example corresponds to the third singular case as pointed in [2, 23]. In the example the coordinate of the camera point is $(1, 1, 1)$, and the coordinates of the four control points are $(-1, 1, 0)$, $(-1, -1, 0)$, $(1, -1, 0)$ and $(1, 1, 0)$ respectively. The corresponding *4-point pose estimation equation system* is:

$$\begin{cases} p_1 := x_2{}^2 - 2.0\,x_1{}^2 - 0.666667\,x_2 + 1.78885\,x_1 - 1.0, \\ p_2 := x_3{}^2 - x_1{}^2 - 0.894427\,x_3 + 0.894427\,x_1, \\ p_3 := x_2{}^2 - 1.49071\,x_1\,x_2 + 0.894427\,x_1 - 1.0, \\ p_4 := -x_1{}^2 + x_3{}^2 - 0.4\,x_1\,x_3 + 1.78885\,x_1 - 2.0, \\ p_5 := x_2{}^2 + x_3{}^2 - 1.49071\,x_2\,x_3 - x_1{}^2 + 0.894427\,x_1 - 1.0. \end{cases} \tag{10}$$

We show how our symbolic-numeric method can be used to solve (10). Under the bijection $\phi : x_i \leftrightarrow \frac{\partial}{\partial x_i}$ where $i = 1, 2, 3$, the system is equivalent to the PDE system R:

$$\begin{cases} \phi(p_1)u := \frac{\partial^2 u}{\partial x_2{}^2} - 2.0\frac{\partial^2 u}{\partial x_1{}^2} - 0.666667\frac{\partial u}{\partial x_2} + 1.78885\frac{\partial u}{\partial x_1} - 1.0u, \\ \phi(p_2)u := \frac{\partial^2 u}{\partial x_3{}^2} - \frac{\partial^2 u}{\partial x_1{}^2} - 0.894427\frac{\partial u}{\partial x_3} + 0.894427\frac{\partial u}{\partial x_1}, \\ \phi(p_3)u := \frac{\partial^2 u}{\partial x_2{}^2} - 1.49071\frac{\partial^2 u}{\partial x_1 \partial x_2} + 0.894427\frac{\partial u}{\partial x_1} - 1.0u, \\ \phi(p_4)u := -\frac{\partial^2 u}{\partial x_1{}^2} + \frac{\partial^2 u}{\partial x_3{}^2} - 0.4\frac{\partial^2 u}{\partial x_1 \partial x_3} + 1.78885\frac{\partial u}{\partial x_1} - 2.0u, \\ \phi(p_5)u := \frac{\partial^2 u}{\partial x_2{}^2} + \frac{\partial^2 u}{\partial x_3{}^2} - 1.49071\frac{\partial^2 u}{\partial x_2 \partial x_3} - \frac{\partial^2 u}{\partial x_1{}^2} + 0.894427\frac{\partial u}{\partial x_1} - 1.0u. \end{cases} \tag{11}$$

Applying the symbolic-numeric completion method to R with tolerance 10^{-9}, we obtain the table of dimensions below:

Table 1. Dim $(\hat{E}^l D^k R)$ for (11)

	$k = 0$	$k = 1$	$k = 2$	$k = 3$	$k = 4$
$l = 0$	5	2	2	2	2
$l = 1$	4	2	2	2	2
$l = 2$	1	2	2	2	2
$l = 3$		1	2	2	2
$l = 4$			1	2	2
$l = 5$				1	2
$l = 6$					1

We seek the smallest k such that there exists an $l = 0, ..., k$ with $\hat{E}^l D^k R$ approximately involutive. Passing the approximate projected elimination test amounts to test looking in the table for the first column with an equal entry in the next column on the downwards sloping diagonal (with both entries being on or above the main diagonal $k = l$). This first occurs for $k = 1$ and $l = 0, 1, 2$.

Applying the approximate version of the projected involutive symbol test to the example, shows that it is passed for $k = 1$, $l = 0$, and $l = 1$, so we choose the largest l ($l = 1$), yielding $\hat{E} D R$ as the sought after approximately involutive system.

The involutive system has $\dim(\hat{E} D R) = 2$ and so by the bijection the polynomial system (10) has 2 solutions up to multiplicity. In the following, we apply an eigenvalue method to solve (10).

1. Compute an approximate basis of the null space of DR, denoted by a 20×2 matrix B. Since $\dim(DR) = \dim(\hat{E} D R) = \dim(\hat{E}^2 D R) = 2$, the 4×2 submatrix B_1 and 10×2 submatrix B_2 of B by deleting entries corresponding to the second and third degree monomials are bases of null spaces of $\hat{E}^2 D R$ and $\hat{E} D R$ respectively.
2. Consider the set of all monomials of degree less than or equal to 1:

$$\mathcal{N} = [x_1, x_2, x_3, 1].$$

For numerical stability, we compute the singular value decomposition of B_1

$$U, S, V := \text{SingularValues}(B_1).$$

The first two columns of U form the 2×4 matrix U_s, and guarantee a stable linear polynomial set $\mathcal{N}_p = U_s^T \cdot \mathcal{N}^T$ for computing multiplication matrices.
3. The multiplication matrix of x_i with respect to $\mathcal{N}_p$ can be formed as

$$M_{x_i} = U_s^T \cdot B_{x_i} \cdot V^T \cdot S_i$$

where $B_{x_1}, B_{x_2}, B_{x_3}$ are the $[1, 2, 3, 7], [2, 4, 5, 8]$ and $[3, 5, 6, 9]$ rows of B_2 respectively, and S_i is a diagonal matrix with elements which are inversions of the first two elements of S: $1.95588, 111.524$.
4. The coordinates x_i of the double root can be found as the average of the eigenvalues of M_{x_i} for $i = 1, 2, 3$:

$$x_1 = 2.23607, x_2 = 3.0, x_3 = 2.23607. \tag{12}$$

Substituting the solution (12) into (10), we find $|p_i(\xi_1, \xi_2, \xi_3)| < 0.42 \cdot 10^{-7}$ for $i = 1, 2, \ldots, 5$. If one substitutes the positive solution (12) to the Jacobian matrix

$$\begin{bmatrix} \frac{\partial p_1}{\partial x_1} & \frac{\partial p_1}{\partial x_2} & \frac{\partial p_1}{\partial x_3} \\ \frac{\partial p_2}{\partial x_1} & \frac{\partial p_2}{\partial x_2} & \frac{\partial p_2}{\partial x_3} \\ \vdots & \vdots & \vdots \\ \frac{\partial p_5}{\partial x_1} & \frac{\partial p_5}{\partial x_2} & \frac{\partial p_5}{\partial x_3} \end{bmatrix} \tag{13}$$

then the singular values of the transpose of the Jacobian matrix are

$$11.8865, 5.42001, 0.109804 \cdot 10^{-8}.$$

The Jacobian matrix is near singular. This tells us that the solution is quite unstable for any small perturbations. Suppose we perturb (10) by errors of order 10^{-6}, the number of solutions read from the dimension table will generally become 1.

In general, we obtain the following table:

Table 2. Dim $(\hat{E}^l D^k R)$ for near critical case

	$k = 0$	$k = 1$	$k = 2$	$k = 3$	$k = 4$
$l = 0$	5	2	1	1	1
$l = 1$	4	2	1	1	1
$l = 2$	1	2	1	1	1
$l = 3$		1	1	1	1
$l = 4$			1	1	1
$l = 5$				1	1
$l = 6$					1

Applying the projected elimination and involutive symbol tests shows that $\hat{E}^2 D^2 R$ is approximately involutive. The computed positive root has backward error of order $10^{-6} \sim 10^{-9}$ in general.

In order to compare the difference between general cases, critical cases and near critical cases, in the below, we also show the dimension table corresponding to the general cases.

Table 3. Dim $(\hat{E}^l D^k R)$ for general case

	$k = 0$	$k = 1$	$k = 2$	$k = 3$	$k = 4$
$l = 0$	5	1	1	1	1
$l = 1$	4	1	1	1	1
$l = 2$	1	1	1	1	1
$l = 3$		1	1	1	1
$l = 4$			1	1	1
$l = 5$				1	1
$l = 6$					1

From the three different dimension tables, it is easy to deduce the following conclusions. Firstly, in the general case, the unique solution can be recovered from the null vector of the 20×20 matrix generated by $p_i, x_i\, p_j$ for $i, j = 1, 2, 3$. Secondly, if the four points are on the critical configuration, we have to deal it with eigenvalue method after forming the multiplication matrix with respect to x_1, x_2, x_3 separately. Finally, if the points are near the critical configuration, then the solution should be found stably from the null vector of the 35×50

matrix generated by $p_i, x_i p_j, x_i x_j p_k$ for $i, j, k = 1, 2, 3$. The main reason is due to that the dimension of the null space of the 20×20 matrix is two from table 2 in near degenerate cases.

4   Experimental Results

Based on the linear symbolic-numeric method, we may have the following algorithm for the *4-point pose estimation problem*:

- Compute the c_{ij} from the image points and the camera calibration matrix K.
- Compute the inter-point distances $|P_i P_j|$ from the reference points.
- Compute the solution x_1, x_2, x_3 of the polynomial system (4) using the symbolic-numeric method [23].
- Recover the camera-point distances X_1, X_2, X_3, X_4 from the equivalence correspondence (3).
- Estimate the coordinates of the 3D reference points in a camera-centered 3D frame: $\bar{P}_i = X_i K^{-1} U_i$.
- Compute the camera rotation and translation using the absolute orientation [9, 20, 21].

The following experiments are done with Maple 8 in the default setting of digits (Digits=10).

The first experiment is to show the accuracy and stability of the algorithm for the general *4-point pose estimation*. The optical center is located at the origin and the matrix of camera's intrinsic parameters is assumed to be the identity matrix. At each trial, four noncoplanar control points are generated at random within a cube centered at $(0, 0, 50)$ and of dimension $60 \times 60 \times 60$. The orientation Euler angles of the camera are positioned randomly. The control points are projected onto an image plane using the camera pose and internal parameters. We carry out one hundred trials and generate 100 sets of control points randomly for each trial. For a set of solutions, we substitute them into (1) and check the backward error. The backward error of the experimental results is generally less than 10^{-8}.

We check the stability of the algorithm. The relative error of the estimated translation t_i w.r.t. the true t is measured by $2|t_i - t|/(|t_i| + |t|)$. The relative error of the estimated rotation R_i w.r.t. the true R is measured by the sum of the absolute values of the three Euler angles of the relative rotation $R_i R^T$ (Fig. 2). We also check the failure rate defined as the percentage of total trials where either the rotation error or the translation error is over 0.5 (Fig. 3).

The second experiment is to show the accuracy and the stability of the algorithm in determining the solutions for the critical configurations. As mentioned in the introduction, the pose problem has some computationally troublesome singular cases. Fig. 4 and Fig. 5 show the relative error and the failure rate for one such critical configuration using our symbolic-numeric linear method.

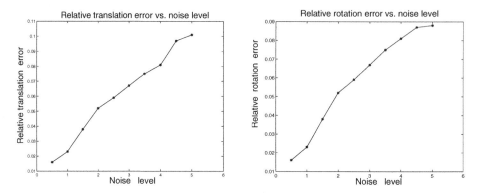

Fig. 2. Relative errors vs. noise level

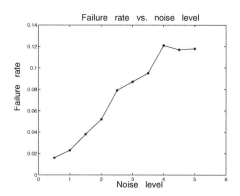

Fig. 3. Failure rate vs. noise level

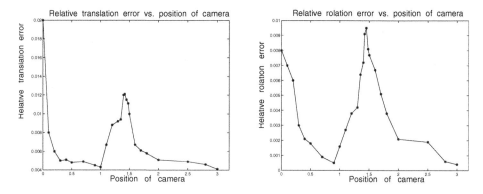

Fig. 4. Relative translation errors vs. noise level for the critical configurations

The data is 4 coplanar points in a square $[-1, 1] \times [-1, 1]$ and the camera starts at position=0, at a singular point directly above their center $(0.5 < h < 1.5)$,

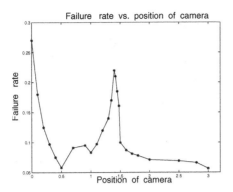

Fig. 5. Failure rate for the critical configurations

where h is the height of the camera. The camera then moves sideways parallel to one edge of the square. At position$=\sqrt{2}$ units it crosses the side of the vertical circular cylinder through the 4 data points, where another singularity occurs. From Fig. 4 and Fig. 5, the relative error and especially the failure rate of the algorithm are significantly lower compared with the algorithm in [2]. The relative error and the failure rate of our algorithm are also acceptable. It is natural that the error and failure rate near the position 0 and $\sqrt{2}$ are a little higher than at other positions.

It is clear that the experimental results are very similar to those we have presented in [23]. However, the computation is simpler due to smaller size of the polynomial system.

5   Conclusion

In this paper, we present a stable algorithm to find the numeric solution for *4-point pose estimation*. Although stated here in terms of derivatives, via the bijection, all the steps of the symbolic-numeric method, can be carried out by equivalent manipulations on the corresponding monomials, and the vector spaces they generate. The algorithm gives a unique solution whenever the control points are not sitting on one of the known critical configurations. When the control points are sitting on or near some known critical configurations, the algorithm also obtains reliable solutions. Compared with other algorithms, the main advantage of our linear algorithm is that it can recognize the critical and near critical cases and deal with different cases in different ways. The matrices in our approach are only bigger than those used in other approaches when the points are near critical configurations. The experiments show that the new simple polynomial system for *4-point pose estimation* can be robustly and reliably solved by our symbolic-numeric method in non-singular, singular and nearly singular configurations.

Acknowledgments

The authors are grateful to X.S. Gao, D.M. Wang for valuable discussions and to Marc-André Ameller, Bill Triggs and Long Quan for sending us their experimental data.

References

1. Abidi, M. A., Chandra, T., *A New Efficient and Direct Solution for Pose Estimation Using Quadrangular Targets: Algorithm and Evaluation, IEEE Transaction on Pattern Analysis and Machine Intelligence*, **17**(5),534-538, 1995.
2. Ameller, M.A., Triggs, B., and Quan, L., *Camera Pose Revisited - New Linear Algorithms*, private communication.
3. Bonasia, J., Reid, G.J., Zhi, L.H., *Determination of approximate symmetries of differential equations*, In Gomez-Ullate, Winternitz, editor, CRM Proceedings and Lecture Notes, *Amer. Math. Soc.* 39, 233-249, 2004.
4. Buchberger, B., *An Algorithm for Finding a Basis for the Residue Class Ring of a Zero-Dimensional Polynomial Ideal*, PhD. Thesis, Univ. of Innsbruck Math. Inst. 1965.
5. J.C. Faugére, *A New Efficient Algorithm for Computing Gröbner Bases without Reduction to Zero(F5)*, Proc. ISSAC, T. Mora, ed., New York, ACM Press, 75-83, 2002.
6. Gao, X.S., Hou, X.R., Tang, J.L. and Cheng, H., *Complete Solution Classification for the Perspective-Three-Point Problem, IEEE Tran. on Pattern Analysis and Machine Intelligence*, **25**(8), 534-538, 2003.
7. X.-S. Gao, J.L. Tang, *On the Solution Number of Solutions for the P4P Problem, Mathematics-Mechanization Research Center Preprints Preprint*, **21**, 64-76, 2002.
8. Horaud, R., Conio, B. and Leboulleux, O., *An Analytic Solution for the Perspective 4-Point Problem. CVGIP* **47**, 33-44, 1989.
9. Horn, B.K.P., *Closed Form Solution of Absolute Orientation Using Unit Quaternions, Journal of the Optical Society of America*, **5**(7), 1127-1135, 1987.
10. M. Kuranishi, *On E. Cartan's Prolongation Theorem of Exterior Differential Systems, Amer. J. Math*, **79**, 1-47, 1957.
11. Z.Y. Hu and F.C. Wu, *A Note on the Number Solution of the Non-coplanar P4P Problem, IEEE Transaction on Pattern Analysis and Machine Intelligence*, **24**(4), 550-555, April 2002.
12. D. Lazard, *Gaussian Elimination and Resolution of Systems of Algebraic Equations, Proc. EUROCAL 83*, 146-157, 1993.
13. Macaulay, F.S., *The Algebraic Theory of Modular Systems*, Cambridge Univ. Press **19** Cambridge tracts in Math. and Math. Physics, 1916.
14. B. Mourrain, *Computing the Isolated Roots by Matrix Methods, J. Symb. Comput.*, **26**, 715-738, 1998.
15. B. Mourrain and Ph. Trébuchet, *Solving Projective Complete Intersection Faster, Proc. ISSAC*, C. Traverso, ed., New York, ACM Press, 430-443, 2000.
16. H.M. Möller, T. Sauer, *H-bases for polynomial interpolation and system solving. Advances Comput. Math.*, to appear.
17. B. Mourrain, *A New Criterion for Normal Form Algorithms. Proc. AAECC*, Fossorier, M.Imai, H.Shu Lin and Poli, A., eds., LNCS, **1719**, Springer, Berlin, 430-443, 1999.

18. Ph. Trébuchet, *Vers une Résolution Stable et Rapide des Équations Algébriques.* Ph.D. Thesis, Université Pierre et Marie Curie, 2002.
19. J.F. Pommaret, *Systems of Partial Differential Equations and Lie Pseudogroups,* Gordon and Breach Science Publishers, 1978.
20. Quan, L. and Lan, Z., *Linear N-Point Camera Pose Determination, IEEE Transaction on PAMI,* **21**(8), 774-780, 1999.
21. P. Rives, P. Bouthémy, B. Prasada, and E. Dubois, *Recovering the Orientation and the Position of a Rigid Body in Space from a Single View, Technical Report,* INRS-Télécommunications, Quebec, Canada, 1981.
22. Reid, G.J., Lin, P. and Wittkopf, A.D., *Differential elimination-completion algorithms for DAE and PDAE, Studies in Applied Mathematics,* **106**(1), 1-45,2001.
23. Reid, G.J., Tang, J. and Lihong Zhi, *A complete symbolic-numeric linear method for camera pose determination, Proceedings of the 2003 International Symposium on Symbolic and Algebraic Computation,* Scotland, ACM Press, 215-223, 2003.
24. Reid, G.J., Smith, C. and Verschelde, J., *Geometric Completion of Differential Systems using Numeric-Symbolic Continuation, SIGSAM Bulletin* **36**(2), 1-17, 2002.
25. H.J. Stetter, *Numerical Polynomial Algebra,* SIAM, 2004.
26. Auzinger, W., Stetter, H., *An Elimination Algorithm for the Computation of All Zeros of a System of Multivariate Polynomial Equations, Numerical Mathematics Proceedings of the International Conference,* Singapore, 11-30, 1988.
27. Seiler, W.M., *Analysis and Application of the Formal Theory of Partial Differential Equations,* PhD. Thesis, Lancaster University, 1994.
28. Sommese, A.J., Verschelde, J. and Wampler, C.W., *Numerical decomposition of the solution sets of polynomial systems into irreducible components, SIAM J. Numer. Anal.* **38**(6), 2022-2046, 2001.
29. Wang, D., *Characteristic Sets and Zero Structures of Polynomial Sets, Preprint RISC-LINZ,* 1989.
30. Wittkopf, A.D. and Reid, G.J., *Fast Differential Elimination in C: The CDiffElim Environment, Comp. Phys. Comm.* **139**(2) 192-217, 2001.
31. Wu, W. T., *Basic principles of mechanical theorem proving in geometries Volume I: Part of Elementary Geometries,* Science Press Beijing(in Chinese) (1984), English Version Springer-Verlag 1995.

Some Necessary Conditions on the Number of Solutions for the $P4P$ Problem

Jianliang Tang

Department of Mathematics, College of Science, Shenzhen University,
Shenzhen 518060, P.R.China
jtang@szu.edu.cn

Abstract. The perspective-n-point (PnP) problem is to find the position and orientation of a camera with respect to a scene object from n correspondence points and is a widely used technique for pose determination in the computer vision community. Finding out geometric conditions of multiple solutions is the ultimate and most desirable goal of the multi-solution analysis, a key research issue of the problem in the literature. In this paper, we study the multi-solution phenomenon of the $P4P$ problem and give some necessary conditions under which there are five positive solutions for the $P4P$ problem. Moreover, we give a geometric configuration for the five solutions.

1   Introduction

Given a set of correspondences between 3D reference points and their images, *camera pose determination* is to determine the position and orientation of the calibrated camera with respect to the known reference points. The problem is called *PnP* problem in mathematics. It is a classical problem in computer vision, photogrammetry and mathematics. This problem has many interesting applications in robotics and cartography within many important fields, such as computer vision, automation, image analysis, automated cartography, photogrammetry, robotics and model based machine vision system, etc. It was first formally introduced by Fischler and Bolles in 1981 [6], and later extensively studied by others [1, 4, 8, 9, 14, 18]. The problem was summarized as follows:

> " Given the relative spatial locations of n control points, and given the angle to every pair of control points from an additional point called the Center of Perspective (C_P), find the lengths of the line segments joining C_P to each of the control points."

The perspective three points is the smallest subset of control points that yields a finite number of solutions. In the literature, the P3P problem has been extensively studied. Su et al. [21] showed the necessary and sufficient condition for the P3P problem. Fischler and Bolles [6] showed that the P3P problem has at most four positive solutions and this upper bound is also attainable via a concrete example. Haralick et al. [10] reviewed six different direct approaches to solve the

H. Li, P. J. Olver and G. Sommer (Eds.): IWMM-GIAE 2004, LNCS 3519, pp. 56–64, 2005.

P3P problem and carefully examined their numerical stabilities under different order of substitution and elimination. In the paper [8], Gao et al. gave a complete solution classification of the P3P problem and the necessary and sufficient algebraic conditions under which the P3P problem could have one, two, three, and four solutions.

Three point methods intrinsically give multiple solutions. In order to obtain a unique solution, we need to add more information and conditions. One of the natural way is to add one control point to consider a $P4P$ problem. For the $P4P$ problem, Rivers [20] gave a set of six quadratic equations with four variables. Fischler and Bolles attacked the problem by finding solutions associated with subsets of three points and selecting the solutions that they have in common. Horaud et al. [12] converted the $P4P$ into a special 3-line problem and obtained a fourth degree polynomial equation showing that this equation system has at most four solutions. In [18], Quan and Lan presented a linear algebraic algorithm which finds the unique solution in the generic case. Reid G. L. et al. [19] gave a complete linear algorithm to solve the $P4P$ problem including the critical configurations and the experiments showed that the algorithm is stable. In [9], Gao & Tang gave the triangular decomposition and closed form solution for this problem. When the control points are coplanar, the $P4P$ problem has a unique solution [1]. If the control points are not coplanar, the $P4P$ problem could have five solutions [14]. It had been proved that the $P5P$ problem could have two solutions [15]. For $n \geq 6$, the PnP problem has one solution and can be solved with the DLT method [2].

In this paper, we study the multi-solution phenomenon for the $P4P$ problem and give some necessary algebraic conditions under which the problem has five positive solutions. In particular, we give a concrete example indicating the $P4P$ problem may have five positive solutions.

The rest of the paper is organized as follows. In section 2, we introduce the $P4P$ problem and give a simplified form of the $P4P$ equation system. In section 3, we present the main result. In section 4, we give the conclusion.

2   The $P4P$ Equation System

Given a calibrated camera centered at P and correspondence between four 3D reference points P_i and their images U_i, each pair of correspondence i and j gives a constraint on the unknown camera-point distances $X_i = |P_i - P|$ (cf. Fig. 1):

$$P_{ij}(X_i, X_j) \equiv X_i^2 + X_j^2 + c_{ij}X_iX_j - d_{ij}^2 = 0$$

$$c_{ij} \equiv -2\cos\theta_{ij}$$

where $d_{ij} = |P_i - P_j|$ is the known inter-point distance between the i-th and j-th reference points and θ_{ij} is the 3D viewing angle subtended at the camera center by the i-th and j-th points. In the following, we will mainly discuss the case: $n = 4$. Let P be the center of perspective, and A, B, C, D be the

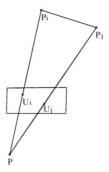

Fig. 1. The basic geometry of camera pose determination for each pair of correspondences

control points. Let $p = 2\cos\angle(BPC), q = 2\cos\angle(APC), r = 2\cos\angle(APB), s = 2\cos\angle(CPD), t = 2\angle(APD), u = 2\cos\angle(BPD), l_1 = |AB|^2, l_2 = |AC|^2, l_3 = |BC|^2, l_4 = |AD|^2, l_5 = |CD|^2, l_6 = |BD|^2$. From triangles PAB, PAC, PBC, PAD, PBD and PCD, we obtain the *P4P equation system*:

$$\begin{cases} p_1 = X^2 + Y^2 - XYr - l_1 = 0 \\ p_2 = X^2 + L^2 - XLq - l_2 = 0 \\ p_3 = Y^2 + L^2 - YLp - l_3 = 0 \\ p_4 = X^2 + Z^2 - XZs - l_4 = 0 \\ p_5 = Z^2 + L^2 - ZLt - l_5 = 0 \\ p_6 = Y^2 + Z^2 - YZu - l_6 = 0 \end{cases} \qquad (1)$$

We need to find the *positive solutions* for X, Y, Z, L. Since $L = |PC|$ is positive, we may make the following variable changes. Let $l_1 = awL^2, l_2 = bwL^2, l_3 = wL^2, l_4 = cwL^2, l_5 = dwL^2, l_6 = ewL^2, X = xL, Y = yL, Z = zL$. Equation system (1) becomes the following equivalent equation system:

$$\begin{cases} q_1 = x^2 + y^2 - xyr - aw = 0 \\ q_2 = x^2 + 1 - xq - w = 0 \\ q_3 = y^2 + 1 - yp - bw = 0 \\ q_4 = x^2 + z^2 - xzs - cw = 0 \\ q_5 = z^2 + 1 - zt - dw = 0 \\ q_6 = y^2 + z^2 - yzu - ew = 0 \end{cases} \qquad (2)$$

From $q_2 = x^2 + 1 - xq - w = 0$ and $|q| < 2$ (see (4)), we have $w = x^2 + 1 - xq = (x - q/2)^2 + 1 - q^2/4 > 0$. L can be uniquely determined by $L = |AC|/\sqrt{w}$. And the equivalent correspondence is:

$$(x_1, x_2, x_3, \sqrt{w}) \xleftarrow{\quad} \overset{|P_1P_4|=\sqrt{w}X_4}{----------} \xrightarrow{\quad} (x_1, x_2, x_3, X_4)$$

$$\overset{X_1=x_1X_4, X_2=x_2X_4, X_3=x_3X_4}{\xleftarrow{\quad}----------------\quad} \xrightarrow{\quad} (X_1, X_2, X_3, X_4)$$

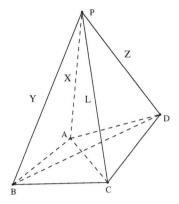

Fig. 2. The P4P problem

Substituting w into (2), we have the following equivalent equation system:

$$\begin{cases} h_1 = x^2 + y^2 - xyr - a(x^2 + 1 - xq) = 0 \\ h_2 = y^2 + 1 - yp - b(x^2 + 1 - xq) = 0 \\ h_3 = x^2 + z^2 - xzs - c(x^2 + 1 - xq) = 0 \\ h_4 = z^2 + 1 - zt - d(x^2 + 1 - xq) = 0 \\ h_5 = y^2 + z^2 - yzu - e(x^2 + 1 - xq) = 0 \end{cases} \quad (3)$$

The above equation system is simper than the original *P4P equation system*. This is a parametric equation system with three variables x, y, z and eleven parameters $\mathcal{U} = \{a, b, c, d, e, p, q, r, s, t, u\}$. It is clear that we need to add the following "reality conditions", which are assumed through out the paper.

$$\begin{cases} x > 0, y > 0, z > 0, a > 0, b > 0, c > 0, d > 0, e > 0 \\ -2 < p < 2, -2 < q < 2, -2 < r < 2, -2 < s < 2, -2 < t < 2, -2 < u < 2 \\ \text{Triangles } ABC, ACD, ABD, BCD \text{ do not degenerate to lines.} \end{cases} \quad (4)$$

solutions. Consider a P3P problem with a perspective center P and control points A, B, C. Assume that A, B, C are not on the same line. It is known that the P3P problem P-ABC has an infinite number of solutions if and only if P is on the circumscribed circle of the triangle ABC [8]. Since we assumed that the triangles ABC, ACD, ABD, BCD in Figure 2 would not degenerate to lines, the $P4P$ problem has an infinite number of solutions if and only if point P is on the circumscribed circles of these four triangles, which could happen if and only if points A, B, C, D and P are on the same circle.

3   Number of Solutions for *P4P* Problem

In [14], the author gave a theorem which showed that the upper bound of the positive solutions of the noncoplanar $P4P$ problem is five and also obtained some sufficient conditions for multiple positive solutions of the noncoplanar $P4P$ problem. In the following Theorem 2, we obtained some algebraic expressions as the necessary conditions for five positive solutions of the noncoplanar $P4P$ problem.

Lemma 1. *If the four control points of the P4P problem are not on the same plane then it could only have a finite number of solutions.*

Proof. Let the four control points be A,B,C and D. The $P3P$ problem with center of perspective P and control points A,B,C has an infinite number of solutions if and only if P is on the circumscribed circle of triangle ABC. Thus the $P4P$ problem has an infinite number of solutions if and only if point P is on the circumscribed circles of triangles ABC, ABD, ACD, BCD, which could happen if and only if points A,B,C,D and P are on the same circle. This contradicts to the fact that A,B,C,D are not co-planar.

Lemma 2. *Let P and Q be solutions of the P4P problem with non-planar control points A,B,C and D. If P and Q are symmetric with respect to plane ABC, then D must be on line PQ.*

Proof. Since the angles between line PQ and the line PA, PB and PC are given, PD is the intersection line of three cones with PA, PB and PC as the central axes respectively. PD is the unique intersection line of three cones Q_A, Q_B, Q_C with QA, QB, QC as central axes and passing through point D respectively. Suppose that D is not on line PQ. Let line PD meet plane ABC at point $\bar{D}$. Since A,B,C,D are not planar, $\bar{D} \neq D$. Since D is not on the line PQ, QD and $Q\bar{D}$ are different lines. Because P and Q are symmetric with plane ABC, we have $\angle AQ\bar{D} = \angle APD = \angle APD = \angle AQD$. Similarly, $\angle BQ\bar{D} = \angle BQD$, $\angle CQ\bar{D} = \angle CQD$. In other words, $Q\bar{D}$ is also the intersection of the three cones Q_A, Q_B, Q_C. This contradicts to the uniqueness of the intersection line. Therefore, D must be on line PQ.

Theorem 1. *The P4P problem with control points A, B, C and D could have up to five solutions. If the P4P problem has five solutions, then there exists one pair of solutions which are symmetric with respect to one of the plane ABC, ABD, ACD or BCD and among the other three solutions there exist no two which are symmetric with this plane.*

Proof. By Lemma 1, the $P4P$ problem could have a finite number of solutions. There are at most four solutions for the P3P problem with control points A, B and C on one side of plane ABC. If we consider both sides of plane ABC, the P3P problem could have eight solutions, of which four solutions are above the plane ABC and four solutions are under the plane ABC, are both solutions of the $P4P$ problem, then point D is on line PQ. It is clear that all other solutions cannot be on line PQ. It is impossible for another pair of solutions symmetric with plane ABC of the $P3P$ problem to exist since point D is already on line PQ. Then there is at most one pair of solutions symmetric with plane ABC. As a consequence, the $P4P$ problem has at most five solutions. In Table 1, we give

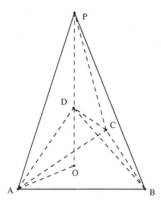

Fig. 3. The figure for multiple positive solutions

a concrete $P4P$ problem which has five solutions. Therefore, the $P4P$ problem could have up to five solutions.

Theorem 2. *Use the notations introduced above and the Fig. 3. Let $|DO| = h$. The equations*

$$\begin{cases} (l_4^2 - h^2)s^2(4 - s^2) = (l_6^2 - h^2)s^2(4 - t^2) \\ (l_4^2 - h^2)t^2(4 - u^2) = (l_5^2 - h^2)u^2(4 - t^2) \\ (l_5^2 - h^2)u^2(4 - s^2) = (l_6^2 - h^2)s^2(4 - u^2) \end{cases} \tag{5}$$

indicate the necessary conditions for multiple positive solutions of the noncoplanar $P4P$ problem.

Proof. *From the proof of above Theorem 1, we know that there exist five solutions if and only if point D is on the line PQ. Suppose that the point D is on the PQ. From the triangles PAO and DAO, we have the following expression: $|PO|^2 = (|AD|^2 - |DO|^2)\frac{s^2}{4-s^2}$. For the triangles PBO, DBO and the triangles PCO and DCO, there are similar results. Using the notations introduced above, we get the expressions of equations system (5).*

The following Table shows that the $P4P$ problem may have five solutions. Therefore, the $P4P$ problem could have up to five solutions. The coordinates for A, B, C, D are $(1/2, 0, 0), (-1/4, -\sqrt{3}/4, 0), (-1/4, \sqrt{3}/4, 0), (0, 0, -\sqrt{3}/12)$

Table 1. Parametric values for which the $P4P$ problem could have 5 solutions

Parameter	—AB—	—AC—	—BC—	—AD—	—CD—	—BD—
value	1	1	1	$\frac{\sqrt{13}}{6}$	$\frac{\sqrt{13}}{6}$	$\frac{\sqrt{13}}{6}$
Parameter	$\cos \angle BPC$	$\cos \angle APC$	$\cos \angle APB$	$\cos \angle CPD$	$\cos \angle APD$	$\cos \angle BPD$
value	$\frac{5}{8}$	$\frac{5}{8}$	$\frac{5}{8}$	$\frac{\sqrt{3}}{2}$	$\frac{\sqrt{3}}{2}$	$\frac{\sqrt{3}}{2}$

respectively. The coordinates for P_1, P_2, P_3, P_4, P_5 are $(0, 0, 7\sqrt{3}/12)$, $(0, 0, -7\sqrt{3}/12)$, $(5/8, 0, \sqrt{3}/8)$, $(-5/16, -5\sqrt{3}/16, \sqrt{3}/8)$, $(-5/16, 5\sqrt{3}/16, \sqrt{3}/8)$ respectively and are showed by following figures.

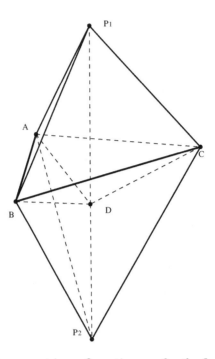

Fig. 4. The geometric configuration one for the five solutions

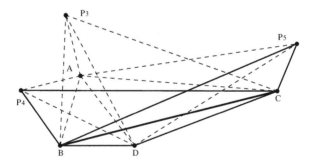

Fig. 5. The geometric configuration two for the five solutions

Remarks:

1. In this paper, we give a pure geometric proof that the $P4P$ problem could have five solutions. The proof is simple. Theorem 1 also gives clear characterization for the geometrical configurations of the solutions which can bring some new insights into a better understanding of the multi-solution problem.

2. The algebraic equations system (5) in the Theorem 2 gives some necessary conditions under which the problem has five solutions. Furthermore, we can obtain more algebraic conditions similar to the algebraic equations system (5).

4   Conclusion

The Perspective-n-Point (PnP) problem originated from camera calibration. Also known as pose estimation, it is to determine the position and orientation of the camera with respect to a scene object from n corresponding points and extensively studied in computer vision and mathematics. For the $P4P$ problem, there are many algorithms to solve the $P4P$ equation system including the critical configurations. From the geometrical view, we study the multi-solution phenomenon and give a geometrical proof that the $P4P$ problem has at most five solutions. From the process of proof, we obtain some necessary algebraic conditions under which the noncoplanar $P4P$ problem has five solutions. Moreover, we give a concrete example to show our results.

Acknowledgements

The author is grateful to professor X. S. Gao as supervisor and to Hongbo Li, D. K. Wang and L. H. Zhi for valuable suggestions.

References

1. Abidi, M. A., Chandra, T.: A New Efficient and Direct Solution for Pose Estimation Using Quadrangular Targets:Algorithm and Evaluation. IEEE Transaction on Pattern Analysis and Machine Intelligence **17**(5) (1995) 534-538
2. Abdel-Aziz, Y. I. and Karara, H. M.: Direct Linear Transformation into Object Space Coordinates in Close-range Photogrammetry. Proc. Symp. Close-range Photogrammetry (1971) 1-18
3. Ansar, A. and Daniilidis, K.: Linear Pose Estimation from Points and Lines. IEEE Transaction on Pattern Analysis and Machine Intelligence **25**(5) (2003) 578-589
4. DeMenthon, D. and Davis, L. S.: Exact and Approximate Solutions of the Perspective-Three-Point Problem. IEEE Transaction on Pattern Analysis and Machine Intelligence **14**(11) (1992) 1100-1105
5. DeMenthon, D. F. and Davis, L. S.: Model-Based Object Pose in 25 Lines of Code. International Journal of Computer Vision **15** (1995) 123-141
6. Fischler, M. A., Bolles, R. C.: Random Sample Consensus: A Paradigm for Model Fitting with Applications to Image Analysis and Automated Cartomated Cartography. Communications of the ACM **24**(6) (1981) 381-395
7. Ganapathy, S.: Decomposition of Transformtion Matrices for Robot Vision. Proc. IEEE Con. Robotics and Automation IEEE Press (1984) 130-139
8. Gao, X. S., Hou, X. R., Tang, J. L. and Cheng, H.: Complete Solution Classification for the Perspective-Three-Point Problem. IEEE Transaction on Pattern Analysis and Machine Intelligence **25**(8) (2003) 534-538

9. Gao, X. S., Tang, J. L.: On the Solution Number of Solutions for the $P4P$ Problem. Mathematics-Mechanization Research Center Preprints Preprint **21** (2002) 64-76
10. Haralick, R. M., Lee, C., Ottenberg, K. and Nolle, M.: Analysis and Solutions of the Three Point Perspective Pose Estimation Problem. Proc. of the Int. Conf. on Computer Vision and Pattern Recognition (1991) 592-598
11. Hartley, R. I. and Zisserman, A.: Multiple view geometry in computer vision. Cambridge University Press (2000)
12. Horaud, R., Conio, B., and Leboulleux, O.: An Analytic Solution for the Perspective 4-Point Problem. Computer Vision, Graphics and Image Processing **47** (1989) 33-44
13. Horn, B. K. P.: Closed Form Solution of Absolute Orientation Using Unit Quaternions. Journal of the Optical Society of America. **5**(7) (1987) 1127-1135
14. Hu, Z. Y., Wu, F. C.: A Note on the Number Solution of the Non-coplanar $P4P$ Problem. IEEE Transactions on Pattern Analysis and Machine Intelligence. **24**(4) (2002) 550-555
15. Hu, Z. Y., Wu F. C.: A Study on the $P5P$ Problem. Chinese Journal of Software **12**(5) (2001) 768-775(in Chinese)
16. Mishra, B.: Algorithmic Algebra. Springer-Verlag New York (1993)
17. Pehkonen, K., Harwood, D., and Davis, L. S.: Parallel Calculation of 3-D Pose of a Known Object in a Single View. Pattern Recognition Lett. **12** (1991) 353-361
18. Quan, L. and Lan, Z.: Linear N-Point Camera Pose Determination. IEEE Transaction on Pattern Analysis and Machine Intelligence **21**(8) (1999) 774-780
19. Reid, G. J., Tang, J. and Lihong Zhi: A complete symbolic-numeric linear method for camera pose determination. Proceedings of the 2003 International Symposium on Symbolic and Algebraic Computation Scotland ACM Press (2003)
20. Rives, P., Bouthémy, P., Prasada, B. and Dubois, E.: Recovering the Orientation and the Position of a Rigid Body in Space from a Single View. Technical Report INRS-Télécommunications place du commerce Ile-des-Soeurs Verdun H3E 1H6 Quebec Canada **3** (1981)
21. Su, C., Xu, Y., Li, H. and Liu, S.: Necessary and Sufficient Condition of Positive Root Number of P3P Problem (in Chinese). Chinese Journal of Computer Sciences **21** (1998) 1084-1095
22. Su, C., Xu, Y., Li, H. and Liu, S.: Application of Wu's Method in Computer Animation. The Fifth Int. Conf. on Computer Aided Design/Computer Graphics **1** (1997) 211-215
23. Dingkang Wang and Xiao-Shan Gao: Counting the Number of Solution for Algebraic Parametric Equation Systems. Mathematics-Mechanization Research Center Priprints **20** (2001) 209-220
24. Wolfe, W. J. and Jones, K.: Camera Calibration Using the Perspective View of a Triangle. Proc. SPIE Conf. Auto. Inspection Measurement **730** (1986) 47-50
25. Wu, W. T.: Basic principles of mechanical theorem proving in geometries. Volume I: Part of Elementary Geometries Science Press Beijing(in Chinese) (1984) English Version Springer-Verlag (1995)
26. Wu, W. T.: Mathematics Mechanization. Science Press Beijing(in Chinese) (2000) English Version Kluwer Aacademic Publishers London (2000)
27. Yuan, J. S. C.: A General Photogrammetric Method for Determining Object Position and Orientation. IEEE Transaction on Robotics and Automation **5**(2) (1989) 129-142

A Generalization of Xie-Nie Stability Criterion[*]

Xiaorong Hou and Xuemin Wang

Institute of Information and Computing Sciences,
Ningbo University, Ningbo 315211, China

Abstract. In this paper, we obtain a sufficient condition for a polynomial with real positive coefficients to be stable, which generalizes Xie-Nie stability criterion.

Keywords: Hurwitzean polynomial; stability criterion.

1 Introduction

One of the fundamental problems in the stability theory is to determine whether a polynomial is Hurwitzean or not. The well-known Routh-Hurwitz criterion, which gives a sufficient and necessary condition for being Hurwitzean polynomial, is algorithmically not a practical method in most cases. Hence it is necessary to find more efficient, simpler conditions for being Hurwitzean polynomial. There have been many papers devoted to the study of this problem.

In [1], X. Xie presented a sufficient condition for being Hurwitzean polynomial. Then Y. Nie [2] improved the result as follows: Let H be the set of all Hurwitzean polynomials, then any polynomial with real positive coefficients,

$$f(x) = a_n x^n + a_{n-1} x^{n-1} + \dots + a_1 x + a_0, \tag{1}$$

is in H if

$$b_i \leq \delta_0, \quad \text{for } i = 1, \dots, n-2, \tag{2}$$

where $\delta_0 \approx 0.4655712319$ is the unique real root of polynomial $\delta^3 + 2\delta^2 + \delta - 1$, called *Xie-Nie constant*, and

$$b_i = \frac{a_{i-1} a_{i+2}}{a_i a_{i+1}}. \tag{3}$$

Among other approaches, V.V. Maslennikhov [6] proposed a beautiful conjecture: $f(x) \in H$ if

$$b_1 + b_2 + \dots + b_{n-2} < 1, \tag{4}$$

which was proved to be true by A. F. Kleptsyn[1]. Recently, A. Borobia, S. Dormido [3] and X. Yang [4,5] obtained some necessary conditions for Hurwitzean polynomials.

---

[*] Supported by the NNSF of China (No. 60273095), NKBRSF-2004CB318003 and NSF of Zhejiang (No. Y604089)
[1] Informed by an anonymous referee.

H. Li, P. J. Olver and G. Sommer (Eds.): IWMM-GIAE 2004, LNCS 3519, pp. 65–71, 2005.

In this paper, we obtain the following **Main Theorem**:

Theorem 1. *Let $f(x)$ and b_i be defined as in (1) and (3). Then $f(x) \in H$ if*

$$\begin{cases} b_{i-1} + b_i \leq \sigma, & i = 2, 3, \cdots, n-2, \\ b_l \leq \beta, & l = 1, 2, \cdots, n-2, \end{cases} \tag{5}$$

where σ, β satisfy

$$\begin{cases} (\sigma^2 + 4\beta)^2 - 16\beta = 0, \\ \beta \leq \sigma \leq 2\beta. \end{cases} \tag{6}$$

The equal signs in (5) should be omitted when $n = 5$.

Our result is a significant generalization of the Xie-Nie stability criterion (2). This can be seen also from the following figure, in which the bold curve shows the range of parameters β and σ. The special case $\sigma = 2\beta$ and $\beta = \delta_0$ where δ_0 is the Xie-Nie constant, is just Xie-Nie's stability criterion.

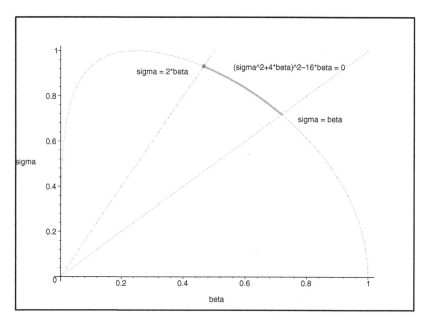

Fig. 1. Our criterion (bold curve) vs. Xie-Nie's criterion (upper-left end of the curve)

2   Lemmas

Definition 1. *Let $L(x)$ and $M(x)$ be polynomials of degree l and m respectively, where $l - 1 \leq m \leq l$. Let $u_l, u_{l-1}, \cdots, u_1$ be the roots of $L(x)$, and $v_m, v_{m-1}, \cdots, v_1$ the roots of $M(x)$. Then $(L(x), M(x))$ is called a **positive***

pair *if* (1) $u_{i+1} < v_i < u_i$ *for* $i = 1, 2, \cdots$, (2) $u_l < u_{l-1} < \cdots < u_1 < 0$, (3) $v_m < v_{m-1} < \cdots < v_1 < 0$.

Similarly, $(L(x), M(x))$ is called a **negative pair** if (1) $u_{i+1} > v_i > u_i$ for $i = 1, 2, \cdots$, (2) $u_l > u_{l-1} > \cdots > u_1 > 0$, (3) $v_m > v_{m-1} > \cdots > v_1 > 0$.

In (1), (3), $a_3, a_4, ..., a_n$ can be expressed in terms of a_0, a_1, a_2, b_i as follows:

$$a_{2s-1} = B_1 B_3 ... B_{2s-3} \, a_1 \left(\frac{a_2}{a_0}\right)^{s-1}, \quad a_{2s} = B_2 B_4 ... B_{2s-2} \, a_0 \left(\frac{a_2}{a_0}\right)^s, \tag{7}$$

where

$$B_i = b_1 b_2 \cdots b_i, \quad \text{for } i = 1, ..., n-2. \tag{8}$$

Let

$$H(\eta) = \sum_{k=0}^{p} (-1)^k \left(\prod_{i=0}^{k} B_{2i-2}\right) \eta^{p-k}, \quad G(\eta) = \sum_{k=0}^{q} (-1)^k \left(\prod_{i=0}^{k} B_{2i-1}\right) \eta^{q-k}, \tag{9}$$

where $p = [n/2]$, $q = [(n-1)/2]$ and $B_{-2} = B_{-1} = B_0 = 1$.

Lemma 1. *Let the notations be as above. Polynomial $f(x)$ in (1) is stable if and only if $(H(\eta), G(\eta))$ is a negative pair.*

Proof. Let

$$F(x) = x^n f\left(\frac{1}{x}\right) = h(x^2) + xg(x^2) = \sum_{i=0}^{n} a_i x^{n-i},$$

where

$$h(x^2) = \frac{F(x) + F(-x)}{2}, \quad g(x^2) = \frac{F(x) - F(-x)}{2x}.$$

Obviously, $f(x)$ is stable if and only if $F(x)$ is stable. According to the Hermite-Biehler criterion, $F(x)$ is stable if and only if $(h(u), g(u))$ is a positive pair. Now we need to prove that $(h(u), g(u))$ is a positive pair if and only if $(H(\eta), G(\eta))$ is a negative pair.

Without loss of generality, we assume that $F(x)$ is a polynomial of even degree. Computing the coefficients of h, g, we get

$$\begin{cases} h(u) = a_0 u^p + a_2 u^{p-1} + B_2 a_0 \left(\frac{a_2}{a_0}\right)^2 u^{p-2} + \cdots + B_2 B_4 \cdots B_{2k-2} a_0 \left(\frac{a_2}{a_0}\right)^k u^{p-k} + \\ \qquad\qquad\qquad\qquad\qquad\qquad\qquad\qquad\qquad\qquad\qquad\qquad + \cdots, \\ g(u) = a_1 u^q + B_1 a_1 \left(\frac{a_2}{a_0}\right) u^{q-1} + \cdots + B_1 B_3 \cdots B_{2k-1} a_1 \left(\frac{a_2}{a_0}\right)^k u^{q-k} + \cdots. \end{cases} \tag{10}$$

Let $\eta = -\frac{a_0}{a_2} u$. It is easy to verify that

$$H(\eta) = (-1)^p h\left(-\frac{a_2}{a_0}\eta\right) / a_0 \left(\frac{a_2}{a_0}\right)^p, \quad G(\eta) = (-1)^q g\left(-\frac{a_2}{a_0}\eta\right) / a_1 \left(\frac{a_2}{a_0}\right)^q. \tag{11}$$

Thus $(H(\eta), G(\eta))$ is a negative pair if and only if $(h(x), g(x))$ is a positive pair. $\qquad\square$

Let $\gamma_0 = 1$, $\gamma_{2p-1} = B_{2p-2}$, and for $k = 1, 2, \cdots, p-1$,

$$\gamma_{2k-1} = \frac{B_{2k-2}(1 + \sqrt{1 - 4b_{2k-1}b_{2k}})}{2}, \quad \gamma_{2k} = \frac{B_{2k-2}(1 - \sqrt{1 - 4b_{2k-1}b_{2k}})}{2}. \tag{12}$$

Lemma 2. *Let the notations be as above, and let $h_1, h_2, \cdots, h_p$ be the p simple real positive roots of $H(\eta)$. If for $p \geq 3$,*

$$0 < b_{i-1} + b_i < 1 \text{ for } i = 2, \cdots, n-2, \tag{13}$$

then $1 = \gamma_0 > \gamma_1 > ... > \gamma_{2p-1} > 0$, and for $k = 1, \cdots, p$, $h_k \in (\gamma_{2k-1}, \gamma_{2k-2})$.

Proof. (i) First we prove that for $p \geq 3$ and $k = 1, 2, \cdots, p-1$,

$$B_{2k} < \gamma_{2k} < \gamma_{2k-1} < B_{2k-2}. \tag{14}$$

That $\gamma_{2k} < \gamma_{2k-1} < B_{2k-2}$ is obvious. Since $\sqrt{1-x} < 1 - x/2$ for $x \in (0, 1)$, we have

$$B_{2k} = \frac{B_{2k-2}(1 - (1 - 2b_{2k-1}b_{2k}))}{2} < \frac{B_{2k-2}(1 - \sqrt{1 - 4b_{2k-1}b_{2k}})}{2} = \gamma_{2k}.$$

This proves (14). From (12) and (14), we can see that $\gamma_i \in (0, 1]$ for $i = 0, ..., 2p-1$, and $\gamma_0, \gamma_1, ..., \gamma_{2p-1}$ form a strictly descending sequence.

(ii) Next we prove that for $p \geq 3$,

$$\begin{cases} H(\gamma_0) > 0, \\ (-1)^p H(\gamma_{2p-1}) > 0, \\ (-1)^k H(\eta_k) > 0, \text{ for all } \eta_k \in [\gamma_{2k}, \gamma_{2k-1}]. \end{cases} \tag{15}$$

The first two inequalities are trivial. Now consider $\eta_k \in [\gamma_{2k}, \gamma_{2k-1}]$. We have

$$(\eta_k - \gamma_{2k})(\gamma_{2k-1} - \eta_k) \geq 0. \tag{16}$$

Substituting (12) into (16), we obtain

$$-\eta_k^2 + B_{2k-2}\eta_k - B_{2k-2}B_{2k} \geq 0. \tag{17}$$

Hence

$$(-1)^k H(\eta_k) = \cdots + B_2 B_4 \cdots B_{2k-8}\eta_k^{p-k+2}(-\eta_k + B_{2k-6})$$
$$+ B_2 B_4 \cdots B_{2k-4}\eta_k^{p-k-1}(-\eta_k^2 + B_{2k-2}\eta_k - B_{2k-2}B_{2k})$$
$$+ B_2 B_4 \cdots B_{2k+2}\eta_k^{p-k-3}(\eta_k - B_{2k+4}) + \cdots > 0,$$

and then $h_k \in (\gamma_{2k-1}, \gamma_{2k-2})$.

□

Let $\gamma_0' = B_1$, $\gamma_{2q-1}' = B_{2q-1}$, and for $k = 1, 2, \cdots, q-1$,

$$\gamma_{2k-1}' = \frac{B_{2k-1}(1 + \sqrt{1 - 4b_{2k}b_{2k+1}})}{2}, \quad \gamma_{2k}' = \frac{B_{2k-1}(1 - \sqrt{1 - 4b_{2k}b_{2k+1}})}{2}. \tag{18}$$

Similar to the proof of Lemma 2, we get that for $q \geq 2$,

$$\begin{cases} B_{2k+1} < \gamma_{2k}' < \gamma_{2k-1}' < B_{2k-1} \text{ for } k = 1, \cdots, q-1, \\ G(\gamma_0') > 0, \\ (-1)^q G(\gamma_{2q-1}') > 0, \\ (-1)^k G(\eta_k') \geq 0, \text{ for all } \eta_k' \in [\gamma_{2k}', \gamma_{2k-1}']. \end{cases} \tag{19}$$

Lemma 3. *Let the notations be as above, and let $g_1, g_2, \cdots, g_q$ be the q simple real positive roots of $G(\eta)$. If $q \geq 2$ and*

$$0 < b_{i-1} + b_i < 1 \text{ for } i = 2, ..., n-2, \tag{20}$$

then $1 > \gamma_0' > \gamma_1' > ... > \gamma_{2p-1}' > 0$, and for $k = 1, \cdots, q$, $g_k \in [\gamma_{2k-1}', \gamma_{2k-2}']$.

Lemma 4. *Let the notations be as above. If $n > 5$,*

$$0 < b_{i-1} + b_i < 1, \quad i = 2, ..., n-2,$$

and

$$1 - \sqrt{1 - 4b_{i-1}b_i} \leq b_{i-1}(1 + \sqrt{1 - 4b_ib_{i+1}}), \quad i = 2, 3, \cdots, n-3, \tag{21}$$

then $f(x)$ is stable.

Proof. Let $h_1, h_2, \cdots, h_p$ be the p simple real positive roots of $H(\eta)$, and let $g_1, g_2, \cdots, g_q$ be the q simple real positive roots of $G(\eta)$. By Lemma 2 and Lemma 3, we know that $h_j \in (\gamma_{2j-1}, \gamma_{2j-2}), j = 1, \cdots, p$, and $g_k \in [\gamma_{2k-1}', \gamma_{2k-2}'], k = 1, \cdots, q$.

The condition (21) implies that

$$\gamma_2 \leq \gamma_1', \gamma_2' \leq \gamma_3, \gamma_4 \leq \gamma_3', \gamma_4' \leq \gamma_5, \cdots,$$
$$\gamma_{2k} \leq \gamma_{2k-1}', \gamma_{2k}' \leq \gamma_{2k+1}, \cdots, \begin{cases} \gamma_{2p-2} \leq \gamma_{2p-3}', \text{if } q = p. \\ \gamma_{2q-2}' \leq \gamma_{2q-1}, \text{if } q = p-1. \end{cases}$$

Thus

$$g_1 > h_2 > g_2 > h_3 > \cdots > g_{k-1} > h_k > g_k > \cdots > \begin{cases} h_p & \text{if } q = p. \\ g_q & \text{if } q = p-1. \end{cases} \tag{22}$$

If we can prove that

$$\begin{cases} H(\gamma_0') < 0, \\ (-1)^q G(\gamma_{2p-1}) < 0 \text{ if } q = p, \\ (-1)^p H(\gamma_{2q-1}') < 0 \text{ if } q = p-1, \end{cases} \tag{23}$$

then, by (22) and (23), we will have

$$h_1 > g_1 > h_2 > g_2 > h_3 > \cdots > g_{k-1} > h_k > g_k > \cdots > \begin{cases} g_q & \text{if } q = p, \\ h_p & \text{if } q = p - 1, \end{cases} \tag{24}$$

which means that $(G(\eta), H(\eta))$ is a negative pair. Then by Lemma 1, we see immediately that $f(x)$ is stable.

Below we prove (23). By substituting b_1 for γ_0' in $H(\gamma_0')$, we have

$$
\begin{aligned}
H(\gamma_0') &= b_1^{p-2}(b_1^2 - b_1 + B_2) - B_2 B_4 b_1^{p-4}(b_1 - B_6) - \cdots \\
&= b_1^{p-1}(b_1 + b_2 - 1) - B_2 B_4 b_1^{p-4}(b_1 - B_6) - \cdots < 0.
\end{aligned}
$$

When $p = q \geq 3$, by substituting B_{2p-2} for γ_{2p-1} in $G(\gamma_{2p-1})$, we have

$$
\begin{aligned}
(-1)^q G(\gamma_{2p-1}) &= B_1 B_3 \cdots B_{2q-5}(B_{2q-3}B_{2q-1} - B_{2q-3}B_{2q-2} + B_{2q-2}{}^2) \\
&\quad - B_1 B_3 \cdots B_{2q-9}B_{2q-2}{}^3(B_{2q-7} - B_{2q-2}) - \cdots \\
&= B_1 B_3 \cdots B_{2q-5}B_{2q-3}B_{2q-2}(b_{2q-1} + b_{2q-2} - 1) \\
&\quad - B_1 B_3 \cdots B_{2q-9}B_{2q-2}{}^3(B_{2q-7} - B_{2q-2}) - \cdots < 0.
\end{aligned}
$$

When $q = p - 1 \geq 2$, by substituting B_{2q-1} for γ_{2q-1}' in $H(\gamma_{2q-1}')$, we have

$$
\begin{aligned}
(-1)^p H(\gamma_{2q-1}') &= B_2 B_4 \cdots B_{2p-6}(B_{2p-4}B_{2p-2} - B_{2p-4}B_{2p-3} + B_{2p-3}{}^2) \\
&\quad - B_2 B_4 \cdots B_{2p-10}B_{2p-3}{}^3(B_{2p-8} - B_{2p-3}) - \cdots \\
&= B_2 B_4 \cdots B_{2p-6}B_{2p-4}B_{2p-3}(b_{2p-2} + b_{2p-3} - 1) \\
&\quad - B_2 B_4 \cdots B_{2p-10}B_{2p-3}{}^3(B_{2p-8} - B_{2p-3}) - \cdots < 0.
\end{aligned}
$$

$\square$

3   Proof of the Main Theorem

It is easy to verify that (5) and (6) imply

$$0 < b_{i-1} + b_i < 1, \quad i = 2, \ldots, n - 2. \tag{25}$$

Case 1. $n = 4$.

By (25), we have $b_1 < 1$ and $b_1 + b_2 < 1$, which happen to be Routh-Hurwitz conditions on the stability of polynomial $f(x)$ for $n = 4$.

Case 2. $n > 5$, i.e., $p \geq 3$ and $q \geq 2$.

If we can prove

$$1 - \sqrt{1 - 4b_{i-1}b_i} \leq b_{i-1}(1 + \sqrt{1 - 4b_i b_{i+1}}), \quad i = 2, 3, \cdots, n - 3, \tag{26}$$

then by Lemma 4, we know that $f(x)$ will be stable.

Set $\sigma_i = b_{i-1} + b_i$. Set $\beta = \max_i(b_i)$ and $\sigma = \max_i(\sigma_i)$. (26) is equivalent to

$$b_i \frac{1 - \sqrt{1 - 4b_{i-1}b_i}}{4b_{i-1}b_i} \leq \frac{1 + \sqrt{1 - 4b_i b_{i+1}}}{4}, \quad i = 2, 3, \cdots, n - 3. \tag{27}$$

Note that $\dfrac{1 - \sqrt{1 - x}}{x}$ is strictly increasing, and $\dfrac{1 + \sqrt{1 - x}}{4}$ is strictly decreasing for $x \in (0, 1)$. Since $0 < 4b_{i-1}b_i \leq (b_{i-1} + b_i)^2 = \sigma_i^2 < 1$, we have

$$b_i \frac{1 - \sqrt{1 - 4b_{i-1}b_i}}{4b_{i-1}b_i} \leq b_i \frac{1 - \sqrt{1 - \sigma_i^2}}{\sigma_i^2} \leq \beta \frac{1 - \sqrt{1 - \sigma^2}}{\sigma^2}, \tag{28}$$

and

$$\frac{1 + \sqrt{1 - \sigma^2}}{4} \leq \frac{1 + \sqrt{1 - \sigma_{i+1}^2}}{4} \leq \frac{1 + \sqrt{1 - 4b_i b_{i+1}}}{4}. \tag{29}$$

If

$$\beta \frac{1 - \sqrt{1 - \sigma^2}}{\sigma^2} \leq \frac{1 + \sqrt{1 - \sigma^2}}{4}, \tag{30}$$

then, by (28) and (29), (27) and hence (26) will hold. (30) is equivalent to

$$(\sigma^2 + 4\beta)^2 - 16\beta \leq 0, \tag{31}$$

which is in our condition (6) already.

Case 3. $n = 5$.

The roots of $H(\eta)$ and $G(\eta)$ are respectively γ_1, γ_2 and γ_1', γ_2', which are alternate if

$$1 - \sqrt{1 - 4b_1 b_2} < b_1(1 + \sqrt{1 - 4b_2 b_3}). \tag{32}$$

Therefore, the equal signs in (5) should be omitted in this case.

This finishes the proof.

References

[1] Xukai Xie, A new criterion of linear system stability, Special Issue on Basic Theory of Transaction of Northeast Institute of Technology of China, 1(1963),26-30. (in Chinese)

[2] Yiyong Nie, A new criterion of Hurwitz polynomial, Mechanics, 2(1976), 110-116.(in Chinese)

[3] Alberto Borobia, Sebastian Dormido, Three coefficients of a polynomial can determine its instability, Linear Algebra and its Applications, 338(2001), 67-76.

[4] Xiaojing Yang, Necessary conditions of Hurwitz polynomials, Linear Algebra and its Applications, 359(2003), 21-27.

[5] Xiaojing Yang, Some necessary conditions for Hurwitz stability, Automatica, 40 (2004), 527-529.

[6] V.V. Maslennikhov, The hypothesis on existence of a simple analytical sufficient stability condition, Automatica Telemehanika, 1984, No.2, 160-161. (in Russian)

[7] Lin Huang, Basic theory of stability and robust, Science Press, 2003. (in Chinese)

Formal Power Series and Loose Entry Formulas for the Dixon Matrix

Wei Xiao and Eng-Wee Chionh

School of Computing, National University of Singapore, Singapore 117543
{xiaowei, chionhew}@comp.nus.edu.sg

Abstract. Formal power series are used to derive four entry formulas for the Dixon matrix. These entry formulas have uniform and simple summation bounds for the entire Dixon matrix. When corner cutting is applied to the monomial support, each of the four loose entry formulas simplifies greatly for some rows and columns associated with a particular corner, but still maintains the uniform and simple summation bounds. Uniform summation bounds make the entry formulas loose because redundant brackets that eventually vanish are produced. On the other hand, uniform summation bounds reveal valuable information about the properties of the Dixon matrix for a corner-cut monomial support.

1   Introduction

Resultants are powerful in solving polynomial systems by elimination [9, 18]. The Dixon bracket method is a well-known technique for constructing resultants [10]. Recent research in Dixon resultants includes [17, 14, 7]. An entry formula allows the Dixon matrix to be computed efficiently [5, 3] and is indispensable in deriving properties of the Dixon matrix [12, 13, 14]. While the concise entry formula given in [1] is good for computing the Dixon matrix, it is not as well suited for theoretical exploration because, to be concise, each entry has distinct and complicated summation bounds, and this obscures rather than reveals useful information. It would greatly simplify derivation if the summation bounds can be the same for the entire matrix, or at least for those rows or columns that are of interest.

This paper answers the above need by presenting four loose entry formulas. These entry formulas have uniform and simple summation bounds for the entire matrix for a canonical, or uncut, bidegree monomial support. For corner-cut monomial supports, each of these entry formulas becomes even simpler for some rows and columns but the summation bounds remain uniform and simple. The tradeoff is that these formulas are loose rather than concise because they produce redundant brackets, that is, brackets that either self-vanish or are mutually cancelled. It is gratifying that these loose entry formulas can be obtained quite easily, all we have to do is to expand a formal power series a little bit differently.

This paper consists of five sections. Section 2 quickly reviews the construction of the Dixon matrix and defines the notation used in the paper. Section 3 presents

H. Li, P. J. Olver and G. Sommer (Eds.): IWMM-GIAE 2004, LNCS 3519, pp. 72–82, 2005.

the four loose entry formulas for the Dixon matrix in two theorems. Section 4 customizes the entry formulas for some rows and columns when the monomial support undergoes corner cutting. Section 5 ends the paper with a discussion on the concise entry formulas in [1] and the formulas in this paper.

2   Preliminaries

To be self-contained we first describe the Dixon quotient, the Dixon polynomial, and the Dixon matrix. Some notations used in this paper are also defined.

Let $a..b$ denote the set of consecutive integers from a to b inclusive. The uncut bidegree monomial support $\mathcal{A}_{m,n}$ can be written as a Cartesian product

$$\mathcal{A}_{m,n} = 0..m \times 0..n = \{0, \cdots, m\} \times \{0, \cdots, n\} = \{(0,0), \cdots, (m,n)\} \ . \qquad (1)$$

Let f, g, h be bivariate polynomials on the monomial support $\mathcal{A} \subseteq \mathcal{A}_{m,n}$:

$$f(s,t) = \sum_{(i,j)\in\mathcal{A}} f_{i,j}s^i t^j, \ g(s,t) = \sum_{(i,j)\in\mathcal{A}} g_{i,j}s^i t^j, \ h(s,t) = \sum_{(i,j)\in\mathcal{A}} h_{i,j}s^i t^j \ . \qquad (2)$$

Their Dixon quotient is

$$\Delta(f(s,t), g(\alpha,t), h(\alpha,\beta)) = \frac{\begin{vmatrix} f(s,t) & g(s,t) & h(s,t) \\ f(\alpha,t) & g(\alpha,t) & h(\alpha,t) \\ f(\alpha,\beta) & g(\alpha,\beta) & h(\alpha,\beta) \end{vmatrix}}{(s-\alpha)(t-\beta)} \ . \qquad (3)$$

The quotient actually divides the numerator and becomes the Dixon polynomial

$$\Delta(f(s,t), g(\alpha,t), h(\alpha,\beta)) = \sum_{\sigma,\tau,a,b} \Delta_{\sigma,\tau,a,b} s^\sigma t^\tau \alpha^a \beta^b. \qquad (4)$$

By writing the Dixon polynomial in the matrix form

$$\Delta = \left[\cdots s^\sigma t^\tau \cdots\right] D \left[\cdots \alpha^a \beta^b \cdots\right]^T, \qquad (5)$$

we obtain the Dixon matrix for f, g, h:

$$D = (\Delta_{\sigma,\tau,a,b}) \ . \qquad (6)$$

The monomials $s^\sigma t^\tau$ and $\alpha^a \beta^b$ that appear in the Dixon polynomial are called respectively the row and column indices of D. Furthermore, the monomial support of Δ considered as a polynomial in s, t or α, β, is called the row support $\mathcal{R}$ or column support $\mathcal{C}$ of D respectively.

The classical Dixon resultant is the determinant $|D|$ when $\mathcal{A} = \mathcal{A}_{m,n}$. The row and column supports of the classical Dixon matrix are

$$\mathcal{R}_{m,n} = 0..m-1 \times 0..2n-1, \quad \mathcal{C}_{m,n} = 0..2m-1 \times 0..n-1 \ . \qquad (7)$$

The coefficient $\Delta_{\sigma,\tau,a,b}$ in the Dixon polynomial is a sum of brackets, which are 3×3 determinants whose entries are coefficients of f, g, h. To be concise brackets are denoted by 6-tuples:

$$(i, j, k, l, p, q) = \begin{vmatrix} f_{i,j} & g_{i,j} & h_{i,j} \\ f_{k,l} & g_{k,l} & h_{k,l} \\ f_{p,q} & g_{p,q} & h_{p,q} \end{vmatrix} . \tag{8}$$

To save space in the examples below we omit the punctuations in the brackets, e.g., we use the shorthand notation

$$(1, 2, 4, 3, 0, 5) = 124305. \tag{9}$$

Example 1. Consider the monomial support $\mathcal{A}_{1,1} = 0..1 \times 0..1 = \{(0,0), (0,1), (1,0), (1,1)\}$:

Its Dixon polynomial is

$$\Delta = \mathcal{R}D\mathcal{C} = \begin{bmatrix} 1 & t \end{bmatrix} \begin{bmatrix} 001001 & 001011 \\ 001101 & 011011 \end{bmatrix} \begin{bmatrix} 1 \\ \alpha \end{bmatrix} . \tag{10}$$

Finally, the Minkowski sum of two ordered pairs is

$$(a, b) \oplus (c, d) = (a + c, b + d) . \tag{11}$$

3   Four Loose Entry Formulas for Uncut Monomial Supports

The reciprocal of the denominator of the Dixon quotient (3) can be regarded as a formal power series. Four ways of expanding this formal power series lead to four equivalent loose entry formulas that are different in appearance.

Theorem 1. *The Dixon matrix entry indexed by $(s^\sigma t^\tau, \alpha^a \beta^b)$ is*

$$\Delta_{\sigma,\tau,a,b} = \sum_{u=0}^{\infty} \sum_{v=0}^{\infty} \sum_{k=0}^{m} \sum_{l=0}^{n} B \tag{12}$$

where the summand is any of the following:

$$B = (\sigma + u + 1, \tau + v + 1 - l, k, l, a - u - k, b - v), \; or \tag{13}$$
$$B = -(\sigma - u, \tau + v + 1 - l, k, l, a + u + 1 - k, b - v), \; or \tag{14}$$
$$B = (\sigma - u, \tau - v - l, k, l, a + u + 1 - k, b + v + 1), \; or \tag{15}$$
$$B = -(\sigma + u + 1, \tau - v - l, k, l, a - u - k, b + v + 1). \tag{16}$$

Proof. The entry formulas can be derived in parallel easily: simply expand $\frac{1}{(s-\alpha)(t-\beta)}$ in the Dixon quotient in four ways in terms of $\frac{s}{\alpha}$ or $\frac{\alpha}{s}$ and $\frac{t}{\beta}$ or $\frac{\beta}{t}$:

$$
\begin{aligned}
\frac{1}{(s-\alpha)(t-\beta)} &= -\sum_{u=0}^{\infty} \frac{\alpha^u}{s^{u+1}} \sum_{v=0}^{\infty} \frac{t^v}{\beta^{v+1}} = \sum_{u=0}^{\infty} \frac{s^u}{\alpha^{u+1}} \sum_{v=0}^{\infty} \frac{t^v}{\beta^{v+1}} \\
&= \sum_{u=0}^{\infty} \frac{\alpha^u}{s^{u+1}} \sum_{v=0}^{\infty} \frac{\beta^v}{t^{v+1}} = -\sum_{u=0}^{\infty} \frac{s^u}{\alpha^{u+1}} \sum_{v=0}^{\infty} \frac{\beta^v}{t^{v+1}} \; .
\end{aligned}
\tag{17}
$$

By

$$
\begin{vmatrix}
f(s,t) & g(s,t) & h(s,t) \\
f(\alpha,t) & g(\alpha,t) & h(\alpha,t) \\
f(\alpha,\beta) & g(\alpha,\beta) & h(\alpha,\beta)
\end{vmatrix}
= \sum_{i,j,k,l,p,q} (i,j,k,l,p,q) s^i t^{j+l} \alpha^{k+p} \beta^q,
\tag{18}
$$

the Dixon polynomial can be written in any of the four expansions:

$$
\begin{aligned}
\Delta &= \sum_{i,j,k,l,p,q} \sum_{u,v} (i,j,k,l,p,q) s^{i-u-1} t^{j+l-v-1} \alpha^{k+p+u} \beta^{q+v} \\
&= -\sum_{i,j,k,l,p,q} \sum_{u,v} (i,j,k,l,p,q) s^{i+u} t^{j+l-v-1} \alpha^{k+p-u-1} \beta^{q+v} \\
&= \sum_{i,j,k,l,p,q} \sum_{u,v} (i,j,k,l,p,q) s^{i+u} t^{j+l+v} \alpha^{k+p-u-1} \beta^{q-v-1} \\
&= -\sum_{i,j,k,l,p,q} \sum_{u,v} (i,j,k,l,p,q) s^{i-u-1} t^{j+l+v} \alpha^{k+p+u} \beta^{q-v-1} \; .
\end{aligned}
$$

By comparing the coefficients of

$$
\Delta = \sum_{\sigma,\tau,a,b} \Delta_{\sigma,\tau,a,b} s^{\sigma} t^{\tau} \alpha^a \beta^b
\tag{19}
$$

with each of the four expansions, we obtain respectively the four equations given by (13), (14), (15), (16). □

The above entry formulas will generate three types of redundant brackets:

1. self-vanishing brackets such as (i,j,i,j,p,q),
2. mutually cancelled brackets such as $(i,j,k,l,p,q) + (i,j,p,q,k,l)$,
3. out of range brackets (i,j,k,l,p,q), with (i,j), (k,l), or $(p,q) \notin \mathcal{A}_{m,n}$. A bracket involving out of range indices is zero since $(i,j) \notin \mathcal{A}_{m,n}$ means $f_{i,j} = g_{i,j} = h_{i,j} = 0$.

For practical computation, we have to shrink the ranges of u and v in the above entry formulas:

Theorem 2. *The Dixon matrix entry indexed by $(s^\sigma t^\tau, \alpha^a \beta^b)$ is*

$$\Delta_{\sigma,\tau,a,b} = \sum_{u=0}^{m-1} \sum_{v=0}^{n-1} \sum_{k=0}^{m} \sum_{l=0}^{n} B \tag{20}$$

where either $B = (i,j,k,l,p,q)$ which can be (13) or (15), or $B = -(i,j,k,l,p,q)$ which can be (14) or (16).

Proof. When $i = \sigma + u + 1$, to have $(i,j) \in \mathcal{A}_{m,n}$ we need

$$\sigma + u + 1 \leq m \Rightarrow u \leq m - 1 . \tag{21}$$

When $i = \sigma - u$, to have $(i,j) \in \mathcal{A}_{m,n}$, we need

$$\sigma - u \geq 0 \Rightarrow u \leq \sigma \leq m - 1 . \tag{22}$$

Consequently, the upper bound of u is reduced from ∞ to $m - 1$.
Similarly, when $q = b + v + 1$, to have $(p,q) \in \mathcal{A}_{m,n}$ we need

$$b + v + 1 \leq n \Rightarrow v \leq n - 1 . \tag{23}$$

When $q = b - v$, to have $(p,q) \in \mathcal{A}_{m,n}$ we need

$$b - v \geq 0 \Rightarrow v \leq b \leq n - 1 . \tag{24}$$

Consequently, the upper bound of v is reduced from ∞ to $n - 1$.

$\square$

Example 2. We use Theorem 2 to compute the Dixon matrix D of Example 1 in four ways.
    Using (13), (14), (15), (16), we have respectively

$$D_1 = \begin{bmatrix} \sum_{k=0}^{1} \sum_{l=0}^{1}(1, 1-l, k, l, -k, 0) & \sum_{k=0}^{1} \sum_{l=0}^{1}(1, 1-l, k, l, 1-k, 0) \\ \sum_{k=0}^{1} \sum_{l=0}^{1}(1, 2-l, k, l, -k, 0) & \sum_{k=0}^{1} \sum_{l=0}^{1}(1, 2-l, k, l, 1-k, 0) \end{bmatrix}$$
$$= \begin{bmatrix} 100100 & 110010 \\ 110100 & 110110 \end{bmatrix} , \tag{25}$$

$$D_2 = \begin{bmatrix} \sum_{k=0}^{1} \sum_{l=0}^{1} -(0, 1-l, k, l, 1-k, 0) & \sum_{k=0}^{1} \sum_{l=0}^{1} -(0, 1-l, k, l, 2-k, 0) \\ \sum_{k=0}^{1} \sum_{l=0}^{1} -(0, 2-l, k, l, 1-k, 0) & \sum_{k=0}^{1} \sum_{l=0}^{1} -(0, 2-l, k, l, 2-k, 0) \end{bmatrix}$$
$$= \begin{bmatrix} -011000 & -001110 \\ -011100 & -011110 \end{bmatrix} , \tag{26}$$

$$D_3 = \begin{bmatrix} \sum_{k=0}^{1} \sum_{l=0}^{1}(0, -l, k, l, 1-k, 1) & \sum_{k=0}^{1} \sum_{l=0}^{1}(0, -l, k, l, 2-k, 1) \\ \sum_{k=0}^{1} \sum_{l=0}^{1}(0, 1-l, k, l, 1-k, 1) & \sum_{k=0}^{1} \sum_{l=0}^{1}(0, 1-l, k, l, 2-k, 1) \end{bmatrix}$$
$$= \begin{bmatrix} 001001 & 001011 \\ 001101 & 011011 \end{bmatrix} , \tag{27}$$

$$D_4 = \begin{bmatrix} \sum_{k=0}^{1} \sum_{l=0}^{1} -(1, -l, k, l, -k, 1) & \sum_{k=0}^{1} \sum_{l=0}^{1} -(1, -l, k, l, 1-k, 1) \\ \sum_{k=0}^{1} \sum_{l=0}^{1} -(1, 1-l, k, l, -k, 1) & \sum_{k=0}^{1} \sum_{l=0}^{1} -(1, 1-l, k, l, 1-k, 1) \end{bmatrix}$$

$$= \begin{bmatrix} -100001 & -100011 \\ -110001 & -100111 \end{bmatrix} . \tag{28}$$

From the properties of determinants, it is obvious that $D = D_1 = D_2 = D_3 = D_4$.

4   Corner-Specific Simplification of the Loose Entry Formulas for Corner-Cut Monomial Supports

Each of the four loose entry formulas can be simplified greatly for a particular corner when the monomial support undergoes corner cutting. To make clear the applicable corner (i, j), $i = 0, m$ and $j = 0, n$, of $\mathcal{A}_{m,n}$ at first glance, we use the following notation

If $\begin{array}{|c|c|} \hline P_{0,n} & P_{m,n} \\ \hline P_{0,0} & P_{m,0} \\ \hline \end{array}$ then $\begin{array}{|c|c|} \hline Q_{0,n} & Q_{m,n} \\ \hline Q_{0,0} & Q_{m,0} \\ \hline \end{array}$

to denote that if $P_{i,j}$ then $Q_{i,j}$ for $i = 0, m$ and $j = 0, n$.

The following theorems describe the corner-specific simplification effects for the entries of some rows and columns when corner cutting is applied.

Theorem 3. *Let $(x, y) \in \mathcal{A}_{m,n}$, and let cutting C be applied to $\mathcal{A}_{m,n}$ such that $(x, y) \in \mathcal{A} \subseteq \mathcal{A}_{m,n} \setminus C$, with C given by*

$$\begin{array}{|c|c|} \hline 0..x \times y..n \setminus \{(x,y)\} & x..m \times y..n \setminus \{(x,y)\} \\ \hline 0..x \times 0..y \setminus \{(x,y)\} & x..m \times 0..y \setminus \{(x,y)\} \\ \hline \end{array} . \tag{29}$$

Then the entries of the row indexed by (x', y') and given by

$$\begin{array}{|c|c|} \hline (x,y) \oplus (0, n-1) & (x,y) \oplus (-1, n-1) \\ \hline (x,y) \oplus (0,0) & (x,y) \oplus (-1,0) \\ \hline \end{array} \tag{30}$$

are the following $\Delta_{x',y',a,b}$ given by

$$\begin{array}{|c|c|} \hline -\sum_{k=0}^{m}(x,y,k,n,a+1-k,b) & \sum_{k=0}^{m}(x,y,k,n,a-k,b) \\ \hline \sum_{k=0}^{m}(x,y,k,0,a+1-k,b+1) & -\sum_{k=0}^{m}(x,y,k,0,a-k,b+1) \\ \hline \end{array} . \tag{31}$$

Proof. Apply the entry formula in Theorem 2 and choose the bracket B in the formula to be

$$\begin{array}{|c|c|} \hline \text{Equation (14)} & \text{Equation (13)} \\ \hline \text{Equation (15)} & \text{Equation (16)} \\ \hline \end{array} . \tag{32}$$

After substituting $(\sigma, \tau) = (x', y')$ into B, the first ordered pair of B becomes

$$
\begin{array}{|c|c|}
\hline
(x - u, y + n + v - l) & (x + u, y + n + v - l) \\
\hline
(x - u, y - v - l) & (x + u, y - v - l) \\
\hline
\end{array}
\tag{33}
$$

To have this ordered pair not in C, we need

$$
\begin{array}{|c|c|}
\hline
-u \geq 0, n + v - l \leq 0 & u \leq 0, n + v - l \leq 0 \\
\hline
-u \geq 0, -v - l \geq 0 & u \leq 0, -v - l \geq 0 \\
\hline
\end{array}
\tag{34}
$$

Thus,

$$
\begin{array}{|c|c|}
\hline
u = v = 0, l = n & u = v = 0, l = n \\
\hline
u = v = l = 0 & u = v = l = 0 \\
\hline
\end{array}
\tag{35}
$$

Substituting the values of u, v, l into the entry formula we obtain the row entry formulas (31). $\qquad\square$

Theorem 4. *Let $(x, y) \in \mathcal{A}_{m,n}$. If the same corner cutting C as (29) is applied, then the entries of the column indexed by (x'', y'') and given by*

$$
\begin{array}{|c|c|}
\hline
(x, y) \oplus (0, -1) & (x, y) \oplus (m - 1, -1) \\
\hline
(x, y) \oplus (0, 0) & (x, y) \oplus (m - 1, 0) \\
\hline
\end{array}
\tag{36}
$$

are the following $\Delta_{\sigma, \tau, x'', y''}$ given by

$$
\begin{array}{|c|c|}
\hline
-\displaystyle\sum_{l=0}^{n} (\sigma + 1, \tau - l, 0, l, x, y) & \displaystyle\sum_{l=0}^{n} (\sigma, \tau - l, m, l, x, y) \\
\hline
\displaystyle\sum_{l=0}^{n} (\sigma + 1, \tau + 1 - l, 0, l, x, y) & -\displaystyle\sum_{l=0}^{n} (\sigma, \tau + 1 - l, m, l, x, y) \\
\hline
\end{array}
\tag{37}
$$

Proof. Apply the entry formula in Theorem 2 and choose the bracket B in the formula to be

$$
\begin{array}{|c|c|}
\hline
\text{Equation (16)} & \text{Equation (15)} \\
\hline
\text{Equation (13)} & \text{Equation (14)} \\
\hline
\end{array}
\tag{38}
$$

After substituting $(a, b) = (x'', y'')$ into B, the last ordered pair of B becomes

$$
\begin{array}{|c|c|}
\hline
(x - u - k, y + v) & (x + u + m - k, y + v) \\
\hline
(x - u - k, y - v) & (x + u + m - k, y - v) \\
\hline
\end{array}
\tag{39}
$$

To have this ordered pair not in C, we need

$$
\begin{array}{|c|c|}
\hline
-u - k \geq 0, v \leq 0 & u + m - k \leq 0, v \leq 0 \\
\hline
-u - k \geq 0, -v \geq 0 & u + m - k \leq 0, -v \geq 0 \\
\hline
\end{array}
\tag{40}
$$

Thus,

$$
\begin{array}{|c|c|}
\hline
u = v = k = 0 & u = v = 0, k = m \\
\hline
u = v = k = 0 & u = v = 0, k = m \\
\hline
\end{array}
\tag{41}
$$

Substituting the values of u, v, k into the entry formula, we obtain the column entry formulas (37). □

The following example illustrates the row and column loose entry formulas (31), (37) for a corner-cut monomial support.

Example 3. Consider the following monomial support $\mathcal{A} \subseteq \mathcal{A}_{3,3}$ and its row and column supports:

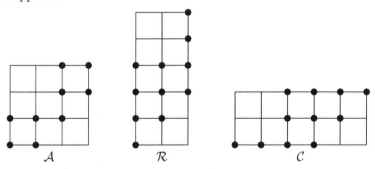

$\mathcal{A}$ $\mathcal{R}$ $\mathcal{C}$

It is easy to check that the sets $\{(1,0), (2,1), (3,2)\}$ and $\{(0,1), (2,3)\}$ satisfy the cutting conditions (29) for the bottom right and top left corners respectively. Thus, the 3×2 sub-matrix formed with the rows indexed by

$$
\{(1,0), (2,1), (3,2)\} \oplus (-1,0) = \{(0,0), (1,1), (2,2)\}
\tag{42}
$$

and the columns indexed by

$$
\{(0,1), (2,3)\} \oplus (0,-1) = \{(0,0), (2,2)\}
$$

can be computed in two ways.

Using the row entry formula (31) for the bottom right corner, we have:

$$
S_1 = \begin{bmatrix}
-\sum_{k=0}^{3}(1,0,k,0,-k,1) & -\sum_{k=0}^{3}(1,0,k,0,2-k,3) \\
-\sum_{k=0}^{3}(2,1,k,0,-k,1) & -\sum_{k=0}^{3}(2,1,k,0,2-k,3) \\
-\sum_{k=0}^{3}(3,2,k,0,-k,1) & -\sum_{k=0}^{3}(3,2,k,0,2-k,3)
\end{bmatrix}.
\tag{43}
$$

Using the column entry formula (37) for the top left corner, we have:

$$S_2 = \begin{bmatrix} -\sum_{l=0}^{3}(1,0-l,0,l,0,1) & -\sum_{l=0}^{3}(1,0-l,0,l,2,3) \\ -\sum_{l=0}^{3}(2,1-l,0,l,0,1) & -\sum_{l=0}^{3}(2,1-l,0,l,2,3) \\ -\sum_{l=0}^{3}(3,2-l,0,l,0,1) & -\sum_{l=0}^{3}(3,2-l,0,l,2,3) \end{bmatrix}. \tag{44}$$

A simple calculation shows that

$$S_1 = \begin{bmatrix} -100001 & -100023 \\ -210001 & -210023 \\ -320001 & -320023 \end{bmatrix} = S_2. \tag{45}$$

5   Discussion

The loose entry formulas presented in the paper have very simple summation bounds:

$$\Delta_{\sigma,\tau,a,b} = \sum_{u=0}^{m-1}\sum_{v=0}^{n-1}\sum_{k=0}^{m}\sum_{l=0}^{n} B,$$

where the bracket B can be any one of (13), (14), (15), (16); this is in sharp contrast with the complicated summation bounds in the concise entry formula given in [1]:

$$\Delta_{\sigma,\tau,a,b} = \sum_{u=0}^{\min(a,m-1-\sigma)}\sum_{v=0}^{\min(b,2n-1-\tau)}\sum_{k=\max(0,a-u-\sigma)}^{\min(m,a-u)}\sum_{l=\max(b+1,\tau+1+v-b)}^{\min(n,\tau+1+v)} B$$

$$+ \sum_{u=0}^{\min(a,m-1-\sigma)}\sum_{v=0}^{\min(b,2n-1-\tau)}\sum_{k=\max(0,a-u-m)}^{\min(\sigma,a-u)}\sum_{l=\max(b+1,\tau+1+v-n)}^{\min(n,\tau+v-b)} B,$$

where B is given by formula (13). Even more significant is that when the cutting of the types in Theorems 3, 4 are applied, the entry formulas for certain rows and columns can be further simplified, such that only one summation bound, instead of four, remains.

This single-uniform summation bound form is very helpful for discovering properties of the Dixon matrix. For example, with Theorem 3, we immediately see that the row associated with the monomial point (x,y) defined by (30) contains the monomial point (x,y) in every bracket of the sum. This observation, together with a single-uniform summation bound, leads to important conclusions concerning the linear dependence of the bracket factors produced by some rows and columns. It is beyond the scope of this paper to provide more detail and we refer the reader to [19] instead.

The loose entry formulas are very convenient for deriving theoretical results, but when computing the Dixon entry it is better to use the concise entry formula as the latter produces no redundant brackets. The total number of brackets in the Dixon matrix for bidegree polynomial [1] is

$$\frac{m(m+1)^2(m+2)n(n+1)^2(n+2)}{36},$$

but with the loose entry formulas the total number of brackets produced is

$$4m^3(m+1)n^3(n+1) \ .$$

Thus it is almost one hundred times faster to compute the Dixon matrix using the concise entry formula than using a loose entry formula.

References

1. E.W. Chionh: Concise Parallel Dixon Determinant. Computer Aided Geometric Design, **14** (1997) 561-570
2. E.W. Chionh: Rectangular Corner Cutting and Dixon $\mathcal{A}$-resultants. J. Symbolic Computation, **31** (2001) 651-669
3. E.W. Chionh: Parallel Dixon Matrices by Bracket. Advances in Computational Mathematics **19** (2003) 373-383
4. E.W. Chionh, M. Zhang, and R.N. Goldman: Implicitization by Dixon $\mathcal{A}$-resultants. In Proceedings of Geometric Modeling and Processing (2000) 310-318
5. E.W. Chionh, M. Zhang, and R.N. Goldman: Fast Computations of the Bezout and the Dixon Resultant Matrices. Journal of Symbolic Computation, **33** (2002) 13-29.
6. A.D. Chtcherba, D. Kapur: On the Efficiency and Optimality of Dixon-based Resultant Methods. ISSAC (2002) 29-36
7. A.D. Chtcherba, D. Kapur: Resultants for Unmixed Bivariant Polynomial Systems using the Dixon formulation. Journal of Symbolic Computation, **38** (2004) 915-958
8. A.D. Chtcherba: A New Sylvetser-type Resultant Method Based on the Dixon-Bézout Formulation. Ph.d. Dissertation, The University of New Mexico 2003
9. D. Cox, J. Little and D. O'Shea: Using Algebraic Geometry. Springer-Verlag, New York (1998)
10. A.L. Dixon: The Eliminant of Three Quantics in Two Independent Variables. Proc. London Math. Soc. Second Series, **7** (1909) 49-69, 473-492
11. I.Z. Emiris, B. Mourrain: Matrices in Elimination Theory. Journal of Symbolic Computation, **28** (1999) 3-44
12. M.C. Foo, E.W. Chionh: Corner Point Pasting and Dixon $\mathcal{A}$-Resultant Quotients. Asian Symposium on Computer Mathematics (2003) 114-127
13. M.C. Foo, E.W. Chionh: Corner Edge Cutting and Dixon $\mathcal{A}$-Resultant Quotients. J. Symbolic Computation, **37** (2004) 101-119
14. M.C. Foo, E.W. Chionh: Dixon $\mathcal{A}$-Resultant Quotients for 6-Point Isoceles Triangular Corner Cutting. Geometric Computation, Lecture Notes Series on Computing **11** (2004) 374-395
15. M.C. Foo: Master's thesis. National Unviversity of Singapore (2003)
16. D. Kapur, T. Saxena: Comparison of Various Multivariate Resultants. In ACM ISSAC, Montreal, Canada (1995)

17. D. Kapur, T. Saxena: Sparsity Considerations in the Dixon Resultant Formulation. In Proc. ACM Symposium on Theory of Computing, Philadelphia (1996)
18. D.M. Wang: Elimination Methods. Springer-Verlag, New York (2001)
19. W. Xiao: Master's thesis. National University of Singapore (2004)

Constructive Theory and Algorithm for Blending Several Implicit Algebraic Surfaces

Yurong Li[1], Na Lei[2], Shugong Zhang[2,3], and Guochen Feng[2,3]

[1] The Information Electronic Engineering School,
Shandong Institute of Business and Technology, 264005,Yantai, China
lyry@263.net
[2] Key Laboratory of Symbolic Computation and Knowledge Engineering in JLU
Ministry of Education, 130012, Changchun, China
leina@email.jlu.edu.cn
[3] Institute of Mathematics, Jilin University, 130012,Changchun, China
{sgzh, fgc}@mail.jlu.edu.cn

Abstract. Blending implicit algebraic surfaces has long been a hard work due to the complex calculations and the lack of effective algorithms. In this paper, we present a recursive method to derive the existence conditions and expressions of the blending surface for an arbitrary number of quadratic surfaces based on the blending surface for few quadratic surfaces. The existence conditions can be described in terms of geometric parameters of the given quadratic surfaces, which makes them easy to check. This greatly simplifies the calculations. Finally, some examples are presented.

1   Introduction

By blending surfaces we mean constructing a smooth transitional surface among given surfaces. Blending surfaces is a fundamental task in CAGD (Computer Aided Geometric Design), see [1],[2],[3] and [4]. It has attracted much attention due to not only its theoretical significance, but also its wide applications in engineering. There have been some methods to solve problems of this type, but most of them deal with parametric surfaces. In recent years, taking account of the advantages of implicit algebraic surfaces (defined by an implicit equation $f(x, y, z) = 0$), such as the closure property under some geometric operations (intersection, union, offset etc.), people began to pay much attention to the study on them. In practice, most problems we meet concern blending implicit quadratic surfaces, for instance, the problems arisen in pipe connection, limb design, and other engineering fields.

The theory on blending implicit algebraic surfaces developed slowly in the past decades even though it has been widely applied in CAGD. Blending surfaces is a hard mathematical task yet. In fact it is a typical algebraic geometry problem. Because constructive theory in classical algebraic geometry did not well developed, we met many difficulties in applications when approximation methods were used.

H. Li, P. J. Olver and G. Sommer (Eds.): IWMM-GIAE 2004, LNCS 3519, pp. 83–96, 2005.
© Springer-Verlag Berlin Heidelberg 2005

With the fast development of the theory on constructive algebraic geometry recently, certain methods have been presented. In 1989, J.Warren[5] converted the problem into seeking the lowest degree member of some ideal. Then he was able to obtain the blending surface by computing a Groebner basis. But a well known fact is that the upper bound of the degrees of the polynomials in a Groebner basis for an ideal generated by polynomials $f_1, \cdots, f_s$ can take the form $O(d^{2^n})$, in which $d = \max\{\deg(f_i)|1 \le i \le s\}$ and the number of the variables n. Due to the huge calculating amount, the method mentioned above is therefore difficult to realize. In 1993, Wu Wen-Tsun[6] converted this problem into calculating an irreducible ascending set of the related polynomials. For the case that the axes of three quadratic surfaces intersect vertically in one point and that the clipping planes vertical to the relative axis, Wu Wen-Tsun obtained the blending conditions as the Wu-formula for three pipes. Wu's method involves complicated calculation and is difficult to apply in general blending of algebraic surfaces. Therefore further improvement of the theory and algorithm is required.

In practice, we hope not only to find all clipping planes which can ensure the existence of the blending surface but also to analyze the geometric position of these clipping planes. Then an appropriate one may be chosen according to the requirement in engineering. There have been some complete results about blending two implicit algebraic surfaces in [8],[9] and [10]. The existence conditions of the blending surface for three implicit algebraic surfaces was also obtained in [11], but these conditions are too complicated to analyze.

In this paper, we will derive the existence conditions and the expressions of the blending surface for n quadratic surfaces based on the relative results for two or three quadratic surfaces. These conditions can be formulated in terms of the geometric parameters of the given quadratic surfaces, so it is easy to check whether the conditions are satisfied. Therefore, the amount of calculation is greatly reduced and the method can be readily applied to work out the blending surface in realtime.

2   Foundation

In the following, let $g_1, \cdots, g_n, h_1, \cdots, h_n \in R[x, y, z]$ denote polynomials, and $V(g_i), V(h_i)$ denote the algebraic surfaces determined by g_i and h_i. In [5], Warren proved that

Theorem 1. *[5] If the algebraic surfaces $V(f)$ and $V(g_i)$ meet with G^k continuity(geometric continuity, see [3]) along the curves $V(g_i, h_i)$ respectively, then*

$$f \in \langle g_1, h_1^{k+1} \rangle \cap \cdots \cap \langle g_n, h_n^{k+1} \rangle, \tag{1}$$

where $\langle g_i, h_i^{k+1} \rangle$ denotes the ideal generated by g_i and h_i^{k+1}. Consequently,

$$f = u_i g_i + a_i h_i^{k+1}, \quad i = 1, 2, \cdots, n. \tag{2}$$

Condition (2) is also sufficient if the u_i does not identically vanish on $V(g_i, h_i)$.

We only discuss the case of blending quadratic surfaces $V(g_1), \cdots, V(g_n)$, with pair-wisely different planes $V(h_1), \cdots, V(h_n)$, and always assume that $V(g_i, h_i)$ are irreducible planar quadratic curves for all $i, 1 \leq i \leq n$. Then we have

Proposition 1. *[7] If g_i, h_i are quadratic and linear polynomials respectively and $V(g_i, h_i)$ are irreducible planar quadratic curves, then the u_i and a_i in (2) satisfy*

$$\deg(u_i) \leq \deg(f) - 2, \quad \deg(a_i) \leq \deg(f) - (k+1). \tag{3}$$

By Theorem 1 and Proposition 1, we get

Proposition 2. *[7] Suppose that g_i, h_i are quadratic and linear polynomials respectively and that $V(g_i, h_i)$ are irreducible planar quadratic curves. A quadratic G^0 blending surface $V(g)$ ($g \in R[x, y, z]$) for $V(g_i)$ along $V(g_i, h_i)$ exists if and only if there exist real numbers λ_i and polynomials a_i such that*

$$g = \lambda_1 g_1 + a_1 h_1 = \lambda_2 g_2 + a_2 h_2 = \ldots = \lambda_n g_n + a_n h_n, \tag{4}$$
$$\lambda_i \neq 0, \quad \deg(a_i) \leq 1, \quad i = 1, 2, \cdots, n.$$

The quadratic G^0 blending surface $V(g)$ is called a ruled surface (for $V(g_i)$, $i = 1, \cdots, n$).

Theorem 2. *[7] If the ruled surface $V(g)$ exists, then*

$$\langle g_1, h_1 \rangle \cap \cdots \cap \langle g_n, h_n \rangle = \langle g, h_1 \cdots h_n \rangle. \tag{5}$$

And for arbitrary $f \in \langle g, h_1 h_2 \cdots h_n \rangle$,

$$\begin{cases} f = u\,g + a h_1 h_2 \cdots h_n, \\ \deg(u) \leq \deg(f) - 2, \quad \deg(a) \leq \deg(f) - n. \end{cases} \tag{6}$$

Corollary 1. *In the case of $n = 2$, if $V(g)$ is a ruled surface, then $V(g')$ is also a ruled surface if and only if there is a real number α, such that $g' = g + \alpha h_1 h_2$. In the case of $n \geq 3$, the ruled surface is unique if it exists.*

Proof. It is a direct conclusion of Theorem 2.

Notice that if the ruled surface exists, then

$$\bigcap_{i=1}^{n} \langle g_i, h_i^2 \rangle \subset \bigcap_{i=1}^{n} \langle g_i, h_i \rangle = \langle g, h_1 h_2 \cdots h_n \rangle.$$

For the sake of convenience, we naturally hope to choose G^1 blending surface, f, from those polynomials that are in the form of (6). That is to find $u, a \in R[x, y, z]$ such that

$$u\,g + a h_1 h_2 \cdots h_n \in \bigcap_{i=1}^{n} \langle g_i, h_i^2 \rangle, \tag{7}$$

which implies that $f = ug + ah_1h_2 \cdots h_n$ blends $V(g_i)$ more smoothly than g does.

In [9] and [10], one author of this paper studied the existence of G^1 blending surfaces in the form of (7) with $u = \omega_1 h_2 + \omega_2 h_1 + \beta$ in the case of $n = 2$. It is proved that in almost all cases one cannot get a G^1 blending surface if $\beta \neq 0$. Therefore hereafter we will take u in the form of

$$u = \sum_{i=1}^{n} \omega_i \prod_{j \neq i} h_j, \quad \omega_i \neq 0. \tag{8}$$

The polynomial u defined in (8) is called smoothing factor. In the following, we will always refer to G^1 blending surfaces $V(f)$ for $V(g_i)$ with $f = ug + ah_1h_2 \cdots h_n$, in which the smoothing factors u take the form of (8).

Theorem 3. *Suppose that there exits a ruled surface $V(g)$*

$$g = \lambda_1 g_1 + a_1 h_1 = \cdots = \lambda_n g_n + a_n h_n. \tag{9}$$

Let $f = ug + vh_1h_2 \cdots h_n$ be a polynomial of degree $n+1$ with u in the form of (8) and v a linear polynomial. Then surfaces $V(f)$ and $V(g_i)$ meet with G^1 continuity along the curves $V(g_i, h_i)$ if and only if there exist real numbers $\varepsilon_i, \omega_i \neq 0$ $(i = 1, 2, \cdots, n)$ such that

$$\varepsilon_1 h_1 - \omega_1 a_1 = \varepsilon_2 h_2 - \omega_2 a_2 = \cdots = \varepsilon_n h_n - \omega_n a_n, \tag{10}$$

and

$$\begin{cases} u = \sum_{i=1}^{n} \omega_i \prod_{j \neq i} h_j, \\ v = \varepsilon_i h_i - \omega_i a_i, \quad \omega_i \neq 0. \end{cases} \tag{11}$$

Proof. The sufficiency has been proved in [7], we therefore only prove the necessity.

Suppose that there is a G^1 blending surface $V(f)$ with

$$f = ug + v \prod_{k=1}^{n} h_k = g \sum_{j=1}^{n} \omega_j \prod_{k \neq j} h_k + v \prod_{k=1}^{n} h_k.$$

Substituting $g = \lambda_i g_i + a_i h_i$ into f, we get

$$f = u(\lambda_i g_i + a_i h_i) + v \prod_{k=1}^{n} h_k$$

$$= \lambda_i u g_i + a_i u h_i + v \prod_{k=1}^{n} h_k$$

$$= \lambda_i u g_i + a_i h_i \sum_{j=1}^{n} \omega_j \prod_{k \neq j} h_k + v \prod_{k=1}^{n} h_k$$

$$= \lambda_i u g_i + \left(a_i \sum_{j=1}^{n} \omega_j \prod_{k \neq j} h_k + v \prod_{k \neq i} h_k \right) h_i$$

$$= \lambda_i u g_i + \left(a_i \sum_{j=1, j \neq i}^{n} \omega_j \prod_{k \neq j} h_k + a_i \omega_i \prod_{k \neq i} h_k + v \prod_{k \neq i} h_k \right) h_i$$

$$= \lambda_i u g_i + a_i h_i^2 \sum_{j=1, j \neq i}^{n} \omega_j \prod_{k \neq i, j} h_k + h_i (a_i \omega_i + v) \prod_{k \neq i} h_k.$$

The fact that $f \in \langle g_i, h_i^2 \rangle$ implies $h_i(a_i\omega_i + v) \prod_{k \neq i} h_k \in \langle g_i, h_i^2 \rangle$. Obviously, $\langle g_i, h_i^2 \rangle$ is a primary ideal since $V(g_i, h_i)$ is irreducible. Then from $h_i \notin \langle g_i, h_i^2 \rangle$ we infer that for some positive integer m

$$\left((a_i\omega_i + v) \prod_{k \neq i} h_k \right)^m \in \langle g_i, h_i^2 \rangle \subset \langle g_i, h_i \rangle.$$

Notice that $\langle g_i, h_i \rangle$ is prime ideal, we can therefore deduce that $(a_i\omega_i + v) \in \langle g_i, h_i \rangle$ for $h_k \notin \langle g_i, h_i \rangle, k \neq i$. Because $a_i\omega_i + v$ is a linear polynomial, there must exist real numbers ε_i such that

$$a_i\omega_i + v = \varepsilon_i h_i,$$

that is

$$v = \varepsilon_i h_i - a_i\omega_i \quad i = 1, \cdots, n.$$

The necessity of Theorem 3 is significant: it implies that if there exists a G^1 blending surface of degree $n+1$ for n quadratic surfaces, then for any subset of $m(1 < m < n)$ quadratic surfaces there exists a relative G^1 blending surface of degree $m + 1$. By Proposition 2, the same conclusion holds true for the ruled surface. Therefore, if the existence conditions of the G^1 blending surface and the ruled surface are obtained for a smaller number of quadratic surfaces, then the existence conditions of the G^1 blending surface for more surfaces can be derived.

Theorem 4. *Suppose that*
1. There exists a ruled surface $V(g)$ for $V(g_i)$ with

$$g = \lambda_i g_i + a_i h_i, \lambda_i \neq 0,$$
$$\deg(a_i) \leq 1, i = 1, \cdots, n.$$

2. There exists an order (maybe a rearrangement is required) such that the G^1 blending surface $V(f^i)$ of degree 4 exists for arbitrary three adjacent quadratic surfaces $V(g_{i-1})$, $V(g_i)$ and $V(g_{i+1})$ $(i = 2, \cdots, n-1)$, where

$$\begin{cases} f^i = u^i g + v^i h_{i-1} h_i h_{i+1}, \\ u^i = \omega_{i-1}^i h_i h_{i+1} + \omega_i^i h_{i-1} h_{i+1} + \omega_{i+1}^i h_{i-1} h_i, \\ v^i = \varepsilon_{i-1}^i h_{i-1} - \omega_{i-1}^i a_{i-1} = \varepsilon_i^i h_i - \omega_i^i a_i = \varepsilon_{i+1}^i h_{i+1} - \omega_{i+1}^i a_{i+1}, \\ \omega_{i-1}^i \omega_i^i \omega_{i+1}^i \neq 0. \end{cases} \quad (12)$$

If there exist nonzero real numbers $\omega_1, \omega_2, \cdots, \omega_n$ *such that*

$$\frac{\omega_{i-1}^i}{\omega_{i-1}} = \frac{\omega_i^i}{\omega_i} = \frac{\omega_{i+1}^i}{\omega_{i+1}} = \alpha_i, \tag{13}$$

then the G^1 *blending surface* $V(f)$ *of degree* $n+1$ *for* $V(g_i)(i=1,\ldots,n)$ *exists and the following holds*

$$\begin{cases} f = ug + vh_1h_2\cdots h_n, \\ u = \sum_{i=1}^n \omega_i \prod_{k=1, k\neq i}^n h_i, \\ v = \varepsilon_i h_i - \omega_i a_i, \quad i = 1, 2, \cdots, n, \\ \varepsilon_i = \frac{\varepsilon_i^i}{\alpha_i}, i = 2, \cdots, n-1, \varepsilon_1 = \frac{\varepsilon_1^2}{\alpha_2}, \varepsilon_n = \frac{\varepsilon_n^{n-1}}{\alpha_{n-1}}. \end{cases} \tag{14}$$

Proof. From (12) and (13), we get

$$\begin{aligned} &\bar{\varepsilon}_j^j h_i - \omega_i a_i = \bar{\varepsilon}_j^j h_j - \omega_j a_j = \bar{\varepsilon}_k^j h_k - \omega_k a_k, \\ &\bar{\varepsilon}_j^k h_j - \omega_j a_j = \bar{\varepsilon}_k^k h_k - \omega_k a_k = \bar{\varepsilon}_l^k h_l - \omega_l a_l, \\ &\bar{\varepsilon}_m^j = \varepsilon_m^j/\alpha_j, \quad m = i, j, k, \\ &\bar{\varepsilon}_m^k = \varepsilon_m^k/\alpha_k, \quad m = j, k, l, \end{aligned} \tag{15}$$

with adjacent $1 < i < j < k < l < n$. Subtracting the second equation from the first one in (15), we get

$$(\bar{\varepsilon}_j^j - \bar{\varepsilon}_j^k)h_j = (\bar{\varepsilon}_k^j - \bar{\varepsilon}_k^k)h_k.$$

Since $V(h_j)$ and $V(h_k)$ are different, we have $\bar{\varepsilon}_j^j = \bar{\varepsilon}_j^k$, $\bar{\varepsilon}_k^j = \bar{\varepsilon}_k^k$.
With $\bar{\varepsilon}_1^2 = \varepsilon_1, \bar{\varepsilon}_i^i = \varepsilon_i, \bar{\varepsilon}_n^{n-1} = \varepsilon_n$, we get from (15)

$$\varepsilon_1 h_1 - \omega_1 a_1 = \varepsilon_2 h_2 - \omega_2 a_2 = \cdots = \varepsilon_n h_n - \omega_n a_n.$$

The conclusion then follows from Theorem 3.

Theorem 5. *Suppose that* $V(g_i)$ *and* $V(h_i)$ *intersect transversally along irreducible curves* $V(g_i, h_i)(i = 1, 2, 3)$, *and the different planes* $V(h_1), V(h_2)$ *and* $V(h_3)$ *neither contain the same line nor are all parallel to the same plane. If*

1. *There exists a ruled surface* $V(g)$ *for* $V(g_i)(i = 1, 2, 3)$ *with*

$$g = \lambda_1 g_1 + a_1 h_1 = \lambda_2 g_2 + a_2 h_2 = \lambda_3 g_3 + a_3 h_3. \tag{16}$$

2. *For arbitrary* $1 \leq i \neq j \leq 3$, *there exists a cubic* G^1 *blending surface* $V(f^{(ij)})$ *for* $V(g_i)$ *and* $V(g_j)$, *where*

$$\begin{cases} f^{(ij)} = u^{(ij)}g + v^{(ij)}h_ih_j, \\ u^{(ij)} = \mu^{(ij)}h_i + h_j, \\ v^{(ij)} = \varepsilon_i^{(ij)}h_i - a_i = \varepsilon_j^{(ij)}h_j - \mu^{(ij)}a_j, \end{cases} \tag{17}$$

and

$$\mu^{(12)}\mu^{(23)}\mu^{(31)} = 1, \tag{18}$$

then there is a G^1 *blending surface* $V(f)$ *of degree 4 for the* $V(g_i)(i = 1, 2, 3)$,

$$\begin{cases} f = u\,g + v h_1 h_2 h_3, \\ u = h_2 h_3 + \mu^{(12)} h_1 h_3 + 1/\mu^{(31)} h_1 h_2, \\ v = \varepsilon_1 h_1 - a_1 = \varepsilon_2 h_2 - \mu^{(12)} a_2 = \varepsilon_3 h_3 - 1/\mu^{(31)} a_3, \\ \varepsilon_1 = \varepsilon_1^{(12)}, \quad \varepsilon_2 = \mu^{(12)} \varepsilon_2^{(23)}, \quad \varepsilon_3 = \varepsilon_3^{(31)}/\mu^{(31)}. \end{cases} \tag{19}$$

Proof. According to the assumption about the $V(h_i)$, there exists an $h_i \notin \langle h_j, h_k \rangle$, $1 \le i \ne j \ne k \le 3$, without loss of generality, suppose $h_1 \notin \langle h_2, h_3 \rangle$. We obtain from (17)

$$\begin{aligned} \varepsilon_1^{(12)} h_1 - a_1 &= \varepsilon_2^{(12)} h_2 - \mu^{(12)} a_2, \\ \varepsilon_2^{(23)} h_2 - a_2 &= \varepsilon_3^{(23)} h_3 - \mu^{(23)} a_3, \\ \varepsilon_3^{(31)} h_3 - a_3 &= \varepsilon_1^{(31)} h_1 - \mu^{(31)} a_1. \end{aligned}$$

With $\omega_1 = 1$, $\omega_2 = \mu^{(12)}$, $\omega_3 = \mu^{(12)} \mu^{(23)} = 1/\mu^{(31)}$, we get

$$\begin{aligned} \varepsilon_1 h_1 - \omega_1 a_1 &= \varepsilon_2' h_2 - \omega_2 a_2, \\ \varepsilon_2 h_2 - \omega_2 a_2 &= \varepsilon_3' h_3 - \omega_3 a_3, \\ \varepsilon_3 h_3 - \omega_3 a_3 &= \varepsilon_1' h_1 - \omega_1 a_1, \end{aligned} \tag{20}$$

where

$$\begin{aligned} \varepsilon_1 &= \varepsilon_1^{(12)}, \quad \varepsilon_2 = \mu^{(12)} \varepsilon_2^{(23)}, \quad \varepsilon_3 = \varepsilon_3^{(31)}/\mu^{(31)}, \\ \varepsilon_1' &= \varepsilon_2^{(12)}, \quad \varepsilon_2' = \mu^{(12)} \varepsilon_3^{(23)}, \quad \varepsilon_3' = \varepsilon_1^{(31)}/\mu^{(31)}. \end{aligned}$$

Adding up the equations in (20), we get

$$(\varepsilon_1 - \varepsilon_1') h_1 + (\varepsilon_2 - \varepsilon_2') h_2 + (\varepsilon_3 - \varepsilon_3') h_3 = 0.$$

Noticing $h_1 \notin \langle h_2, h_3 \rangle$, we have $\varepsilon_1 - \varepsilon_1' = 0$, and further $\varepsilon_2 = \varepsilon_2'$, $\varepsilon_3 = \varepsilon_3'$ because $V(h_2), V(h_3)$ are different. Therefore equation (20) can be rewritten as

$$\varepsilon_1 h_1 - \omega_1 a_1 = \varepsilon_2 h_2 - \omega_2 a_2 = \varepsilon_3 h_3 - \omega_3 a_3.$$

By Theorem 3, the statement is verified.

By Theorems 4 and 5, we can derive the existence conditions of the blending surface for n quadratic surfaces from those for three or even two quadratic surfaces. In the next section we will present the existence condition of the G^1 blending surface and its expression for three given quadratic surfaces.

For given quadratic surfaces, the manipulation for calculating the ruled surface gets much easier. In some cases, we can deduce the existence of the ruled surface for n quadratic surfaces from the existence of the ruled surface for three quadratic surfaces.

Theorem 6. *Suppose that any three of the different planes $V(h_i)(i = 1, 2, \cdots, n)$ neither contain the same line nor are all parallel to the same plane. If every three of the quadratic surfaces $V(g_i)(i = 1, 2, \cdots, n)$ have a ruled surface, then the ruled surface for all $V(g_i)$ exists.*

Proof. In the case of $n = 4$, assume that the quadratic surfaces $V(g_i), V(g_j)$ and $V(g_k)$ have a ruled surface $V(g_{ijk})$, and that $V(g_j), V(g_k)$ and $V(g_l)$ have a ruled surface $V(g_{jkl})$, $(1 \le i, j, k, l \le 4)$. Notice that both $V(g_{ijk})$ and $V(g_{jkl})$ are the ruled surfaces for $V(g_j)$ and $V(g_k)$. By Corollary 1 there exists a real number α_{jk} such that $g_{ijk} - g_{jkl} = \alpha_{jk} h_j h_k$. Hence

$$\alpha_{23} h_2 h_3 + \alpha_{34} h_3 h_4 + \alpha_{14} h_1 h_4 + \alpha_{12} h_1 h_2 = 0.$$

If some of α_{jk} are nonzero, we get from the above equation

$$(\alpha_{23} h_2 + \alpha_{34} h_4) h_3 + (\alpha_{14} h_4 + \alpha_{12} h_2) h_1 = 0,$$

or

$$(\alpha_{14} h_1 + \alpha_{34} h_3) h_4 + (\alpha_{12} h_1 + \alpha_{23} h_3) h_2 = 0.$$

Since $V(h_i)$ and $V(h_j)$ are different and irreducible, we can deduce that $h_1, h_3 \in \langle h_2, h_4 \rangle$ or $h_2, h_4 \in \langle h_1, h_3 \rangle$, which contradicts the assumption about $V(h_i)$. So

$$\alpha_{23} = \alpha_{34} = \alpha_{14} = \alpha_{12} = 0.$$

Hence, $g_{123} = g_{234}$, i.e. there is a ruled surface for all the 4 quadratic surfaces.

If $n > 4$, according to the proof above every four of the quadratic surfaces have a ruled surface. Then there is a ruled surface $V(g_{ijkl})$ for $V(g_i), V(g_j), V(g_k)$ and $V(g_l)$, and there is a ruled surface $V(g_{jklm})$ for $V(g_j), V(g_k), V(g_l)$ and $V(g_m)$. Notice that both $V(g_{ijkl})$ and $V(g_{jklm})$ are the ruled surfaces for $V(g_i)$, $V(g_j)$ and $V(g_k)$. We get from Corollary 1 that $V(g_{ijkl}) = V(g_{jklm})$.

3   Existence Conditions of Smooth Blending Surface

For a set of given quadratic surfaces $V(g_i)$, our aim is to find appropriate clipping planes $V(h_i)$ and a relatively low degree surface $V(f)$ to meet the quadratic surfaces $V(g_i)$ along $V(g_i, h_i)$ smoothly. Based on the discussion above, we only need to find the existence conditions of the blending surface for two or three quadratic surfaces. There are already complete results for the existence conditions of the blending surface for two quadratic surfaces in [9] and [10]. So we now only discuss the instance for three quadratic surfaces.

We assume from now on that three different clipping planes $V(h_1), V(h_2)$ and $V(h_3)$ neither contain the same line nor are all parallel to the same plane. Under this assumption, there are only two cases that need to discuss:

Case 1. The three clipping planes intersect in one common point;

Case 2. $V(h_1)$ and $V(h_2)$ intersect but the three planes do not have any common point.

In Case 2, there must exist real numbers s_1, s_2, s_0 such that $h_3 = s_1 h_1 + s_2 h_2 + s_0$, where $s_0 \ne 0$ and $s_1^2 + s_2^2 \ne 0$. We may choose $s_0 = 1$ because $V(f) = V(\alpha f)$ for arbitrary $\alpha \ne 0$.

In Case 1, let $p_1 = h_1, p_2 = h_2, p_3 = h_3$. We may choose $V(p_1), V(p_2)$ and $V(p_3)$ as coordinate planes. In Case 2, let $p_1 = h_1, p_2 = h_2$ and we select another plane $V(p_3)$ such that the three planes $(V(p_1), V(p_2)$ and $V(p_3))$ constitute a coordinate system. In this coordinate system, a given g_i can be expressed as

$$g_i = g_{i,200}p_1^2 + g_{i,020}p_2^2 + g_{i,002}p_3^2 + g_{i,110}p_1p_2 + g_{i,101}p_1p_3 + g_{i,011}p_2p_3 \qquad (21)$$
$$+ g_{i,100}p_1 + g_{i,010}p_2 + g_{i,001}p_3 + g_{i,000}, \quad i = 1, 2, 3,$$

where (l, m, k) in $g_{i,lmk}$ is the degree tuple of the term $p_1^l p_2^m p_3^k$.

We always suppose that $V(g_1, h_1)$, $V(g_2, h_2)$ and $V(g_3, h_3)$ are irreducible quadratic curves. Then the existence condition of the ruled surface is

$$g = \lambda_1 g_1 + a_1 h_1 = \lambda_2 g_2 + a_2 h_2 = \lambda_3 g_3 + a_3 h_3. \qquad (22)$$

In the sequel we will denote

$$\begin{aligned}
g_{12,lmk} &= \lambda_1 g_{1,lmk} - \lambda_2 g_{2,lmk}, \\
g_{23,lmk} &= \lambda_2 g_{2,lmk} - \lambda_3 g_{3,lmk}, \\
g_{31,lmk} &= \lambda_3 g_{3,lmk} - \lambda_1 g_{1,lmk}.
\end{aligned} \qquad (23)$$

Substituting (21) into (22) and comparing the coefficients, we get

Proposition 3. *The ruled surface $V(g)$ exists if and only if there are nonzero real numbers λ_1, λ_2 and λ_3 such that the $g_{\alpha\beta,lmk}$ defined in equation (23) satisfy respectively*

For Case 1:

$$\begin{cases}
g_{12,002} = 0, & g_{12,001} = 0, & g_{12,000} = 0, \\
g_{23,200} = 0, & g_{23,100} = 0, & g_{23,000} = 0, \\
g_{31,020} = 0, & g_{31,010} = 0, & g_{31,000} = 0,
\end{cases} \qquad (24)$$

where

$$\begin{aligned}
a_1 &= -g_{12,200}p_1 + g_{31,110}p_2 - g_{12,101}p_3 - g_{12,100}, \\
a_2 &= -g_{23,110}p_1 - g_{23,020}p_2 + g_{12,011}p_3 - g_{23,010}, \\
a_3 &= g_{23,101}p_1 - g_{31,011}p_2 - g_{31,002}p_3 - g_{31,001}, \\
g &= \lambda_2 g_{2,200}p_1^2 + \lambda_1 g_{1,020}p_2^2 + \lambda_1 g_{1,002}p_3^2 + \lambda_3 g_{3,110}p_1p_2 + \lambda_1 g_{1,011}p_2p_3 \\
&\quad + \lambda_2 g_{2,101}p_1p_3 + \lambda_2 g_{2,100}p_1 + \lambda_1 g_{1,010}p_2 + \lambda_1 g_{1,001}p_3 + \lambda_1 g_{1,000}.
\end{aligned}$$

For Case 2:

$$\begin{cases}
g_{12,00i} = 0, \quad i = 0, 1, 2, \\
g_{31,011} - s_2 g_{31,001} = 0, \quad g_{23,002} = 0, \\
g_{23,101} - s_1 g_{23,001} = 0, \quad g_{31,002} = 0, \\
g_{23,200} - s_1 g_{23,100} + s_1^2 g_{23,000} = 0, \\
g_{31,020} - s_2 g_{31,010} + s_2^2 g_{31,000} = 0,
\end{cases} \qquad (25)$$

where

$$a_1 = (g_{31,110} + s_2 g_{23,100} - s_1 g_{31,010} + 2s_1 s_2 g_{31,000})p_2$$
$$\quad - g_{12,200}p_1 - g_{12,101}p_3 - g_{12,100},$$
$$a_2 = g_{12,020}p_2 + g_{12,011}p_3 + g_{12,010}$$
$$\quad - (g_{23,110} - s_2 g_{23,100} + s_1 g_{31,010} + 2s_1 s_2 g_{31,000})p_1,$$
$$a_3 = (g_{23,100} + s_1 g_{31,000})p_1 - g_{31,001}p_3 - (g_{31,010} - s_2 g_{31,000})p_2 - g_{31,000},$$
$$g = \lambda_2 g_{2,200}p_1^2 + \lambda_1 g_{1,020}p_2^2 + \lambda_1 g_{1,002}p_3^2 + \lambda_2 g_{2,101}p_1 p_3 + \lambda_1 g_{1,011}p_2 p_3$$
$$\quad + \lambda_2 g_{2,100}p_1 + (\lambda_3 g_{3,110} + 2s_1 s_2 g_{31,000} + s_2 g_{23,100} - s_1 g_{31,010})p_2 p_1$$
$$\quad + \lambda_1 g_{1,010}p_2 + \lambda_1 g_{1,001}p_3 + \lambda_1 g_{1,000}.$$

By Theorem 5, we get

Proposition 4. *If there exist nonzero real numbers λ_1, λ_2 and λ_3 such that equations in (22) hold, then the smooth blending surface $V(f)$ for $V(g_i)(i = 1, 2, 3)$ exists if and only if there are nonzero real numbers $\mu_{23}, \mu_{31}, \mu_{12}$ satisfying $\mu_{23}\mu_{31}\mu_{12} = 1$ and respectively*

For Case 1:

$$\begin{cases} \dfrac{g_{12,100}}{g_{12,010}} = \dfrac{g_{12,101}}{g_{12,011}} = -\mu_{12}, \\[2mm] \dfrac{g_{23,010}}{g_{23,001}} = \dfrac{g_{23,110}}{g_{23,101}} = -\mu_{23}, \\[2mm] \dfrac{g_{31,001}}{g_{31,100}} = \dfrac{g_{31,011}}{g_{31,110}} = -\mu_{31}. \end{cases}$$

For Case 2:

$$\begin{cases} \dfrac{g_{12,100}}{g_{12,010}} = \dfrac{g_{12,101}}{g_{12,011}} = -\mu_{12}, \\[2mm] \dfrac{g_{12,011}}{g_{31,001}} = \dfrac{g_{23,110} - g_{23,010}s_1 - g_{23,100}s_2 + 2\,g_{31,000}s_2 s_1}{g_{23,100} + 2\,g_{31,000}s_1} = -\mu_{23}, \\[2mm] \dfrac{g_{31,001}}{g_{12,101}} = \dfrac{g_{31,010} - 2\,g_{31,000}s_2}{-g_{31,110} + g_{31,010}s_1 + g_{31,100}s_2 - 2\,g_{31,000}s_2 s_1} = -\mu_{31}. \end{cases}$$

Subsequently, the definition polynomial of the blending surface can be expressed as

$$f = (\omega_1 h_2 h_3 + \omega_2 h_1 h_3 + \omega_3 h_1 h_2)g + v h_1 h_2 h_3,$$

where $V(g)$ is the ruled surface and ω_1 is an arbitrary nonzero real number, $\omega_2 = \mu_{12}\omega_1$, $\omega_3 = \omega_1/\mu_{31}$, and respectively
In Case 1:

$$v = \omega_2 g_{23,110}p_1 + \omega_3 g_{31,011}p_2 + \omega_1 g_{12,101}p_3 + \omega_1 g_{12,100}.$$

In Case 2:

$$v = \omega_2(g_{23,110} + g_{31,010}s_1 - g_{23,100}s_2 - 2g_{31,000}s_2 s_1)p_1 + \omega_1 g_{12,100}$$
$$\quad - \omega_1(g_{31,110} - g_{31,010}s_1 + g_{23,100}s_2 + 2g_{31,000}s_2 s_1)p_2 + \omega_1 g_{12,101}p_3.$$

4   Examples

The main task of this paper is to convert the problem of blending several quadratic surfaces to that of blending two or three quadratic surfaces. The problem of blending two quadratic surfaces has been completely solved in [9], so we only need to verify equations (13) and (18) in Theorems 4 and 5.

For given quadratic surfaces, practical problems always demand us to find appropriate clipping planes such that a minimal degree G^1 blending surface exists. It is very convenient for applications to be able to analyze the geometric position of clipping planes. The existence conditions of the blending surface for two quadratic surfaces are relatively simple and easy to analyze. With the help of those results and methods for blending two quadratic surfaces, it is possible to analyze the existence conditions of the blending surface for n quadratic surfaces.

This recursive method simplifies the problem. In [9], one of us showed that a G^1 blending surface almost never exists in case that the axes of two quadratic surfaces are not coplanar. So we only consider the instances that the axes of quadratic surfaces intersect in one common point.

For example, we obtained interesting results about blending three cylinders:

Case I. The axes of the three cylinders are coplanar and intersect in a common point O. Let L denotes the line which is perpendicular to the plane containing the axes and passes through the point O. The G^1 blending surface will exist only if the clipping planes are all parallel to the line L or all intersect L at the same point. If the radiuses of the cylinders r_1, r_2 and r_3 do not satisfy $r_1 = r_2 = r_3$, the G^1 blending surface of the three cylinders exists only if the clipping planes are all perpendicular to the axes.

Case II. The axes of the three cylinders intersect in a common point O and are not coplanar. If the radiuses of the cylinders do not satisfy $r_1 = r_2 = r_3$, then the G^1 blending surface for three cylinders exists if and only if the clipping planes are perpendicular to the axes and the following equation holds:

$$r_1^2 + d_1^2 = r_2^2 + d_2^2 = r_3^2 + d_3^2,$$

where d_i denote the distances from the intersection point O to the clipping planes $V(h_i)$ (see Fig.1) .

In [12], one author of this paper did a detailed discussion concerning the existence conditions of the blending surface for quadratic surfaces of all kinds. Here we only propose the conditions in the case that the clipping planes are all perpendicular to the axes.

For the case that the axes of quadratic surfaces vertically intersect in one point and that the clipping planes are vertical to the axes, Wu Wen-Tsun obtained the blending conditions i.e. the Wu-formula for three pipes [6]. Here we get rid of the restriction that the axes are perpendicular, and derive the existence conditions of the blending surface for several quadratic surfaces. The Wu-formula is thereby enriched. So the following conditions can be regarded as the generalized Wu-formulas.

In the following formulas, d_i denote the distances from the intersection point of the axes to the clipping planes.

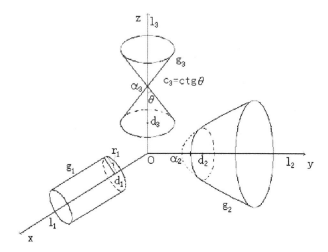

Fig. 1. The parameters in the formulas

Cylinders: Let r_i denote the radiuses of cylinders. For arbitrary three cylinders, the existence condition of a G^1 blending surface is, according to Theorem 5,

$$r_i^2 + d_i^2 = r_j^2 + d_j^2 = r_k^2 + d_k^2.$$

We can therefore get the existence condition of a G^1 blending surface for n cylinders from Theorem 4:

$$r_1^2 + d_1^2 = r_2^2 + d_2^2 = \cdots = r_n^2 + d_n^2.$$

Circular paraboloids: Let p_i denote the focuses of circular paraboloids, α_i denote the distances between the vertexes of the circular paraboloids and the intersection point of the axes (see Fig. 1). The existence condition of a G^1 blending surface for arbitrary three circular paraboloids is, according to Theorem 5,

$$2\,p_i(\alpha_i - d_i) - d_i^2 = 2\,p_j(\alpha_j - d_j) - d_j^2 = 2\,p_k(\alpha_k - d_k) - d_k^2.$$

We can therefore get the existence condition of a G^1 blending surface for n circular paraboloids from Theorem 4:

$$2\,p_1(\alpha_1 - d_1) - d_1^2 = 2\,p_2(\alpha_2 - d_2) - d_2^2 = \cdots = 2\,p_n(\alpha_n - d_n) - d_n^2.$$

Circular cones: Let c_i denote the slopes of the circular cones' generatrices, α_i denote the distances between the vertexes of the circular cones and the intersection point of the axes (see Fig. 1). For arbitrary three circular cones, the existence condition of a G^1 blending surface is, according to Theorem 5,

$$c_i^2(\alpha_i - d_i)^2 + d_i^2 = c_j^2(\alpha_j - d_j)^2 + d_j^2 = c_k^2(\alpha_k - d_k)^2 + d_k^2.$$

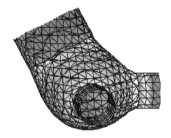 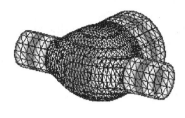

Fig. 2. Blending three cylinders

 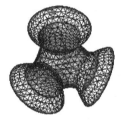

Fig. 3. Blending three quadratic surfaces

We can therefore get the existence condition of a G^1 blending surface for n circular cones from Theorem 4:

$$c_1^2(\alpha_1 - d_1)^2 + d_1^2 = c_2^2(\alpha_2 - d_2)^2 + d_2^2 = \cdots = c_n^2(\alpha_n - d_n)^2 + d_n^2.$$

Circular hyperboloids of two sheets: Let α_i denote the distances between the centers of the hyperboloids and the intersection point of the axes, and a_i, b_i denote the real and imaginary radiuses of the hyperboloids respectively. For n circular hyperboloids of two sheets, the existence condition of a G^1 blending surface of degree $n + 1$ is

$$a_1^2 - d_1^2 - \frac{a_1^2(d_1 - \alpha_1)^2}{b_1^2} = a_2^2 - d_2^2 - \frac{a_2^2(d_2 - \alpha_2)^2}{b_2^2}$$
$$= \cdots = a_n^2 - d_n^2 - \frac{a_n^2(d_n - \alpha_n)^2}{b_n^2}.$$

Cylinder, Circular paraboloid and Circular cone: Using the above definitions of the parameters we get the existence condition of a G^1 blending surface:

$$r_1^2 + d_1^2 = d_2^2 + 2p_2(d_2 - \alpha_2) = d_3^2 + c_3^2(d_3 - \alpha_3)^2.$$

5   Conclusion

In this paper, we first propose the theoretical basis to find all clipping planes which ensure the existence of the blending surface, then present an approach to construct the blending surface for n quadratic surfaces based on the results for two or three quadratic surfaces. In terms of geometric parameters of the given quadratic surfaces, we describe the existence conditions of the blending surface and reveal their geometric meanings. Such existence conditions therefore can be easily checked and the method can be readily applied to compute smooth blending surfaces in realtime.

References

1. Rossignac, J. R. and Requicha, A. G. : Constant radius blending in solid modeling, Comput. Mech. Engrg., **3**(1984), 65–73
2. Bajaj, C. L. and Ihm, I. : C^1 smoothing of polyhedra with implicit algebraic splines,   Comput. Graphics, **26** (1992), 79–88
3. Garrity, T. and Warren, J. : Geometric continuity, Comput. Aided Geom. Design, **8** (1991), 51–65
4. Changsong Chen, Falai Chen, and Yuyu Feng: Blending quadric surfaces with piecewise algebraic surfaces, Graphical Models, **63** (2001), 212–227
5. Warren, J. : Blending algebraic surfaces, ACM Transactions on Graphics, **8** (1989), 263–278
6. Wen-tsun Wu: On surface-fitting problem in CAGD, MM-Res. Preprints, **10** (1993), 1–11
7. Tieru Wu, Yunshi Zhou: On blending of several quadratic algebraic surfaces, Computer Aided Geometric Design, **17** (2000), 759–766
8. Tieru Wu, Yunshi Zhou and Guochen Feng: Blending two quadratic algebraic surfaces with cubic surfaces,   Proceeding of ASCM'96. Kobe, Japan, (1996), 73–79
9. Na Lei: Blending of two quadratic surfaces and Wu Wen-Tsun formulae, Ph.D. Dissertation, Jilin University, P.R.China, (2002)
10. Tieru Wu, Na Lei and Jinsan Cheng: Wu Wen-tsun formulae for the blending of pipe surfaces, Northeast. Math. J. **17** (2001), 383–386
11. Guochen Feng , Hongju Ren and Yunshi Zhou: Blending several implicit algebraic surfaces, Mathematics Mechanization and Applications, Academic Press, NewYork, (2000), 461–489
12. Yurong Li: The constructive theory and algorithm for blending several implicit surfaces with the lowest degree surface, Ph.D. Dissertation, Jilin University, P. R. China, (2003)

Minimum-Cost Optimization in Multicommodity Logistic Chain Network

Hongxia Li[1], Shuicheng Tian[2,3], Yuan Pan[1], Xiao Zhang[1], and Xiaochen Yu[2]

[1] School of Management, Xi'an University of Science and Technology,
Xi'an 710054, P. R. China
[2] School of Energy, Xi'an University of Science and Technology,
Xi'an 710054, P. R. China
[3] Key Laboratory of Mathematics Mechanization, Chinese Academy of Sciences,
Beijing 100080, P. R. China

Abstract. This paper presents a method of modeling and solving a very complicated real logistic problem in the management of transportation and sales. The problem to be addressed is a large-scale multicommodity, multi-source and multi-sink network flow optimization, of 12 types of coal from 29 mines, through over 200 railway stations along 5 railroad arteries, in Chongqing Coal Industry Company of China. A minimum-cost flow model is established for the network system, and several maximal-flow algorithms are implemented to produce an optimal scheme.

Keywords: Transportation Network Management; Network Operations; Multicommodity Flow; Maximal-Flow Algorithm; Minimum-Cost Optimization.

1   Introduction

In coal transportation management, the basic logistic problem is to efficiently transport one or several types of commodities, in different ways of transportation, from multiple origins to one or several destinations. This is one of the most complicated large-scale network flow problems in operations research, in that the constraints in optimization involve the transportation capacity of a complex system of railways and highways, the capacity of cargo storing and unloading at the nodes of transportation, the time-dependent demands of the customs, and the time-varying modes of transportation.

In this paper, we report part of our work in recent years on developing a decision-making system for multicommodity suppliers in their logistic transportation. It is part of a research project towards developing a decision support system for Chongqing Municipal Administration in emergency response management. In particular, we concern the problem of *minimum-cost multicommodity flow optimization* (MCMF).

Multicommodity flow can be used to model various management problems arising from transportation, distribution, multiperiod assignment, equipment replacement, disaster relief, communication, etc. Its algorithmic study dates back

H. Li, P. J. Olver and G. Sommer (Eds.): IWMM-GIAE 2004, LNCS 3519, pp. 97–104, 2005.

to the 1970's. Several classes of integral multicommodity networks are described in [1], [4], [6], [8], [10], [11], [12], [17], [18], [26], [29], [30]. Approaches like the simplex method [17], bottleneck method [12], and dynamic programming [8] are proposed. Minimum-cost multicommodity flow was used in [10] to solve the *equal-flow problem*, in which the flows through some given sets of arcs are required to take equal values.

In this paper, we formulate the minimum-cost multicommodity flow problem as a model of maximal-flow programming in an enlarged network. We implement different algorithms to solve the maximal-flow problem in this special network. For real data provided by the Chongqing Coal Industry Company (CCIC), our system turns out to be very effective. It is being in use by CCIC in their management of coal transportation and sales. A related project has won a Science and Technology Award from the Chongqing Municipal Administration.

The paper is organized as follows. In Section 2 we model the MCMF problem as a weighted maximal-flow programming problem. In Section 3 we discuss different maximal-flow algorithms performing on real data from the department of coal transportation and sales of CCIC.

2   Modeling Minimum-Cost Multicommodity Flows

Let there be R types of commodities, labeled by $r = 1, \ldots, R$. Let the transportation be within a network represented by an undirected graph $G = (V, E)$, where V is the nodes and E is the arcs. Let the number of nodes and arcs be n, m respectively. Assume that $m > n$, and the graph be connected and without parallel arcs. When the network is in use for transportation, the graph becomes directed.

For each arc (i, j) directed from node i to node j, let the flow capacity be $d(i, j)$, and let the cost per unit flow be $c(i, j)$. Assume that commodity r is located at s_r different nodes $\mathcal{S}^r$ indexed by $i_1^r, \ldots, i_{s_r}^r$, with amount of supply $\lambda_{i_k^r}$ at node i_k^r. The transportation of commodity r is towards t_r different nodes $\mathcal{T}^r$ indexed by $j_1^r, \ldots, j_{t_r}^r$, with amount of demand $\mu_{j_k^r}$ at node j_k^r. Let the nodes outside sets $\mathcal{S}^r, \mathcal{T}^r$ be $\mathcal{M}^r$. At any node in $\mathcal{M}^r$, the input of any commodity r equals the output.

Similar to [1] and [29], we introduce the so-called "supersource" S^r and "supersink" T^r to change the multiple origins $\mathcal{S}^r$ and multiple terminations $\mathcal{T}^r$ into single origin S^r and single termination T^r for each commodity r. The new node S^r is connected only to the nodes in $\mathcal{S}^r$, and the new node T^r is connected only to the nodes in $\mathcal{T}^r$. For each node $i_k^r \in \mathcal{S}^r$, we have

$$d(S^r, i_k^r) = \lambda_{i_k^r}, \quad d(i_k^r, S^r) = c(i_k^r, S^r) = c(S^r, i_k^r) = 0. \tag{1}$$

For each node $j_k^r \in \mathcal{T}^r$, we have

$$d(j_k^r, T^r) = \mu_{j_k^r}, \quad d(T^r, j_k^r) = c(T^r, j_k^r) = c(j_k^r, T^r) = 0. \tag{2}$$

The *enlarged network* with the $2R$ supersources and supersinks is denoted by $GE = (VE, EE)$.

The following figure displays an enlarged network of two commodities to be transported from both origins $1, 2$ to terminations $3, 5$ and $4, 5$ respectively.

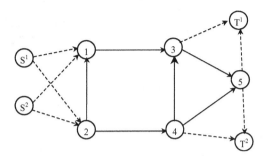

Fig. 1. Enlarged network of two commodities

Let $x^r(i,j)$ be the flow of commodity r from node i to node j, where the nodes belong to the enlarged network. At any node $i \in V$, the input and output are balanced:

$$\sum_{j:\,(j,i)\in EE} x^r(j,i) = \sum_{j:\,(i,j)\in EE} x^r(i,j). \tag{3}$$

At nodes S^r and T^r,

$$x^r(S^r, i) = \lambda_i^r; \quad x^r(j, T^r) = \mu_j^r;$$
$$x^r(i, S^r) = x^r(T^r, j) = 0 \text{ for any } i \in \mathcal{S}^r, \; j \in \mathcal{T}^r. \tag{4}$$

The total cost C^r of commodity r is

$$C^r = \sum_{(i,j)\in EE} c(i,j) x^r(i,j). \tag{5}$$

The MCMF problem is to determine the flows $x^r(i,j)$ for each commodity at each arc, so that the sum of total costs $C^1 + C^2 + \cdots C^R$ is minimized under the equality constraints (3), (4) and the following inequality constraints:

$$\sum_{r=1}^{R} x^r(i,j) \le d(i,j); \quad x^r(i,j) \ge 0. \tag{6}$$

This is essentially a weighted maximal-flow problem.

3 Algorithm and Practice

In CCIC's coal transportation system, there are 29 mines producing 12 types of coal, to supply a population of more than 100 million in the area of Chongqing

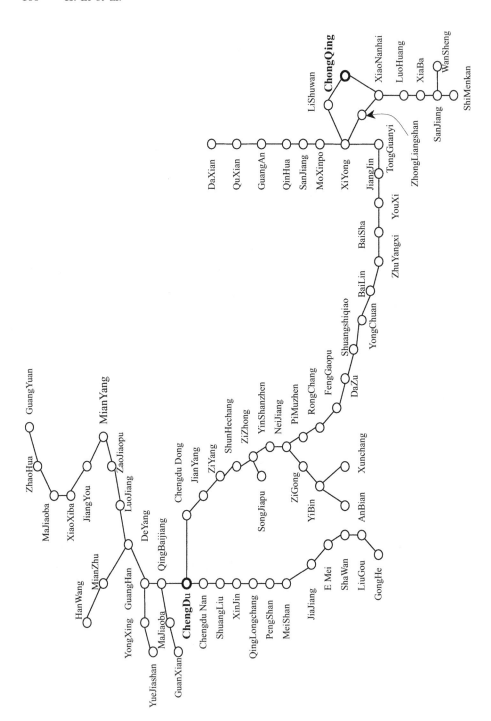

Fig. 2. A portion of CCIC's coal transportation network

Commodity of coal	Number of nodes passed through	Source	Sink	Distance /KM	Flow / Ton	T·KM
Mixture Coal	4	Wansheng	Shimenkan	86	10600	911600
Mixture Coal	23	Wansheng	Zigong	407	6000	2442000
Mixture Coal	35	Wansheng	Meishan	671	15000	100653000
Mixture Coal	33	Wansheng	Qinglongchang	647	2000	1294000
Mixture Coal	32	Wansheng	Guanxian	686	7900	5419400
Mixture Coal	31	Wansheng	Pengxian	656	1000	656000
Mixture Coal	30	Wansheng	Qinbaijiang	626	1000	626000
Mixture Coal	28	Wansheng	Chengdudong	583	21000	12243000
Mixture Coal	26	Wansheng	Ziyan	465	24000	11160000
Mixture Coal	23	Wansheng	Yinshanzhen	383	10000	3830000
Mixture Coal	22	Wansheng	Neijiang	369	187300	69113700
Mixture Coal	4	Wansheng	Xiaba	70	17000	1190000
Mixture Coal	3	Wansheng	Sanjiang	32	35700	1142400
Mixture Coal	19	Moxinpo	Neijiang	271	218700	59267700
Mixture Coal	5	Sanhuiba	Chongqing	79	152000	12008000
Mixture Coal	6	Sanhuiba	Chongqing	105	575000	60375000
Mixture Coal	20	Sanhuiba	Neijiang	342	14000	4788000
Mixture Coal	10	Sanhuiba	Bailin	192	9500	1824000
Mixture Coal	9	Sanhuiba	Zhuyangxi	174	600	104400
Mixture Coal	6	Sanhuiba	Jiangjin	115	41500	4772500
Mixture Coal	5	Sanhuiba	Tongguanyi	93	4400	409200
...	...	...	...	...	...	...
Top-grade coal	22	Rongchang	Emei	439	12000	5268000
...	...	...	...			
Block coal						
...						
Block coal of top-grade	10	Yongchuan	Chongqing	165	198100	32686500
...	...	...	...	...	...	...
Powder coal	14	Longchang	Guanxian	356	3300	1174800
...	...	...	...	...	...	...
Anthracite coal	37	Suimengkan	Emei	757	9300	7040100
...	...	...	...	...	...	...

Fig. 3. Part of an optimal scheme for CCIC

City and Sichuan Province, through over 200 railway stations along 5 railroad arteries. A portion of the transportation network is shown in Figure 2.

For this large-scale programming problem, some simplifications are necessary. We assume that all the cost functions are linear, and all the supplies, demands and capacities in the network take integer values. Then the maximal-flow programming is simplified to mixed integer linear programming. Below we discuss the algorithm aspect of the problem.

Maximal-flow programming is one of the oldest and well-studied problems in operations research. In its 50 years history, Ford and Fulkerson's "augmenting path" algorithm [19] was one of the first algorithms to solve the problem. Subsequently, Edmonds and Karp [16] proposed two labeling algorithms. Dinic [15] introduced the concept "layered networks" into the problem solving, and Karzanov [27] made further improvement by introducing the concept "preflow". Until 1985, Dinic's and Karzanov's algorithms were widely considered to be the fastest algorithms for solving the maximum-flow problem.

In 1988, Goldberg and Tarjan [22] proposed a new algorithm called "push-relabel method". Derigs and Meier [14] found that Goldberg and Tarjan's algorithm is substantially faster than Dinic's and Karzanov's algorithms. On the other hand, by Bertsekas' comparison [9], among the auction algorithm, the highest-label preflow-push algorithm and Derigs and Meier's algorithm, the first one outperformed the other two by a constant factor for most classes of problems. Recently, Ahuja et al. [2] tested some of the major algorithmic ideas to assess their utility on the empirical front, including the shortest augmenting path algorithm, Dinic's algorithm, Karzanov's algorithm, Goldberg-Tarjan's algorithm, etc. They found that the time to perform relabeling operations is at least as much as that for augmentations and/or pushes.

Being unable to determine beforehand what algorithms are best suited for our special network, we have to implement different maximal-flow algorithms in our system. For real data provided by CCIC, different algorithms cost much the same amount of computing time, suggesting that our network may be an equilibrium case for the algorithms participating the performance contest.

The computing results are given by labels within the MCMF model. As an illustration, some labels in an optimal scheme are shown in Figure 3. This scheme reduces the transportation load by 8,663,400 Ton-Kilometer per year, and saves at least 2,290,000 RMB's transportation expenses per year.

4   Conclusion

We have introduced a system of modeling and computing the multicommodity minimum cost network flow problem, and its application in the operational decision of the coal transportation and sales of Chongqing Coal Industry Company. The system is currently in use by the Company and has brought considerable economic benefits.

References

1. A. K. Aggarwal, M. Oblak, R. R. Vemuganti, A heuristic solution procedure for multicommodity integer flows, *Computer Operation Research* 22(10), 1075-1087, 1995.

2. R. K. Ahuja, M. Kodialam, A. K. Mishra, J. B. Orlin, Computational investigations of maximum flow algorithms, *European J. of Operational Research* 97, 509-54, 1997.

3. Anderson, R. J. and Sctubal, J. C., Parallel and sequential implementations of maximum flow algorithms, In: D. S. Johnson and C. C. McGeoch (eds.), *Network Flows and Matching: First DIMACS Implementation Challenge*, AMS, 1993.

4. R. Aringhierie and R. Cordone, The Multicommodity Multilevel Bottleneck Assignment Problem, *Electronic Notes in Discrete Mathematics* 17, 35-40, 2004.

5. T. Asano and Y. Asano, Recent developments in maximum flow algorithms, *Journal of the Operations Research Society of Japan*, Vol. 43, No. 1, 2000, 2-31.

6. A. Assad, Multicommodity network flows, a survey, *Networks* 8, 37-91, 1978.

7. Badics, T., Boros, E., and Cepek, O., Implementing a new maximum flow algorithm, In: D. S. Johnson and C. C. McGeoch (eds.), *Network Flows and Matching: First DIMACS Implementation Challenge*, AMS, l993.

8. J. E. Beasley and B. Cao, A Dynamic Programming based algorithm for the Crew Scheduling Problem, *Computers & Operations Research* 53, 567-582, 1998.

9. Bertsekas, D. P., An auction algorithm for the max-flow problem, *IEEE Conference on the Foundations of Computer Science*, pp. 118-123, 1994.

10. H. I. Calvete, Network simplex algorithm for the general equal flow problem, *European J. of Operational Research* 150 (2003) 585-600.

11. P. Cappanera and G. Gallo, On the Airline Crew Rostering Problem, *European J. of Operational Research* 150 (2003).

12. P. Carraresi and G. Gallo, A Multi-level Bottleneck Assignment Approach to the bus drivers' roistering problem, *European J. of Operational Res.* 16, 163-173, 1984.

13. N. D. Curet, J. DeVinney, M. E. Gaston, An efficient network flow code for finding all minimum cost s-t cutsets, *Computers & Operations Research* 29 (2002) 205-219.

14. Derigs, U., and Meier, W., Implementing Goldberg's max-flow algorithm: A computational investigation, *Zeitschrift fur Operations Research* 33, 383-403, 1989.

15. E. A. Dinic, Algorithm for solution of a problem of maximum flow in networks with power estimation. *Soviet Math. Doklady* 11, 1277-1280, 1970.

16. J. Edmonds and R. M. Karp, Theoretical improvements in algorithmic efficiency for network flow problems. *Journal of the ACM* 19, 248-264, 1972.

17. J. R. Evans, The simplex method for integral multicommodity networks. *Naval Res. Logistics* 4, 31-37, 1978.

18. J. R. Evans, Reducing computational effort in detecting integral multicommodity networks. *Computation Operation Research* 7, 261-265, 1980.

19. L. R. Ford, Jr. and D. R. Fulkerson, Maximal flow through a network. *Canad. J. Math.* 8, 399-404, 1956.

20. V. Gabrela, A. Knippelb, M. Minouxb, Exact solution of multicommodity network optimization problems with general step cost functions, *Operations Research Letters* 25 (1999) 15-23.

21. Gabow, H. N., Scaling algorithms for network flow problems, *Journal of Computer and System Sciences* 31, 148-168, 1985.

22. A. V. Goldberg and R. E. Tarjan, A new approach to the maximum flow problem. *Journal of the ACM* 35, 921-940, 1988.

23. Goldfarb, D. and Grigoriadis, M. D., A computational comparison of the Dinic and network simplex methods for maximum flow, *Annals of Operations Research* 13, 83-123, 1988.
24. Goldfarb, D. and Hao, J., On strongly polynomial variants of the network simplex algorithm for the maximum flow problem, *Operations Res. Letters* 10, 383-387, 1991.
25. A. Haghani, S. Oh, Formulation and solution of a multicommodity, multi-modal network flow model for disaster relief operations, *Transportation Research* 30(3), 231-250, 1996.
26. R. Hassin, On multicommodity flows in planar graphs. *Networks* 14, 225-235, 1984.
27. A. V. Karzanov, Determining the maximal flow in a network by the method of preflows. *Soviet Math. Doklady* 15, 434-437, 1974.
28. J. Kennington and R. Helgason, *Algorithms for Network Programming*. John Wiley & Sons, New York, 1980.
29. H.-X. Li, A Study on Optimization of Transportation and sale of A Variety of Coal in Chongqing Area, *Master's Degree Thesis*, Xi'an Univ. of Science & Tech., 1993.
30. M. V. Lomonosov, On the planar integer two-flow problem, *Combinatorica* 3, 207-218, 1983.
31. Nguyen, Q. C. and Venkateshwaran, V., Implementations of the Goldberg Tarjan maximum flow algorithm, In: D. S. Johnson and C. C. McGeoch (eds.), *Network Flows and Matching: First DIMACS Implementation Challenge*, AMS, 1993.

A Survey of Moving Frames

Peter J. Olver

School of Mathematics, University of Minnesota,
Minneapolis, MN 55455, USA
olver@umn.edu
http://www.math.umn.edu/~olver

Abstract. This paper surveys the new, algorithmic theory of moving frames developed by the author and M. Fels. Applications in geometry, computer vision, classical invariant theory, the calculus of variations, and numerical analysis are indicated.

1 Introduction

According to Akivis, [1], the idea of moving frames can be traced back to the method of moving trihedrons introduced by the Estonian mathematician Martin Bartels (1769–1836), a teacher of both Gauß and Lobachevsky. The modern method of moving frames or repères mobiles[1] was primarily developed by Élie Cartan, [22, 23], who forged earlier contributions by Cotton, Darboux, Frenet and Serret into a powerful tool for analyzing the geometric properties of submanifolds and their invariants under the action of transformation groups.

In the 1970's, several researchers, cf. [29, 42, 44, 53], began the attempt to place Cartan's intuitive constructions on a firm theoretical foundation. I've been fascinated by the power of the method since my student days, but, for many years, could not see how to release it from its rather narrow geometrical confines, e.g. Euclidean or equiaffine actions on submanifolds of Euclidean space. The crucial conceptual leap is to decouple the moving frame theory from reliance on any form of frame bundle or connection, and define a moving frame as an equivariant map from the manifold or jet bundle back to the transformation group. In other words,

$$\text{Moving frames} \;\neq\; \text{Frames}\,!$$

A careful study of Cartan's analysis of the case of projective curves, [22], reveals that Cartan was well aware of this viewpoint; however, this important and instructive example did not receive the attention it deserved. Once freed from the confining fetters of frames, Mark Fels and I, [39, 40], were able to formulate

---

[1] In French, the term "repère mobile" refers to a temporary mark made during building or interior design, and so a more accurate English translation might be "movable landmarks".

H. Li, P. J. Olver and G. Sommer (Eds.): IWMM-GIAE 2004, LNCS 3519, pp. 105–138, 2005.

a new, powerful, constructive approach to the equivariant moving frame theory that can be systematically applied to general transformation groups. All classical moving frames can be reinterpreted in this manner, but the equivariant approach applies in far broader generality.

Cartan's construction of the moving frame through the normalization process is interpreted with the choice of a cross-section to the group orbits. Building on these two simple ideas, one may algorithmically construct equivariant moving frames and, as a result, complete systems of invariants for completely general group actions. The existence of a moving frame requires freeness of the underlying group action. Classically, non-free actions are made free by prolonging to jet space, leading to differential invariants and the solution to equivalence and symmetry problems via the differential invariant signature. More recently, the moving frame method was also applied to Cartesian product actions, leading to classification of joint invariants and joint differential invariants, [86]. Recently, a seamless amalgamation of jet and Cartesian product actions dubbed *multi-space* was proposed in [88] to serve as the basis for the geometric analysis of numerical approximations, and, via the application of the moving frame method, to the systematic construction of invariant numerical algorithms.

New and significant applications of these results have been developed in a wide variety of directions. In [84, 6, 58, 59], the theory was applied to produce new algorithms for solving the basic symmetry and equivalence problems of polynomials that form the foundation of classical invariant theory. The moving frame method provides a direct route to the classification of joint invariants and joint differential invariants, [40, 86, 10], establishing a geometric counterpart of what Weyl, [108], in the algebraic framework, calls the first main theorem for the transformation group. In computer vision, joint differential invariants have been proposed as noise-resistant alternatives to the standard differential invariant signatures, [14, 21, 33, 79, 105, 106]. The approximation of higher order differential invariants by joint differential invariants and, generally, ordinary joint invariants leads to fully invariant finite difference numerical schemes, [9, 18, 19, 88, 57]. In [19, 5, 9], the characterization of submanifolds via their differential invariant signatures was applied to the problem of object recognition and symmetry detection, [12, 13, 15, 92]. A complete solution to the calculus of variations problem of directly constructing differential invariant Euler-Lagrange equations from their differential invariant Lagrangians was given based on the moving frame construction of the invariant variational bicomplex, [62].

As these methods become more widely disseminated, many additional applications are being pursued by a number of research groups, and include the computation of symmetry groups and classification of partial differential equations [69, 80]; projective and conformal geometry of curves and surfaces, with applications in Poisson geometry and integrable systems, [71, 72]; recognition of polygons and point configurations, with applications in image processing, [11, 54]; classification of projective curves in visual recognition, [48]; classification of the invariants and covariants of Killing tensors arising in general relativity and geometry, with applications to separation of variables and Hamiltonian systems, [32, 75]; and the development of noncommutative Gröbner basis methods, [50, 70]. Finally, in recent work with Pohjanpelto, [89, 90, 91], the theory has recently been extended to the vastly more complicated case of infinite-dimensional Lie pseudo-groups.

2   Moving Frames

We begin by outlining the basic moving frame construction in [40]. Let G be an r-dimensional Lie group acting smoothly on an m-dimensional manifold M. Let $G_S = \{\, g \in G \,|\, g \cdot S = S \,\}$ denote the *isotropy subgroup* of a subset $S \subset M$, and $G_S^* = \bigcap_{z \in S} G_z$ its *global isotropy subgroup*, which consists of those group elements which fix all points in S. We always assume, without any significant loss of generality, that G acts *effectively on subsets*, and so $G_U^* = \{e\}$ for any open $U \subset M$, i.e., there are no group elements other than the identity which act completely trivially on an open subset of M.

Definition 1.   A *moving frame* is a smooth, G-equivariant map $\rho : M \to G$.

The group G acts on itself by left or right multiplication. If $\rho(z)$ is any right-equivariant moving frame then $\widetilde{\rho}(z) = \rho(z)^{-1}$ is left-equivariant and conversely. All classical moving frames are left equivariant, but, in many cases, the right versions are easier to compute. In many geometrical situations, one can identify our left moving frames with the usual frame-based versions, but these identifications break down for more general transformation groups.

Theorem 2.   *A moving frame exists in a neighborhood of a point $z \in M$ if and only if G acts freely and regularly near z.*

Recall that G acts *freely* if the isotropy subgroup of each point is trivial, $G_z = \{e\}$ for all $z \in M$. This implies that the orbits all have the same dimension as G itself. *Regularity* requires that, in addition, each point $x \in M$ has a system of arbitrarily small neighborhoods whose intersection with each orbit is connected, cf. [82].

The practical construction of a moving frame is based on Cartan's method of *normalization*, [56, 22], which requires the choice of a (local) *cross-section* to the group orbits.

Theorem 3.   *Let G act freely and regularly on M, and let $K \subset M$ be a cross-section. Given $z \in M$, let $g = \rho(z)$ be the unique group element that maps z to the cross-section: $g \cdot z = \rho(z) \cdot z \in K$. Then $\rho : M \to G$ is a right moving frame for the group action.*

Given local coordinates $z = (z_1, \dots, z_m)$ on M, let $w(g, z) = g \cdot z$ be the explicit formulae for the group transformations. The right[2] moving frame $g = \rho(z)$

---

[2] The left version can be obtained directly by replacing g by g^{-1} throughout the construction.

associated with a *coordinate cross-section* $K = \{z_1 = c_1, \ldots, z_r = c_r\}$ is obtained by solving the *normalization equations*

$$w_1(g, z) = c_1, \qquad \cdots \qquad w_r(g, z) = c_r, \qquad (2.1)$$

for the group parameters $g = (g_1, \ldots, g_r)$ in terms of the coordinates $z = (z_1, \ldots, z_m)$. Substituting the moving frame formulae into the remaining transformation rules leads to a complete system of invariants for the group action.

Theorem 4.   *If $g = \rho(z)$ is the moving frame solution to the normalization equations* (2.1), *then the functions*

$$I_1(z) = w_{r+1}(\rho(z), z), \qquad \cdots \qquad I_{m-r}(z) = w_m(\rho(z), z), \qquad (2.2)$$

form a complete system of functionally independent invariants.

Definition 5.   The *invariantization* of a scalar function $F \colon M \to \mathbb{R}$ with respect to a right moving frame ρ is the the invariant function $I = \iota(F)$ defined by $I(z) = F(\rho(z) \cdot z)$.

Invariantization amounts to restricting F to the cross-section, $I \,|\, K = F \,|\, K$, and then requiring that I be constant along the orbits. In particular, if $I(z)$ is an invariant, then $\iota(I) = I$, so invariantization defines a projection, depending on the moving frame, from functions to invariants. Thus, a moving frame provides a canonical method of associating an invariant with an arbitrary function.

Of course, most interesting group actions are *not* free, and therefore do not admit moving frames in the sense of Definition 1. There are two basic methods for converting a non-free (but effective) action into a free action. The first is to look at the product action of G on several copies of M, leading to joint invariants. The second is to prolong the group action to jet space, which is the natural setting for the traditional moving frame theory, and leads to differential invariants. Combining the two methods of prolongation and product will lead to joint differential invariants. In applications of symmetry constructions to numerical approximations of derivatives and differential invariants, one requires a unification of these different actions into a common framework, called multispace, [57, 88].

3   Prolongation and Differential Invariants

Traditional moving frames are obtained by prolonging the group action to the n-th order (extended) jet bundle $\mathrm{J}^n = \mathrm{J}^n(M, p)$ consisting of equivalence classes of p-dimensional submanifolds $S \subset M$ modulo n-th order contact at a single point; see [82–Chapter 3] for details. Since G preserves the contact equivalence relation, it induces an action on the jet space J^n, known as its n-th order *prolongation* and denoted by $G^{(n)}$.

An *n-th order moving frame* $\rho^{(n)} : \mathrm{J}^n \to G$ is an equivariant map defined on an open subset of the jet space. In practical examples, for n sufficiently large, the prolonged action $G^{(n)}$ becomes regular and free on a dense open subset $\mathcal{V}^n \subset \mathrm{J}^n$, the set of *regular jets*. It has been rigorously proved that, for $n \gg 0$ sufficiently large, if G acts effectively on subsets, then $G^{(n)}$ acts locally freely on an open subset $\mathcal{V}^n \subset \mathrm{J}^n$, [85].

Theorem 6.   *An n-th order moving frame exists in a neighborhood of a point $z^{(n)} \in \mathrm{J}^n$ if and only if $z^{(n)} \in \mathcal{V}^n$ is a regular jet.*

Our normalization construction will produce a moving frame and a complete system of differential invariants in the neighborhood of any regular jet. Local coordinates $z = (x, u)$ on M — considering the first p components $x = (x^1, \ldots, x^p)$ as independent variables, and the latter $q = m - p$ components $u = (u^1, \ldots, u^q)$ as dependent variables — induce local coordinates $z^{(n)} = (x, u^{(n)})$ on J^n with components u_J^α representing the partial derivatives of the dependent variables with respect to the independent variables, [82, 83]. We compute the prolonged transformation formulae

$$w^{(n)}(g, z^{(n)}) = g^{(n)} \cdot z^{(n)}, \qquad \text{or} \qquad (y, v^{(n)}) = g^{(n)} \cdot (x, u^{(n)}),$$

by implicit differentiation of the v's with respect to the y's. For simplicity, we restrict to a coordinate cross-section by choosing $r = \dim G$ components of $w^{(n)}$ to normalize to constants:

$$w_1(g, z^{(n)}) = c_1, \qquad \cdots \qquad w_r(g, z^{(n)}) = c_r. \tag{3.3}$$

Solving the normalization equations (3.3) for the group transformations leads to the explicit formulae $g = \rho^{(n)}(z^{(n)})$ for the right moving frame. As in Theorem 4, substituting the moving frame formulae into the unnormalized components of $w^{(n)}$ leads to the *fundamental n-th order differential invariants*

$$I^{(n)}(z^{(n)}) = w^{(n)}(\rho^{(n)}(z^{(n)}), z^{(n)}) = \rho^{(n)}(z^{(n)}) \cdot z^{(n)}. \tag{3.4}$$

Once the moving frame is established, the *invariantization* process will map general differential functions $F(x, u^{(n)})$ to differential invariants $I = \iota(F) = F \circ I^{(n)}$. As before, invariantization defines a projection, depending on the moving frame, from the space of differential functions to the space of differential invariants. The fundamental differential invariants $I^{(n)}$ are obtained by invariantization of the coordinate functions

$$\begin{aligned}
H^i(x, u^{(n)}) &= \iota(x^i) = y^i(\rho^{(n)}(x, u^{(n)}), x, u), \\
I_K^\alpha(x, u^{(k)}) &= \iota(u_J^\alpha) = v_K^\alpha(\rho^{(n)}(x, u^{(n)}), x, u^{(k)}).
\end{aligned} \tag{3.5}$$

In particular, those corresponding to the normalization components (3.3) of $w^{(n)}$ will be constant, and are known as the *phantom differential invariants*.

Theorem 7.   *Let $\rho^{(n)} : \mathrm{J}^n \to G$ be a moving frame of order $\leq n$. Every n-th order differential invariant can be locally written as a function $J = \Phi(I^{(n)})$ of the fundamental n-th order differential invariants (3.5). The function Φ is unique provided it does not depend on the phantom invariants.*

Example 8.   Let us begin with a very simple, classical example: curves in the Euclidean plane. The orientation-preserving Euclidean group SE(2) acts on $M = \mathbb{R}^2$, mapping a point $z = (x, u)$ to

$$y = x \cos \theta - u \sin \theta + a, \qquad v = x \sin \theta + u \cos \theta + b. \qquad (3.6)$$

For a general parametrized[3] curve $z(t) = (x(t), u(t))$, the prolonged group transformations

$$v_y = \frac{dv}{dy} = \frac{\dot{x} \sin \theta + \dot{u} \cos \theta}{\dot{x} \cos \theta - \dot{u} \sin \theta}, \qquad v_{yy} = \frac{d^2 v}{dy^2} = \frac{\dot{x} \ddot{u} - \ddot{x} \dot{u}}{(\dot{x} \cos \theta - \dot{u} \sin \theta)^3}, \qquad (3.7)$$

and so on, are found by successively applying the implicit differentiation operator

$$\frac{d}{dy} = \frac{1}{\dot{x} \cos \theta - \dot{u} \sin \theta} \frac{d}{dt} \qquad (3.8)$$

to v. The classical Euclidean moving frame for planar curves, [46], follows from the cross-section normalizations

$$y = 0, \qquad v = 0, \qquad v_y = 0. \qquad (3.9)$$

Solving for the group parameters $g = (\theta, a, b)$ leads to the right-equivariant moving frame

$$\theta = -\tan^{-1} \frac{\dot{u}}{\dot{x}}, \qquad a = -\frac{x\dot{x} + u\dot{u}}{\sqrt{\dot{x}^2 + \dot{u}^2}} = -\frac{z \cdot \dot{z}}{\| \dot{z} \|}, \qquad b = \frac{x\dot{u} - u\dot{x}}{\sqrt{\dot{x}^2 + \dot{u}^2}} = \frac{z \wedge \dot{z}}{\| \dot{z} \|}. \qquad (3.10)$$

The inverse group transformation $g^{-1} = (\tilde{\theta}, \tilde{a}, \tilde{b})$ is the classical left moving frame, [22, 46]: one identifies the translation component $(\tilde{a}, \tilde{b}) = (x, u) = z$ as the point on the curve, while the columns of the rotation matrix $\tilde{R}(\tilde{\theta}) = (\mathbf{t}, \mathbf{n})$ are the unit tangent and unit normal vectors. Substituting the moving frame normalizations (3.10) into the prolonged transformation formulae (3.7), results in the fundamental differential invariants

$$v_{yy} \longmapsto \kappa = \frac{\dot{x} \ddot{u} - \ddot{x} \dot{u}}{(\dot{x}^2 + \dot{u}^2)^{3/2}} = \frac{\dot{z} \wedge \ddot{z}}{\| \dot{z} \|^3},$$

$$v_{yyy} \longmapsto \frac{d\kappa}{ds}, \qquad v_{yyyy} \longmapsto \frac{d^2\kappa}{ds^2} + 3\kappa^3, \qquad (3.11)$$

where $d/ds = \| \dot{z} \|^{-1} d/dt$ is the arc length derivative — which is itself found by substituting the moving frame formulae (3.10) into the implicit differentiation operator (3.8). A complete system of differential invariants for the planar

---

[3] While the local coordinates $(x, u, u_x, u_{xx}, \ldots)$ on the jet space assume that the curve is given as the graph of a function $u = f(x)$, the moving frame computations also apply, as indicated in this example, to general parametrized curves. Two parametrized curves are equivalent if and only if one can be mapped to the other under a suitable reparametrization.

Euclidean group is provided by the curvature and its successive derivatives with respect to arc length: $\kappa, \kappa_s, \kappa_{ss}, \ldots$.

The one caveat is that the first prolongation of SE(2) is only locally free on J^1 since a 180° rotation has trivial first prolongation. The even derivatives of κ with respect to s change sign under a 180° rotation, and so only their absolute values are fully invariant. The ambiguity can be removed by including the second order constraint $v_{yy} > 0$ in the derivation of the moving frame. Extending the analysis to the full Euclidean group E(2) adds in a second sign ambiguity which can only be resolved at third order. See [86] for complete details.

Example 9.   Let $n \neq 0, 1$. In classical invariant theory, the planar actions

$$y = \frac{\alpha x + \beta}{\gamma x + \delta}, \qquad \bar{u} = (\gamma x + \delta)^{-n} u, \qquad (3.12)$$

of $G = \mathrm{GL}(2)$ play a key role in the equivalence and symmetry properties of binary forms, when $u = q(x)$ is a polynomial of degree $\leq n$, [49, 84, 6]. We identify the graph of the function $u = q(x)$ as a plane curve. The prolonged action on such graphs is found by implicit differentiation:

$$v_y = \frac{\sigma u_x - n\gamma u}{\Delta \sigma^{n-1}}, \qquad v_{yy} = \frac{\sigma^2 u_{xx} - 2(n-1)\gamma\sigma u_x + n(n-1)\gamma^2 u}{\Delta^2 \sigma^{n-2}},$$

$$v_{yyy} = \frac{\sigma^3 u_{xxx} - 3(n-2)\gamma\sigma^2 u_{xx} + 3(n-1)(n-2)\gamma^2\sigma u_x - n(n-1)(n-2)\gamma^3 u}{\Delta^3 \sigma^{n-3}},$$

and so on, where $\sigma = \gamma p + \delta$, $\Delta = \alpha\delta - \beta\gamma \neq 0$. On the regular subdomain

$$\mathcal{V}^2 = \{uH \neq 0\} \subset J^2, \qquad \text{where} \qquad H = uu_{xx} - \frac{n-1}{n}u_x^2$$

is the classical Hessian covariant of u, we can choose the cross-section defined by the normalizations

$$y = 0, \qquad v = 1, \qquad v_y = 0, \qquad v_{yy} = 1.$$

Solving for the group parameters gives the right moving frame formulae[4]

$$\begin{aligned} \alpha &= u^{(1-n)/n}\sqrt{H}, & \beta &= -x\,u^{(1-n)/n}\sqrt{H}, \\ \gamma &= \tfrac{1}{n} u^{(1-n)/n} u_x, & \delta &= u^{1/n} - \tfrac{1}{n} x\,u^{(1-n)/n} u_x. \end{aligned} \qquad (3.13)$$

Substituting the normalizations (3.13) into the higher order transformation rules gives us the differential invariants, the first two of which are

$$v_{yyy} \longmapsto J = \frac{T}{H^{3/2}}, \qquad v_{yyyy} \longmapsto K = \frac{V}{H^2}, \qquad (3.14)$$

---

[4] See [6] for a detailed discussion of how to resolve the square root ambiguities.

where

$$T = u^2 u_{xxx} - 3\,\frac{n-2}{n}\,uu_x u_{xx} + 2\,\frac{(n-1)(n-2)}{n^2}\,u_x^3,$$

$$V = u^3 u_{xxxx} - 4\,\frac{n-3}{n}\,u^2 u_x u_{xxx} + 6\,\frac{(n-2)(n-3)}{n^2}\,uu_x^{\,2} u_{xx} -$$

$$-\,3\frac{(n-1)(n-2)(n-3)}{n^3}\,u_x^4,$$

and can be identified with classical covariants, which may be constructed using the basic transvectant process of classical invariant theory, cf. [49, 84]. Using $J^2 = T^2/H^3$ as the fundamental differential invariant will remove the ambiguity caused by the square root. As in the Euclidean case, higher order differential invariants are found by successive application of the normalized implicit differentiation operator $D_s = uH^{-1/2}D_x$ to the fundamental invariant J.

4   Equivalence and Signatures

The moving frame method was developed by Cartan expressly for the solution to problems of equivalence and symmetry of submanifolds under group actions. Two submanifolds $S, \overline{S} \subset M$ are said to be *equivalent* if $\overline{S} = g \cdot S$ for some $g \in G$. A *symmetry* of a submanifold is a group transformation that maps S to itself, and so is an element $g \in G_S$. As emphasized by Cartan, [22], the solution to the equivalence and symmetry problems for submanifolds is based on the functional interrelationships among the fundamental differential invariants restricted to the submanifold.

Suppose we have constructed an n-th order moving frame $\rho^{(n)} \colon \mathrm{J}^n \to G$ defined on an open subset of jet space. A submanifold S is called *regular* if its n-jet $\mathrm{j}_n S$ lies in the domain of definition of the moving frame. For any $k \geq n$, we use $J^{(k)} = I^{(k)} \,|\, S = I^{(k)} \circ \mathrm{j}_k S$ to denote the k-th order *restricted differential invariants*. The k-th order *signature* $\mathcal{S}^{(k)} = \mathcal{S}^{(k)}(S)$ is the set parametrized by the restricted differential invariants; S is called *fully regular* if $J^{(k)}$ has constant rank $0 \leq t_k \leq p = \dim S$ for all $k \geq n$. In this case, $\mathcal{S}^{(k)}$ forms a submanifold of dimension t_k — perhaps with self-intersections. In the fully regular case,

$$t_n < t_{n+1} < t_{n+2} < \cdots < t_s = t_{s+1} = \cdots = t \leq p,$$

where t is the *differential invariant rank* and s the *differential invariant order* of S.

Theorem 10.   *Two fully regular p-dimensional submanifolds $S, \overline{S} \subset M$ are (locally) equivalent, $\overline{S} = g \cdot S$, if and only if they have the same differential invariant order s and their signature manifolds of order $s + 1$ are identical: $\mathcal{S}^{(s+1)}(\overline{S}) = \mathcal{S}^{(s+1)}(S)$.*

Since symmetries are the same as self-equivalences, the signature also determines the symmetry group of the submanifold.

Theorem 11.   *If $S \subset M$ is a fully regular p-dimensional submanifold of differential invariant rank t, then its symmetry group G_S is an $(r-t)$-dimensional subgroup of G that acts locally freely on S.*

A submanifold with maximal differential invariant rank $t = p$, and hence only a discrete symmetry group, is called *nonsingular*. The number of symmetries is determined by the *index* of the submanifold, defined as the number of points in S map to a single generic point of its signature:

$$\operatorname{ind} S = \min \ \left\{ \ \# \, (J^{(s+1)})^{-1}\{\zeta\} \ \Big| \ \zeta \in \mathcal{S}^{(s+1)} \ \right\}.$$

Theorem 12.   *If S is a nonsingular submanifold, then its symmetry group is a discrete subgroup of cardinality $\# \, G_S = \operatorname{ind} S$.*

At the other extreme, a rank 0 or *maximally symmetric* submanifold has all constant differential invariants, and so its signature degenerates to a single point.

Theorem 13.   *A regular p-dimensional submanifold S has differential invariant rank 0 if and only if its symmetry group is a p-dimensional subgroup $H = G_S \subset G$ and an H–orbit: $S = H \cdot z_0$.*

Remark: "Totally singular" submanifolds may have even larger, non-free symmetry groups, but these are not covered by the preceding results. See [85] for details and precise characterization of such submanifolds.

Example 14.   The *Euclidean signature* for a curve in the Euclidean plane is the planar curve $\mathcal{S}(C) = \{(\kappa, \kappa_s)\}$ parametrized by the curvature invariant κ and its first derivative with respect to arc length. Two planar curves are equivalent under oriented rigid motions if and only if they have the same signature curves. The maximally symmetric curves have constant Euclidean curvature, and so their signature curve degenerates to a single point. These are the circles and straight lines, and, in accordance with Theorem 13, each is the orbit of its one-parameter symmetry subgroup of SE(2). The number of Euclidean symmetries of a curve is equal to its index — the number of times the Euclidean signature is retraced as we go around the curve.

An example of a Euclidean signature curve is displayed in figure 1. The first figure shows the curve, and the second its Euclidean signature; the axes are κ and κ_s in the signature plot. Note in particular the approximate three-fold symmetry of the curve is reflected in the fact that its signature has winding number three. If the symmetries were exact, the signature would be exactly retraced three times on top of itself. The final figure gives a discrete approximation to the signature which is based on the invariant numerical algorithms to be discussed below.

In figure 2 we display some signature curves computed from an actual medical image — a 70×70, 8-bit gray-scale image of a cross section of a canine heart, obtained from an MRI scan. We then display an enlargement of the left ventricle. The boundary of the ventricle has been automatically segmented through use of

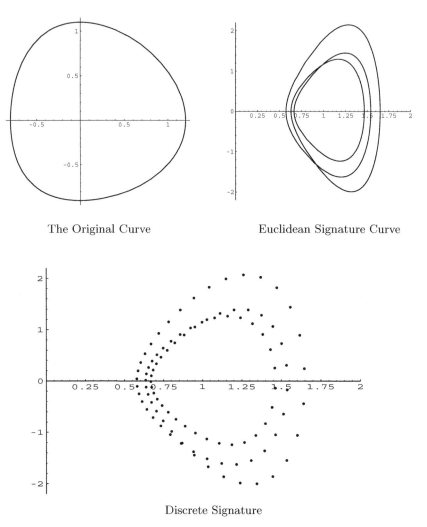

The Original Curve   Euclidean Signature Curve

Discrete Signature

Fig. 1. The Curve $x = \cos t + \frac{1}{5} \cos^2 t$, $y = \sin t + \frac{1}{10} \sin^2 t$

the conformally Riemannian moving contour or snake flow that was proposed in
[55] and successfully applied to a wide variety of 2D and 3D medical imagery,
including MRI, ultrasound and CT data, [109]. Underneath these images, we display the ventricle boundary curve along with two successive smoothed versions
obtained application of the standard Euclidean-invariant curve shortening procedure. Below each curve is the associated spline-interpolated discrete signature
curves for the smoothed boundary, as computed using the invariant numerical
approximations to κ and κ_s discussed below. As the evolving curves approach
circularity the signature curves exhibit less variation in curvature and appear
to be winding more and more tightly around a single point, which is the signature of a circle of area equal to the area inside the evolving curve. Despite

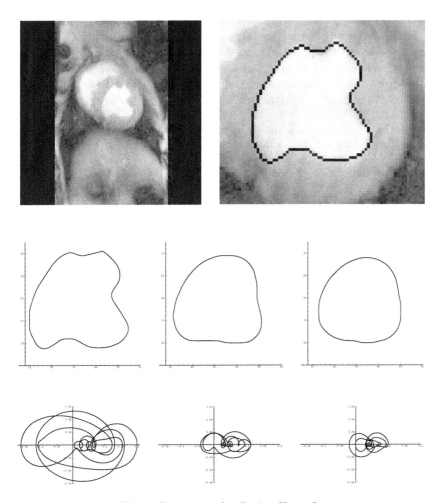

Fig. 2. Signature of a Canine Heart Image

the rather extensive smoothing involved, except for an overall shrinking as the contour approaches circularity, the basic qualitative features of the different signature curves, and particularly their winding behavior, appear to be remarkably robust.

Thus, the signature curve method has the potential to be of practical use in the general problem of object recognition and symmetry classification. It offer several advantages over more traditional approaches. First, it is purely local, and therefore immediately applicable to occluded objects. Second, it provides a mechanism for recognizing symmetries and approximate symmetries of the object. The design of a suitably robust "signature metric" for practical comparison of signatures is the subject of ongoing research. See the contribution by Shakiban and Lloyd, [97], in these proceedings for recent developments in this direction.

Example 15.   Let us next consider the equivalence and symmetry problems for binary forms. According to the general moving frame construction in Example 9, the signature curve $\mathcal{S} = \mathcal{S}(q)$ of a function (polynomial) $u = q(x)$ is parametrized by the covariants J^2 and K, as given in (3.14). The following solution to the equivalence problem for complex-valued binary forms, [6, 81, 84], is an immediate consequence of the general equivalence Theorem 10.

Theorem 16.   *Two nondegenerate complex-valued forms $q(x)$ and $\bar{q}(x)$ are equivalent if and only if their signature curves are identical: $\mathcal{S}(q) = \mathcal{S}(\bar{q})$.*

All equivalence maps $\bar{x} = \varphi(x)$ solve the two rational equations

$$J(x)^2 = \bar{J}(\bar{x})^2, \qquad K(x) = \overline{K}(\bar{x}). \tag{4.15}$$

In particular, the theory guarantees φ is necessarily a linear fractional transformation!

Theorem 17.   *A nondegenerate binary form $q(x)$ is maximally symmetric if and only if it satisfies the following equivalent conditions*:

a) *q is complex-equivalent to a monomial x^k, with $k \neq 0, n$.*
b) *The covariant T^2 is a constant multiple of $H^3 \not\equiv 0$.*
c) *The signature is just a single point.*
d) *q admits a one-parameter symmetry group.*
e) *The graph of q coincides with the orbit of a one-parameter subgroup of $\mathrm{GL}(2)$.*

A binary form $q(x)$ is nonsingular if and only if it is not complex-equivalent to a monomial if and only if it has a finite symmetry group.

The symmetries of a nonsingular form can be explicitly determined by solving the rational equations (4.15) with $\bar{J} = J$, $\overline{K} = K$. See [6] for a MAPLE implementation of this method for computing discrete symmetries and classification of univariate polynomials. In particular, we obtain the following useful bounds on the number of symmetries.

Theorem 18.   *If $q(x)$ is a binary form of degree n which is not complex-equivalent to a monomial, then its projective symmetry group has cardinality*

$$k \leq \begin{cases} 6n - 12 & \text{if } V = cH^2 \text{ for some constant } c, \text{ or} \\ 4n - 8 & \text{in all other cases.} \end{cases}$$

In her thesis, Kogan, [58], extends these results to forms in several variables. In particular, a complete signature for ternary forms, [59], leads to a practical algorithm for computing discrete symmetries of, among other cases, elliptic curves.

5   Joint Invariants and Joint Differential Invariants

One practical difficulty with the differential invariant signature is its dependence upon high order derivatives, which makes it very sensitive to data noise. For this reason, a new signature paradigm, based on joint invariants, was proposed in [86]. We consider now the joint action

$$g \cdot (z_0, \ldots, z_n) = (g \cdot z_0, \ldots, g \cdot z_n), \qquad g \in G, \quad z_0, \ldots, z_n \in M. \tag{5.16}$$

of the group G on the $(n+1)$-fold Cartesian product $M^{\times(n+1)} = M \times \cdots \times M$. An invariant $I(z_0, \ldots, z_n)$ of (5.16) is an $(n+1)$-*point joint invariant* of the original transformation group. In most cases of interest, although not in general, if G acts effectively on M, then, for $n \gg 0$ sufficiently large, the product action is free and regular on an open subset of $M^{\times(n+1)}$. Consequently, the moving frame method outlined in section 1 can be applied to such joint actions, and thereby establish complete classifications of joint invariants and, via prolongation to Cartesian products of jet spaces, joint differential invariants. We will discuss two particular examples — planar curves in Euclidean geometry and projective geometry, referring to [86] for details.

Example 19.   *Euclidean joint differential invariants.* Consider the proper Euclidean group SE(2) acting on oriented curves in the plane $M = \mathbf{R}^2$. We begin with the Cartesian product action on $M^{\times 2} \simeq \mathbf{R}^4$. Taking the simplest cross-section $x_0 = u_0 = x_1 = 0, u_1 > 0$ leads to the normalization equations

$$y_0 = x_0 \cos \theta - u_0 \sin \theta + a = 0, \qquad v_0 = x_0 \sin \theta + u_0 \cos \theta + b = 0,$$
$$y_1 = x_1 \cos \theta - u_1 \sin \theta + a = 0. \tag{5.17}$$

Solving, we obtain a right moving frame

$$\theta = \tan^{-1}\left(\frac{x_1 - x_0}{u_1 - u_0}\right), \qquad a = -x_0 \cos \theta + u_0 \sin \theta, \qquad b = -x_0 \sin \theta - u_0 \cos \theta, \tag{5.18}$$

along with the fundamental interpoint distance invariant

$$v_1 = x_1 \sin \theta + u_1 \cos \theta + b \qquad \longmapsto \qquad I = \| z_1 - z_0 \|. \tag{5.19}$$

Substituting (5.18) into the prolongation formulae (3.7) leads to the the normalized first and second order joint differential invariants

$$\frac{dv_k}{dy} \quad \longmapsto \quad J_k = -\frac{(z_1 - z_0) \cdot \dot{z}_k}{(z_1 - z_0) \wedge \dot{z}_k},$$
$$\frac{d^2 v_k}{dy^2} \quad \longmapsto \quad K_k = -\frac{\| z_1 - z_0 \|^3 (\dot{z}_k \wedge \ddot{z}_k)}{[(z_1 - z_0) \wedge \dot{z}_0]^3}, \tag{5.20}$$

for $k = 0, 1$. Note that

$$J_0 = -\cot \phi_0, \qquad J_1 = +\cot \phi_1, \tag{5.21}$$

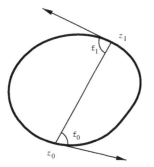 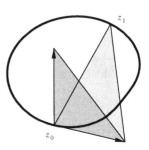

Fig. 3. First and Second Order Joint Euclidean Differential Invariants

where $\phi_k = \sphericalangle(z_1 - z_0, \dot{z}_k)$ denotes the angle between the chord connecting z_0, z_1 and the tangent vector at z_k, as illustrated in figure 3. The modified second order joint differential invariant

$$\widehat{K}_0 = -\| z_1 - z_0 \|^{-3} K_0 = \frac{\dot{z}_0 \wedge \ddot{z}_0}{\left[(z_1 - z_0) \wedge \dot{z}_0 \right]^3} \tag{5.22}$$

equals the ratio of the area of triangle whose sides are the first and second derivative vectors $\dot{z}_0, \ddot{z}_0$ at the point z_0 over the *cube* of the area of triangle whose sides are the chord from z_0 to z_1 and the tangent vector at z_0; see figure 3.

On the other hand, we can construct the joint differential invariants by invariant differentiation of the basic distance invariant (5.19). The normalized invariant differential operators are

$$\mathcal{D}_{y_k} \longmapsto \mathcal{D}_k = -\frac{\| z_1 - z_0 \|}{(z_1 - z_0) \wedge \dot{z}_k} \, \mathcal{D}_{t_k}. \tag{5.23}$$

Proposition 20. *Every two-point Euclidean joint differential invariant is a function of the interpoint distance $I = \| z_1 - z_0 \|$ and its invariant derivatives with respect to (5.23).*

A generic product curve $\mathbf{C} = C_0 \times C_1 \subset M^{\times 2}$ has joint differential invariant rank $2 = \dim \mathbf{C}$, and its joint signature $\mathcal{S}^{(2)}(\mathbf{C})$ will be a two-dimensional submanifold parametrized by the joint differential invariants I, J_0, J_1, K_0, K_1 of order ≤ 2. There will exist a (local) syzygy $\Phi(I, J_0, J_1) = 0$ among the three first order joint differential invariants.

Theorem 21. *A curve C or, more generally, a pair of curves $C_0, C_1 \subset \mathbf{R}^2$, is uniquely determined up to a Euclidean transformation by its reduced joint signature, which is parametrized by the first order joint differential invariants I, J_0, J_1. The curve(s) have a one-dimensional symmetry group if and only if their signature is a one-dimensional curve if and only if they are orbits of a common* *one-parameter subgroup (i.e., concentric circles or parallel straight lines); otherwise the signature is a two-dimensional surface, and the curve(s) have only discrete symmetries.*

For $n > 2$ points, we can use the two-point moving frame (5.18) to construct the additional joint invariants

$$y_k \longmapsto H_k = \| z_k - z_0 \| \cos \psi_k, \qquad v_k \longmapsto I_k = \| z_k - z_0 \| \sin \psi_k,$$

where $\psi_k = \sphericalangle(z_k - z_0, z_1 - z_0)$. Therefore, a complete system of joint invariants for SE(2) consists of the angles ψ_k, $k \geq 2$, and distances $\| z_k - z_0 \|$, $k \geq 1$. The other interpoint distances can all be recovered from these angles; vice versa, given the distances, and the sign of one angle, one can recover all other angles. In this manner, we establish a "First Main Theorem" for joint Euclidean differential invariants.

Theorem 22. *If $n \geq 2$, then every n-point joint E(2) differential invariant is a function of the interpoint distances $\| z_i - z_j \|$ and their invariant derivatives with respect to (5.23). For the proper Euclidean group SE(2), one must also include the sign of one of the angles, say $\psi_2 = \sphericalangle(z_2 - z_0, z_1 - z_0)$.*

Generic three-pointed Euclidean curves still require first order signature invariants. To create a Euclidean signature based entirely on joint invariants, we take four points z_0, z_1, z_2, z_3 on our curve $C \subset \mathbf{R}^2$. As illustrated in figure 4, there are six different interpoint distance invariants

$$\begin{aligned} a = \| z_1 - z_0 \|, \qquad b = \| z_2 - z_0 \|, \qquad c = \| z_3 - z_0 \|, \\ d = \| z_2 - z_1 \|, \qquad e = \| z_3 - z_1 \|, \qquad f = \| z_3 - z_2 \|, \end{aligned} \qquad (5.24)$$

which parametrize the joint signature $\widehat{\mathcal{S}} = \widehat{\mathcal{S}}(C)$ that uniquely characterizes the curve C up to Euclidean motion. This signature has the advantage of requiring no differentiation, and so is not sensitive to noisy image data. There are two local syzygies

$$\Phi_1(a, b, c, d, e, f) = 0, \qquad \Phi_2(a, b, c, d, e, f) = 0, \qquad (5.25)$$

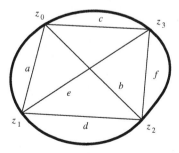

Fig. 4. Four-Point Euclidean Curve Invariants

among the the six interpoint distances. One of these is the universal *Cayley–Menger syzygy* which is valid for all possible configurations of the four points, and is a consequence of their coplanarity, cf. [8, 77]. The second syzygy in (5.25) is curve-dependent and serves to effectively characterize the joint invariant signature. Euclidean symmetries of the curve, both continuous and discrete, are characterized by this joint signature. For example, the number of discrete symmetries equals the signature index — the number of points in the original curve that map to a single, generic point in $\mathcal{S}$.

A wide variety of additional cases, including curves and surfaces in two and three-dimensional space under the Euclidean, equi-affine, affine and projective groups, are investigated in detail in [86].

6   Multi-space for Curves

In modern numerical analysis, the development of numerical schemes that incorporate additional structure enjoyed by the problem being approximated have become quite popular in recent years. The first instances of such schemes are the symplectic integrators arising in Hamiltonian mechanics, and the related energy conserving methods, [27, 65, 104]. The design of symmetry-based numerical approximation schemes for differential equations has been studied by various authors, including Shokin, [98], Dorodnitsyn, [34, 35], Axford and Jaegers, [52], and Budd and Collins, [16]. These methods are closely related to the active area of geometric integration of differential equations, [17, 47, 73]. In practical applications of invariant theory to computer vision, group-invariant numerical schemes to approximate differential invariants have been applied to the problem of symmetry-based object recognition, [9, 19, 18].

In this section, we outline the basic construction of multi-space that forms the foundation for the study of the geometric properties of discrete approximations to derivatives and numerical solutions to differential equations; see [88] for more details. We will only discuss the case of curves, which correspond to functions of a single independent variable, and hence satisfy ordinary differential equations. The more difficult case of higher dimensional submanifolds, corresponding to functions of several variables that satisfy partial differential equations, relies on a new approach to multi-dimensional interpolation theory, [87].

Numerical finite difference approximations to the derivatives of a function $u = f(x)$ rely on its values $u_0 = f(x_0), \ldots, u_n = f(x_n)$ at several distinct points $z_i = (x_i, u_i) = (x_i, f(x_i))$ on the curve. Thus, discrete approximations to jet coordinates on J^n are functions $F(z_0, \ldots, z_n)$ defined on the $(n+1)$-fold Cartesian product space $M^{\times(n+1)} = M \times \cdots \times M$. In order to seamlessly connect the jet coordinates with their discrete approximations, then, we need to relate the jet space for curves, $\mathrm{J}^n = \mathrm{J}^n(M, 1)$, to the Cartesian product space $M^{\times(n+1)}$. Now, as the points $z_0, \ldots, z_n$ coalesce, the approximation $F(z_0, \ldots, z_n)$ will not be well-defined unless we specify the "direction" of convergence. Thus, strictly speaking, F is not defined on all of $M^{\times(n+1)}$, but, rather, on the "off-diagonal" part, by which we mean the subset

$$M^{\diamond(n+1)} = \left\{ (z_0, \ldots, z_n) \mid z_i \neq z_j \text{ for all } i \neq j \right\} \subset M^{\times(n+1)}$$

consisting of all *distinct* $(n+1)$-tuples of points. As two or more points come together, the limiting value of $F(z_0, \ldots, z_n)$ will be governed by the derivatives (or jet) of the appropriate order governing the direction of convergence. This observation serves to motivate our construction of the *n-th order multi-space* $M^{(n)}$, which shall contain both the jet space J^n and the off-diagonal Cartesian product space $M^{\diamond(n+1)}$ in a consistent manner.

Definition 23.   An $(n+1)$-*pointed curve* $\mathbf{C} = (z_0, \ldots, z_n; C)$ consists of a smooth curve C and $n+1$ not necessarily distinct points $z_0, \ldots, z_n \in C$ thereon. Given $\mathbf{C}$, we let $\#i = \#\{j \mid z_j = z_i\}$ Two $(n+1)$-pointed curves $\mathbf{C} = (z_0, \ldots, z_n; C)$, $\widetilde{\mathbf{C}} = (\widetilde{z}_0, \ldots, \widetilde{z}_n; \widetilde{C})$, have *n-th order multi-contact* if and only if

$$z_i = \widetilde{z}_i, \quad \text{and} \quad j_{\#i-1} C|_{z_i} = j_{\#i-1} \widetilde{C}|_{z_i}, \quad \text{for each} \quad i = 0, \ldots, n.$$

Definition 24.   The n-th order *multi-space*, denoted $M^{(n)}$ is the set of equivalence classes of $(n+1)$-pointed curves in M under the equivalence relation of n-th order multi-contact. The equivalence class of an $(n+1)$-pointed curves $\mathbf{C}$ is called its n-th order *multi-jet*, and denoted $\mathbf{j}_n \mathbf{C} \in M^{(n)}$.

In particular, if the points on $\mathbf{C} = (z_0, \ldots, z_n; C)$ are all distinct, then $\mathbf{j}_n \mathbf{C} = \mathbf{j}_n \widetilde{\mathbf{C}}$ if and only if $z_i = \widetilde{z}_i$ for all i, which means that $\mathbf{C}$ and $\widetilde{\mathbf{C}}$ have all $n+1$ points in common. Therefore, we can identify the subset of multi-jets of multi-pointed curves having distinct points with the off-diagonal Cartesian product space $M^{\diamond(n+1)} \subset J^n$. On the other hand, if all $n+1$ points coincide, $z_0 = \cdots = z_n$, then $\mathbf{j}_n \mathbf{C} = \mathbf{j}_n \widetilde{\mathbf{C}}$ if and only if $\mathbf{C}$ and $\widetilde{\mathbf{C}}$ have n-th order contact at their common point $z_0 = \widetilde{z}_0$. Therefore, the multi-space equivalence relation reduces to the ordinary jet space equivalence relation on the set of coincident multi-pointed curves, and in this way $J^n \subset M^{(n)}$. These two extremes do not exhaust the possibilities, since one can have some but not all points coincide. Intermediate cases correspond to "off-diagonal" Cartesian products of jet spaces

$$J^{k_1} \diamond \cdots \diamond J^{k_i} \equiv \left\{ (z_0^{(k_1)}, \ldots, z_i^{(k_i)}) \in J^{k_1} \times \cdots \times J^{k_i} \mid \pi(z_\nu^{(k_\nu)}) \text{ are distinct} \right\},$$
$$(6.26)$$

where $\sum k_\nu = n$ and $\pi \colon J^k \to M$ is the usual jet space projection. These *multi-jet spaces* appear in the work of Dhooghe, [33], on the theory of "semi-differential invariants" in computer vision.

Theorem 25.   *If M is a smooth m-dimensional manifold, then its n-th order multi-space $M^{(n)}$ is a smooth manifold of dimension $(n+1)m$, which contains the off-diagonal part $M^{\diamond(n+1)}$ of the Cartesian product space as an open, dense submanifold, and the n-th order jet space J^n as a smooth submanifold.*

The proof of Theorem 25 requires the introduction of coordinate charts on $M^{(n)}$. Just as the local coordinates on J^n are provided by the coefficients of

Taylor polynomials, the local coordinates on $M^{(n)}$ are provided by the coefficients of interpolating polynomials, which are the classical divided differences of numerical interpolation theory, [78, 93].

Definition 26.   Given an $(n+1)$-pointed graph $\mathbf{C} = (z_0, \ldots, z_n; C)$, its divided differences are defined by $[z_j]_C = f(x_j)$, and

$$[z_0 z_1 \cdots z_{k-1} z_k]_C = \lim_{z \to z_k} \frac{[z_0 z_1 z_2 \cdots z_{k-2} z]_C - [z_0 z_1 z_2 \cdots z_{k-2} z_{k-1}]_C}{x - x_{k-1}}.$$

(6.27)

When taking the limit, the point $z = (x, f(x))$ must lie on the curve C, and take limiting values $x \to x_k$ and $f(x) \to f(x_k)$.

In the non-confluent case $z_k \neq z_{k-1}$ we can replace z by z_k directly in the difference quotient (6.27) and so ignore the limit. On the other hand, when all $k + 1$ points coincide, the k-th order confluent divided difference converges to

$$[z_0 \cdots z_0]_C = \frac{f^{(k)}(x_0)}{k!}.$$

(6.28)

Remark: Classically, one employs the simpler notation $[u_0 u_1 \ldots u_k]$ for the divided difference $[z_0 z_1 \ldots z_k]_C$. However, the classical notation is ambiguous since it assumes that the mesh $x_0, \ldots, x_n$ is fixed throughout. Because we are regarding the independent and dependent variables on the same footing — and, indeed, are allowing changes of variables that scramble the two — it is important to adopt an unambiguous divided difference notation here.

Theorem 27.   *Two $(n+1)$-pointed graphs $\mathbf{C}, \widetilde{\mathbf{C}}$ have n-th order multi-contact if and only if they have the same divided differences:*

$$[z_0 z_1 \ldots z_k]_C = [z_0 z_1 \ldots z_k]_{\widetilde{C}}, \qquad k = 0, \ldots, n.$$

The required local coordinates on multi-space $M^{(n)}$ consist of the independent variables along with all the divided differences

$$x_0, \ldots, x_n, \quad \begin{matrix} u^{(0)} = u_0 = [z_0]_C, & u^{(1)} = [z_0 z_1]_C, \\ u^{(2)} = 2\,[z_0 z_1 z_2]_C & \cdots & u^{(n)} = n!\,[z_0 z_1 \ldots z_n]_C, \end{matrix}$$

(6.29)

prescribed by $(n + 1)$-pointed graphs $\mathbf{C} = (z_0, \ldots, z_n; C)$. The $n!$ factor is included so that $u^{(n)}$ agrees with the usual derivative coordinate when restricted to J^n, cf. (6.28).

7   Invariant Numerical Methods

To implement a numerical solution to a system of differential equations

$$\Delta_1(x, u^{(n)}) = \cdots = \Delta_k(x, u^{(n)}) = 0.$$

(7.30)

by finite difference methods, one relies on suitable discrete approximations to each of its defining differential functions Δ_ν, and this requires extending the differential functions from the jet space to the associated multi-space, in accordance with the following definition.

Definition 28. An $(n+1)$-*point numerical approximation of order* k to a differential function $\Delta \colon J^n \to \mathbf{R}$ is an function $F \colon M^{(n)} \to \mathbf{R}$ that, when restricted to the jet space, agrees with Δ to order k.

The simplest illustration of Definition 28 is provided by the divided difference coordinates (6.29). Each divided difference $u^{(n)}$ forms an $(n+1)$-point numerical approximation to the n-th order derivative coordinate on J^n. According to the usual Taylor expansion, the order of the approximation is $k = 1$. More generally, any differential function $\Delta(x, u, u^{(1)}, \ldots, u^{(n)})$ can immediately be assigned an $(n+1)$-point numerical approximation $F = \Delta(x_0, u^{(0)}, u^{(1)}, \ldots, u^{(n)})$ by replacing each derivative by its divided difference coordinate approximation. However, these are by no means the only numerical approximations possible.

Now let us consider an r-dimensional Lie group G which acts smoothly on M. Since G evidently maps multi-pointed curves to multi-pointed curves while preserving the multi-contact equivalence relation, it induces an action on the multi-space $M^{(n)}$ that will be called the n-th *multi-prolongation* of G and denoted by $G^{(n)}$. On the jet subset $J^n \subset M^{(n)}$ the multi-prolonged action reduced to the usual jet space prolongation. On the other hand, on the off-diagonal part $M^{\diamond(n+1)} \subset M^{(n)}$ the action coincides with the $(n+1)$-fold Cartesian product action of G on $M^{\times(n+1)}$.

We define a *multi-invariant* to be a function $K \colon M^{(n)} \to \mathbf{R}$ on multi-space which is invariant under the multi-prolonged action of $G^{(n)}$. The restriction of a multi-invariant K to jet space will be a differential invariant, $I = K \mid J^n$, while restriction to $M^{\diamond(n+1)}$ will define a joint invariant $J = K \mid M^{\diamond(n+1)}$. Smoothness of K will imply that the joint invariant J is an *invariant n-th order numerical approximation to the differential invariant* I. Moreover, every invariant finite difference numerical approximation arises in this manner. Thus, the theory of multi-invariants *is* the theory of invariant numerical approximations!

Furthermore, the restriction of a multi-invariant to an intermediate multi-jet subspace, as in (6.26), will define a joint differential invariant, [86] — also known as a semi-differential invariant in the computer vision literature, [33, 79]. The approximation of differential invariants by joint differential invariants is, therefore, based on the extension of the differential invariant from the jet space to a suitable multi-jet subspace (6.26). The invariant numerical approximations to joint differential invariants are, in turn, obtained by extending them from the multi-jet subspace to the entire multi-space. Thus, multi-invariants also include invariant semi-differential approximations to differential invariants as well as joint invariant numerical approximations to differential invariants and semi-differential invariants — all in one seamless geometric framework.

Effectiveness of the group action on M implies, typically, freeness and regularity of the multi-prolonged action on an open subset of $M^{(n)}$. Thus, we can apply the basic moving frame construction. The resulting *multi-frame* $\rho^{(n)} \colon M^{(n)} \to G$ will lead us immediately to the required multi-invariants and hence a general, systematic construction for invariant numerical approximations to differential invariants. Any multi-frame will evidently restrict to a classical moving frame $\rho^{(n)} \colon J^n \to G$ on the jet space along with a suitably compatible product frame $\rho^{\diamond(n+1)} \colon M^{\diamond(n+1)} \to G$.

In local coordinates, we use $w_k = (y_k, v_k) = g \cdot z_k$ to denote the transformation formulae for the individual points on a multi-pointed curve. The multi-prolonged action on the divided difference coordinates gives

$$y_0, \ldots, y_n, \qquad \begin{aligned} v^{(0)} &= v_0 = [\,w_0\,], & v^{(1)} &= [\,w_0 w_1\,], \\ v^{(2)} &= [\,w_0 w_1 w_2\,], & \cdots & \quad v^{(n)} &= n!\,[\,w_0, \ldots, w_n\,], \end{aligned} \qquad (7.31)$$

where the formulae are most easily computed via the difference quotients

$$[\,w_0 w_1 \ldots w_{k-1} w_k\,] = \frac{[\,w_0 w_1 w_2 \ldots w_{k-2} w_k\,] - [\,w_0 w_1 w_2 \ldots w_{k-2} w_{k-1}\,]}{y_k - y_{k-1}}, \qquad (7.32)$$

$$[\,w_j\,] = v_j,$$

and then taking appropriate limits to cover the case of coalescing points. Inspired by the constructions in [40], we will refer to (7.31) as the *lifted divided difference invariants*.

To construct a multi-frame, we need to normalize by choosing a cross-section to the group orbits in $M^{(n)}$, which amounts to setting $r = \dim G$ of the lifted divided difference invariants (7.31) equal to suitably chosen constants. An important observation is that in order to obtain the limiting differential invariants, we must require our local cross-section to pass through the jet space, and define, by intersection, a cross-section for the prolonged action on J^n. This compatibility constraint implies that we are only allowed to normalize the first lifted independent variable $y_0 = c_0$.

With the aid of the multi-frame, the most direct construction of the requisite multi-invariants and associated invariant numerical differentiation formulae is through the invariantization of the original finite difference quotients (6.27). Substituting the multi-frame formulae for the group parameters into the lifted coordinates (7.31) provides a complete system of multi-invariants on $M^{(n)}$; this follows immediately from Theorem 4. We denote the fundamental multi-invariants by

$$y_i \;\longmapsto\; H_i = \iota(x_i), \qquad v^{(n)} \;\longmapsto\; K^{(n)} = \iota(u^{(n)}), \qquad (7.33)$$

where ι denotes the invariantization map associated with the multi-frame. The fundamental differential invariants for the prolonged action of G on J^n can all be obtained by restriction, so that $I^{(n)} = K^{(n)} \,|\, J^n$. On the jet space, the points are coincident, and so the multi-invariants H_i will all restrict to the *same* differential invariant $c_0 = H = H_i \,|\, J^n$ — the normalization value of y_0. On the other hand, the fundamental joint invariants on $M^{\diamond(n+1)}$ are obtained by restricting the multi-invariants $H_i = \iota(x_i)$ and $K_i = \iota(u_i)$. The multi-invariants can computed by using a multi-invariant divided difference recursion

$$[I_j] = K_j = \iota(u_j)$$

$$[I_0 \ldots I_k] = \iota([z_0 z_1 \ldots z_k]) = \frac{[I_0 \ldots I_{k-2} I_k] - [I_0 \ldots I_{k-2} I_{k-1}]}{H_k - H_{k-1}}, \qquad (7.34)$$

and then relying on continuity to extend the formulae to coincident points. The multi-invariants

$$K^{(n)} = n! \, [I_0 \ldots I_n] = \iota(u^{(n)}) \qquad (7.35)$$

define the fundamental first order invariant numerical approximations to the differential invariants $I^{(n)}$. Higher order invariant approximations can be obtained by invariantization of the higher order divided difference approximations. The moving frame construction has a significant advantage over the infinitesimal approach used by Dorodnitsyn, [34, 35], in that it does not require the solution of partial differential equations in order to construct the multi-invariants.

Given a regular G-invariant differential equation

$$\Delta(x, u^{(n)}) = 0, \qquad (7.36)$$

we can invariantize the left hand side to rewrite the differential equation in terms of the fundamental differential invariants:

$$\iota(\Delta(x, u^{(n)})) = \Delta(H, I^{(0)}, \ldots, I^{(n)}) = 0.$$

The invariant finite difference approximation to the differential equation is then obtained by replacing the differential invariants $I^{(k)}$ by their multi-invariant counterparts $K^{(k)}$:

$$\Delta(c_0, K^{(0)}, \ldots, K^{(n)}) = 0. \qquad (7.37)$$

Example 29. Consider the elementary action

$$(x, u) \quad \longmapsto \quad (\lambda^{-1} x + a, \lambda u + b)$$

of the three-parameter similarity group $G = \mathbf{R}^2 \, \mathbf{n} \, \mathbf{R}$ on $M = \mathbf{R}^2$. To obtain the multi-prolonged action, we compute the divided differences (7.31) of the basic lifted invariants

$$y_k = \lambda^{-1} x_k + a, \qquad v_k = \lambda u_k + b.$$

We find

$$v^{(1)} = [w_0 w_1] = \frac{v_1 - v_0}{y_1 - y_0} = \lambda^2 \frac{u_1 - u_0}{x_1 - x_0} = \lambda^2 [z_0 z_1] = \lambda^2 u^{(1)}.$$

More generally,

$$v^{(n)} = \lambda^{n+1} u^{(n)}, \qquad n \geq 1. \qquad (7.38)$$

Note that we may compute the multi-space transformation formulae assuming initially that the points are distinct, and then extending to coincident cases by continuity. (In fact, this gives an alternative method for computing the standard jet space prolongations of group actions!) In particular, when all the points coincide, each $u^{(n)}$ reduces to the n-th order derivative coordinate, and (7.38) reduces to the prolonged action of G on J^n. We choose the normalization cross-section defined by

$$y_0 = 0, \qquad v_0 = 0, \qquad v^{(1)} = 1,$$

which, upon solving for the group parameters, leads to the basic moving frame

$$a = -\sqrt{u^{(1)}}\, x_0, \qquad b = -\frac{u_0}{\sqrt{u^{(1)}}}, \qquad \lambda = \frac{1}{\sqrt{u^{(1)}}}, \qquad (7.39)$$

where, for simplicity, we restrict to the subset where $u^{(1)} = [\,z_0 z_1\,] > 0$. The fundamental joint similarity invariants are obtained by substituting these formulae into

$$y_k \longmapsto H_k = (x_k - x_0)\sqrt{u^{(1)}} = (x_k - x_0)\sqrt{\frac{u_1 - u_0}{x_1 - x_0}},$$

$$v_k \longmapsto K_k = \frac{u_k - u_0}{\sqrt{u^{(1)}}} = (u_k - u_0)\sqrt{\frac{x_1 - x_0}{u_1 - u_0}},$$

both of which reduce to the trivial zero differential invariant on J^n. Higher order multi-invariants are obtained by substituting (7.39) into the lifted invariants (7.38), leading to

$$K^{(n)} = \frac{u^{(n)}}{(u^{(1)})^{(n+1)/2}} = \frac{n!\,[\,z_0 z_1 \ldots z_n\,]}{[\,z_0 z_1 z_2\,]^{(n+1)/2}}.$$

In the limit, these reduce to the differential invariants $I^{(n)} = (u^{(1)})^{-(n+1)/2}\,u^{(n)}$, and so $K^{(n)}$ give the desired similarity-invariant, first order numerical approximations. To construct an invariant numerical scheme for any similarity-invariant ordinary differential equation

$$\Delta(x, u, u^{(1)}, u^{(2)}, \ldots u^{(n)}) = 0,$$

we merely invariantize the defining differential function, leading to the general similarity–invariant numerical approximation

$$\Delta(0, 0, 1, K^{(2)}, \ldots, K^{(n)}) = 0.$$

Example 30. For the action (3.6) of the proper Euclidean group of SE(2) on $M = \mathbf{R}^2$, the multi-prolonged action is free on $M^{(n)}$ for $n \geq 1$. We can thereby determine a first order multi-frame and use it to completely classify Euclidean multi-invariants. The first order transformation formulae are

$$y_0 = x_0 \cos\theta - u_0 \sin\theta + a, \qquad v_0 = x_0 \sin\theta + u_0 \cos\theta + b,$$

$$y_1 = x_1 \cos\theta - u_1 \sin\theta + a, \qquad v^{(1)} = \frac{\sin\theta + u^{(1)} \cos\theta}{\cos\theta - u^{(1)} \sin\theta}, \qquad (7.40)$$

where $u^{(1)} = [z_0 z_1]$. Normalization based on the cross-section $y_0 = v_0 = v^{(1)} = 0$ results in the right moving frame

$$a = -x_0 \cos\theta + u_0 \sin\theta = -\frac{x_0 + u^{(1)} u_0}{\sqrt{1 + (u^{(1)})^2}},$$

$$\tan\theta = -u^{(1)}. \qquad (7.41)$$

$$b = -x_0 \sin\theta - u_0 \cos\theta = \frac{x_0 u^{(1)} - u_0}{\sqrt{1 + (u^{(1)})^2}},$$

Substituting the moving frame formulae (7.41) into the lifted divided differences results in a complete system of (oriented) Euclidean multi-invariants. These are easily computed by beginning with the fundamental joint invariants $I_k = (H_k, K_k) = \iota(x_k, u_k)$, where

$$y_k \longmapsto H_k = \frac{(x_k - x_0) + u^{(1)}(u_k - u_0)}{\sqrt{1 + (u^{(1)})^2}} = (x_k - x_0)\frac{1 + [z_0 z_1][z_0 z_k]}{\sqrt{1 + [z_0 z_1]^2}},$$

$$v_k \longmapsto K_k = \frac{(u_k - u_0) - u^{(1)}(x_k - x_0)}{\sqrt{1 + (u^{(1)})^2}} = (x_k - x_0)\frac{[z_0 z_k] - [z_0 z_1]}{\sqrt{1 + [z_0 z_1]^2}}.$$

The multi-invariants are obtained by forming divided difference quotients

$$[I_0 I_k] = \frac{K_k - K_0}{H_k - H_0} = \frac{K_k}{H_k} = \frac{(x_k - x_1)[z_0 z_1 z_k]}{1 + [z_0 z_k][z_0 z_1]},$$

where, in particular, $I^{(1)} = [I_0 I_1] = 0$. The second order multi-invariant

$$I^{(2)} = 2[I_0 I_1 I_2] = 2\frac{[I_0 I_2] - [I_0 I_1]}{H_2 - H_1} = \frac{2[z_0 z_1 z_2]\sqrt{1 + [z_0 z_1]^2}}{(1 + [z_0 z_1][z_1 z_2])(1 + [z_0 z_1][z_0 z_2])}$$

$$= \frac{u^{(2)}\sqrt{1 + (u^{(1)})^2}}{[1 + (u^{(1)})^2 + \frac{1}{2}u^{(1)}u^{(2)}(x_2 - x_0)][1 + (u^{(1)})^2 + \frac{1}{2}u^{(1)}u^{(2)}(x_2 - x_1)]}$$

provides a Euclidean–invariant numerical approximation to the Euclidean curvature:

$$\lim_{z_1, z_2 \to z_0} I^{(2)} = \kappa = \frac{u^{(2)}}{(1 + (u^{(1)})^2)^{3/2}}.$$

Similarly, the third order multi-invariant

$$I^{(3)} = 6[I_0 I_1 I_2 I_3] = 6\frac{[I_0 I_1 I_3] - [I_0 I_1 I_2]}{H_3 - H_2}$$

will form a Euclidean–invariant approximation for the normalized differential invariant $\kappa_s = \iota(u_{xxx})$, the derivative of curvature with respect to arc length, [19, 40].

To compare these with the invariant numerical approximations proposed in [18, 19], we reformulate the divided difference formulae in terms of the geometrical configurations of the four distinct points z_0, z_1, z_2, z_3 on our curve. We find

$$H_k = \frac{(z_1 - z_0) \cdot (z_k - z_0)}{\| z_1 - z_0 \|} = r_k \cos \phi_k,$$

$$K_k = \frac{(z_1 - z_0) \wedge (z_k - z_0)}{\| z_1 - z_0 \|} = r_k \sin \phi_k,$$

$$[\, I_0 I_k\,] = \tan \phi_k,$$

where

$$r_k = \| z_k - z_0 \|, \qquad \phi_k = \sphericalangle (z_k - z_0, z_1 - z_0),$$

denotes the distance and the angle between the indicated vectors. Therefore,

$$I^{(2)} = 2 \, \frac{\tan \phi_2}{r_2 \cos \phi_2 - r_1},$$

$$I^{(3)} = 6 \, \frac{(r_2 \cos \phi_2 - r_1) \tan \phi_3 - (r_3 \cos \phi_3 - r_1) \tan \phi_2}{(r_2 \cos \phi_2 - r_1)(r_3 \cos \phi_3 - r_1)(r_3 \cos \phi_3 - r_2 \cos \phi_2)}.$$

(7.42)

Interestingly, $I^{(2)}$ is *not* the same Euclidean approximation to the curvature that was used in [19, 18]. The latter was based on the Heron formula for the radius of a circle through three points:

$$I^\star = \frac{4 \Delta}{abc} = \frac{2 \sin \phi_2}{\| z_1 - z_2 \|}.$$

(7.43)

Here Δ denotes the area of the triangle connecting z_0, z_1, z_2 and

$$a = r_1 = \| z_1 - z_0 \|, \qquad b = r_2 = \| z_2 - z_0 \|, \qquad c = \| z_2 - z_1 \|,$$

are its side lengths. The ratio tends to a limit $I^\star / I^{(2)} \to 1$ as the points coalesce. The geometrical approximation (7.43) has the advantage that it is symmetric under permutations of the points; one can achieve the same thing by symmetrizing the divided difference version $I^{(2)}$. Furthermore, $I^{(3)}$ is an invariant approximation for the differential invariant κ_s, that, like the approximations constructed by Boutin, [9], converges properly for arbitrary spacings of the points on the curve.

Recently, Pilwon Kim and I have been developing the invariantization techniques to a variety of numerical integrators, e.g., Euler and Runge–Kutta, for ordinary differential equations with symmetry, with sometimes striking results, [57]. In preparation for extending these methods to functions of several variables and partial differential equations, I have recently formulated a new approach to the theory of multivariate interpolation based on noncommutative quasi-determinants, [87].

8   Invariant Variational Problems

In the fundamental theories of modern physics, [7, 43], one begins by postulating an underlying symmetry group (e.g., conformal invariance, Poincaré invariance,

supersymmetry, etc.), and then seeks a suitably invariant Lagrangian or variational principle. The governing field equations are the Euler–Lagrange equations,
which retain the invariance properties of the underlying pseudo-group. As first
recognized by Lie, [67], under appropriate regularity assumptions, all invariant
differential equations and variational problems can be written in terms of the
differential invariants. Surprisingly, though, complete classifications of differential invariants remain, for the most part, unknown, even for some of the most
basic cases in physics, e.g., the full Poincaré group. A principal aim of the moving frame approach is to provide the necessary mathematical tools for resolving
such fundamental issues.

In this direction, Irina Kogan and I, [61, 62], extended the invariantization
process to formulate an invariant version of the *variational bicomplex*. In particular, our results solve the previously outstanding problem of directly constructing
the differential invariant form of the Euler-Lagrange equations from that of the
underlying variational problem. Previously, only a handful of special examples
were known, [2, 45].

Example 31. To illustrate, the simplest example is that of plane curves in
Euclidean geometry. Any Euclidean-invariant variational problem

$$\mathcal{I}[u] = \int \widetilde{L}(\kappa, \kappa_s, \kappa_{ss}, \ldots) \, ds$$

can be written in terms of the Euclidean curvature differential invariant κ and
its successive derivatives $\mathcal{D}^n \kappa = D_s^n \kappa$ with respect to arc length ds. The associated Euler-Lagrange equation is Euclidean-invariant, and so is equivalent to an
ordinary differential equation of the form

$$F(\kappa, \kappa_s, \kappa_{ss}, \ldots) = 0.$$

The basic problem is to go directly from the invariant form of the variational
problem to the invariant form of its Euler-Lagrange equation. The correct formula for the Euler-Lagrange equation is

$$(\mathcal{D}^2 + \kappa^2) \, \mathcal{E}(\widetilde{L}) + \kappa \, \mathcal{H}(\widetilde{L}) = 0,$$

where

$$\mathcal{E}(\widetilde{L}) = \sum_n (-\mathcal{D})^n \frac{\partial \widetilde{L}}{\partial \kappa_n}, \qquad \mathcal{H}(\widetilde{L}) = \sum_{i>j} \kappa_{i-j}(-\mathcal{D})^j \frac{\partial \widetilde{L}}{\partial \kappa_i} - \widetilde{L},$$

are, respectively, the *invariant Euler-Lagrange expression* (or *Eulerian*), and the
invariant Hamiltonian of the *invariant Lagrangian* $\widetilde{L}$.

Kogan and I proved that, in general, the invariant Euler–Lagrange formula assumes an analogous form

$$\mathcal{A}^* \mathcal{E}(\widetilde{L}) - \mathcal{B}^* \mathcal{H}(\widetilde{L}) = 0,$$

where $\mathcal{E}(\widetilde{L})$ is the invariantized Eulerian, $\mathcal{H}(\widetilde{L})$ an invariantized Hamiltonian tensor, [95], based on the invariant Lagrangian of the problem, while $\mathcal{A}^*, \mathcal{B}^*$ are certain invariant differential operators, which we name the *Eulerian* and *Hamiltonian operators*. The precise forms of these operators follows from the recurrence formulae for the moving frame on the invariant variational bicomplex, which, as they rely solely on linear differential algebraic formulae, can be readily implemented in computer algebra systems. Complete details on the construction and applications can be found in our papers [61, 62].

9   Lie Pseudo–Groups

With the moving frame constructions for finite-dimensional Lie group actions taking more or less final form, my attention has shifted to developing a comparably powerful theory that can be applied to infinite-dimensional Lie pseudo-groups. The subject is classical: Lie, [66], and Medolaghi, [76], classified all planar pseudo-groups, and gave applications to Darboux integrable partial differential equations, [4, 100]. Cartan's famous classification of transitive simple pseudo-groups, [24], remains a milestone in the subject. Remarkably, despite numerous investigations, there is still no entirely satisfactory abstract object that will properly represent a Lie pseudo-group, cf. [64, 99, 101, 94].

Pseudo-groups appear in a broad range of physical and geometrical contexts, including gauge theories in physics, [7]; canonical and area-preserving transformations in Hamiltonian mechanics, [82]; conformal symmetry groups on two-dimensional surfaces, [37]; foliation-preserving groups of transformations, with the associated characteristic classes defined by certain invariant forms, [41]; symmetry groups of both linear and nonlinear partial differential equations appearing in fluid and plasma mechanics, such as the Euler, Navier-Stokes and boundary layer equations, [20, 82], in meteorology, such as semi-geostrophic models, [96], and in integrable (soliton) equations in more than one space dimension such as the Kadomtsev–Petviashvili (KP) equation, [31]. Applications of pseudo-groups to the design of geometric numerical integrators are being emphasized in recent work of McLachlan and Quispel, [73, 74].

Juha Pohjanpelto and I, [89, 90, 91], recently announced a breakthrough in the development of a practical moving frame theory for general Lie pseudo-group actions. (A more abstract version was concurrently developed by my former student Vladimir Itskov, [51].) Just as in the finite-dimensional theory, the new methods lead to general computational algorithms for (*i*) determining complete systems of differential invariants, invariant differential operators, and invariant differential forms, (*ii*) complete classifications of syzygies and recurrence formulae relating the differentiated invariants and invariant forms, (*iii*) a general algorithm for computing the Euler–Lagrange equations associated with an invariant variational problem. Further extensions — pseudo-group algorithms for joint invariants and joint differential invariants, invariant numerical approximations, and so on — are also evident.

Our approach rests on an amalgamation of two powerful, general modern theories: *groupoids*, [68, 107], which generalize the concept of transformation groups, and the *variational bicomplex*, [2, 83, 103], which underlies the modern geometric approach to differential equations and the calculus of variations. Groupoids (first formalized by Ehresmann, [36], for precisely these purposes) are required because there is no underlying global geometric object to represent the (local) pseudo-group. The simplest case, and one that must be fully understood from the start, is the pseudo-group of all local diffeomorphisms. Their jets (Taylor series) naturally form a groupoid, because one can only compose two Taylor series if the target (or sum) of the first matches the source (or base point) of the second. Thus, the first item of business is to adapt the Lie group moving frame constructions to the groupoid category.

On an infinite jet bundle, the variational bicomplex, [2, 103], follows from the natural splitting of the space of differential one-forms into contact forms and horizontal forms, [83]. Our constructions involve two infinite jet bundles and their associated variational bicomplexes: the first is the groupoid of infinite jets of local diffeomorphisms; the second is the space of jets of submanifolds (or graphs of functions or sections). This seriously complicates the analysis (and the notation), but not beyond the range of being forged into a practical, algorithmic method.

The next challenge is the construction of the Maurer–Cartan forms and the associated structure equations for the pseudo-group. For finite-dimensional Lie groups, the pull-back action of the moving frame on the Maurer–Cartan forms is used to construct the basic recurrence formulae that relate the differentiated invariants and differential forms, [40, 62]. The recurrence formulae are the foundation for all the advanced computational algorithms, including classification of differential invariants and their syzygies, the general invariantization procedure, and the applications in the calculus of variations.

In the case of the diffeomorphism pseudo-group, the Maurer–Cartan forms are the invariant contact forms on the diffeomorphism jet groupoid, and can be explicitly constructed, completely avoiding the more complicated inductive procedure advocated by Cartan, [25]. Let $z = (z^1, \ldots, z^m)$, $Z = (Z^1, \ldots, Z^m)$ be, respectively, the source and target coordinates on M. The induced coordinates on the diffeomorphism jet bundle $\mathcal{D}^\infty \subset \mathrm{J}^\infty(M, M)$ are denoted by Z_J^a, $a = 1, \ldots, m$, $\#J \geq 0$, representing all derivatives of the target coordinates. The space of invariant contact forms on $\mathcal{D}^\infty$ has basis elements μ_J^a, $a = 1, \ldots, m$, $\#J \geq 0$, whose explicit formulas can be found in [89]. Utilizing the variational bicomplex machinery, we readily establish the explicit formulae for the structure equations for the diffeomorphism pseudo-group by equating coefficients in the formal power series formula

$$d\mu[\![H]\!] = \nabla_H \mu[\![H]\!] \wedge \left(\mu[\![H]\!] - dZ \right). \tag{9.44}$$

Here, $\mu[\![H]\!]$ is the vector-valued formal power series depending on the parameters $H = (H^1, \ldots, H^m)$, with entries

$$\mu^a[\![H]\!] = \sum_{\#J \geq 0} \mu_J^a \, H^J,$$

$\nabla_H \mu[\![H]\!]$ is its formal Jacobian matrix, while $dZ = (dZ^1, \ldots, dZ^m)^T$. Given a Lie pseudo-group $\mathcal{G}$ acting on M, let

$$L(z, \ldots, \zeta_J^a, \ldots) = 0 \tag{9.45}$$

denote the involutive system of *determining equations* for its infinitesimal generators $\mathbf{v} = \sum_{a=1}^m \zeta^a \partial_{z^a}$, where $\zeta_J^a = \partial^J \zeta^a / \partial z^J$ stand for the corresponding derivatives (or jets) of the vector field coefficients. For example, if $\mathcal{G}$ is a symmetry group of a system of partial differential equations, then (9.46) are (the involutive completion) of the classical Lie determining equations for its infinitesimal symmetries, [82]. The remarkable fact, proved in [89], is that Maurer–Cartan forms for the pseudo-group, which are obtained by restricting the diffeomorphism Maurer–Cartan forms μ_J^a to the pseudo-group jet subbundle $\mathcal{G}^{(\infty)} \subset \mathcal{D}^{(\infty)}$, satisfy the *exact same* linear relations[5]:

$$L(Z, \ldots, \mu_J^a, \ldots) = 0. \tag{9.46}$$

Therefore, a basis for the solution space to the infinitesimal determining equations (9.46) prescribes the complete system of independent Maurer–Cartan forms for the pseudo-group. Furthermore, the all-important pseudo-group structure equations are obtained by restricting the diffeomorphism structure equations (9.44) to the linear subspace spanned by the pseudo-group Maurer–Cartan forms, i.e., to the space of solutions to (9.46). The result is a direct computational procedure for passing directly from the infinitesimal determining equations to the structure equations for the pseudo-group relying on just linear differential algebra.

With the Maurer–Cartan forms and structure equations in hand, we are now in a position to implement to moving frame method. The primary focus is on the action of the pseudo-group on submanifolds of a specified dimension. There is an induced prolonged action of the (finite dimensional) n-th order pseudo-group jet groupoid on the n-th order submanifold jet bundle. A straightforward adaptation of the general normalization procedure will produce the n-th order moving frame map. The consequent invariantization process is used to produce the complete system of n-th order differential invariants, invariant differential forms, and, when combined with the Maurer–Cartan structure equations, the required recurrence formulae. In [28], these algorithms were applied to the symmetry groups of the Korteweg–deVries and KP equations arising in soliton theory, and general packages for effecting these computations are being developed. More substantial examples, arising as symmetry pseudo-groups of nonlinear partial differential equations such as the KP equation and the equations in fluid mechanics and meteorology, are in the process of being investigated.

---

[5] With the source coordinates z replaced by target coordinates Z

10   Implementation

A noteworthy feature of both the finite-dimensional and infinite-dimensional moving frame methods is that most of the computations rely on purely linear algebra techniques. In particular, the structure of the pseudo-group, the fundamental differential invariants, and the recurrence formulae, syzygies and commutation relations all follow from the infinitesimal determining equations. Only the explicit formulas for the differential invariants requires the nonlinear pseudo-group transformations, coupled with elimination of the normalization equations. The efficiency of the moving frame approach is underscored by the fact that we can replace the complicated Spencer-based analysis of Tresse's prototypical example in [63] by a few lines of easy hand computation, [90]. More substantial examples, such as the symmetry groups of nonlinear partial differential equations, that were previously unattainable are now well within our computational grasp. However, large-scale applications, such as those in Mansfield, [69], will require the development of a suitable noncommutative Gröbner basis theory for such algebras, complicated by the noncommutativity of the invariant differential operators and the syzygies among the differentiated invariants.

Owing to the overall complexity of larger scale computations, any serious application of the methods discussed here will, ultimately, rely on computer algebra, and so the development of appropriate software packages is a significant priority. The moving frame algorithms point to significant weaknesses in current computer algebra technology, particularly when manipulating the rational algebraic functions which inevitably appear within the normalization formulae. Following some preliminary work by the author in MATHEMATICA, Irina Kogan, [60], has implemented the finite-dimensional moving frame algorithms on Ian Anderson's general purpose MAPLE package VESSIOT, [3]. As part of his Ph.D. thesis, Jeongoo Cheh is implementing the full pseudo-group algorithms for symmetry groups of partial differential equations in MATHEMATICA.

References

1. Akivis, M.A., and Rosenfeld, B.A., *Élie Cartan (1869-1951)*, Translations Math. monographs, vol. 123, American Math. Soc., Providence, R.I., 1993.
2. Anderson, I.M., *The Variational Bicomplex*, Utah State Technical Report, 1989, http://math.usu.edu/~fg_mp.
3. Anderson, I.M., *The Vessiot Handbook*, Technical Report, Utah Sate University, 2000.
4. Anderson, I.M., and Kamran, N., The variational bicomplex for second order scalar partial differential equations in the plane, *Duke Math. J.* **87** (1997), 265–319.
5. Bazin, P.–L., and Boutin, M., Structure from motion: theoretical foundations of a novel approach using custom built invariants, *SIAM J. Appl. Math.* **64** (2004), 1156–1174.
6. Berchenko, I.A., and Olver, P.J., Symmetries of polynomials, *J. Symb. Comp.* **29** (2000), 485–514.

7. Bleecker, D., *Gauge Theory and Variational Principles*, Addison–Wesley Publ. Co., Reading, Mass., 1981.

8. Blumenthal, L.M., *Theory and Applications of Distance Geometry*, Oxford Univ. Press, Oxford, 1953.

9. Boutin, M., Numerically invariant signature curves, *Int. J. Computer Vision* **40** (2000), 235–248.

10. Boutin, M., On orbit dimensions under a simultaneous Lie group action on n copies of a manifold, *J. Lie Theory* **12** (2002), 191–203.

11. Boutin, M., Polygon recognition and symmetry detection, *Found. Comput. Math.* **3** (2003), 227–271.

12. Bruckstein, A.M., Holt, R.J., Netravali, A.N., and Richardson, T.J., Invariant signatures for planar shape recognition under partial occlusion, *CVGIP: Image Understanding* **58** (1993), 49–65.

13. Bruckstein, A.M., and Netravali, A.N., On differential invariants of planar curves and recognizing partially occluded planar shapes, *Ann. Math. Artificial Intel.* **13** (1995), 227–250.

14. Bruckstein, A.M., Rivlin, E., and Weiss, I., Scale space semi-local invariants, *Image Vision Comp.* **15** (1997), 335–344.

15. Bruckstein, A.M., and Shaked, D., Skew-symmetry detection via invariant signatures, *Pattern Recognition* **31** (1998), 181–192.

16. Budd, C.J., and Collins, C.B., Symmetry based numerical methods for partial differential equations, in: *Numerical analysis 1997*, D.F. Griffiths, D.J. Higham and G.A. Watson, eds., Pitman Res. Notes Math., vol. 380, Longman, Harlow, 1998, pp. 16–36.

17. Budd, C.J., and Iserles, A., Geometric integration: numerical solution of differential equations on manifolds, *Phil. Trans. Roy. Soc. London A* **357** (1999), 945–956.

18. Calabi, E., Olver, P.J., and Tannenbaum, A., Affine geometry, curve flows, and invariant numerical approximations, *Adv. in Math.* **124** (1996), 154–196.

19. Calabi, E., Olver, P.J., Shakiban, C., Tannenbaum, A., and Haker, S., Differential and numerically invariant signature curves applied to object recognition, *Int. J. Computer Vision* **26** (1998), 107–135.

20. Cantwell, B.J., *Introduction to Symmetry Analysis*, Cambridge University Press, Cambridge, 2003.

21. Carlsson, S., Mohr, R., Moons, T., Morin, L., Rothwell, C., Van Diest, M., Van Gool, L., Veillon, F., and Zisserman, A., Semi-local projective invariants for the recognition of smooth plane curves, *Int. J. Comput. Vision* **19** (1996), 211–236.

22. Cartan, É., *La Méthode du Repère Mobile, la Théorie des Groupes Continus, et les Espaces Généralisés*, Exposés de Géométrie No. 5, Hermann, Paris, 1935.

23. Cartan, É., *La Théorie des Groupes Finis et Continus et la Géométrie Différentielle Traitées par la Méthode du Repère Mobile*, Cahiers Scientifiques, Vol. 18, Gauthier–Villars, Paris, 1937.

24. Cartan, É., Les sous-groupes des groupes continus de transformations, in: *Oeuvres Complètes*, part. II, vol. 2, Gauthier–Villars, Paris, 1953, pp. 719–856.

25. Cartan, É., Sur la structure des groupes infinis de transformations, in: *Oeuvres Complètes*, Part. II, Vol. 2, Gauthier–Villars, Paris, 1953, pp. 571–714.

26. Cartan, É., La structure des groupes infinis, in: *Oeuvres Complètes*, part. II, vol. 2, Gauthier–Villars, Paris, 1953, pp. 1335–1384.

27. Channell, P.J., and Scovel, C., Symplectic integration of Hamiltonian systems, *Nonlinearity* **3** (1990), 231–259.

28. Cheh, J., Olver, P.J., and Pohjanpelto, J., Maurer–Cartan equations for Lie symmetry pseudo-groups of differential equations, *J. Math. Phys.*, to appear.

29. Chern, S.-S., Moving frames, in: *Élie Cartan et les Mathématiques d'Aujourn'hui*, Soc. Math. France, Astérisque, numéro hors série, 1985, pp. 67–77.

30. Cox, D., Little, J., and O'Shea, D., *Ideals, Varieties, and Algorithms*, 2nd ed., Springer–Verlag, New York, 1996.

31. David, D., Kamran, N., Levi, D., and Winternitz, P., Subalgebras of loop algebras and symmetries of the Kadomtsev-Petviashivili equation, *Phys. Rev. Lett.* **55** (1985), 2111–2113.

32. Deeley, R.J., Horwood, J.T., McLenaghan, R.G., and Smirnov, R.G., Theory of algebraic invariants of vector spaces of Killing tensors: methods for computing the fundamental invariants, *Proc. Inst. Math. NAS Ukraine* **50** (2004), 1079–1086.

33. Dhooghe, P.F., Multilocal invariants, in: *Geometry and Topology of Submanifolds, VIII*, F. Dillen, B. Komrakov, U. Simon, I. Van de Woestyne, and L. Verstraelen, eds., World Sci. Publishing, Singapore, 1996, pp. 121–137.

34. Dorodnitsyn, V.A., Transformation groups in net spaces, *J. Sov. Math.* **55** (1991), 1490–1517.

35. Dorodnitsyn, V.A., Finite difference models entirely inheriting continuous symmetry of original differential equations, *Int. J. Mod. Phys. C* **5** (1994), 723–734.

36. Ehresmann, C., Introduction à la théorie des structures infinitésimales et des pseudo-groupes de Lie, in: *Géometrie Différentielle*, Colloq. Inter. du Centre Nat. de la Rech. Sci., Strasbourg, 1953, pp. 97–110.

37. Fefferman, C., and Graham, C.R., Conformal invariants, in: *Élie Cartan et les Mathématiques d'aujourd'hui*, Astérisque, hors série, Soc. Math. France, Paris, 1985, pp. 95–116.

38. Faugeras, O., Cartan's moving frame method and its application to the geometry and evolution of curves in the euclidean, affine and projective planes, in: *Applications of Invariance in Computer Vision*, J.L. Mundy, A. Zisserman, D. Forsyth (eds.), Springer–Verlag Lecture Notes in Computer Science, Vol. 825, 1994, pp. 11–46.

39. Fels, M., and Olver, P.J., Moving coframes. I. A practical algorithm, *Acta Appl. Math.* **51** (1998), 161–213.

40. Fels, M., and Olver, P.J., Moving coframes. II. Regularization and theoretical foundations, *Acta Appl. Math.* **55** (1999), 127–208.

41. Fuchs, D.B., Gabrielov, A.M., and Gel'fand, I.M., The Gauss–Bonnet theorem and Atiyah–Patodi–Singer functionals for the characteristic classes of foliations, *Topology* **15** (1976), 165–188.

42. Green, M.L., The moving frame, differential invariants and rigidity theorems for curves in homogeneous spaces, *Duke Math. J.* **45** (1978), 735–779.

43. Greene, B. , *The Elegant Universe: Superstrings, Hidden Dimensions, and the Quest for the Ultimate Theory*, W.W. Norton, New York, 1999.

44. Griffiths, P.A., On Cartan's method of Lie groups and moving frames as applied to uniqueness and existence questions in differential geometry, *Duke Math. J.* **41** (1974), 775–814.

45. Griffiths, P.A., *Exterior Differential Systems and the Calculus of Variations*, Progress in Math. vol. 25, Birkhäuser, Boston, 1983.

46. Guggenheimer, H.W., *Differential Geometry*, McGraw–Hill, New York, 1963.

47. Hairer, E., Lubich, C., and Wanner, G., *Geometric Numerical Integration*, Springer– Verlag, New York, 2002.

48. Hann, C.E., and Hickman, M.S., Projective curvature and integral invariants, *Acta Appl. Math.* **74** (2002), 177–193.

49. Hilbert, D., *Theory of Algebraic Invariants*, Cambridge Univ. Press, New York, 1993.

50. Hubert, E., Differential algebra for derivations with nontrivial commutation rules, preprint, INRIA Research Report 4972, Sophis Antipolis, France, 2003.
51. Itskov, V., *Orbit Reduction of Exterior Differential Systems*, Ph.D. Thesis, University of Minnesota, 2002.
52. Jaegers, P.J., *Lie group invariant finite difference schemes for the neutron diffusion equation*, Ph.D. Thesis, Los Alamos National Lab Report, LA–12791–T, 1994.
53. Jensen, G.R., *Higher order contact of submanifolds of homogeneous spaces*, Lecture Notes in Math., No. 610, Springer–Verlag, New York, 1977.
54. Kemper, G., and Boutin, M., On reconstructing n-point configurations from the distribution of distances or areas, *Adv. App. Math.* **32** (2004), 709–735.
55. Kichenassamy, S., Kumar, A., Olver, P.J., Tannenbaum, A., and Yezzi, A., Conformal curvature flows: from phase transitions to active vision, *Arch. Rat. Mech. Anal.* **134** (1996), 275–301.
56. Killing, W., Erweiterung der Begriffes der Invarianten von Transformationgruppen, *Math. Ann.* **35** (1890), 423–432.
57. Kim, P., and Olver, P.J., Geometric integration via multi-space, *Regular and Chaotic Dynamics* **9** (2004), 213–226.
58. Kogan, I.A., *Inductive approach to moving frames and applications in classical invariant theory*, Ph.D. Thesis, University of Minnesota, 2000.
59. Kogan, I.A., and Moreno Maza, M., Computation of canonical forms for ternary cubics, in: *Proceedings of the 2002 International Symposium on Symbolic and Algebraic Computation*, T. Mora, ed., The Association for Computing Machinery, New York, 2002, pp. 151–160.
60. Kogan, I.A., personal communication, 2004.
61. Kogan, I.A., and Olver, P.J., The invariant variational bicomplex, *Contemp. Math.* **285** (2001), 131–144.
62. Kogan, I.A., and Olver, P.J., Invariant Euler-Lagrange equations and the invariant variational bicomplex, *Acta Appl. Math.* **76** (2003), 137–193.
63. Kumpera, A., Invariants différentiels d'un pseudogroupe de Lie, *J. Diff. Geom.* **10** (1975), 289–416.
64. Kuranishi, M., On the local theory of continuous infinite pseudo groups I, *Nagoya Math. J.* **15** (1959), 225–260.
65. Lewis, D., and Simo, J.C., Conserving algorithms for the dynamics of Hamiltonian systems on Lie groups, *J. Nonlin. Sci.* **4** (1994), 253–299.
66. Lie, S., Über unendlichen kontinuierliche Gruppen, *Christ. Forh. Aar.* **8** (1883), 1–47; also *Gesammelte Abhandlungen*, Vol. 5, B.G. Teubner, Leipzig, 1924, pp. 314–360.
67. Lie, S., Über Integralinvarianten und ihre Verwertung für die Theorie der Differentialgleichungen, *Leipz. Berichte* **49** (1897), 369–410; also *Gesammelte Abhandlungen*, vol. 6, B.G. Teubner, Leipzig, 1927, pp. 664–701.
68. Mackenzie, K., *Lie Groupoids and Lie Algebroids in Differential Geometry*, London Math. Soc. Lecture Notes, vol. 124, Cambridge University Press, Cambridge, 1987.
69. Mansfield, E.L., Algorithms for symmetric differential systems, *Found. Comput. Math.* **1** (2001), 335–383.
70. Mansfield, E.L., personal communication, 2003.
71. Marí Beffa, G., Relative and absolute differential invariants for conformal curves, *J. Lie Theory* **13** (2003), 213–245.
72. Marí–Beffa, G., and Olver, P.J., Differential invariants for parametrized projective surfaces, *Commun. Anal. Geom.* **7** (1999), 807–839.

73. McLachlan, R.I., and Quispel, G.R.W., Six lectures on the geometric integration of ODEs, in: *Foundations of Computational Mathematics*, R. DeVore, A. Iserles and E. Suli, eds., London Math. Soc. Lecture Note Series, vol. 284, Cambridge University Press, Cambridge, 2001, pp. 155–210.

74. McLachlan, R.I., and Quispel, G.R.W., What kinds of dynamics are there? Lie pseudogroups, dynamical systems and geometric integration, *Nonlinearity* **14** (2001), 1689–1705.

75. McLenaghan, R.G., Smirnov, R.G., and The, D., An extension of the classical theory of algebraic invariants to pseudo-Riemannian geometry and Hamiltonian mechanics, *J. Math. Phys.* **45** (2004), 1079–1120.

76. Medolaghi, P., Classificazione delle equazioni alle derivate parziali del secondo ordine, che ammettono un gruppo infinito di trasformazioni puntuali, *Ann. Mat. Pura Appl.* **1** (3) (1898), 229–263.

77. Menger, K., Untersuchungen über allgemeine Metrik, *Math. Ann.* **100** (1928), 75–163.

78. Milne–Thompson, L.M., *The Calculus of Finite Differences*, Macmilland and Co., Ltd., London, 1951.

79. Moons, T., Pauwels, E., Van Gool, L., and Oosterlinck, A., Foundations of semi-differential invariants, *Int. J. Comput. Vision* **14** (1995), 25–48.

80. Morozov, O., Moving coframes and symmetries of differential equations, *J. Phys. A* **35** (2002), 2965–2977.

81. Olver, P.J., Classical invariant theory and the equivalence problem for particle Lagrangians. I. Binary Forms, *Adv. in Math.* **80** (1990), 39–77.

82. Olver, P.J., *Applications of Lie Groups to Differential Equations*, Second Edition, Graduate Texts in Mathematics, vol. 107, Springer–Verlag, New York, 1993.

83. Olver, P.J., *Equivalence, Invariants, and Symmetry*, Cambridge University Press, Cambridge, 1995.

84. Olver, P.J., *Classical Invariant Theory*, London Math. Soc. Student Texts, vol. 44, Cambridge University Press, Cambridge, 1999.

85. Olver, P.J., Moving frames and singularities of prolonged group actions, *Selecta Math.* **6** (2000), 41–77.

86. Olver, P.J., Joint invariant signatures, *Found. Comput. Math.* **1** (2001), 3–67.

87. Olver, P.J., On multivariate interpolation, preprint, University of Minnesota, 2003.

88. Olver, P.J., Geometric foundations of numerical algorithms and symmetry, *Appl. Alg. Engin. Commun. Comput.* **11** (2001), 417–436.

89. Olver, P.J., and Pohjanpelto, J., Moving frames for pseudo–groups. I. The Maurer–Cartan forms, preprint, University of Minnesota, 2004.

90. Olver, P.J., and Pohjanpelto, J., Moving frames for pseudo–groups. II. Differential invariants for submanifolds, preprint, University of Minnesota, 2004.

91. Olver, P.J., and Pohjanpelto, J., Regularity of pseudogroup orbits, in: *Symmetry and Perturbation Theory*, G. Gaeta, ed., to appear.

92. Pauwels, E., Moons, T., Van Gool, L.J., Kempenaers, P., and Oosterlinck, A., Recognition of planar shapes under affine distortion, *Int. J. Comput. Vision* **14** (1995), 49–65.

93. Powell, M.J.D., *Approximation theory and Methods*, Cambridge University Press, Cambridge, 1981.

94. Robart, T., and Kamran, N., Sur la théorie locale des pseudogroupes de transformations continus infinis I, *Math. Ann.* **308** (1997), 593–613.

95. Rund, H., *The Hamilton-Jacobi Theory in the Calculus of Variations*, D. Van Nostrand Co. Ltd., Princeton, N.J., 1966.

96. Salmon, R., *Lectures on Geophysical Fluid Dynamics*, Oxford Univ. Press, Oxford, 1998.

97. Shakiban, C., and Lloyd, P., Classification of signature curves using latent semantic analysis, preprint, University of St. Thomas, 2004.

98. Shokin, Y.I., *The Method of Differential Approximation*, Springer–Verlag, New York, 1983.

99. Singer, I.M., and Sternberg, S., The infinite groups of Lie and Cartan. Part I (the transitive groups), *J. Analyse Math.* **15** (1965), 1–114.

100. Sokolov, V.V., and Zhiber, A.V., On the Darboux integrable hyperbolic equations, *Phys. Lett. A* **208** (1995), 303–308.

101. Spencer, D.C., Deformations of structures on manifolds defined by transitive pseudo-groups I, II, *Ann. Math.* **76** (1962), 306–445.

102. Tresse, A., Sur les invariants différentiels des groupes continus de transformations, *Acta Math.* **18** (1894), 1–88.

103. Tsujishita, T., On variational bicomplexes associated to differential equations, *Osaka J. Math.* **19** (1982), 311–363.

104. van Beckum, F.P.H., and van Groesen, E., Discretizations conserving energy and other constants of the motion, in: *Proc. ICIAM 87*, Paris, 1987, pp. 17–35 .

105. Van Gool, L., Brill, M.H., Barrett, E.B., Moons, T., and Pauwels, E., Semi-differential invariants for nonplanar curves, in: *Geometric Invariance in Computer Vision*, J.L. Mundy and A. Zisserman, eds., The MIT Press, Cambridge, Mass., 1992, pp. 293–309.

106. Van Gool, L., Moons, T., Pauwels, E., and Oosterlinck, A., Semi-differential invariants, in: *Geometric Invariance in Computer Vision*, J.L. Mundy and A. Zisserman, eds., The MIT Press, Cambridge, Mass., 1992, pp. 157–192.

107. Weinstein, A., Groupoids: unifying internal and external symmetry. A tour through some examples, *Contemp. Math.* **282** (2001), 1–19.

108. Weyl, H., *Classical Groups*, Princeton Univ. Press, Princeton, N.J., 1946.

109. Yezzi, A., Kichenassamy, S., Kumar, A., Olver, P.J., and Tannenbaum, A., A geometric snake model for segmentation of medical imagery, *IEEE Trans. Medical Imaging* **16** (1997), 199–209.

Invariant Geometric Motions of Space Curves

Changzheng Qu

Center for Nonlinear Studies,
Department of Mathematics,
Northwest University, Xi'an 710069,
P. R. China

Abstract. Motions of space curves in similarity and centro-affine geometries are studied. It is shown that the geometric motions of inextensible space curves in similarity and centro-affine geometries are closely related to integrable equations. Several gauge equivalent integrable equations are derived through the relationship between the curvature and graph of the curves. Motion of space curves corresponding to travelling wave solutions to the KdV flow in centro-affine geometry is discussed in detail.

Keywords: Motion of curve, similarity geometry, centro-affine geometry, integrable equation, travelling wave.

1   Introduction

The problem of motion for curves and surface is important in a wide range of applications. Many interesting nonlinear evolution equations have been shown to be related to motions of curves in certain geometries. For instances, Mullins's nonlinear diffusion model of groove development [1] describes the curve shortening problem, and it has been studied in detail from different points of view (see [1-5] and references therein). More interestingly, a number of 1+1-dimensional integrable equations arise naturally from motions of inextensible curves in certain geometries. For examples, in the earlier works of Hasimoto [6], he showed that the Schrödinger equation arises from motion of inextensible curves in $\mathbb{R}^3$, the Schrödinger hierarchy was also obtained by Langer and Perline [7], they provided a geometrical explanation to the recursion operator of the Schrödinger equation. Using the Hasimoto transformation, Lamb [8] obtained the mKdV and sine-Gordon equations from the motion of curves in $\mathbb{R}^3$. The Heisenberg spin chain model which is gauge equivalent to the Schrödinger equation was derived by Lakshmanan [9]. Doliwa and Santini [10] discovered that the NLS hierarchy and complex mKdV equation arise from motions on $S^3(R)$ where the radius R plays the role of the spectral parameter. Schief and Rogers [11] derived an extended Harry-Dym equation and sine-Gordon equation from binormal motions of curves with constant curvature or torsion. Recently, Nakayama [12] showed that the defocusing nonlinear Schrödinger equation, the Regge-Lund equation,

H. Li, P. J. Olver and G. Sommer (Eds.): IWMM-GIAE 2004, LNCS 3519, pp. 139–151, 2005.

a coupled of system of KdV equations and their hyperbolic type arise from motions of curves in hyperboloids in the Minkowski space. In [13] he realized the full AKNS scheme in a hyperboloid in $M^{3,1}$.

As compared to the motions of curves in space, there are very few works considering motions of curves in the plane. In an intriguing paper of Goldstein and Petrich [14], they showed that the mKdV hierarchy comes naturally from the motion of non-stretching plane curves in Euclidean space $\mathbb{R}^2$. After that, Nakayama, Segur and Wadati [15] obtained the sine-Gordon equation by considering a nonlocal motion of curves in $\mathbb{R}^2$, they also pointed out that the Frenet-Serret equations for curves in $\mathbb{R}^2$ and $\mathbb{R}^3$ are equivalent to the AKNS spectral problem without spectral parameter [15, 16]. More recently, we found that many 1+1-dimensional integrable equations including KdV, Sawada-Kotera, Burgers, Harry-Dym, Kaup-Kupershmidt and Camassa-Holm equations naturally arise from motions of plane curves in centro-affine, similarity, affine and fully affine geometries [17-19]. Motion of plane curves in the Minkowski space $M^{2,1}$ was also investigated by Gürses [20].

In this paper, we are mainly concerned with the motions of space curves in similarity and centro-affine geometries as well as the relationship with integrable systems. The outline of this paper is as follows: In Section 2, we study motion of curves in similarity geometry P^3. Motion of space curves in centro-affine is investigated in Sections 3. In Section 4, we study motions of curves corresponding to travelling wave solutions of the KdV flow in centro-affine geometry. Section 5 contains the concluding remarks on this work.

2   Motion of Space Curves in Similarity Geometry

The problem on motion of curves in similarity geometry has been studied in Olver, Sapiro and Tannenbaum [21, 22]. The isometry group for similarity geometry is a composition of Euclidean motions and dilatations [23]. For the three dimensional similarity geometry P^3, the corresponding Lie algebras of the isometry groups are generated by $\{\partial_x, \partial_y, \partial_u, x\partial_u - u\partial_x, x\partial_y - y\partial_x, y\partial_u - u\partial_y, x\partial_x + y\partial_y + u\partial_u\}$. Similar to the Euclidean and affine geometries, we can define curvatures and arc-length in the similarity geometry, they are given by differential invariants and invariant one-form of the isometry group. More precisely, let κ, τ be the curvature and torsion of a curve in Euclidean space $\mathbb{R}^3$, then they are differential invariants of 3-dimensional Euclidean motion. One can readily verify that $\tilde{\kappa} = \kappa_s/\kappa^2$ and $\tilde{\tau} = \tau/\kappa$ are differential invariants under the similarity motion, we define them to be curvatures of curves in P^3, where s is the arc-length of curves in Euclidean space. In addition, $d\theta = \kappa ds$ is an invariant one-form, where θ is the angle between the tangent and a fixed direction, we define it to be the arc-length of a curve in P^3. After that we can define its frame vectors $\mathbf{t}_i$ in terms of Euclidean's: $\mathbf{t}_1 = \kappa\mathbf{t}$, $\mathbf{t}_2 = \kappa\mathbf{n}$, $\mathbf{t}_3 = \kappa\mathbf{b}$. Using them we can represent the geometric motion of curves in P^3 in the form

$$\gamma_t = A\mathbf{t}_1 + B\mathbf{t}_2 + C\mathbf{t}_3, \tag{1}$$

where A, B and C are respectively the tangent, normal and binormal velocities which depend on the curvatures $\tilde{\kappa}$, $\tilde{\tau}$ and their derivatives with respect to the arc-length θ. Using the Frenet-Serret formulae in Euclidean geometry

$$\begin{pmatrix} \mathbf{t} \\ \mathbf{n} \\ \mathbf{b} \end{pmatrix}_s = \begin{pmatrix} 0 & \kappa & 0 \\ -\kappa & 0 & \tau \\ 0 & -\tau & 0 \end{pmatrix} \begin{pmatrix} \mathbf{t} \\ \mathbf{n} \\ \mathbf{b} \end{pmatrix}, \tag{2}$$

one obtains the Frenet-Serret formulae in the similarity geometry P^3:

$$\begin{pmatrix} \mathbf{t}_1 \\ \mathbf{t}_2 \\ \mathbf{t}_3 \end{pmatrix}_\theta = \begin{pmatrix} -\tilde{\kappa} & 1 & 0 \\ -1 & -\tilde{\kappa} & \tilde{\tau} \\ 0 & -\tilde{\tau} & -\tilde{\kappa} \end{pmatrix} \begin{pmatrix} \mathbf{t}_1 \\ \mathbf{t}_2 \\ \mathbf{t}_3 \end{pmatrix}, \tag{3}$$

where $\tilde{\kappa}$ and $\tilde{\tau}$ are respectively the curvature and torsion of curves in P^3, $\mathbf{t}_i$, $i = 1, 2, 3$ are the frame vector in P^3, and are related to the Euclidean tangent $\mathbf{t}$, normal $\mathbf{n}$ and binormal $\mathbf{b}$ by

$$\mathbf{t}_1 = \frac{\mathbf{t}}{\kappa}, \quad \mathbf{t}_2 = \frac{\mathbf{n}}{\kappa}, \quad \mathbf{t}_3 = \frac{\mathbf{b}}{\kappa}. \tag{4}$$

It is possible to relate (1) with the Euclidean motion in $\mathbb{R}^3$ [6, 15]

$$\gamma_t = W\mathbf{t} + U\mathbf{n} + V\mathbf{b}, \tag{5}$$

with $W = A/\kappa$, $U = B/\kappa$ and $V = C/\kappa$. Using the formulae in Euclidean space [15]

$$\mathbf{t}_t = (U_s - \tau V + \kappa W)\mathbf{n} + (V_s + \tau U)\mathbf{b},$$

and

$$\kappa_t = (\mathbf{n}, \mathbf{t}_{st}) = (\mathbf{n}, \mathbf{t}_{ts}) - \frac{s_t}{s}\kappa,$$

we obtain

$$\kappa_t + \frac{s_t}{s}\kappa = (\mathbf{n}, \mathbf{t}_{ts}),$$

which gives time variation of the perimeter $L = \oint d\theta = \oint \kappa ds$ for a closed curve

$$\begin{aligned} \frac{dL}{dt} &= \oint (\kappa_t + \frac{s_t}{s}\kappa)ds \\ &= \oint \left[\frac{1}{\kappa}(U_s - \tau V + \kappa W)_s - \frac{\tau}{\kappa}(V_s + \tau U) \right] d\theta \\ &= \oint [(U_s - \tau V + \kappa W)_\theta - \tilde{\tau}(V_s + \tau U)]d\theta \\ &= \oint [(B_\theta - \tilde{\kappa}B - \tilde{\tau}C + A)_\theta - \tilde{\tau}(C_\theta - \tilde{\kappa}C + \tilde{\tau}B)]d\theta. \end{aligned}$$

So the inextensibility condition means

$$\oint \tilde{\tau}(C_\theta - \tilde{\kappa}C + \tilde{\tau}B)d\theta = 0,$$
$$(B_\theta - \tilde{\kappa}B - \tilde{\tau}C + A)_\theta = \tilde{\tau}(C_\theta - \tilde{\kappa}C + \tilde{\tau}B). \tag{6}$$

The first equation in (6) means that the perimeter of a closed curve is invariant under the motion (1), and the second one means that the arc-length θ commutes with time t. In terms of (1), we obtain the time evolution for the frame

$$\begin{pmatrix} \mathbf{t}_1 \\ \mathbf{t}_2 \\ \mathbf{t}_3 \end{pmatrix}_t = \begin{pmatrix} F_1 & G_1 & H_1 \\ -G_1 & F_1 & H_2 \\ -H_1 & -H_2 & F_1 \end{pmatrix} \begin{pmatrix} \mathbf{t}_1 \\ \mathbf{t}_2 \\ \mathbf{t}_3 \end{pmatrix}, \tag{7}$$

where

$$F_1 = A_\theta - \tilde{\kappa}A - B,$$
$$G_1 = B_\theta - \tilde{\kappa}B - \tilde{\tau}C + A,$$
$$H_1 = C_\theta - \tilde{\kappa}C + \tilde{\tau}B,$$
$$H_2 = H_{1\theta} + \tilde{\tau}G_1.$$

The compatibility condition between (3) and (7) implies $\tilde{\kappa}$ and $\tilde{\tau}$ satisfying

$$\begin{pmatrix} \tilde{\tau} \\ \tilde{\kappa} \end{pmatrix}_t = \begin{pmatrix} \Omega_1\tilde{\tau} & \Omega_1(\partial_\theta - \tilde{\kappa}) \\ \Omega_2(\partial_\theta^2 - \partial_\theta\tilde{\kappa} - \tilde{\tau}^2) + \partial_\theta & -\Omega_2(\partial_\theta\tilde{\tau} + \tilde{\tau}\partial_\theta - \tilde{\kappa}\tilde{\tau}) \end{pmatrix} \begin{pmatrix} B \\ C \end{pmatrix}, \tag{8}$$

where $\Omega_1 = \partial_\theta^2 + \tilde{\tau}^2 + \tilde{\tau}_\theta\partial_\theta^{-1}\tilde{\tau} + 1$ and $\Omega_2 = \partial_\theta - \tilde{\kappa} - \tilde{\kappa}_\theta\partial_\theta^{-1}$ are respectively the recursion operators of the mKdV and Burgers equation. In particular, setting

$$C = -\tilde{\tau}, \quad B = -\tilde{\kappa}, \quad A = \tilde{\kappa}_\theta - \tilde{\kappa}^2 - \frac{3}{2}\tilde{\tau}^2,$$

so that (6) holds. Then (7) becomes

$$\tilde{\tau}_t + \tilde{\tau}_{\theta\theta\theta} + \frac{3}{2}\tilde{\tau}^2\tilde{\tau}_\theta + \tilde{\tau}_\theta = 0,$$
$$\tilde{\kappa}_t + \tilde{\kappa}_{\theta\theta\theta} - 3(\tilde{\kappa}\tilde{\kappa}_\theta)_\theta + 3\tilde{\kappa}^2\tilde{\kappa}_\theta + \frac{3}{2}(\tilde{\tau}^2\tilde{\kappa})_\theta - 3(\tilde{\tau}\tilde{\tau}_\theta)_\theta + \tilde{\kappa}_\theta = 0. \tag{9}$$

It is easy to see that this system is integrable. Indeed, letting $\tilde{\kappa} = -\mu_\theta/\mu$, μ satisfies

$$\mu_t + \mu_{\theta\theta\theta} + \frac{3}{2}(\tilde{\tau}^2\mu)_\theta + \mu_\theta = 0.$$

Therefore the motion of curves in P^3 yields the Burgers-mKdV hierarchy. Noting that for curves satisfying $\tilde{\kappa} = -\tilde{\tau}_\theta/\tilde{\tau}$, the system (9) is reduced to the mKdV equation (the first one of (9)). In terms of the local graph $(x, u(x,t), v(x,t))$, the Burgers-mKdV flow (9) can be written as

$$u_t = -\frac{(1 + u_x^2 + v_x^2)[(1 + u_x^2 + v_x^2)u_{xxx} - 3u_{xx}(u_x u_{xx} + v_x v_{xx})]}{[u_{xx}^2 + v_{xx}^2 + (u_x v_{xx} - u_{xx} v_x)^2]^{\frac{3}{2}}},$$

$$v_t = -\frac{(1 + u_x^2 + v_x^2)[(1 + u_x^2 + v_x^2)v_{xxx} - 3v_{xx}(u_x u_{xx} + v_x v_{xx})]}{[u_{xx}^2 + v_{xx}^2 + (u_x v_{xx} - u_{xx} v_x)^2]^{\frac{3}{2}}}.$$

Remarks: With the local graph of curves, we found that the mKdV flow in the Euclidean space $\mathbb{R}^2$ [14], the KdV flow in centro-affine geometry [17, 18], the Sawada-Kotera flow in affine geometry [18] and and the mKdV flow in Euclidean space $\mathbb{R}^3$ [15, 24] are gauge equivalent respectively to the WKI equation [25]

$$u_t = \left[\frac{u_{xx}}{(1 + u_x^2)^{\frac{3}{2}}}\right]_x,$$

the affine KdV equation

$$u_t = \left[\frac{u_{xx}}{(xu_x - u)^3}\right]_x,$$

the affine Sawada-Kotera equation [26]

$$u_t = (u_{xx}^{-\frac{5}{3}} u_{xxxx} - \frac{5}{3} u_{xx}^{-\frac{8}{3}} u_{xxx}^2)_x$$

and the two-component WKI equation [27]

$$u_t = -\left[\frac{u_{xx}}{[(1 + u_x^2 + v_x^2)^{\frac{3}{2}}}\right]_x,$$

$$v_t = -\left[\frac{v_{xx}}{(1 + u_x^2 + v_x^2)^{\frac{3}{2}}}\right]_x.$$

3 Motion of Curves in Centro-Affine Geometry

Centro-affine geometry is obtained by deleting translations from the affine geometry; see books about affine geometry [28, 29] for details. The isometries of centro-affine geometry are the linear transformations $x' = Ax$, $A \in SL(3, \mathbf{R})$. For a general curve γ satisfying $[\gamma, \gamma_p, \gamma_{pp}] \neq 0$ along the curve, one may reparametrize the curve by a special parameter σ satisfying

$$[\gamma, \gamma_\sigma, \gamma_{\sigma\sigma}] = 1, \tag{10}$$

everywhere. In terms of an arbitrary parameter p, the centro-affine arc-length is given by

$$d\sigma = [\gamma, \gamma_p, \gamma_{pp}]^{\frac{1}{3}} dp.$$

The centro-affine Serret-Frenet equation is

$$\begin{pmatrix} \gamma \\ \gamma_\sigma \\ \gamma_{\sigma\sigma} \end{pmatrix}_\sigma = \begin{pmatrix} 0 & 1 & 0 \\ 0 & 0 & 1 \\ \alpha & \beta & 0 \end{pmatrix} \begin{pmatrix} \gamma \\ \gamma_\sigma \\ \gamma_{\sigma\sigma} \end{pmatrix}. \tag{11}$$

Now the motion of curves in centro-affine geometry is governed by

$$\gamma_t = F\gamma + G\gamma_\sigma + H\gamma_{\sigma\sigma}, \tag{12}$$

where the velocities F, G and H depend on the centro-affine curvatures α and β and their derivatives with respect to the arc-length σ. The inextensibility condition $(d/dt)[\gamma, \gamma_p, \gamma_{pp}] = 0$ yields

$$F + G_\sigma + \frac{2}{3}\beta H + \frac{1}{3}H_{\sigma\sigma} = 0. \tag{13}$$

Under the motion (12), the compatibility condition $\gamma_{\sigma\sigma t} = \gamma_{t\sigma\sigma}$ implies α and β satisfying

$$\begin{aligned} \alpha_t &= [F_{\sigma\sigma} + \alpha(G + 2H_\sigma) + \alpha_\sigma H]_\sigma + 2\alpha G_\sigma - \beta F_\sigma + \alpha H_{\sigma\sigma}, \\ \beta_t &= [3F_\sigma + G_{\sigma\sigma} + \beta(G + 2H_\sigma) + (\alpha + \beta_\sigma)H]_\sigma + 2\alpha H_\sigma \\ &\quad + \beta H_{\sigma\sigma} + \beta G_\sigma + \alpha_\sigma H. \end{aligned} \tag{14}$$

Some integrable equations are obtained as follows:

Case 1. $\alpha = (1/2)\beta_\sigma$, $H = 0$, $F = -G_\sigma$. Then (14) is reduced to

$$\beta_t = -2(D_\sigma^2 - \beta - \frac{1}{2}\beta_\sigma \partial_\sigma^{-1})G_\sigma = -2\Omega_1 G_\sigma.$$

Example 1. Choosing $G_\sigma = -(1/2)\Omega_1^{n-1}\beta_\sigma$, we get the KdV hierarchy

$$\beta_t = \Omega_1^n \beta_\sigma.$$

Example 2. Taking $G_\sigma = [\beta^{-3/2}\beta_\sigma \partial_\sigma^{-1}(\beta q) - 2\beta^{1/2}q]$, $q = \Omega_2^{n-1}(\psi^3 \psi_{\sigma\sigma\sigma})$, $\beta = \psi^{-2}$. We obtain the Harry Dym hierarchy

$$\psi_t = \Omega_2^n \beta_\sigma,$$

where $\Omega_2 = \psi^2 D_\sigma^2 - \psi\psi_\sigma D_\sigma + \psi\psi_{\sigma\sigma} + \psi^3 \psi_{\sigma\sigma\sigma}\partial_\sigma^{-1}\psi^{-2}$ is the recursion operator of the Harry Dym equation

$$\psi_t + \psi^3 \psi_{\sigma\sigma\sigma} = 0.$$

Example 3. Taking $G_\sigma = -(1/2)(u_{\sigma\sigma} - (1/4)\beta u)$, where $u = (D_\sigma^2 - (1/4)\beta - (1/4)\beta_\sigma \partial_\sigma^{-1})\Omega_3^{n-1}\beta_\sigma$, we get the Sawada-Kotera hierarchy

$$\beta_t = \Omega_3^n \beta_\sigma,$$

where

$$\Omega_3 = (D_\sigma^3 - \frac{1}{2}\beta D_\sigma - \frac{1}{2}D_\sigma\beta)[D_\sigma^3 - \frac{1}{4}D_\sigma^2\beta\partial_\sigma^{-1} - \frac{1}{4}\partial_\sigma^{-1}\beta D_\sigma^2 + \frac{1}{32}(\beta^2\partial_\sigma^{-1} + \partial_\sigma^{-1}\beta^2)]$$

is the recursion operator of the Sawada-Kotera equation [30-32].

Case 2. $F = -(2/3)\beta$, $G = 0$, $H = 1$. We get the Boussinesq equation

$$\alpha_t = 2\alpha_\sigma - \beta_{\sigma\sigma},$$
$$\beta_t = -\frac{2}{3}\beta_{\sigma\sigma\sigma} + \alpha_{\sigma\sigma} + \frac{2}{3}\beta\beta_\sigma.$$

Case 3. $\alpha = (1/2)\beta_\sigma + \lambda$, $F = (1/2)\beta_{\sigma\sigma\sigma} - 2\beta\beta_\sigma + 6\lambda\beta$, $G = \beta^2 - (1/2)\beta_{\sigma\sigma}$, $H = 9\lambda$. We derive the Kaup-Kupershmidt equation

$$\beta_t = \beta_5 - 5\beta\beta_3 - \frac{25}{2}\beta_1\beta_2 + 5\beta^2\beta_1.$$

Case 4. $\alpha = \lambda$, $F = 6\lambda\beta$, $G = \beta_{\sigma\sigma} + \beta^2$, $H = -3\beta_\sigma - 9\lambda$. We obtain the Sawada-Kotera equation again.

Case 5. $\alpha = -[(u_{\sigma\sigma} - (1/2)u_\sigma^2)_\sigma + \lambda]$, $\beta = u_\sigma^2 - 2u_{\sigma\sigma}$, $F = -\lambda^{-1}e^{-2u}u_{\sigma\sigma}$, $G = -\lambda^{-1}e^{-2u}u_\sigma$, $H = -\lambda^{-1}e^{-2u}$. We get the Tzitzéica equation [33, 34]

$$u_{\sigma t} + e^u + e^{-2u} = 0.$$

Case 6. $\alpha = \lambda$, $\beta = 6u_\sigma$, $F = 0$, $G = -(2/\lambda)u_{\sigma t}$, $H = (2/\lambda)u_t$. We have the Hirota-Satsuma equation [36]

$$u_{\sigma\sigma t} + 6u_\sigma u_t = 0.$$

4   The KdV Flow in Centro-Affine Geometry

In this section, we consider motion of space curves corresponding to the travelling waves of the KdV equation in the centro-affine geometry. In general, given any two functions α and β, we solve the equation

$$y_{\sigma\sigma\sigma} - \beta y_\sigma - \alpha y = 0,$$

to obtain three independent solutions y_1, y_2 and y_3. Then for any $A \in SL(3)$, the curve $\gamma(\sigma) = AY$, $Y^t = (y_1, y_2, y_3)$, takes σ and α, β to be its respective arc-length and curvatures. Just as the Euclidean curvature determines the curve up to a rigid motion, the centro-affine curvatures determine the curve up to an centro-affine motion.

Now, suppose the curve moves according to the KdV equation

$$\beta_t = \beta_{\sigma\sigma\sigma} - \frac{3}{2}\beta\beta_\sigma, \tag{15}$$

and $\alpha = \frac{1}{2}\beta_\sigma$. Then from Case 1 in Section 3, we know that the curve satisfies

$$\gamma_{\sigma\sigma\sigma} - \beta\gamma_\sigma - \frac{1}{2}\beta_\sigma\gamma = 0. \tag{16}$$

It determines a family of curves $\gamma(\sigma,t) = A(t)Y(\sigma,t)$, where $[Y(0,t), Y_\sigma(0,t), Y_{\sigma\sigma}(0,t)]$ may be taken to be the identity matrix. By substituting $\gamma(\sigma,t)$ into the KdV flow

$$\gamma_t = \frac{1}{2}(\beta_\sigma\gamma - \beta\gamma_\sigma), \tag{17}$$

we obtain

$$A'Y + AY_t = \frac{1}{2}(\beta_\sigma\gamma - \beta\gamma_\sigma). \tag{18}$$

It follows from (16) that Y_t satisfies

$$Y_{t\sigma\sigma\sigma} - \beta Y_{t\sigma} - \frac{1}{2}\beta_\sigma Y_t = \beta_t Y_\sigma + \frac{1}{2}\beta_{t\sigma}Y,$$
$$Y_t(0,t) = Y_{\sigma t}(0,t) = Y_{\sigma\sigma t} = 0.$$

We know that the general solution of a third-order linear inhomogeneous ODE can be expressed in terms of the three independent solutions y_1, y_2 and y_3 of its homogeneous equation by

$$y = \sum_{i=1}^{3} c_i y_i + y_3 \int_0^\sigma JW(y_1, y_2)d\sigma - y_2 \int_0^\sigma JW(y_1, y_3)d\sigma + y_1 \int_0^\sigma JW(y_2, y_3)d\sigma,$$

where $W(y_i, y_j)$ is the Wronskian determinant, and J is the inhomogeneous term of the ODE. Using this formula, we may express Y_t in terms of β, Y and their derivatives with respect to σ. Putting this expression into (18), we obtain, after performing integration by parts and a lengthy computation,

$$A' + AC(t) = 0, \tag{19}$$

where

$$C(t) = \begin{pmatrix} \frac{1}{2}\beta_\sigma(0,t) & \frac{1}{2}\beta_{\sigma\sigma}(0,t) & \frac{1}{2}(\beta_{\sigma\sigma\sigma} - \frac{1}{2}\beta\beta_\sigma)(0,t) \\ -\frac{1}{2}\beta(0,t) & 0 & \frac{1}{2}(\beta_{\sigma\sigma} - \beta^2)(0,t) \\ 0 & -\frac{1}{2}\beta(0,t) & -\frac{1}{2}\beta_\sigma(0,t) \end{pmatrix}.$$

Thus the KdV flow is given by

$$\gamma(s,t) = A(t)Y(\sigma,t),$$

where A satisfies (19).

Suppose now that $\beta(\sigma, t) = \hat{\beta}(\sigma + ct)$ is a travelling wave of (15). The corresponding flow is given by

$$\gamma(\sigma, t) = A(t)\hat{\gamma}(\sigma + ct), \qquad (20)$$

where A satisfies (19) with $C(t)$ replaced by the matrix

$$\hat{C}(t) = \begin{pmatrix} \frac{1}{2}\hat{\beta}'(ct) & \frac{1}{2}\hat{\beta}''(ct) & \frac{1}{2}(\hat{\beta}''' - \frac{1}{2}\hat{\beta}\hat{\beta}')(ct) \\ -\frac{1}{2}\hat{\beta}(ct) & 0 & \frac{1}{2}(\hat{\beta}'' - \hat{\beta}^2)(ct) \\ 0 & -\frac{1}{2}\hat{\beta}(ct) & -\frac{1}{2}\hat{\beta}'(ct) \end{pmatrix},$$

where $\hat{\gamma}$ is the curve determined by $\hat{\beta}$ (taking $A = id$). So the curve evolves under an centro-affine motion for every t. For convenience, letting $\hat{\beta} = -4\tilde{\beta}$, it follows from (15) that $\tilde{\beta}$ satisfies

$$\tilde{\beta}_{zz} + 3\tilde{\beta}^2 - c\tilde{\beta} = D, \quad z = \sigma + ct,$$

and

$$\frac{1}{2}\tilde{\beta}_z^2 + \tilde{\beta}^3 - \frac{1}{2}c\tilde{\beta}^2 - D\tilde{\beta} - E = 0,$$

where D and E are arbitrary constants. Only when the equation $\rho^3 - \frac{1}{2}c\rho^2 - D\rho - E = 0$ has three real roots, the solution is bounded. Let's consider the generic case when there are three distinct roots c_1, c_2 and c_3, $c_1 < c_2 < c_3$. The solution ($u = \beta$ here) is given by

$$u = u_0 + A\operatorname{cn}^2 az, \qquad (21)$$

where

$$u_0 = \frac{1}{6}c_2, \ A = \frac{1}{6}(c_3 - c_2), \ c = \frac{1}{3}(c_1 + c_2 + c_3), \ a^2 = \frac{1}{12}(c_3 - c_1), \ k^2 = \frac{c_3 - c_2}{c_3 - c_1},$$

and k is the modulus of the Jacobi elliptic function $\operatorname{sn} z$. So the curve $\hat{\gamma}$ satisfies

$$\hat{\gamma}_{zzz} + 4u\hat{\gamma}_z + 2u_z\hat{\gamma} = 0, \qquad (22)$$

To solve (22), we use the following fact: If ϕ_1 and ϕ_2 are two independent solutions of the Hill's equation satisfying $\phi_1(0) = 1$, $\phi_1'(0) = 0$, $\phi_2(0) = 0$, $\phi_2'(0) = 1$,

$$\phi_{zz} + u\phi = 0, \qquad (23)$$

then ϕ_1^2, ϕ_2^2 and $\phi_1\phi_2$ are three independent solutions of (22). In terms of ϕ_1 and ϕ_2, one may express the curve by

$$\gamma = A(t)Y, \quad Y^t = (\phi_1^2(z), \frac{1}{2}\phi_1(z)\phi_2(z), \phi_2^2(z)),$$

where $A(t)$ satisfies (19) with $C(t)$ replaced by

$$\hat{C}(t) = \begin{pmatrix} \frac{1}{2}\hat{\beta}'(ct) & (\frac{1}{2}\hat{\beta}^2 - \hat{\beta}'')(ct) & \frac{1}{4}(\hat{\beta}''' - \frac{3}{2}\hat{\beta}\hat{\beta}')(ct) \\ -\frac{1}{4}\hat{\beta}(ct) & 0 & \frac{1}{8}(\hat{\beta}'' - \frac{1}{2}\hat{\beta}^2)(ct) \\ 0 & 2\hat{\beta}(ct) & -\frac{1}{2}\hat{\beta}'(ct) \end{pmatrix},$$

Hence the problem is reduced to solve Hill's equation (23). Letting $\hat{\phi}(z)$ be any component of $\phi(z/a)$, we have

$$\hat{\phi}_{zz} + (\lambda - 2k^2\mathrm{sn}^2 z)\hat{\phi} = 0, \quad \lambda = \frac{2c_3}{c_3 - c_1}. \tag{24}$$

Equation (24) is precisely Lamé's equation with $m = 1$, [35]. We can describe all its periodic solutions as follows. Given any $m, n \geq 1$, $(m, n) = 1$, we solve

$$\frac{\Delta(\lambda)}{2} = \cos(\frac{2n\pi}{m}),$$

for λ. For each root λ we can solve (24) to obtain a closed curve $\hat{\gamma}$ with m leaves. More precisely, there are exactly two stable intervals of the form $I_0 = (\lambda_0, \lambda_1')$ and $I_1 = (\lambda_2', \infty)$ so that when $|\cos(\frac{2n\pi}{m})| < 1$, there is exactly one root in I_0 and all other roots lie in I_1. When $|\cos(\frac{2n\pi}{m})| = 1$, all roots lie inside I_1. Several examples of closed $\hat{\gamma}$ for $\lambda \in I_0$ and $k^2 = 1/2$ can be found in Figure 1. Closed $\hat{\gamma}$'s for $\lambda \in I_1$ and $k^2 = 1/2$ can be found in Figure 2.

A limit case of the above travelling wave is $c_1 = c_2 < c_3$, where one obtains the one-soliton of the KdV equation

$$\tilde{\beta} = \frac{c}{2}\mathrm{sech}^2 \frac{\sqrt{c}}{2}z, \quad z = \sigma + ct.$$

Solving the Hill's equation with u replaced by the one-soliton, we obtain two independent solutions

$$\phi_1 = 2 - \theta\tanh\frac{\theta}{2}, \quad \phi_2 = \tanh\frac{\theta}{2}, \quad \theta = \sqrt{c}(s + ct).$$

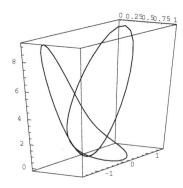

Fig. 1. A space curve for $k = 1/\sqrt{2}$, $I_0 = (0.5, 1)$, $\lambda = 0.788$ with $m = 3$ in centro-affine geometry

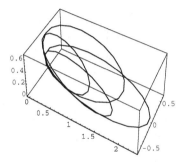

Fig. 2. A space curve for $k = 1/\sqrt{2}$, $I_1 = (1.5, \infty)$, λ=1.873 with m=3 in centro-affine geometry

So the space curve $\hat{\gamma}$ is given by

$$\hat{\gamma}(\sigma, t) = t(t)(2 - \theta \tanh \frac{\theta}{2})^2 + n(t) \tanh \frac{\theta}{2}(2 - \theta \tanh \frac{\theta}{2}) + b(t) \tanh^2 \frac{\theta}{2}.$$

Substituting it into the KdV flow (17), we find

$$t' = 0, \quad n' - 2c^{\frac{3}{2}}t = 0, \quad b' - c^{\frac{3}{2}}n = 0,$$

which yields

$$t(t) = t_0,$$
$$n(t) = 2c^{\frac{3}{2}}tt_0 + n_0,$$
$$b(t) = c^3 t^2 t_0 + c^{\frac{3}{2}}tn_0 + b_0,$$

where t_0, n_0 and b_0 are constant vectors. Taking $t_0 = (1,0,0)$, $n_0 = (0,0,1)$ and $b_0 = (0,0,1)$, we get the curve given by

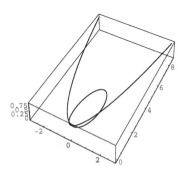

Fig. 3. A space curve with one loop corresponding to the one-soliton of the KdV equation in centro-affine geometry

$$\gamma(\sigma, t) = \begin{pmatrix} 1 & 2c^{\frac{3}{2}}t & c^3t^2 \\ 0 & 1 & c^{\frac{3}{2}}t \\ 0 & 0 & 1 \end{pmatrix} \begin{pmatrix} (2 - \theta \tanh\frac{\theta}{2})^2 \\ \tanh\frac{\theta}{2}(2 - \theta \tanh\frac{\theta}{2}) \\ \tanh^2\frac{\theta}{2} \end{pmatrix}.$$

So γ moves under centro-affine actions. The curve corresponding to the one-soliton has one loop (see Figure 3), which is analogous to the mKdV flow in the Euclidean geometry that one-soliton gives one-loop curve [16, 37].

5 Concluding Remarks

We have shown that several well-known integrable equations including KdV, Sawada-Kotera, Harry Dym hierarchies and Boussinesq, Kaup-Kupershmidt, Tzitz-éica, Hirota-Satsuma equations and Burgers-mKdV system arise naturally from motions of space curves in centro-affine and similarity geometries. It is worthy of mentioning that the mKdV equation in Euclidean spaces $\mathbf{E}^2$ and $\mathbf{E}^3$, the KdV equation in centro-affine geometry and the Sawada-Kotera equation in affine geometry $SA(3)$ are obtained by choosing the normal velocity to be the derivative of the group curvature with respect to the group arc-length. More interestingly, we found that the N-solitons of the mKdV and Sawada-Kotera equations give N loop curves respectively in Euclidean and affine geometries, but in centro-affine geometry, N-solitons of the KdV equation gives $N - 1$ loop curves. These analogies suggest that the KdV equation and Sawada-Kotera equation are respectively the centro-affine version and affine version of the mKdV equation.

It is of great interest to study the motion of discrete curves in centro-affine, affine and similarity geometries. We are currently investigating this issue and our findings will appear in future papers.

Acknowledgments

This work was supported by the National NSF (Grant No. 10371098) of China, NSF of Shaan Xi Province and Excellent Youth Teacher's Foundation of China Ministry of Education.

References

1. W.W. Mullins, J. Appl. Phys. **28** (1957) 333-339.
2. H. Gage and R. Hamilton, J. Diff. Geom. **23** (1986) 69-96.
3. M. Grayson, J. Diff. Geom. **26** (1987) 285-314.
4. C.Z. Qu, IMA J. Appl. Math. **62** (1999) 283-302.

5. C.Z. Qu, J. Phys. Soc. Jpn. **69** (2000) 1307-1312.
6. H. Hasimoto, J. Fluid. Mech. **51** (1972) 477-485.
7. J.S. Langer and R. Perline, J. Nonlin. Sci. **1** (1991) 71-93.
8. G.L. Lamb, J. Math. Phys. **18** (1977) 1654-1661.
9. M. Lakshmanan, Th. W. Ruijgrok and C.J. Thompson, Physica A **84** (1976) 577-84.
10. A. Doliwa and P.M. Santini, Phys. Lett. A **85** (1994) 373-84.
11. W.K. Schief and C. Rogers, Proc. Roy. Soc. Lond. A **455** (1999) 3163-88.
12. K. Nakayama, J. Phys. Soc. Jpn. **67** (1998) 3031-3037.
13. K. Nakayama, J. Phys. Soc. Jpn. **68** (1999) 3214-3218.
14. R.E. Goldstein and D.M. Petrich, Phys. Rev. Lett. **67** (1991) 3203-3206.
15. K. Nakayama, H. Segur and M. Wadati, Phys. Rev. Lett. **69** (1992) 2603-2606.
16. K. Nakayama and M. Wadati, J. Phys. Soc. Jpn. **62** (1993) 473-479.
17. K. S. Chou and C. Z. Qu, J. Phys. Soc. Jpn. **70** (2001) 1912-1916.
18. K. S. Chou and C. Z. Qu, Physica D, **162** (2002) 9-33.
19. K. S. Chou and C. Z. Qu, J. Nonlin. Sci. **13** (2003) 487-517.
20. M. Gürses, Phys. Lett. A **241** (1998) 329-334.
21. P. J. Olver, G. Sapiro and A. Tannenbaum, Differential invariant signatures and flows in computer vision: A symmetry group approach, Geometry Driven Diffusion in Computer Vision, B. M. Ter Haar Romeny, Ed., 205-306, Kluwer, Dordrecht, 1994.
22. G. Sapiro and A. Tannenbaum, IEEE Trans. PAMI. **17** (1995) 67-72.
23. P. J. Olver, Equivalence, Invariants and Symmetry, Cambridge University Press, Cambridge, 1995.
24. J.S. Langer and R. Perline, Phys. Lett.A **239** (1998) 36-40.
25. M. Wadati, K. Konno and Y.H. Ichikawa, J. Phys. Soc. Jpn. **47** (1979) 1698-1700.
26. C.Z. Qu and Y.Q. Si and R.C. Liu, Chaos, Solitons and Fractals, **15** (2003) 131-139.
27. C.Z. Qu, R. X. Yao and R.C. Liu, Phys. Lett. A, in press, 2004.
28. H.G. Guggenheimer, Differential Geometry, Dover, New York, 1963.
29. K. Nomizu and T. Sasaki, Affine Differential Geometry, Cambridge University Press, Cambridge, 1994.
30. K. Sawada and T. Kotera, Prog. Theor. Phys. **51** (1974) 1335-1367.
31. P.J. Caudrey, R.K. Dodd and J.D. Gibbon, Proc. R. Soc. London A **351** (1976) 407-422.
32. B. Fuchssteiner and W. Oevel, J. Math. Phys. **23** (1982) 358-363.
33. G. Tzitzéica, C. R. Acad. Sci. Paris **150** (1910) 955-956.
34. R.K Dodd and R.K. Bullough, Proc. R. Soc. Lond. A **352** (1977) 481-503.
35. W. Magnus and S. Winkler, Hill's Equations, Dover, New York, 1979.
36. R. Hirota and J. Satsuma, J. Phys. Soc. Jpn. **40** (1976) 611-612.
37. K. Konno, Y.H. Ichikawa and M. Wadati, J. Phys. Soc. Jpn. **50** (1981) 1025-1026.

Classification of Signature Curves Using Latent Semantic Analysis

Cheri Shakiban and Ryan Lloyd

Department of Mathematics,
University of St. Thomas
St. Paul, MN, 55105, USA

Abstract. In this paper we describe the Euclidean signature curves for two dimensional closed curves in the plane and will give a discrete numerical method for finding such invariant curves. Further we describe an analog of Latent Semantic Analysis (LSA) and present data and noise reduction techniques as well as an optimal combination of normalizing transformations to categorize signature curves. We will then introduce a system for determining the correct category for a new object from a pre-existing database of information on objects and give an example for sorting out leaves of two types of trees regardless of their orientation using their signature curves.

Keywords: computer vision, pattern recognition, object recognition, signature curves, Latent Semantic Analysis (LSA), weight functions, categorization.

1   Introduction

The Euclidean signature curves are unique closed curves assigned to the boundary of objects in the plane. These unique curves are most useful in the study of computer vision applications because they remain invariant under Euclidean transformations such as rotation. Signature curves can be used in medical imaging devices such as CAT or MRI scans. There are also many military and civil defense systems that can employ the use of signature curves for object recognition purposes. Classification of two dimensional objects is an old problem in computer vision and most common used techniques are based upon Fourier Analysis, however, using signature curves has an advantage over the most commonly used methods as it allows for the transformed (e.g. rotated) objects to be considered for classification without any concern about their angle of rotation.

Although we will only discuss the two dimensional classification techniques in this paper, the method described could easily be extended to the objects that are invariant under affine transformations and space curve transformations. In fact, the classification of space curves is discussed in an upcoming paper, *Latent Semantic Analysis in the Identification of Supercoiled DNA Molecules Generalized by Euclidean Signature Curves*. However, the projective invariant signature curves are still under investigation.

H. Li, P. J. Olver and G. Sommer (Eds.): IWMM-GIAE 2004, LNCS 3519, pp. 152–162, 2005.
© Springer-Verlag Berlin Heidelberg 2005

2   Calculation of Signature Curves

Signature curves can be calculated and graphed for curves that are described by functions algebraically but this process can be difficult as the calculations get very tedious. However they can simply be created numerically by methods described by Calabi et. al., [2]. These numerical methods can also be applied to find the signature curves of objects within an image taken with a camera or other imaging device. One can find the boundary of such an object using a segmentation algorithm such as the method of active contours or snakes described in [4], and thereby create the corresponding signature curve.

Algebraically, the Euclidean signature curve is associated to a planar curve C in $\mathbb{R}^2$ that is parameterized by arclength $s = s(t) = \int_0^t \sqrt{x_t^2 + y_t^2}$, i.e. if the original curve is given by $C = \{x(s), y(s)\}$. Then its signature curve is given by the curve $S = \{\kappa, \kappa_s\}$, where $\kappa = \kappa(s) = x_s y_{ss} - y_s x_{ss}$ is the curvature as a function of arclength and κ_s is its derivative with respect to arclength.

The Euclidean signature curve is then displayed by simply plotting (κ, κ_s). It is important to note however, that this formula for curvature only applies to Euclidean signature curves. In order to plot *affine signature curves*, the affine curvature and arclength must be used; see [2].

Figure 1 shows the graph of a curve C in its parametric form $x(t) = \cos t + \frac{1}{5} \cos^2 t$, $y(t) = \sin t + \frac{1}{10} \sin^2 t$ followed by the graph of its algebraic Euclidean signature curve S. Note that if the original curve is rotated or translated in any direction, the signature curve S remains unchanged.

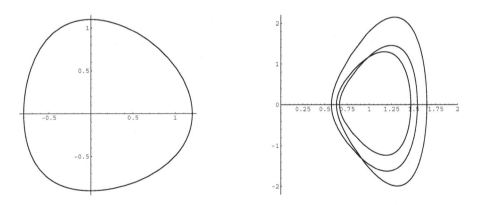

Fig. 1. The Original curve C and its Signature curve S

Although the formulated approach to finding curvature is effective for explicitly parameterized curves, it is not used in computer vision applications as it is not so easy to fit a function to a segmented image. Thus, it becomes important that curvature be approximated by some numerical method that preserves the underlying symmetry group. The process for approximating discrete curvature begins by introducing three successive mesh points on the original curve (image).

Three mesh points is optimal because the Euclidean curvature, as defined above, is a second order differential function. Let A, B, and C denote the three points, and let $a = d(A, B), b = d(B, C), c = d(A, C)$ be the distances between them. The circle passing though the points A, B and C in figure 2 is an approximation of the osculating circle at point B. As a result, $\tilde{\kappa} = \tilde{\kappa}(A, B, C)$, which is the reciprocal of the radius of this circle, acts as an approximation to the curvature at point B. We can apply Heron's formula to find the radius of the circle passing through these three points, and thus find a good approximation of the curvature.

$$\kappa(A, B, C) \approx \tilde{\kappa}(A, B, C) = \frac{4\Delta}{abc} = \pm\frac{4\sqrt{s(s-a)(s-b)(s-c)}}{abc}$$

where $s = \frac{1}{2}(a + b + c)$.

The first derivative of curvature can also be approximated using a centered difference formula.

$$\tilde{\kappa}_s(P_{i-2}, P_{i-1}, P_i, P_{i+1}, P_{i+2}) = (\tilde{\kappa}(P_i, P_{i+1}, P_{i+2}) - \tilde{\kappa}(P_{i-2}, P_{i-1}, P_i))/d(P_{i+1}, P_{i-1}).$$

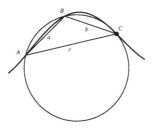

Fig. 2. Original Curve with points A, B, and C and the circle going through these points

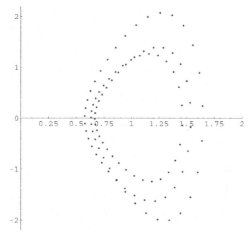

Fig. 3. The discrete version of the signature curve obtained numerically for the curve C of Figure 1

This formula is only accurate when the points are fairly regularly spaced around the curve; a more general formula was found by Mireille Boutin, [1]. Thus, the entire numerically discrete approximation to the Euclidean signature curve can be plotted by:

$$(\tilde{\kappa}(P_{i-1}, P_i, P_{i+1}), \tilde{\kappa}_s(P_{i-2}, P_{i-1}, P_i, P_{i+1}, P_{i+2}))$$

Figure 4 gives the smooth and discrete versions of the signature curve obtained numerically for the curve given in polar form $r = 3 + \frac{1}{10}\cos 3\theta + \frac{1}{40}\cos 7\theta$.

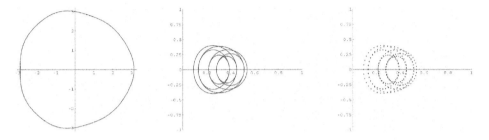

Fig. 4. The original curve and its smooth and discrete signature curves

One difficulty when considering signature curves is how a computer will interpret these curves and decide if they are representing similar or the same objects. This is where Latent Semantic Analysis comes in.

3   Latent Semantic Analysis

Latent Semantic Analysis, (LSA), is a statistical method that produces an algorithm to calculate the similarity between two or more broad objects that contain units that appear with certain frequencies. LSA was introduced in 1990, [3] and remains one of the most commonly used in the statistical examination of texts or documents. For more information on LSA, refer to For more information on LSA, refer to [6], [7]. In the analysis of texts, each category could hold the works of a particular author, or the writings of a particular era. The classes might each have a large number of documents. The frequencies of the terms common within each document are entered into a huge matrix, called term-document matrix (TDM) which then undergoes a series of transformations. The final result in LSA is a correlation matrix with the ij^{th} element representing the similarity between the documents i and j (the correlation values are between 0 and 1). Further, if a new document is added to the system, the computer could determine which category this new document belongs to, if any.

The algorithm is as follows:

Step 1: *Constructing the TDM.* To create the TDM, we let the columns of the matrix represent the documents under investigation and the rows of the matrix represent certain common terms (words). For every term the value of entry (i, j) is the frequency of the term i that appears in the document j.

Step 2: *Normalizing the columns of the TDM.* We will assign a value between 0 and 1 to each entry of the matrix while taking the length of each document into account.

Step 3: *Applying weight functions.* We replace each entry of the matrix with the product of a local weight function (LWF) and a global weight function(GWF). LWF considers the frequency of a word within a particular document and GWF examines a term's frequency across all the documents. For a list of these weight functions see [6].

Step 4: *Applying singular value decomposition, SVD.* We remove noise or infrequent terms that do not effect our calculations in classifying the document by using SVD to factor the $m \times n$ matrix M, (obtained in step 3) into three new matrices, $M = P\Sigma Q$. The center matrix Σ contains the singular values, while the matrices on the left and right are orthogonal. Matrix Σ, enables us to reduce the noise and to obtain a more suitable matrix. For a discussion of SVD see [8].

Step 5: *Building a correlation matrix.* The final step in the LSA is to use the transformed matrix obtained in step 4, $\widetilde{M}$ to calculate the correlation between the ith and j^{th} documents by calculating the cosine of the angle between column vectors $\mathbf{C}_i$ and $\mathbf{C}_j$ in the matrix $\widetilde{M}$:

Example 1.

Suppose that the documents of interest are four short texts, which are listed here:

a) Text #1: "Buy bargain goods here."
b) Text #2: "Buy imported deals."
c) Text #3: "Find best goods, best deals here."
d) Text #4: "Imported goods are bargain. Are best goods bargain?"

The term-document matrix corresponding to this example is given by figure 5. The documents contain nine unique words, and the number in each cell discloses the frequency of a word within a particular document.

After normalizing and applying the local and global transformations and using the singular value decomposition to de-noise the matrix TDM, we arrive

	Text #1	Text #2	Text #3	Text #4
buy	1	1	0	0
bargain	1	0	0	2
goods	1	0	1	2
here	1	0	1	0
imported	0	1	0	1
deals	0	1	1	0
find	0	0	1	0
best	0	0	2	1
are	0	0	0	2

Fig. 5. TDM for the texts

	Text #1	Text #2	Text #3	Text #4
buy	0.25	0.33	0	0
bargain	0.38	0	0	0.38
goods	0.33	0	0.23	0.33
here	0.25	0	0.17	0
imported	0	0.33	0	0.13
deals	0	0.33	0.17	0
find	0	0	0.17	0
best	0	0	0.50	0.20
are	0	0	0	0.50

Fig. 6. Transformed TDM

at the transformed matrix in figure 6. This matrix is then used to describe how similar the documents are by finding the cosine of the angle between two column vectors $\mathbf{C}_i$ and $\mathbf{C}_j$ appearing in the matrix, using the familiar formula

$$\cos(\theta) = \frac{\mathbf{C}_i \cdot \mathbf{C}_j}{\|\mathbf{C}_i\| \|\mathbf{C}_j\|}. \tag{1}$$

For example if we calculate the correlation between the first and the second text, we get 0.2346, which indicates a rather weak correlation between the two texts.

4   Procedure for the Signature Curves

The version of Latent Semantic Analysis that involves signature curves is similar to the traditional version of LSA featuring texts. Many of the same local and global weight functions (LWF,GWF), and normalization routines utilized in the common method of LSA can be carried over to signature curve analysis. The first step in the LSA-signature curve algorithm is the construction of the Square-Object Matrix, SOM which is a variant of TDM. SOM contains the various objects along the columns and a range of infinitesimal square addresses forming a grid that covers all the signature curves of the objects, along the rows as in figure 7. This grid that these signature curves we are investigating lie within, is segmented into several thousand squares. Since most signature curves lie within a n unit by n unit square centered at the origin, the square division only needs to take place in this grid. The length of each infinitesimal square could be any value and depends on n, and the number of points used in creating the signature curves. In example 1, $n = 4$ and the lengths of squares ranged from 0.1 to 0.25 unit.

The ij^{th} element of SOM denotes the number of points that lie in the ith square for the jth object. This square segmentation process gives a digital representation of the general shape of the curve and reveals where the points are concentrated in the coordinate system.

A series of transformations, including LWF and GWF is then applied on this matrix. For a list of the normalization and possible weight functions, refer to [6] and [7]. These transformations highlight the important data in the SOM. The

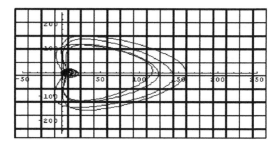

Fig. 7. Creating the SOM

goal of these transformations is to emphasize the differences between the signature curves belonging to opposing categories and to bring out the similarities among the curves in the same group. Information that is unnecessary, such as the squares that consistently have a large amount of points (or no points) for all the signature curves and the squares that have similar frequency values across all categories of items are ignored . We will then apply a trigonometric weight function on the new, simplified matrix. These weight functions give greater values to the point counts in the squares that are far from the origin as the points far from the origin make the various signature curves unique and the modified version of LSA should emphasizes these outer points. The proposed trigonometric weight functions used are shown below:

$$TWF = 0 \quad \longrightarrow \quad T(i,j) = 1$$
$$TWF = 1 \quad \longrightarrow \quad T(i,j) = (a_{ij}^{\,2} + b_{ij}^{\,2})^{1/2}$$
$$TWF = 2 \quad \longrightarrow \quad T(i,j) = (a_{ij}^{\,2} + b_{ij}^{\,2})$$
$$TWF = 3 \quad \longrightarrow \quad T(i,j) = (a_{ij}^{\,2} + b_{ij}^{\,2})^{1/4}$$

Here $TWF = 0$ is the trivial weight function. The remaining functions are applied to the points (a_{ij}, b_{ij})-the center of the ij^{th} square in the grid that covers the signature curve. For $TWF = 1$, $T(i,j)$ is the distance from the origin to the center of the square. $TWF = 2$ is similar to $TWF = 1$ with the exception that the distance is squared. $TWF = 2$ allows new frequency values to grow more rapidly, while $TWF = 1$ yields linear growth. The last trigonometric weight function provides growth that slowly diminishes for squares farther from the origin. Ultimately to obtained the final transformed matrix, we apply SVD on the obtained matrix to de-noise the matrix.

5 Noise Reduction Techniques

Squares that have similar point frequencies across all signature curves create noise and don't highlight differences between signature curves and must be eliminated. Therefore, the rows that have few non-zero frequencies are eliminated.

For example if a SOM has 100 columns, and the third row has only two non-zero values, then the third row is deleted all together to minimize noise. Another phase in this square-filtering endeavor involves erasing squares with low standard deviations. A lower standard deviation means that all the elements in a row are quite similar and could be ignored. The squares with high standard deviations contain data on the differences between different categories and the similarities within the same classes and thus must be highlighted. Besides using the standard deviation as a measure in determining whether or not a row should remain in the SOM, one could utilize the entropy measurement. Recall that entropy is a description of how spread out data is. A low entropy signifies that the vector's information varies drastically in magnitude, while a high entropy value is attached to a vector with rather uniform data. The equation for the entropy of vector $\mathbf{v}$ is:

$$E(\mathbf{v}) = \sqrt{-\sum_{k=1}^{n} \frac{v_k^2}{\|\mathbf{v}\|^2} \log_2 \frac{v_k^2}{\|\mathbf{v}\|^2}}.$$

A routine that filters squares based on entropy should remove the rows with high entropies. These three data reducing methods described above can dramatically decrease the size of database files in computer memory while increasing the accuracy of LSA in categorizing objects. For example, in the maple leaf-buckthorn leaf trial (see Example 2, discussed below), after applying these methods, the size of the SOM reduced from 4,046 kilobytes to just 12 kilobytes(a 99.7% decrease).

6   Modelling the Optimal Transformations

Selecting a model for the combination of the normalization and transformations used to accelerate the algorithm is also essential. One way to achieve this goal is to use the plots of singular values of several transformation combinations. The singular value plot with the lowest condition number, i.e., the ratio of the largest singular value to the smallest singular value, is tied to the optimal transformation, and provides the best transformation that does not create noise in the SOM. Figure 8 displays the singular value plot of an optimal model for the *Maple Leaf-Buckthorn Leaf Trial* , discussed below.

In the final stage in the LSA-signature curve routine we construct a correlation matrix using formula (1), viewing each column in the SOM as a vector. The similarity between i^{th} object and the jth object is the cosine of the angle between the objects' corresponding column vectors $\mathbf{C}_i$ and $\mathbf{C}_j$.

In general a quantity of 1 in the correlation matrix implies that two bodies are identical, while a value around zero reveals that the items are quite dissimilar. The correlation matrix will be an n by n square, where n is the number of objects under investigation. The ij^{th} entry gives the correlation between the ith and the jth signature curves. The matrix is always symmetric with a line of ones running along the main diagonal. If a correlation matrix is represented in color, then red traditionally symbolizes high correlations and blue signifies weak relationships.

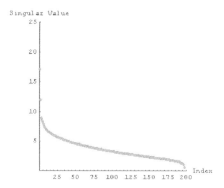

Fig. 8. SDV plot of an optimal Model

In the following example, we will illustrate how the algorithm described above was used to examine LSA's ability to classify signature curves of objects with two different shapes.

Example 2. *Maple Leaf-Buckthorn Leaf Trial:* In this example we used the leaves from two different trees, sugar maple and buckthorn trees and established the correlation matrix among the shape of the leaves. The example featured 100 total images-50 photos for each type of leaf. In the photos, all the leaves were viewed in the same top-view orientation but some of the leaves were rotated differently relative to the viewer, but as expected from the result of section 3, the signature curves were immune to these rotations. An edge detection routine [4] was used to identify the edge points in each picture that belonged to the outline of the leaves of interest. We generated the signature curves for the sets of initial outline points. Each signature curve contained between 10,000 and 20,000 points depending on how large the original photo was. The variables for the signature curves had values ranging from about -50 to 50. Figure 9 shows picture of an example of a maple leaf and figure 10 shows an example of a buckthorn leaf used in our tests, along with their signature curves. We applied the LSA algorithm to all the signature curves by using square partitions of various sizes and in various quantities. We worked with various weighting function combinations until we discovered the best combination to produce a favorable result for the correlation matrix.

A sample correlation matrix based on the information on the 50 maple leaves and 50 buckthorn leaves and using an optimal order of operations discussed in the section is given by figure 11.

Giving a computer the ability to accurately label new objects by appearance is the central goal of this computer vision problem. A computer as it examines a new object in its environment, it builds the corresponding signature curve of that object and determines whether this object is new or old to its knowledge. The correlations between the new object's signature curve and the older signature curves is then calculated. If the new object matches a class that already exists in the database, then the computer labels it and adds the information to the

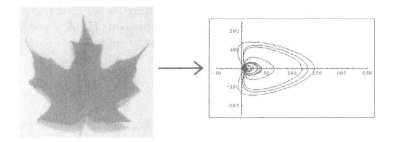

Fig. 9. The Maple Leaf and its Signature Curve

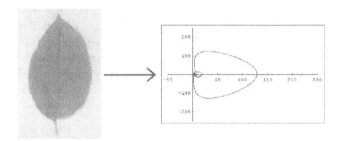

Fig. 10. The Buckthorn Leaf and its Signature Curve

Fig. 11. Correlation Matrix for Maple versus Buckthorn

cluster of pre-existing data on the category. However, if the observed object is foreign, then the machine creates a new cluster in the database for the new object, adds the information to the database, and gives this new object a name. The method for labelling a new photographic image is remarkably similar to the conventional method of categorizing a new document in LSA. The categorization process follows the method discussed in [5]. Essentially, the correlations between the new object's signature curve and all the pre-existing signature curves are calculated. The average correlation between the object in question and all the objects in a group is then found for all the classes. The group with the highest average correlation is the category that the new object belongs to.

This procedure was tested with several signature curves of maple and buck-thorn leaves and also other similar leaves and the method of LSA was able to identify each leaf by its signature curve with a high degree of accuracy (over 90 %).

7   Conclusion

The system using Latent Semantic Analysis to classify signature curves is encouraging. In the experiments performed so far the algorithm has proved to be able to distinguish one category of objects from a second group. The correlation matrices for the experiments with two categories have been nearly perfect. The computer's accuracy in naming new objects has also been extraordinarily high. Overall, LSA has proved that it can classify signature curves belonging to two distinct, clearly defined groups. LSA has even been used to sort a random pile of signature curves into classes. The proposed data or noise reduction methods are promising, and the techniques using SVD for deciding which of the transformations is the best for an application of LSA has greatly improve LSA's speed and precision. To summarize, Latent Semantic Analysis can aid a computer to be trained to identify the patterns of Euclidean signature curves and has potential to be further developed.

References

1. Boutin, M., Numerically invariant signature curves, *Int. J. Computer Vision* **40** (2000), 235–248.
2. Calabi, E., Olver, P.J., Shakiban, C., Tannenbaum, A., and Haker, S., Differential and numerically invariant signature curves applied to object recognition, *Int. J. Computer Vision* **26** (1998), 107–135.
3. Deerwester, S., Dumais, S., and Harshman , R., Indexing by Latent Semantic Analysis , *Journal of the Society for Information Science, 41(6)* **391-407** (1990),. .
4. Kichenassamy, S., Kumar, A., Olver, P.J., Tannenbaum, A., and Yezzi, A., Conformal curvature flows: from phase transitions to active vision, *Arch. Rat. Mech. Anal.* **134** (1996), 275–301.
5. Lloyd, P., and Shakiban, C., Signature Curves Statistics of DNA Supercoils, *in: vol. 5, I.M. Mladenov and A.C. Hirschfeld, eds.*, Softex, Sofia, Bulgaria, 2004, pp. 67–80.
6. Lloyd, R., and Shakiban, C., An improved method in the field of latent semantic analysis, *Amer. J. Undergrad. Research* (2004),.
7. Nakov, P., Popova, A., and Mateev, P., *Weight Functions Impact on LSA Performance*, Sofia: Sofia University Press, 2001.
8. Olver, P.J., and Shakiban, C., *Applied Linear Algebra*, Prentice–Hall, Inc., Upper Saddle River, N.J., 2005.

Hamiltonian System and Algebro-Geometric Solution Associated with Dispersive Long Wave Equation

Engui Fan

Institute of Mathematics, Fudan University,
Shanghai 200433, P. R. China
faneg@fudan.edu.cn

Abstract. By using an iterative algebraic method, we derive from a spectral problem a hierarchy of nonlinear evolution equations associated with dispersive long wave equation. It is shown that the hierarchy is integrable in Liouville sense and possesses bi-Hamiltonian structure. Under a Bargmann constraint the spectral is nonlinearized to a completely integrable finite dimensional Hamiltonian system. By introducing the Abel-Jacobi coordinates, an algebro-geometric solution for the dispersive long wave equation is derived by resorting to the Riemann theta function.

1   Introduction

The development of new integrable system representing well-known physical phenomena has shown to be meaningful and valuable. For instance, the illustration of the bi-Hamiltonian structure of a system of partial differential equations is proven to be a direct and elegant method in establishing the complete integrability of the system [1-5]. If a set of partial differential equations can be formulated as a Hamiltonian system in two distinct but compatible ways, Magri had proven in [1] that this give rise to an infinite sequence of conserved Hamiltonians which are in involution with respect to one of those two symplectic structures. Recently, two constructive approaches that can handle both finite-dimensional and infinite-dimensional integrable Hamiltonian systems were successfuly developed. The first approach is based on the trace identity [6, 7], which is effective in constructing the infinite-dimensional Liouville integrable Hamiltonian systems. Starting from a properly defined isospectral problem, many integrable hierarchies and their Hamiltonian structures (e.g. AKNS, TC, TA, BPT, Yang) had been obtained by applying this method [6-10]. The second one used non-linearization technique [11, 12], which has also been proven to be powerful for obtaining new finite-dimensional integrable Hamiltonian systems from various soliton hierarchies. Under the Bargmann or Neumann constraints on the potentials and the eigenvalues, which play a central role in the process of nonlinearization, the related eigenvalue problem can be nonlinearized as a finite-dimensional completely integrable system. This covers the

H. Li, P. J. Olver and G. Sommer (Eds.): IWMM-GIAE 2004, LNCS 3519, pp. 163–178, 2005.

eigenvalue problems associated the well-known soliton hierarchies such as the KdV, AKNS, Jaulent-Miodek, Kaup-Newell, and etc.[10-17]. An advantage of this method is that its solution to the soliton equation associated with an eigenvalue problem can be reduced to solving a compatible system of nonlinear ordinary differential equations [9-15]. This approach has now been further developed to a general method for handling higher order constraints associated with infinitely many hierarchies of finite-dimensional integrable Hamiltonian systems [18-20].

The Algebro-geometric method is an analogue of inverse scattering transformation. It was first developed by Novikov, Dubrovin, Matveev et al [21-24]. The method can derive an important class of exact solutions, which is called algebro-geometric solution, to the soliton equation such as KdV equation, Sin-Gordon equation, and Schrodinger equation. In the degenerated case the method gives the soliton solution and elliptic function solution [24-28]. Recently, an alternate approach based on the nonlinearization technique of Lax pairs or the restricted flow technique has been proposed. By this method or other methods, algebro-geometric solutions for (1+1)- and (2+1)-dimensional soliton equations can now be obtained [29-32].

The dispersive long wave (DLW) equation

$$u_t = 2uu_x + 2v_x + u_{xx},$$
$$v_t = 2(uv)_x - v_{xx}. \tag{1.1}$$

was first derived by Whitham and Broer for simulating dispersive waves in shallow water [33, 34]. Its symmetries, conservation laws, similarity reductions, painlevé property and soliton solutions had been fully discussed [35-37]. In recent years, the following spectral problem associated with the DLW equation (1.1)

$$\psi_x = U\psi = \begin{pmatrix} -\frac{1}{2}(\lambda - u) & -v \\ 1 & \frac{1}{2}(\lambda - u) \end{pmatrix}\psi, \tag{1.2}$$

has been proposed [13,38]. Under two different constraints between the potentials and eigenfunctions, the nonlinearization of the spectral problem gives two kinds of finite-dimensional completely integrable systems [13].

In this paper, we would investigate some aspects related to the DLW equation (1.1) and its spectral problem (1.2). In the following Section 2, starting from a spectral problem, we first derive a hierarchy associated with the DLW equation by using an iterative algebraic method. It will be shown that the hierarchy is integrable in Liouville sense and its bi-Hamiltonian structure can be established from trace identity. In Section 3, from the nonlinearization of the spectral problem (1.2), we devise a finite dimensional completely integrable Hamiltonian system under a Bargmann constraint. The Abel-Jacobi coordinates are introduced in Section 4 from which the algebro-geometric solutions for the DLW equations will be derived by resorting to the Riemann theta function.

2   DLW Hierarchy and Its Bi-Hamiltonian Structure

Let G be a matrix Lie algebra over the complex field $\mathcal{C}$, and $\widetilde{G} = G \otimes \mathcal{C}(\lambda, \lambda^{-1})$ be its loop algebra, where $\mathcal{C}(\lambda, \lambda^{-1})$ is the set of Laurent polynomials in λ. The gradation of $\widetilde{G}$ is taken by

$$deg(\rho \otimes \lambda^n) = n, \quad \rho \in G$$

Consider in the following a general spectral problem:

$$\psi_x = U(u, \lambda)\psi, \tag{2.1}$$

where $u = (u_1, \cdots, u_p)^T \in S$, S is a Schwarz space, and λ is a spectral parameter. The function $U(u, \lambda)$ can be unified to the following representation:

$$U = R + u_1 e_1 + \cdots + u_p e_p,$$

where $R, e_1, \cdots, e_p \in \widetilde{G}$. We define the ranks for ∂, u, λ, and $\rho \in \widetilde{G}$ in such a way that if ab is well defined for any two entities a and b, then

$$\mathrm{rank}(ab) = \mathrm{rank}(a) + \mathrm{rank}(b).$$

We also define the rank of R in such a way that above element U has homogeneous rank, i.e.,

$$\mathrm{rank}(R) = \mathrm{rank}(u_1 e_1) = \cdots = \mathrm{rank}(u_p e_p).$$

Finally, we define

$$\mathrm{rank}(\rho) = \deg(\rho), \quad \rho \in \widetilde{G}; \quad \mathrm{rank}(\lambda) = \deg(\rho\lambda) - \deg(\rho),$$
$$\mathrm{rank}(u_i) = \alpha - \epsilon_i; \quad \mathrm{rank}(\partial) = \alpha, \quad \mathrm{rank}(\beta) = 0, (\beta = \text{constant}),$$

where $\deg(R) = \alpha$ and $\deg(e_i) = \epsilon_i$.

Denote

$$g_+ = \sum_{n \geq \pi} g_n, \quad g_- = \sum_{n < \pi} g_n,$$

and call g_+ the positive part of g, where $\pi \in Z$ is a properly chosen integer.

The DLW hierarchy corresponding to equation (1.1) derived from the spectral problem (1.2) is now given to be a basis of iterated algebra over $\mathbb{C}$ as follows:

$$h(n) = \begin{pmatrix} \frac{1}{2}\lambda^n & 0 \\ 0 & -\frac{1}{2}\lambda^n \end{pmatrix}, \quad e(n) = \begin{pmatrix} 0 & \lambda^n \\ 0 & 0 \end{pmatrix}, \quad f(n) = \begin{pmatrix} 0 & 0 \\ \lambda^n & 0 \end{pmatrix}, \tag{2.2}$$

such that the following equalities

$$[h(m), e(n)] = e(m + n), \quad [f(m), h(n)] = f(m + n),$$
$$[e(m), f(n)] = 2h(m + n),$$

hold. The gradation for the basis is defined by

$$\deg h(n) = \deg e(n) = \deg f(n) = n.$$

From equations (2.2), the U in (1.2) can be expressed as

$$U = -h(1) + f(0) + uh(0) - ve(0),$$

and we have

$$\deg h(1) = 1, \quad \deg e(0) = \deg f(0) = 0,$$
$$\mathrm{rank}(\lambda) = \mathrm{rank}(\partial) = \mathrm{rank}(u) = \mathrm{rank}(v) = 1.$$

The adjoint equation

$$V_x = [U, V] = UV - VU$$

with

$$V = ah(0) + be(0) - cf(0)$$

can be solved and we obtain

$$a_x = 2vc - 2b, \quad b_x = -b\lambda + ub + va,$$
$$c_x = c\lambda + uc + a. \tag{2.3}$$

Substituting the Laurent expansion $a = \sum a_m \lambda^{-m}, b = \sum b_m \lambda^{-m}, c = \sum c_m \lambda^{-m}$ into (2.3), we obtain the following recursive formulas:

$$a_{mx} = 2vc_m - 2b_m,$$
$$b_{mx} = -b_{m+1} + ub_m + va_m, \tag{2.4}$$
$$c_{mx} = c_{m+1} + uc_m + a_m.$$

Taking $a_0 = 0, c_0 = \alpha, b_0 = \alpha v$, we further derive from (2.4) that

$$\begin{pmatrix} \frac{1}{2}a_{m+1} \\ c_{m+1} \end{pmatrix} = L \begin{pmatrix} \frac{1}{2}a_m \\ c_m \end{pmatrix},$$

where L is the matrix operator defined by

$$L = \begin{pmatrix} -\partial + \partial^{-1}u\partial & \partial^{-1}v\partial + v \\ 2 & \partial + u \end{pmatrix}.$$

Set

$$(\lambda^n V)_+ = \sum_{m=0}^{n} [a_m h(n-m) + b_m e(n-m) - c_m f(n-m)].$$

Then, $V_x = [U, V]$ implies that

$$-(\lambda^n V)_{+x} + [U, (\lambda^n V)_+] = -(\lambda^n V)_{-x} + [U, (\lambda^n V)_-]. \tag{2.5}$$

Notice that all the terms in the left-hand side of (2.5) have degrees ≥ 0, whereas all the terms in the right-hand side of (2.5) have degrees ≤ 0. Therefore, we have

$$-(\lambda^n V)_{+x} + [U, (\lambda^n V)_+] \in \mathbb{C}h(0) + \mathbb{C}e(0) + \mathbb{C}f(0). \tag{2.6}$$

In other words, we only need to calculate those terms that contain $h(0), e(0), f(0)$. Direct calculation shows that

$$-(\lambda^n V)_{+x} + [U, (\lambda^n V)_+] = b_{n+1}e(0) + c_{n+1}f(0).$$

To cancel the term $c_{n+1}f(0)$, we introduce $\Delta_n = c_{n+1}h(0)$. It is then easy to prove that, for $V^{(n)} = (\lambda^n V)_+ + \Delta_n$, we have

$$-(V^{(n)})_{+x} + [U, (V^{(n)})_+] = -c_{n+1x}h(0) - \frac{1}{2}a_{n+1x}e(0).$$

Then zero curvature equation $U_t - (V^{(n)})_{+x} + [U, (V^{(n)})_+]$ gives the following a DLW hierarchy:

$$u_t = c_{n+1x}, \quad v_t = \frac{1}{2}a_{n+1x}, \quad n = 1, 2, \cdots \tag{2.7}$$

We note here that the first two representative systems in the hierarchy (2.7) are

$$u_t = \alpha u_x, \quad v_t = \alpha v_x;$$

$$u_t = \alpha(u_{xx} + 2v_x + 2uu_x), \quad v_t = \alpha(-v_{xx} + 2u_x v + 2uv_x).$$

The second system is exactly the DLW equation (1.1) in the case when $\alpha = 1$.

The following theorem serves to develop the Hamiltonian structure of the hierarchy (2.7).

Theorem 2.1. The hierarchy (2.7) possesses the bi-Hamiltonian structure

$$w_t = J\frac{\delta H_n}{\delta w} = K\frac{\delta H_{n-1}}{\delta w}, \tag{2.8}$$

where the vector function w, the Hamiltonian operators J, K and the Hamiltonian function are given by

$$w = \begin{pmatrix} u \\ v \end{pmatrix}, \quad K = JL = \begin{pmatrix} 2\partial & \partial^2 + \partial u \\ -\partial^2 + u\partial & v\partial + \partial v \end{pmatrix},$$

$$J = \begin{pmatrix} 0 & \partial \\ \partial & 0 \end{pmatrix}, \quad H_n = \frac{1}{2}a_{n+2}, \quad n = 0, 1, \cdots \tag{2.9}$$

Proof. We introduce

$$G_{n+1} = \begin{pmatrix} \frac{1}{2}a_{n+1} \\ c_{n+1} \end{pmatrix}, \tag{2.10}$$

then hierarchy (2.7) can then be rewritten in the form of

$$w_t = JG_{n+1} = KG_n = JL^{n+1}G_0 = J(J^{-1}K)^{n+1}G_0, \quad n = 1, 2, \cdots \tag{2.11}$$

where $G_0 = (0, 1)^T$.

We take the Killing-Cartan form $< A, B >$ to be $tr(AB)$. From direct calculation we then obtain

$$\langle V, \frac{\partial U}{\partial \lambda} \rangle = -\frac{1}{2}a, \quad \langle V, \frac{\partial U}{\partial u} \rangle = \frac{1}{2}a, \quad \langle V, \frac{\partial U}{\partial v} \rangle = c.$$

By using the result of trace identity [6], we have

$$\frac{\delta}{\delta w}(-\frac{1}{2}a) = \lambda^{-\gamma} \frac{\partial}{\partial \lambda}[\lambda^{\gamma}(\frac{1}{2}a, c)^T].$$

Substituting

$$a = \sum_{n=0}^{\infty} a_n \lambda^{-n}, \quad c = \sum_{n=0}^{\infty} c_n \lambda^{-n}$$

into the above equation yields

$$\frac{\delta}{\delta w}(-\frac{1}{2}a_{n+1}) = (\gamma - n)G_n, \tag{2.12}$$

which holds for all $n = 0, 1, 2, \cdots$. The condition $n = 0$ in (2.12) gives $\gamma = 1$. Therefore, we obtain

$$G_{n+1} = \frac{\delta H_n}{\delta w}, \tag{2.13}$$

where H_n is given by (2.9).

Combining (2.11) with (2.13) we then obtain the desired Hamiltonian formulation (2.8) for the DLW hierarchy (2.7).

Since $J^* = -J, JL = L^*J$, we conclude that

Theorem 2.2. The Hamiltonian functions $\{H_n\}_{n=0}^{\infty}$ defined by (2.9) constitute common conserved densities for the whole hierarchy (2.8). It other words, it is an integrable Hamiltonian system in Liouville sense.

3   A Finite Dimensional Hamiltonian System

To construct the algebro-geometric solution of DLW hierarchy, in this section we investigate the finite dimensional Hamiltonian systems associated with the spectral problem (1.2) through a nonlinearization approach [12,13]. Let $\lambda_j, j = 1, \ldots, N$ be N different eigenvalues of equation (1.2), and (p_j, q_j) be the associated eigenfunctions, i.e., we consider the following N eigenvalues

$$\begin{pmatrix} p_j \\ q_j \end{pmatrix}_x = \begin{pmatrix} -\frac{1}{2}(\lambda_j - u) & -v \\ 1 & \frac{1}{2}(\lambda_j - u) \end{pmatrix} \begin{pmatrix} p_j \\ q_j \end{pmatrix}. \tag{3.1}$$

Denote $p = (p_1, p_2, \cdots, p_N)^T, q = (q_1, q_2, \cdots, q_N)^T$, and $\wedge = \text{diag}(\lambda_1, \lambda_2, \cdots, \lambda_N)$. From the Bargmann constraint

$$G_1 = \begin{pmatrix} v \\ u \end{pmatrix} = \sum_{j=1}^{N} \nabla \lambda_j = \begin{pmatrix} < p, q > \\ - < q, q > \end{pmatrix}, \tag{3.2}$$

we then obtain

$$w = \begin{pmatrix} u \\ v \end{pmatrix} = \begin{pmatrix} - <q,q> \\ <p,q> \end{pmatrix} \equiv h(p,q). \tag{3.3}$$

From the constraint (3.3), the equation (3.1) can further be nonlinearized into the following finite-dimensional Hamiltonian system:

$$p_x = -\frac{1}{2} \wedge p - \frac{1}{2} <q,q> p - <p,q> q = -\frac{\partial H}{\partial q},$$

$$q_x = p + \frac{1}{2} \wedge q + \frac{1}{2} <q,q> q = \frac{\partial H}{\partial p}, \tag{3.4}$$

whose Hamiltonian function H is given by

$$H = \frac{1}{2} <p,p> + \frac{1}{2} <\wedge p,q> + \frac{1}{2} <q,q><p,q>.$$

In the following, we proceed to show that the Hamilton system (3.4) is completely integrable in Liouville sense. The Poisson bracket of two functions in symplectic space $(\mathbb{R}^{2N}, dp \wedge dq)$ is defined as

$$(F,G) = \sum_{j=1}^{N} \left(\frac{\partial F}{\partial q_j} \frac{\partial G}{\partial p_j} - \frac{\partial F}{\partial p_j} \frac{\partial G}{\partial q_j} \right) = \left\langle \frac{\partial F}{\partial q}, \frac{\partial G}{\partial p} \right\rangle - \left\langle \frac{\partial F}{\partial p}, \frac{\partial G}{\partial q} \right\rangle,$$

which is skew-symmetric, bilinear, and satisfies the Jacobi identity. In particular, F and G are called involution if $(F,G) = 0$.

Consider a bilinear function $Q_\lambda(p,q) = <(\lambda I - \wedge)^{-1} p, q>$ such that

$$G_\lambda = G_0 + \sum_{j=1}^{N} \frac{\nabla \lambda_j}{\lambda - \lambda_j} = \begin{pmatrix} Q_\lambda(p,q) \\ 1 - Q_\lambda(q,q) \end{pmatrix}. \tag{3.5}$$

From (3.4) and (3.5), we have

$$V_\lambda = V(G_\lambda) = \begin{pmatrix} -\frac{1}{2}\lambda - Q_\lambda(p,q) & - <p,q> +Q_\lambda(p,p) \\ 1 - Q_\lambda(q,q) & \frac{1}{2}\lambda + Q_\lambda(p,q) \end{pmatrix}.$$

Let $F_\lambda = \det V_\lambda$. Then, we have

$$F_\lambda = -\frac{1}{4}\lambda^2 - Q_\lambda(\wedge p,q) + Q_\lambda(p,p) + <p,p> Q_\lambda(q,q)$$

$$+ \begin{vmatrix} Q_\lambda(p,p) & Q_\lambda(p,q) \\ Q_\lambda(p,q) & Q_\lambda(q,q,) \end{vmatrix} = -\frac{1}{4}\lambda^2 + \sum_{m=0}^{\infty} F_m \lambda^{-m-1}, \tag{3.6}$$

where

$$F_0 = 2H,$$

$$F_m = - <\wedge^{m+1} p,p> + <p,q><\wedge^m q,q) + \sum_{m=0}^{m-1} \begin{vmatrix} <\wedge^j p,p> & <\wedge^{m-1-j}p,q> \\ <\wedge^j p,q> & <\wedge^{m-1-j}q,q> \end{vmatrix}.$$

By considering that F_λ is a Hamiltonian in the symplectic space $(\mathbb{R}^{2N}, dp \wedge dq)$, the canonical equations of the F_λ-flow can be computed as:

$$\frac{d}{d\tau_\lambda}\begin{pmatrix} p_k \\ q_k \end{pmatrix} = I\nabla F_\lambda = \begin{pmatrix} -\partial F_\lambda/\partial q_k \\ \partial F_\lambda/\partial p_k \end{pmatrix} = W(\lambda, \lambda_k)\begin{pmatrix} p_k \\ q_k \end{pmatrix},$$

where

$$W(\lambda, \lambda_k) = \frac{2}{\lambda - \lambda_k}V_\lambda + (1 - Q_\lambda(p, q))\begin{pmatrix} 1 & 0 \\ 0 & -1 \end{pmatrix}.$$

Lemma 3.1 The Lax matrix V_μ satisfies the Lax equation (3.7) alone the t_λ-flow:

$$\frac{dV_\mu}{d\tau_\lambda} = [W(\lambda, \mu), V_\mu] \tag{3.7}$$

and

$$(F_\mu, F_\lambda) = 0, \quad \forall \lambda, \mu \in \mathbb{C}, \tag{3.8}$$

$$(F_j, F_k) = 0, \quad \forall j, k = 0, 1, \cdots. \tag{3.9}$$

Proof. Notice that

$$Q_\lambda(\wedge\xi, \eta) = \lambda Q_\lambda(\wedge\xi, \eta) - <\xi, \eta>,$$

$$< (\mu I - \wedge)^{-1}(\lambda I - \wedge)^{-1}\xi, \eta > = \frac{1}{\mu - \lambda}(Q_\lambda(\xi, \eta) - Q_\mu(\xi, \eta)),$$

a direct calculation shows that (3.7) holds. From the Lax equation (3.7) we have

$$(F_\mu, F_\lambda) = \sum_{j=1}^N \left(\frac{\partial F_\mu}{\partial q_j}\frac{\partial F_\lambda}{\partial p_j} - \frac{\partial F_\mu}{\partial p_j}\frac{\partial F_\lambda}{\partial q_j} \right) = \sum_{j=1}^N \left(\frac{\partial F_\mu}{\partial q_j}\frac{dq_j}{dt_\lambda} + \frac{\partial F_\mu}{\partial p_j}\frac{dp_j}{dt_\lambda} \right),$$

$$= \frac{dF_\mu}{d\tau_\lambda} = \frac{d}{d\tau_\lambda}(\det V_\mu) = \frac{d}{d\tau_\lambda}(-\frac{1}{2}\mathrm{tr}V_\mu^2) = -\mathrm{tr}(V_\mu\frac{dV_\mu}{d\tau_\lambda}),$$

$$= -\mathrm{tr}V_\mu\mathrm{tr}[W(\lambda, \mu), V_\mu] = 0.$$

Substituting (3.6) into (3.8) and using the coefficients of λ and μ, we then obtain (3.9).

We conclude the above results in the following theorem.

Theorem 3.1. The finite dimensional Hamiltonian system defined by (3.4) is completely integrable in Liouville sense in the symplectic space $(\mathbb{R}^{2N}, dp \wedge dq)$.

Theorem 3.2. Let (p, q) be a solution of the Hamiltonian system (3.4). Then, u and v defined by (3.3) satisfy the following stationary DLW equation

$$X_N + c_1 X_{N-2} + \cdots + c_{N-1}X_0 = 0 \tag{3.10}$$

with suitably chosen constants $c_1, \cdots, c_{N-1}$.

Introduce the generating function $\{g_k\}$

$$g_\lambda = G_0 + \sum_{k=0}^{\infty} G_k \lambda^{-k}, \tag{3.12}$$

which satisfies

$$(K - \lambda J)g_\lambda = 0,$$

we have

$$\det V(g_\lambda) = -\frac{1}{4}\lambda^2. \tag{3.13}$$

Define a new set of integrals $\{H_k\}$ respectively by

$$H_0 = \frac{1}{2}F_0, \quad H_1 = -\frac{1}{2}F_1, \quad H_2 = -\frac{1}{2}F_2,$$

$$H_m = -\frac{1}{2}F_m + 2\sum_{j=0}^{m-3} H_j H_{m-j-3}, \quad m \geq 3.$$

Equation (3.6) can then be transformed into the following equivalent form

$$-\frac{4}{\lambda^2}F_\lambda = (1 - 4H_\lambda)^2, \tag{3.14}$$

where

$$H_\lambda = \sum_{m=1}^{\infty} H_{m-1}\lambda^{-m-1}.$$

The involutivity of $\{H_k\}$ is based on the equality

$$\{H_\mu, H_\lambda\} = \frac{1}{16\sqrt{F_\lambda F_\mu}}\{F_\mu, F_\lambda\} = 0.$$

By using (3.5) and (3.11), we have

$$G_\lambda = G_0 + \sum_{k=0}^{\infty} \lambda^{-k-1}\sum_{j=1}^{N} \lambda_j^k \nabla \lambda_j,$$

$$= G_0 + \sum_{k=0}^{\infty} c^{-k-1}(G_k + c_1 G_{k-2} + \cdots + c_{k-1}G_0 + c_k G_{-1},$$

$$= c_\lambda g_\lambda + \sum_{k=0}^{\infty} \lambda^{-k-1}c_k G_{-1},$$

where

$$c_\lambda = 1 + \sum_{k=0}^{\infty} c_{k+2}\lambda^{-k-2}. \tag{3.15}$$

Using (3.6) and (3.14) again, we have

$$V_\lambda = V(c_\lambda g_\lambda + \sum_{k=0}^{\infty} \lambda^{-k-1} c_k G_{-1}) = V(c_\lambda g_\lambda).$$

Therefore, we have

$$F_\lambda = \det V_\lambda = -\frac{1}{4} c_\lambda^2 g_\lambda^2. \tag{3.16}$$

From (3.14) and (3.16), we obtain

$$c_\lambda = 1 - 4H_\lambda.$$

Denote the variables of H_λ–flow and H_k–flow by t_λ and t_k respectively. Applying the Leibniz rule of the Poisson bracket, we obtain

$$\frac{1}{2\lambda^2}\{\psi, F_\lambda\} = (1 - 4H_\lambda)\{\psi, H_\lambda\},$$

for any smooth function ψ. Thus,

$$\frac{d}{dt_\lambda} = \frac{1}{2\lambda^2(1 - 4H_\lambda)}\frac{d}{d\tau_\lambda} = \frac{1}{2\lambda^2 c_\lambda}\frac{d}{d\tau_\lambda}. \tag{3.17}$$

For $w = (u, v)^T = h(p, q)$, we have

$$\frac{dw}{d\tau_\lambda} = \begin{pmatrix} -2 < q, \dfrac{dq}{d\tau_\lambda} > \\ < p, \dfrac{dq}{d\tau_\lambda} > + < \dfrac{dp}{d\tau_\lambda}, q > \end{pmatrix} = 2JG_\lambda,$$

$$\frac{dw}{dt_\lambda} = \frac{1}{2\lambda^2 c_\lambda}\frac{dw}{d\tau_\lambda} = \frac{1}{\lambda^2 c_\lambda}JG_\lambda = \frac{1}{\lambda^2}Jg_\lambda,$$

$$= \sum_{k=0}^{\infty} JG_K \lambda^{-k-3} = \sum_{k=0}^{\infty} JG_K \lambda^{-k-3}.$$

Finally, we state the above result in the following theorem.

Theorem 3.3. Let $(p(x, t_k), q(x, t_k))$ be a compatible solution of the h_0- and H_k- flow. Then, $w(x, t_K) = h(p, q)$ solves the kth DLW equation

$$w_t = X_k(w).$$

4   Algebro-Geometric Solutions

Based on results presented in Sections 3 and 4, we give the construction of the algebro-geometric solutions for the hierarchy (2.7) in the following lemmas and theorem.

Lemma 4.1. The solutions u and v for the DLW hierarchy (2.7) can be expressed in the form

$$u = \sum_{k=1}^{N}(\lambda_k - \nu_k), \quad \partial_x \ln v = \sum_{k=1}^{N}(\mu_k - \nu_k), \tag{4.1}$$

where μ_k, ν_k are called the elliptic coordinates of the finite-dimensional Hamiltonian system (3.3).

Proof. We express F_λ in the form

$$F_\lambda = -V_\lambda^{11^2} - V_\lambda^{12}V_\lambda^{21} = -\frac{1}{4}\lambda^2 + \sum_{k=1}^{N}\frac{E_k}{\lambda - \lambda_k},$$

where

$$E_k = -\lambda_k p_k q_k + p_k^2 + < p, q > q_k^2 + \sum_{j=1,j\neq k}^{N}\frac{(p_j q_k - p_k q_j)^2}{\lambda_j - \lambda_k}.$$

Let

$$F_\lambda = -\frac{b(\lambda)}{4a(\lambda)} = -\frac{R(\lambda)}{4a^2(\lambda)}, \tag{4.2}$$

$$V_\lambda^{12} = - < p, q > +Q_\lambda(p, p) = - < p, q > \frac{m(\lambda)}{a(\lambda)},$$

$$V_\lambda^{21} = 1 - Q_\lambda(q, q) = \frac{n(\lambda)}{a(\lambda)}, \tag{4.3}$$

where

$$a(\lambda) = \prod_{k=1}^{N}(\lambda - \lambda_k), \quad b(\lambda) = \prod_{k=1}^{N+2}(\lambda - \lambda_{N+k}),$$

$$m(\lambda) = \prod_{k=1}^{N}(\lambda - \mu_k), \quad n(\lambda) = \prod_{k=1}^{N}(\lambda - \nu_k),$$

$$R(\lambda) = a(\lambda)b(\lambda) = \prod_{k=1}^{2N+2}(\lambda - \lambda_k).$$

Comparing the coefficients of λ^{N-1} in (4.2) and (4.3) gives

$$< q, q >= -\sum_{k=1}^{N}(\lambda_k - \nu_k), \quad \frac{< p, p >}{< p, q >} = \sum_{k=1}^{N}(\mu_k - \lambda_k). \tag{4.4}$$

By using (3.2) and (4.4), we then obtain (4.1).

From the Lax equation (3.7), we have

$$\frac{dV_\mu^{12}}{dt_\lambda} = 2(W^{11}V^{12} - W^{12}V^{11}),$$

$$\frac{dV_\mu^{21}}{dt_\lambda} = 2(W^{21}V^{11} - W^{11}V^{21}).$$

(4.5)

Taking $\lambda = \mu_k, \nu_k$ in (4.2) and (4.3), we get

$$V_{\mu_k}^{11} = \frac{\sqrt{R(\mu_k)}}{2a(\mu_k)}, \quad V_{\mu_k}^{11} = \frac{\sqrt{R(\nu_k)}}{2a(\nu_k)}.$$

Hence, we have

$$\frac{1}{2\sqrt{R(\mu_k)}}\frac{d\mu_k}{d\tau_\lambda} = \frac{m(\lambda)}{a(\lambda)(\lambda - \mu_k)m'(\mu_k)},$$

$$\frac{1}{2\sqrt{R(\nu_k)}}\frac{d\nu_k}{d\tau_\lambda} = -\frac{n(\lambda)}{a(\lambda)(\lambda - \nu_k)n'(\nu_k)}.$$

By using the polynomails interpolation formula, we have

$$\sum_{k=1}^{N} \frac{\mu_k^{N-j}}{2\sqrt{R(\mu_k)}}\frac{d\mu_k}{d\tau_\lambda} = \frac{\lambda^{N-j}}{a(\lambda)}, \quad \sum_{k=1}^{N} \frac{\mu_k^{N-j}}{2\sqrt{R(\nu_k)}}\frac{d\nu_k}{d\tau_\lambda} = -\frac{\lambda^{N-j}}{a(\lambda)}.$$

(4.6)

Define the hyperelliptic curve by

$$\Gamma : \xi^2 - 4R(\lambda) = 0,$$

(4.7)

with genus $g = N$ and the usual holomorphic differentials by

$$\tilde{\omega}_j = \frac{\lambda^{N-j}d\lambda}{2\sqrt{R(\lambda)}}, \quad j = 1, 2, \cdots, N.$$

(4.8)

Denote $P(\mu_k) = (\lambda, \xi = 2\sqrt{R(\lambda)}) \in \Gamma$. Let $P_0 \in \Gamma$ be fixed. Define the quasi-Abel-Jacobi coordinates by

$$\tilde{\phi}_j = \sum_{k=1}^{N} \int_{P_0}^{P(\mu_k)} \tilde{\omega}_j, \quad \tilde{\psi}_j = \sum_{k=1}^{N} \int_{P_0}^{P(\nu_k)} \tilde{\omega}_j, \quad j = 1, 2, \cdots, N.$$

(4.9)

Then, (4.6) can be represented in the form

$$\frac{d\tilde{\phi}_j}{d\tau_\lambda} = \frac{\lambda^{N-j}}{a(\lambda)}, \quad \frac{d\tilde{\psi}_j}{d\tau_\lambda} = -\frac{\lambda^{N-j}}{a(\lambda)}.$$

Let $a_1, b_1, \cdots, a_N, b_N$ be the canonical basis of cycles on the Γ and

$$C = (A_{jk})_{N \times N}^{-1}, \quad A_{jk} = \int_{a_k} \tilde{\omega}_j.$$

Define the normalized holomorphic differential by

$$\omega_s = \sum_{j=1}^{N} C_{sj}\tilde{\omega}_j, \quad \omega = (\omega_1, \cdots, \omega_N)^T = C\tilde{\omega}.$$

Then, we have

$$\int_{a_k} \omega_s = \delta_{sk}, \quad \int_{b_k} \omega_s = B_{sk},$$

where the matrix $B = (B_{sk})$ is symmetric with positive-definite imaginary part. This is used to define the Riemann theta function of Γ as

$$\theta(\zeta) = \sum_{z \in \mathbb{Z}^N} exp(\pi i < Bz, z > + 2\pi i < \zeta, z >), \quad \zeta \in C^N.$$

The Abel map $A(P)$ and the Abel-Jacobi coordinate are defined as

$$A(P) = \int_{P_0}^{P} \omega, \quad A(\sum n_k P_k) = \sum n_k A(P_k),$$

$$\phi = A(\sum_{k=1}^{N} P(\mu_k)) = \sum_{k=1}^{N} \int_{P_0}^{P(\mu_k)} \omega = C\tilde{\phi},$$

$$\psi = A(\sum_{k=1}^{N} P(\nu_k)) = \sum_{k=1}^{N} \int_{P_0}^{P(\nu_k)} \omega = C\tilde{\psi}.$$

Let $S_k = \lambda_1^k + \cdots + \lambda_{2N+2}^k$, and $\tilde{R}(\lambda^{-1}) = \prod_{j=1}^{2N+2}(1 - \lambda_j\lambda^{-1})$. The coefficients in

$$\frac{1}{\sqrt{\tilde{R}(\lambda^{-1})}} = \sum_{k=1}^{\infty} \Lambda_k \lambda^{-k}$$

are then given by

$$\Lambda_0 = 1, \Lambda_1 = \frac{1}{2}S_1, \Lambda_k = \frac{1}{2k}(S_k + \sum_{i+j=k, i,j \geq 1} S_i \Lambda_j).$$

From (4.2), we obtain

$$\sqrt{R(\lambda)} = \lambda a(\lambda)c_\lambda,$$

and the lemma is proved.

Lemma 4.2. Let C_k be the kth column vector of the matrix C. Under the Abel-Jacobi coordinate system, we have straighten out the flows to

$$\frac{d\phi}{dt_\lambda} = \sum_{k=1}^{\infty} \Omega_{k-1}\lambda^{-k-1}, \quad \frac{d\psi}{dt_\lambda} = -\sum_{k=1}^{\infty} \Omega_{k-1}\lambda^{-k-1}, \tag{4.10}$$

$$\frac{d\phi}{dt_k} = \Omega_{k-1}, \quad \frac{d\psi}{dt_k} = -\Omega_{k-1}, \tag{4.11}$$

where

$$\Omega_0 = \frac{1}{2}\Lambda_0 C_1, \Omega_1 = \frac{1}{2}(\Lambda_1 C_1 + \Lambda_0 C_2),$$

$$\Omega_k = \frac{1}{2}(\Lambda_k C_1 + \cdots + \Lambda_0 C_{k+1}), \quad k \le N - 1,$$

$$\Omega_k = \frac{1}{2}(\Lambda_k C_1 + \cdots + \Lambda_{k-N+1} C_N), \quad k \ge N - 1.$$

Proof. Notice that

$$\frac{d\phi}{dt_\lambda} = \frac{1}{2\lambda^2 c_\lambda} \frac{d\phi}{d\tau_\lambda} = \frac{1}{2\lambda^2 c_\lambda} C \frac{d\tilde{\phi}}{d\tau_\lambda},$$

$$= \frac{1}{2}\sum_{k=0}^{\infty} \Lambda_k \lambda^{-k-2} \sum_{j=1}^{N} C_j \lambda^{-j} = \sum_{k=0}^{\infty} \Omega_k \lambda^{-k-3}.$$

Similarly, we can obtain the second formula of (4.10) and compare the coefficients of λ^{-k-1} in (4.10) to obtain (4.11).

From (4.11), we get

$$\phi = \phi_0 + \sum_{k=1}^{\infty} \Omega_{k-1} t_k, \quad \psi = \psi_0 - \sum_{k=1}^{\infty} \Omega_{k-1} t_k.$$

Hence, the evolution picture of the $1 + 1$ flow is given as

$$\phi = \phi_0 + \Omega_0 x + \Omega_1 t, \quad \psi = \psi_0 - \Omega_0 x - \Omega_1 t. \tag{4.12}$$

Since $\deg R = 2N + 2$ on Γ, there are two infinite points ∞_1 and ∞_2 which are not branch points of Γ. From Riemann theorem, there exists two constants $M_1, M_2 \in C^N$ such that $\theta(A(P) - \phi - M_1), \theta(A(P) - \psi - M_2)$ have exactly n zeroes at $\mu_1, \cdots, \mu_N$ and $\nu_1, \cdots, \nu_N$ respectively. That is,

$$\sum_{j=1}^{N} \mu_j = I(\Gamma) - \sum_{j=1}^{2} \mathrm{Res}_{\lambda=\infty_j} \lambda d\ln\theta(A(P) - \phi - M_1), \tag{4.13}$$

$$\sum_{j=1}^{N} \nu_j = I(\Gamma) - \sum_{j=1}^{2} \mathrm{Res}_{\lambda=\infty_j} \lambda d\ln\theta(A(P) - \phi - M_1), \tag{4.14}$$

with the constant

$$I(\Gamma) = \sum_{j=1}^{N} \int_{a_j} \lambda\omega_j.$$

For the same λ, there are two points on the different sheets of the Riemann surface Γ:

$$P_+(\lambda) = (\lambda, \sqrt{R(\lambda)}), \quad P_+(\lambda) = (\lambda, -\sqrt{R(\lambda)}).$$

Under the local coordinate $z = \lambda^{-1}$ at infinity, the hyperelliptic curve $\Gamma : \xi^2 - 4R(\lambda) = 0$ in the neighborhood of infinity is expressed as $\tilde{\xi}^2 - 4\tilde{R}(z) = 0$ with $\tilde{\xi} = z^{N+1}\xi$ and $(z, 2(-1)^{s-1}\sqrt{R(z)})|_{z=0} = (0, 2(-1)^{s-1})$, $s = 1, 2$. Thus, we have

$$\omega = C\tilde{\omega} = (C_1\lambda^{-1} + \cdots + C_N\lambda^{-N})\frac{\lambda^{N-1}d\lambda}{2\sqrt{R(\lambda)}},$$

$$= (C_1\lambda^{-1} + \cdots + C_N\lambda^{-N})\frac{d\lambda}{2\lambda^2\sqrt{\tilde{R}(\lambda^{-1})}}$$

$$= \frac{1}{2}(-1)^{s-1}\sum_{j=0}^{\infty}\Omega_k z^k, dz$$

and

$$A(P(z^{-1})) = \int_{P_0}^{P}\omega = -\int_{\infty_s}^{P_0}\omega + \int_{\infty_s}^{P}\omega$$

$$= -\pi_s + \frac{1}{2}(-)^{s-1}\sum_{k=0}^{\infty}\frac{1}{k+1}\Omega_k z^{k+1}$$

Since the theta function is an even function, we have

$$\theta(A(P(z^{-1})) - \phi - M_1) = \theta(\phi + M_1 + \eta_s) + \frac{1}{2}z(-1)^{s-1}\frac{\partial}{\partial x}\theta(\phi + M_1 + \eta_s) + o(z^2).$$

$$(4.15)$$

From (4.13) and (4.15), we obtain

$$\sum_{j=1}^{N}\mu_j = I(\Gamma) - \frac{1}{2}(-1)^{s-1}\frac{\partial}{\partial x}\theta(\phi + M_1 + \eta_s),$$

$$= I(\Gamma) + \frac{1}{2}\partial_x \ln\frac{\theta(\phi + M_1 + \eta_2)}{\theta(\phi + M_1 + \eta_1)}.$$

$$(4.16)$$

Simiarly, we have

$$\sum_{j=1}^{N}\mu_j = I(\Gamma) + \frac{1}{2}\partial_x \ln\frac{\theta(\psi + M_2 + \pi_1)}{\theta(\psi + M_2 + \pi_2)}. \qquad (4.17)$$

Substituting (4.16) and (4.17) into (4.1), we then obtain the following algebro-geometric solution for the DLW equation

$$u = \sum_{k=1}^{N}\lambda_k - I(\Gamma) - \frac{1}{2}\partial_x \ln\frac{\theta(\Omega_0 x + \Omega_1 t + \alpha_1)}{\theta(\Omega_0 x + \Omega_1 t + \alpha_2)},$$

$$v^2 = \frac{\theta(\Omega_0 x + \Omega_1 t + \alpha_2)\theta(\Omega_0 x + \Omega_1 t + \beta_2)}{\theta(\Omega_0 x + \Omega_1 t + \alpha_1)\theta(\Omega_0 x + \Omega_1 t + \beta_1)}v^2(0, t),$$

where

$$\alpha_l = \phi_0 + M_1 + \pi_l, \quad \beta_l = -\psi_0 - M_2 - \pi_l, l = 1, 2.$$

Acknowledgments

The work described in this paper was partially supported by a grant from the National Science Foundation of China (10371023) and the Shanghai Shuguang Project of China(02SG02).

References

1. F. Magri, J. Math. Phys. **19**, 1156-1162 (1978)
2. F. Magri, *Lectures Notes in Phys.* Vol.120 (Springer, Berlin, 1980)
3. P. J. Olver, *Applications of Lie groups to differential equations* (Berlin, Springer, 1986)
4. I. M. Gel'fand and I.Y. Dorfman, Funct. Anal. Appl. **15** 173-187(1981)
5. Y. Nutku. J. Math. Phys. **28** (1987), 2579
6. G. Z. Tu, J. Math. Phys. **30**(1989), 330
7. G. Z. Tu, J. Phys. A: Math. Gen. **22**(1989), 2375
8. G. Z. Tu, J. Phys. A: Math. Gen. **23**(1990), 3903
9. E. G. Fan, J. Math. Phys. J. Math. Phys. **42** (2001), 4327
10. E. G. Fan, Acta Math. Appl. Sin. **18** (2002), 405
11. E. G. Fan, Phys. Lett. A **274** (2000), 135
12. C. W. Cao, Sci. in China A **33** (1990), 528
13. C. W. Cao and X. G. Geng, J. Phys. A **23**(1990), 4117
14. X. G. Geng, J. Math. Phys. **34**(1993), 805
15. X. G. Geng, Physica A **180**(1992), 241
16. Z. J. Qiao, J. Phys. A **26**(1993), 4407
17. Z. J. Qiao, J. Math. Phys.,**34**(1993), 3110
18. Y. B. Zeng, Phys. Lett. A, **160**(1991), 541
19. Y. B. Zeng, Physica D, **73**(1994), 171
20. Z. B. Zeng, J. Phys. A **30**(1997), 3719
21. S. P. Novikov, Funct. Anal. Appl. **8** (1974) 236.
22. B. A. Dubrovin, Funct. Anal. Appl. **9** (1975) 41.
23. A. Its and V. Matveev, Funct. Anal. Appl. **9** (1975) 69.
24. E. Belokolos, A. Bobenko, V. Enol'skij, A, Its and V. Matveev, *Algebro-Geometrical Approach to Nonlinear Integrable Equations*, Springer, Berlin, 1994.
25. P. L. Christiansen, J. C. Eilbeck, V. Z. Enolskii and N. A. Kostov, Proc. R. Soc. London A Math. **451** (1995) 685
26. M. S. Alber and Y. N. Fedorov, Inverse Probl. **17** (2001) 1017
27. A. V. Porubov and D. F. Paeker, Wave Motion **29**(1999) 97
28. F. Gesztesy and H. Holden, *Soliton Equations and Their Algebro-Geometric Solutions*, Cambridge University Press, 2003.
29. R. G. Zhou, J. Math. Phys. **38** (1997) 2535.
30. C. W. Cao, Y. T. Wu and X. G. Geng, J. Math. Phys. **40** 1999, 3948.
31. Z. J. Qiao, Reviews in Math. Phys. **13** (2001), 545.
32. C. W. Cao, X. G. Geng and H. Y. Wang, J. Math. Phys. **43** 2002, 621.
33. G. B. Whitham, Proc. R. Soc. A **299**(1983), 6
34. L. T. F. Broer, Appl. Sci. Res. **31**(1983), 377
35. B. A. Kupershmidt. Commun. Math. Prhys. **99**(1983), 51
36. M. L. Wang, Y. B. Zhou, Z. B. Li, Phys. Lett. A **216**(1983), 67
37. H. Y. Ruan, S. Y. Lou, Commun. Theor. Phys. **20**(1993), 73
38. D. Levi, A. Sym, S. Wojciechowsk, J. Phys. A **16**(1983),2423

The Painlevé Test of Nonlinear Partial Differential Equations and Its Implementation Using Maple

Gui-qiong Xu[1] and Zhi-bin Li[2]

[1] Department of Information Management, College of International Business, and Management, Shanghai University, Shanghai 201800, China
xuguiqiong@yahoo.com
xugq@staff.shu.edu.cn
[2] Department of Computer Science, East China Normal University, Shanghai 200062, P. R. China

Abstract. The so-called WTC-Kruskal algorithm is presented in order to study the Painlevé integrability of nonlinear partial differential equations, which combines the WTC algorithm and Kruskal's simplification algorithm. Based on the WTC, Kruskal and WTC-Kruskal algorithms, we give an implementation in *Maple* called PDEPtest. The applications of PDEPtest to several nonlinear partial differential equations are also presented and some new results are reported.

1 Introduction

Since more and more problems have to involve nonlinearity, much attention has been focused on the integrability of nonlinear models. There are close relations between the integrability and the Painlevé property of nonlinear partial differential equations (PDEs)[1]-[18]. Therefore, it is of great importance to check whether nonlinear PDEs possess the Painlevé property or not. In order to verify the Painlevé property of nonlinear PDEs (we call it the Painlevé test), one may use different methods, such as the Ablowitz-Ramani-Segur (ARS) method[1], the Weiss-Tabor-Carnevale (WTC) method[2], the Kruskal's simplification method[3], Conte's invariant method[4] and Pickering's approach[5].

The WTC method and Kruskal's simplification method are the most widely applied tools to verify the Painlevé property. The WTC method, apart from its usefulness in proving the Painlevé property, has rather interesting connections with rich integrable properties of nonlinear PDEs, such as Hirota's bilinear forms, symmetries and special solutions[6]-[9]. However, it is rather cumbersome to verify the Painlevé property especially for coupled system of equations or a single equation with large resonance. If one only wants to verify the Painlevé property, then it is sufficient to adopt the Kruskal's simplification for the WTC method. However, the Kruskal's simplification cannot lead to some useful information such as Bäcklund transformation and Lax pairs. Motivated by these facts, we

H. Li, P. J. Olver and G. Sommer (Eds.): IWMM-GIAE 2004, LNCS 3519, pp. 179–190, 2005.

combined them and presented the so-called *WTC-Kruskal algorithm*[10]. The WTC-Kruskal algorithm can not only simplify the Painlevé test, but also obtain some truncated expansions related to integrability at the same time.

It is very tedious to perform the Painlevé test by hand, thus the application of computer algebra can be very helpful in such calculations. Various researchers have developed computer programs for the Painlevé test of nonlinear ODEs[11]-[13]. However, there is little code for nonlinear PDEs. Hereman *et al.* presented two packages in *Macsyma* and *Mathematica* respectively, which are based on the WTC method and the Kruskal's simplification[14]-[16]. Xie *et al.* implemented the WTC method for single PDE, its key step lies in dealing with the compatibility of the resonance equations by the Wu-Ritt elimination method[17]. Recently, we have developed a package **wkptest** written in *Maple*, which is an implementation of the WTC-Kruskal algorithm[18]. **wkptest** has been used to test a large variety of nonlinear PDEs and proved to be very efficient. Some integrable properties such as self-consistent system of equations defining Lax pair can be derived by the WTC method, but they cannot be obtained by using **wkptest**. In addition, some users only want to quickly verify the Painlevé property of nonlinear PDEs. In this paper, as far as the faultiness of **wkptest**, we rewrite a new package PDEPtest(the Painlevé test for nonlinear PDEs) in *Maple*, which integrates the WTC, Kruskal and WTC-Kruskal algorithms. The effectiveness of PDEPtest is illustrated by applying it to a variety of nonlinear PDEs.

2   Algorithm of the Painlevé Test

Consider a system of nonlinear PDE, say in two independent variables x and t

$$H_s(u^{(i)}, u_x^{(i)}, u_t^{(i)}, u_{xt}^{(i)}, u_{xx}^{(i)}, \cdots) = 0, \quad i, s = 1, \cdots, m \tag{1}$$

where $u^{(i)} = u^{(i)}(x, t)$ $(i = 1, \cdots, m)$ are dependent variables, the subscripts denote partial derivatives, $H_s(s = 1, \cdots, m)$ are polynomials about $u^{(i)}$ and their derivatives, maybe after a preliminary change of variables.

Eqs.(1) is said to pass the Painlevé test if its solutions are "single-valued" about arbitrary non-characteristic, movable singularity manifolds. In other words, all solutions of Eqs.(1) can be expressed as Laurent series,

$$u^{(i)} = \sum_{j=0}^{\infty} u_j^{(i)} \phi^{(j+\alpha_i)}, \quad i = 1, \cdots, m \tag{2}$$

with sufficient number of arbitrary functions among $u_j^{(i)}$ in addition to ϕ, α_i should be negative integers. $u_j^{(i)}$ are functions of t only if the Kruskal's simplification is used. The WTC-Kruskal algorithm is made of four steps:

Step a. Leading order analysis

To determine leading order exponents α_i and leading order coefficients $u_0^{(i)}$, letting

$$u^{(i)} = u_0^{(i)} \phi^{\alpha_i}, \quad i = 1, \cdots, m$$

and inserting it into (1), then balancing the minimal power terms, one can obtain all possible $(\alpha_i, u_0^{(i)})$. It may happen that there is more than one solution for $(\alpha_i, u_0^{(i)})$. If the only possible $\alpha_i's$ are not integers, then the algorithm stops. Otherwise, one has to go on with the next step.

Step b. To generate truncated expansions

For each pair of $(\alpha_i, u_0^{(i)})$ from *Step a*, calculate the possible truncated expansions in the form

$$u^{(i)} = u_0^{(i)} \phi^{\alpha_i} + u_1^{(i)} \phi^{\alpha_i+1} + \cdots u_{-\alpha_i}^{(i)} \phi^0, \quad i = 1, \cdots, m.$$

If $u_k^{(i)} (k = 0, \cdots, -\alpha_i - 1)$ cannot be determined, then the series (2) cannot be truncated at constant terms.

Step c. To find the resonances

If all possible $\alpha_i's$ are integers, calculate all resonances r for each pair of $(\alpha_i, u_0^{(i)})$ by inserting

$$u^{(i)} = u_0^{(i)} \phi^{\alpha_i} + u_r^{(i)} \phi^{\alpha_i+r}, \quad i = 1, \cdots, m$$

into (1) and balancing the terms with the minimal power of ϕ.

For a single PDE, all resonances should be distinct integers, whether positive or negative. As for a coupled system, the resonance with equal values (occurring twice or more times) does not necessarily indicate a logarithmic branch point[8]. If a non-integer value is found when solving for r, the algorithm terminates.

Step d. To verify compatibility conditions

The final step is to compute the coefficients at the non-resonances, and verify the compatibility condition at every positive integer resonance. This is done by inserting

$$u^{(i)} = \sum_{j=0}^{rmax} u_j^{(i)} \phi^{j+\alpha_i}, \quad i = 1, \cdots, m$$

into (1) and collecting the terms with the same power of ϕ, where $rmax$ is the largest resonance. If there are compatibility conditions which can not be satisfied, one should turn to verify the compatibility conditions for the next branch.

Eqs.(1) is said to pass the Painlevé test if the above steps can be carried out consistently. It should be pointed out that the singular manifold in *Step c* and *Step d* is defined according to Kruskal's simplification, i.e.

$$\phi(x,t) = x - \psi(t), \quad u_j^{(i)} = u_j^{(i)}(t).$$

Kruskal's simplification is a way of making the computation of the WTC method drastically shorter by separating a variable in the singular manifold and it has been widely used in verifying the Painlevé property in a vast amount of literature[6]-[8].

3   The Package PDEPtest

The algorithm of the Painlevé test mentioned in Section 2, while relatively simple in principle, can be very tedious by hand. We have developed a *Maple* package PDEPtest which is a complete implementation of all the algorithms mentioned above. In PDEPtest, the main procedure is *pltest(< eqnlist >, < algtype >* [, *funclist*]), which contains 15 sub-procedures. The parameter *eqnlist* represents the nonlinear PDEs to be tested, *funclist* denotes the list of unknown functions which appear in *eqnlist*. For the PDEs with constant coefficients, *funclist* is optional. But for the PDEs with variable coefficients, *funclist* should be given explicitly. The second parameter *algtype* can take string values as follows:

algtype = "wtc": perform the Painlevé test according to the WTC algorithm;

algtype = "kruskal": perform the Painlevé test according to Kruskal's simplification algorithm;

algtype = "wtc-kruskal": perform the Painlevé test according to the WTC-Kruskal algorithm.

As an example we consider the (3+1)-dimensional potential-YTSF equation[19]

$$-4\,w_{xt} + w_{xxxz} + 4w_x w_{xz} + 2w_{xx} w_z + 3w_{yy} = 0. \tag{3}$$

To check whether Eq.(3) passes the Painlevé test, one proceeds as follows:

```
> pltest( [-4*diff(w(x,y,z,t),x,t)+diff(w(x,y,z,t),x$3,z)+4*diff(w(x,y,z,t),x)
           *diff(w(x,y,z,t),x,z)+2*diff(w(x,y,z,t),x$2)*diff(w(x,y,z,t),z)
           +3*diff(w(x,y,z,t),y$2)=0], "kruskal" );
```

The procedure `pltest` will generate the following results automatically:
The input equation is:

$$-4\frac{\partial^2}{\partial x \partial t}w(x,y,z,t) + \frac{\partial^4}{\partial x^3 \partial z}w(x,y,z,t) + 4\frac{\partial}{\partial x}w(x,y,z,t)\frac{\partial^2}{\partial x \partial z}w(x,y,z,t)$$

$$+2\frac{\partial}{\partial z}w(x,y,z,t)\frac{\partial^2}{\partial x^2}w(x,y,z,t) + 3\frac{\partial^2}{\partial y^2}w(x,y,z,t) = 0$$

The leading analysis of the first branch gives

$$\alpha_1 = -1, \quad w_0(y,z,t) = 2$$

The resonances of the first branch are:

$$-1, 1, 4, 6$$

The coefficients of the first branch are

$$w_0(y,z,t) = 2, \quad w_1(y,z,t) = w_1(y,z,t),$$

$$w_2(y,z,t) = \frac{2\dfrac{\partial}{\partial z}w_1(y,z,t) + 4\dfrac{\partial}{\partial t}\psi(y,z,t) + 3\left(\dfrac{\partial}{\partial y}\psi(y,z,t)\right)^2}{6\dfrac{\partial}{\partial z}\psi(y,z,t)},$$

$$w_3(y,z,t) = \frac{\dfrac{\partial^2}{\partial y^2}\psi(y,z,t)}{4\dfrac{\partial}{\partial z}\psi(y,z,t)}$$

The compatibility condition(s) cannot hold at resonance r=4 is(are):

$$2\frac{\partial^2}{\partial y^2}\psi(y,z,t)\frac{\partial^2}{\partial z^2}\psi(y,z,t) - 2\frac{\partial}{\partial z}\psi(y,z,t)\frac{\partial^3}{\partial y^2\partial z}\psi(y,z,t) = 0$$

The equation fails the Painleve test in checking conditions.
The running time is: 0.235 seconds.

For Eq.(3), when the second parameter $algtype =$ "wtc", PDEPtest outputs the leading order $\alpha = -1$, $w_0 = 2\phi_x$, the resonances are at $r = -1, 1, 4$ and 6. The coefficients of Laurent series read

$$w_2 = -\frac{1}{6\phi_x^3\phi_z}\,[3\phi_x\phi_y^2 + 4\phi_x^2\phi_z w_{1,x} + 2\phi_x^2\phi_{xxz} + 2\phi_x\phi_z\phi_{3x} - 2\phi_x\phi_{xx}\phi_{xz}$$
$$- \phi_{xx}^2\phi_z - 4\phi_t\phi_x^2 + 2\phi_x^3 w_{1,z}],$$

$$w_3 = -\frac{1}{12\phi_x^5\phi_z^2}\,[(4\phi_{xz}w_{1,z} - 4w_{1,xz}\phi_z)\phi_x^4 + (-4\phi_z\phi_{xxxz} - 2w_{1,z}\phi_z\phi_{xx}$$
$$- 8\phi_t\phi_{xz} + 4\phi_{xz}\phi_{xxz} - 3\phi_{yy}\phi_z - 2\phi_z^2 w_{1,xx} + 8\phi_z\phi_{xt})\phi_x^3 + (-4\phi_{xz}^2\phi_{xx} -$$
$$4\phi_t\phi_z\phi_{xx} - 6\phi_y\phi_{xy}\phi_z + 4\phi_z^2\phi_{xx}w_{1,x} + 6\phi_{xx}\phi_{xxz}\phi_z + 4\phi_{xz}\phi_z\phi_{xxx} - \phi_z^2\phi_{xxxx}$$
$$+ 6\phi_{xz}\phi_y^2)\phi_x^2 + (6\phi_{xx}\phi_z\phi_y^2 + 4\phi_{xx}\phi_{xxx}\phi_z^2 - 6\phi_{xx}^2\phi_z\phi_{xz})\phi_x - 2\phi_{xx}^3\phi_z^2],$$

where w_1 is an arbitrary function of $\{x,y,z,t\}$.

The compatibility condition at the resonance $r = 4$ reduces to

$$-2\,[(\phi_{yyz}\phi_z - \phi_{yy}\phi_{zz})\phi_x^4 + (-2\phi_{yz}\phi_{xy}\phi_z - 2\phi_y\phi_{xyz}\phi_z + \phi_{yy}\phi_{xz}\phi_z$$
$$-\phi_{xyy}\phi_z^2 + 2\phi_y\phi_{xy}\phi_{zz})\phi_x^3 + (2\phi_y\phi_{xxy}\phi_z^2 + 2\phi_{xy}^2\phi_z^2 - \phi_{xx}\phi_y^2\phi_{zz} + \phi_{xxz}\phi_y^2\phi_z$$
$$+2\phi_{xx}\phi_y\phi_{yz}\phi_z)\phi_x^2 - (4\phi_y\phi_{xx}\phi_{xy}\phi_z^2 + \phi_y^2\phi_{xxx}\phi_z^2)\phi_x + 2\phi_y^2\phi_{xx}^2\phi_z^2] = 0,$$

which cannot hold for arbitrary singular manifold $\phi(x,y,z,t)$, therefore, our package also concludes that Eq.(3) fails the Painlevé test. Meanwhile, PDEPtest outputs the truncated expansion,

$$w = 2(\log \phi)_x + w_1. \tag{4}$$

By using (4), the special solutions can be further derived. In order to find solitary wave solutions of (3), one may suppose

$$\phi = 1 + \exp(\theta), \quad \theta = k_1\,x + k_2\,y + k_3\,z + k_4\,t + \xi_0, \tag{5}$$

inserting (4) and (5) into (3), collecting the terms with the same power of ϕ, we get a set of homogeneous PDEs about ϕ. Solving it, the following shock wave solution is obtained,

Table 1. Time for testing examples under three algorithms(seconds)

Name of equations	WTC	Kruskal	WTC − Kruskal	Result	Ref.
Burgers	0.109	0.062	0.093	Pass	[1]
KdV	0.422	0.110	0.172	Pass	[1]
mKdV	0.608	0.234	0.406	Pass	[1]
KdV-Burgers	0.797	0.140	0.281	Fail	[18]
Gen. Kawahara	0.360	0.155	0.703	Fail	[10]
7th-order SK	2294.654	16.357	19.390	Pass	[18]
Coupled mKdV	22.754	2.547	3.594	Pass	[18]
Gen. Ito(I)	> 10000	5.157	5.390	Pass	[25]
Gen. Ito(II)	766.913	3.281	4.358	Pass	[25]
2+1 Dimen. KK	151.904	1.486	1.952	Pass	[10]
2+1 Dimen. Gen. NNV	70.000	2.438	3.733	Pass	[18]
2+1 Dimen. Gen. Hirota	2984.273	13.331	19.503	Pass	[22]
2+1 Dimen. Coupled KP	> 10000	10.251	11.002	Pass	[24]
3+1 Dimen KP	1.126	0.265	0.561	Fail	[18]
3+1 Dimen. YTSF	1.236	0.235	0.311	Fail	[19]
3+1 Dimen. KdV-ZK	1434.771	1.500	2.016	Fail	[26]
3+1 Dimen. mKdV-ZK	91.674	0.952	8.843	Fail	[26]

$$
\begin{aligned}
w &= \frac{2\,k_1\,\exp(k_1\,x + k_2\,y + k_3\,z + k_4\,t + \xi_0)}{1 + \exp(k_1\,x + k_2\,y + k_3\,z + k_4\,t + \xi_0)} \\
&= k_1\left[1 + \tanh(\frac{k_1\,x + k_2\,y + k_3\,z + k_4\,t}{2} + \frac{\xi_0}{2})\right], \quad k_4 = \frac{k_1^3 k_3 + 3k_2^2}{4k_1},
\end{aligned}
\tag{6}
$$

where k_1, k_2, k_3 and ξ_0 are arbitrary constants.

In the package PDEPtest, we apply three algorithms which are WTC, Kruskal, WTC-Kruskal. PDEPtest has been applied to a large variety of nonlinear PDEs, and the involved time of the Painlevé test for 17 examples is listed in Table 1. The tests performed on a Dell Dimension 4600 PC with 2.40GHz Pentium 4 Processor, 256MB of RAM with Maple V.6.0.

From Table 1, we can conclude the following result: for higher order PDEs, higher dimensional PDEs and coupled system in more components, the more time is needed. If one tries to verify the Painlevé property and obtain the truncated expansion form, the WTC-Kruskal algorithm may be the better choice.

4   The Applications of PDEPtest

In what follows, we just consider several higher order equations, coupled system of equations as well as special type of nonlinear PDEs.

Example 1. Let us first consider the Ito's fifth-order mKdV equation[20]

$$
u_t + \left(6u^5 + 10\,p\,(u^2 u_{xx} + u u_x^2) + u_{xxxx}\right)_x = 0,
\tag{7}
$$

where p is a parameter.

When the second parameter *algtype* ="kruskal" PDEPtest outputs that Eq.(7) has four different branches. For these four branches, the leading order exponents for u are both -1. And the corresponding leading coefficients read

(a) $u_0(t) = \dfrac{\sqrt{2\sqrt{25p^2 - 16} - 10p}}{2}$,   (b) $u_0(t) = -\dfrac{\sqrt{2\sqrt{25p^2 - 16} - 10p}}{2}$,

(c) $u_0(t) = \dfrac{\sqrt{-2\sqrt{25p^2 - 16} - 10p}}{2}$,   (d) $u_0(t) = -\dfrac{\sqrt{-2\sqrt{25p^2 - 16} - 10p}}{2}$.

For the former two branches the resonances are given by $r = -1$, 5, 6, and

$$\frac{5 + \sqrt{100p^2 - 20p\sqrt{25p^2 - 16} - 39}}{2}, \frac{5 - \sqrt{100p^2 - 20p\sqrt{25p^2 - 16} - 39}}{2},$$

and for the latter two branches the resonances are given by $r = -1$, 5, 6, and

$$\frac{5 + \sqrt{100p^2 + 20p\sqrt{25p^2 - 16} - 39}}{2}, \frac{5 - \sqrt{100p^2 + 20p\sqrt{25p^2 - 16} - 39}}{2}.$$

It is obvious that there are non-integer resonances, so PDEPtest outputs that Eq.(7) fails the Painlevé test.

Since all the resonances r must be integers in order for Eq.(7) to pass the Painlevé test, we can see that the above resonances are all integers when $100p^2 - 20p\sqrt{25p^2 - 16} = (2m+1)^2 + 39(m = 0, 1, 2, \cdots)$ and $100p^2 + 20p\sqrt{25p^2 - 16} = (2\tilde{m} + 1)^2 + 39(\tilde{m} = 0, 1, 2, \cdots)$ hold simultaneously. It can be shown that the above conditions are both satisfied for $p = -1$ or $p = 1$.

When we take $p = 1$, the procedure **pltest** with *algtype* ="wtc-kruskal" outputs that the the leading order exponent is -1, and the truncated expansions of four branches read

$$(a)\, u = -i\,(\log \phi)_x + u_1, \ (b)\, u = i\,(\log \phi)_x + u_1,$$

$$(c)\, u = -2i\,(\log \phi)_x + u_1, \ (d)\, u = 2i\,(\log \phi)_x + u_1,$$

where $i = \sqrt{-1}$. The resonances for the former two branches arise at $\{-1, 2, 3, 5, 6\}$, and the resonances for the latter two branches occur at $\{-3, -1, 5, 6, 8\}$. Then all compatibility conditions at positive integer resonances are proved to be satisfied identically.

There are close connections between the truncated expansions and integrable properties of nonlinear PDEs. For the above first truncated expansion, we consider the vacuum solution $u_1 = 0$, namely,

$$u = -i(\log \phi)_x. \tag{8}$$

Assuming now $\phi = g/f$ and inserting (8) into (7) and making use of the Hirota D operators[21], we obtain the bilinear form:

$$(D_t + D_x^5)\, g \cdot f = 0, \quad D_x^2\, g \cdot f = 0. \tag{9}$$

By means of Hirota's bilinear method[21], N-soliton solution of Eq.(7) is given by (8) and

$$f = \sum_{\mu=0,1} \exp\left[\sum_{i<j}^{(N)} A_{ij}\mu_i\mu_j + \sum_{j=1}^{N} \mu_j(\eta_j + \frac{\pi}{2}i)\right],$$

$$g = \sum_{\mu=0,1} \exp\left[\sum_{i<j}^{(N)} A_{ij}\mu_i\mu_j + \sum_{j=1}^{N} \mu_j(\eta_j - \frac{\pi}{2}i)\right],$$

where

$$\eta_j = p_j\,x + w_j\,t + \eta_j^0, \; w_j = -p_j^5, \; \exp(A_{ij}) = \frac{(p_i - p_j)^2}{(p_i + p_j)^2}.$$

When we take $p = -1$, the procedure **pltest** with *algtype* ="wtc-kruskal" outputs that the the leading order equals -1, and the truncated expansions of four branches read

$$(a)\, u = -(\log \phi)_x + u_1, \, (b)\, u = (\log \phi)_x + u_1,$$

$$(c)\, u = -2(\log \phi)_x + u_1, \, (d)\, u = 2(\log \phi)_x + u_1.$$

The resonances for the former two branches are at $\{-1, 2, 3, 5, 6\}$, and the resonances for the latter two branches occur at $\{-3, -1, 5, 6, 8\}$. Then it is shown that all compatibility conditions at positive integer resonances are satisfied identically.

With the assistance of the package PDEPtest, we can conclude that Eq.(7) passes the Painlevé test if and only if $p = \pm 1$. It is very interesting that the periodic-wave solutions are obtained just under the same parameter constraint[20].

Example 2. It is well known that physics and engineering often provide special type of nonlinear equations such as sine-Gordon equation and Schrödinger equation. Such equations cannot be directly tested by our package and require some kind of pre-processing technique. Next we consider a new (2+1)-dimensional generalized Hirota equation

$$\begin{cases} i\,\psi_t + \psi_{xy} + i\,\psi_{xxx} + \psi\phi - i\,|\psi|^2\,\psi_x = 0, \\ 3\phi_x + (|\psi|^2)_y = 0, \end{cases} \tag{10}$$

which was studied by Maccari recently[22]. To apply the Painlevé test, we set $\psi = u + i\,v, \phi = w$. Then Eq.(10) becomes

$$\begin{cases} u_t + v_{xy} + u_{xxx} + vw - (u^2 + v^2)u_x = 0, \\ -v_t + u_{xy} - v_{xxx} + uw + (u^2 + v^2)v_x = 0, \\ 3w_x + 2uu_y + 2vv_y = 0. \end{cases} \tag{11}$$

When the second parameter *algtype* ="kruskal" PDEPtest outputs that Eq.(11) has two different branches. For these two branches, the leading order exponents for u, v, w are -1, -1 and -2 respectively. The leading coefficients of the two branches are

$$(a)\ u_0 = \sqrt{6 - v_0^2},\ w_0 = 2\psi_y,\quad (b)\ u_0 = -\sqrt{6 - v_0^2},\ w_0 = 2\psi_y,$$

where v_0 is an arbitrary function. In both cases the resonances are $\{-1, 0, 1, 2, 3, 4, 5\}$. The coefficients of the two branches are too tedious to be listed here. It is shown that the compatibility conditions at all non-negative integer resonances are satisfied identically. Therefore, Eq.(11) passes the Painlevé test.

When the second parameter $algtype$ = "wtc", it takes about 3000 seconds to verify the Painlevé property for Eq.(11). If $algtype$ = "wtc-kruskal", we can not only perform the Painlevé test quickly, but also obtain the truncated Laurent series expansion. The truncated expansion was used to derive Lax pair and Darboux transformation in Ref.[23].

Example 3.   The package PDEPtest can be used to test coupled system in three or more components, such as the following coupled KP system[24]

$$\begin{cases} q_t = \frac{1}{4} \left(q_{xxx} - 6qq_x + 3\partial_x^{-1}q_{yy} + 6(pr)_x \right), \\ p_t = \frac{1}{2} \left(-p_{xxx} + 3qp_x - 3p_{xy} + 3p\partial_x^{-1}q_y \right), \\ r_t = \frac{1}{2} \left(-r_{xxx} + 3qr_x + 3r_{xy} - 3r\partial_x^{-1}q_y \right). \end{cases} \tag{12}$$

The bilinear form, N-soliton solution and algebraic-geometrical solutions have been obtained. Its Painlevé property, as much as we know, has not been proved in literature.

To investigate Eq.(12), let us now apply the transformation $q_y = s_x$, so as to convert it into

$$\begin{cases} q_y = s_x, \\ q_t = \frac{1}{4} \left(q_{xxx} - 6qq_x + 3s_y + 6(pr)_x \right), \\ p_t = \frac{1}{2} \left(-p_{xxx} + 3qp_x - 3p_{xy} + 3p\,s \right), \\ r_t = \frac{1}{2} \left(-r_{xxx} + 3qr_x + 3r_{xy} - 3r\,s \right). \end{cases} \tag{13}$$

When the second parameter $algtype$ = "wtc-kruskal", it is shown that Eq.(13) has two branches. For the first branch, the leading order and coefficients for q, s, p, r read

$$\alpha_1 = \alpha_2 = -2, \alpha_3 = \alpha_4 = -1, q_0 = 2\phi_x^2,\ s_0 = 2\phi_x\phi_y,$$

where p_0, r_0 are arbitrary functions with respect to $\{x, y, t\}$. While for the second branch, the leading order and coefficients for q, s, p, r read

$$\alpha_1 = \alpha_2 = \alpha_3 = \alpha_4 = -2, q_0 = 4\phi_x^2,\ s_0 = 4\phi_x\phi_y,\ r_0 = 4\phi_x^2/p_0,$$

with p_0 being arbitrary function with respect to $\{x, y, t\}$.

The resonances occur at $\{-1, 0, 0, 1, 1, 2, 4, 5, 5, 6\}$ and $\{-1, -2, 0, 2, 2, 3, 4, 6, 7, 8\}$, respectively. Obviously, the first branch is principal one(the branch with the most non-negative integer resonances), and its truncated expansion takes the form

$$q = -2(\log \phi)_{xx} + q_2, \quad s = -2(\log \phi)_{xy} + s_2, \quad p = p_0/\phi + p_1, \quad r = r_0/\phi + r_1. \tag{14}$$

The compatibility conditions at all non-negative integer resonances are satisfied identically, so our package concludes Eq.(13) passes the Painlevé test.

If we consider the trivial solution $q_2 = s_2 = p_1 = r_1 = 0$ in (14) and inserting

$$q = -2(\log f)_{xx}, \quad s = -2(\log f)_{xy}, \quad p = g/f, \quad r = h/f$$

into (13), we can obtain the following bilinear form:

$$
\begin{aligned}
(4\, D_x D_t - D_x^4 - 3\, D_y^2) f \cdot f + 6\, g \cdot h &= 0, \\
(2\, D_t + D_x^3 - 3\, D_x D_y)\, h \cdot f &= 0, \\
(2\, D_t + D_x^3 + 3\, D_x D_y)\, g \cdot f &= 0.
\end{aligned}
\tag{15}
$$

N-soliton solution and Bäcklund transformation can be derived by means of the bilinear form (15).

5   Summary

In this paper, the WTC-Kruskal algorithm is presented, which is based on the WTC algorithm and Kruskal's simplification. Also we have developed a *Maple* package PDEPtest in which the WTC, Kruskal and WTC-Kruskal algorithms are implemented at the same time. Up to now, PDEPtest has been used to verify the Painlevé property of a large variety of nonlinear PDEs. This package can handle a single nonlinear PDE and coupled nonlinear PDEs with finite components, in which every equation is a polynomial (or can be converted to a polynomial) in the unknown functions and their derivatives. However, there exist some limitations in our package. PDEPtest does not deal with the theoretical shortcomings of the standard Painlevé test. Thus far, we have implemented the traditional Painlevé test, but not taken account of the latest advances in Painlevé type methods, such as the poly-Painlevé test, the Fuchs-Painlevé test. Neither did we code the quasi- and weak Painlevé test or other variants. In addition, our package is incapable of checking the compatibility conditions at negative integer resonances, even though we know that negative integer resonances sometimes contain important information regarding the integrability of an equation. The more study on how to develop our package and study other applications of the Painlevé analysis is worthwhile in the future.

References

1. Ablowitz M.J., Clarkson P.A.: Solitons, Nonlinear Evolution Equations and Inverse Scattering, Cambridge University Press, Cambridge(1999).
2. Weiss J., Tabor M. and Carnevale G.: The Painlevé Property of Partial Differential Equations. J. Math. Phys. **24**(1983)522-526.
3. Jimbo M., Kruskal M.D. and Miwa T.: The Painlevé Test for the Self-dual Yang-Mills Equations. Phys. Lett. A. **92**(1982)59.
4. Conte R.: Invariant Painlevé Analysis of Partial Differential Equations. Phys. Lett. A. **140**(1989)383-389.
5. Conte R., Fordy A.P. and Pickering A.: A Perturbation Painlevé Approach to Nonlinear Differential Equations. Phys. D **69**(1993)33-58.
6. Lou S.Y., Chen C.L. and Tang X.Y.: (2+1)-Dimensional (M+N)-Component AKNS System: Painlevé Integrability, Infinitely Many Symmetries and Similarity Reductions. J. Math. Phys. **43**(2002)4078-4109.
7. Estévez P.G., Conde E., Gordoa P.R.: Unified Approach to Miura, Bäcklund and Darboux Transformations for Nonlinear Partial Differential Equations. J. Nonlinear Math. Phys. **5**(1998)82-114.
8. Roy Chowdhury A.: The Painlevé Analysis and Its Applications, Chapman & Hall/CRC, Baton Rouge, Florida(2000).
9. Newell A.C., Tabor M. and Zeng Y.B.: A Unified Approach to Painlevé Expansion. Phys. D **29**(1987)1-68.
10. Xu G.Q., Li Z.B.: A Maple Package for the Painlevé Test of Nonlinear Partial Differential Equations. Chin. Phys. Lett. **20**(2003)975-978.
11. Conte R.(Ed.): The Painlevé Property, One Century Later. Springer Verlag, New York(1999).
12. Hlavatý L.: Test of Resonances in the Painlevé Analysis. Comput. Phys. Commun. **42**(1986)427-433.
13. Scheen C.: Implementation of the Painlevé Test for Ordinary Differential Equation. Theor. Comput. Sci. **187**(1997)87-104.
14. Hereman W., Angenent S.: The Painlevé Test for Nonlinear Ordinary and Partial Differential Equations. MACSYMA Newsletter. **6**(1989)11-18.
15. Hereman W., Göktas Ü., Colagrosso M. et al.: Algorithmic Integrability Tests of Nonlinear Differential and Lattice Equations. Comput. Phys. Commun. **115**(1998)428-446.
16. Baldwin D., Hereman W. and Sayers J.: Symbolic Algorithms for the Painlevé Test, Special Solutions, and Recursion Operators of Nonlinear PDEs, CRM Proceedings and Lecture series **39**, Eds: Winternitz P. and Gomez-Ullate D. American Mathematical Society, Providence, Rhode Island (2004)17-32.
17. Xie F.D., Chen Y.:, An Algorithmic Method in Painlevé Analysis of PDE. Comput. Phys. Commun. **154**(2003)197-204.
18. Xu G.Q., Li Z.B.: Symbolic Computation of the Painlevé Test for Nonlinear Partial Differential Equations using Maple. Comput. Phys. Commun. **161** (2004)65-75.
19. Yan Z.Y.: New Families of Non-travelling Wave Solutions to a New (3+1)-Dimensional Potential-YTSF Equation. Phys. Lett. A. **318**(2003)78-83.
20. Parkes E.J., Duffy B.R. and Abbott P.C.: The Jacobi Elliptic Function Method for Finding Periodic-wave Solutions to Nonlinear Evolution Equations. Phys. Lett. A. **295**(2002)280-286.
21. Hirota R.: Direct Methods in Soliton Theory, in Solitons, Eds: Bullough R.K. and Caudrey P.J. Spinger, Berlin(1980).

22. Maccari A.: A Generalized Hirota Equation in 2+1 Dimensions. J. Math. Phys. **39**(1998)6547-6551.

23. Estévez P.G.: A Nonisospectral Problem in (2+1) Dimensions Derived from KP. Inverse Problems **17**(2001)1043-1052.

24. Geng X.G.: Algebraic-geometrical Solutions of Some Multidimensional Nonlinear Evolution Equations. J. Phys. A: Math. Gen. **36**(2003)2289-2301.

25. Ayse K.K., Atalay K. and Sergei Y.S.: Integrability of a Generalized Ito System: The Painlevé Test. J. Phys. Soc. Jpn. **70**(2001)1165-1166.

26. Das K.P., Verheest F.: Ion-acoustic Solitons in Magnetized Multi-component Plasmas Including Negative Ions. J. Plasma Phys. **41**(1989)139-155.

Hybrid Matrix Geometric Algebra

Garret Sobczyk[1] and Gordon Erlebacher[2]

[1] Universidad de Las Americas - Puebla,
72820 Cholula, Mexico
[2] Florida State University Computer,
Science and Information Technology,
Tallahassee, FL, USA

Abstract. The structures of matrix algebra and geometric algebra are completely compatible and in many ways complimentary, each having their own advantages and disadvantages. We present a detailed study of the hybrid 2×2 matrix geometric algebra $M(2, \boldsymbol{G})$ with elements in the 8 dimensional geometric algebra $\boldsymbol{G} = \boldsymbol{G}_3$ of Euclidean space. The resulting hybrid structure, isomorphic to the geometric algebra $\boldsymbol{G}_{4,1}$ of de Sitter space, combines the simplicity of 2×2 matrices and the clear geometric interpretation of the elements of $\boldsymbol{G}$. It is well known that the geometric algebra $\boldsymbol{G}(4, 1)$ contains the 3-dimensional affine, projective, and conformal spaces of Möbius transformations, together with the 3-dimensional horosphere which has attracted the attention of computer scientists and engineers as well as mathematicians and physicists. In the last section, we describe a sophisticated computer software package, based on Wolfram's Mathematica, designed specifically to facilitate computations in the hybrid algebra.

1 Introduction

The structures of matrix algebra and geometric algebra are completely compatible and in many ways complimentary, each having their own advantages and disadvantages. In this paper, we present a detailed study of the the hybrid matrix geometric algebra $M(2, \boldsymbol{G}_3)$, which exploits the advantages of both structures. Some of the points we wish to make in this paper are the following:

- The geometric algebra $\boldsymbol{G}_3$ is closely related to the familiar Gibbs-Heaviside vector algebra.
- Working with 2×2 matrices simplifies the computer implementation of the resulting structure.
- The resulting hybrid structure is isomorphic both to the algebra of complex 4×4 matrices and to the geometric algebra $\boldsymbol{G}_{4,1}$.
- The geometric algebra $\boldsymbol{G}_{4,1}$ is sufficiently rich to contain the 3-dimensional affine, projective, and conformal spaces which includes Möbius transformations, as well as the 3-dimensional horosphere.
- The matrix geometric $M(2, \boldsymbol{G})$ is the building block for higher dimensional hybrid matrix geometric algebras $M(2^k, \boldsymbol{G}_3)$ for the geometric algebras $\boldsymbol{G}_{3+k,k}$ for $k \geq 1$.

H. Li, P. J. Olver and G. Sommer (Eds.): IWMM-GIAE 2004, LNCS 3519, pp. 191–206, 2005.

2   The Geometric Algebra $\boldsymbol{G}_3$ of Euclidean Space

The associative geometric algebra $\boldsymbol{G}_3 := gen\{e_1, e_2, e_3\}$ is *generated* by taking the geometric sums of products of 3 orthonormal real basis vectors e_1, e_2, e_3, subject to the rules that

$$e_1^2 = e_2^2 = e_3^2 = 1.$$

Products of the basis vectors are denoted by $e_{i...k} := e_i \cdots e_k$. As a linear space, the geometric algebra $\boldsymbol{G}_3$ is spanned by the basis S, where

$$S = \{1, e_1, e_2, e_3, e_{23} = ie_1, e_{31} = ie_2, e_{12} = ie_3, i = e_{123}\}.$$

Consequently, we can express a general element $a \in \boldsymbol{G}_3$ in the form

$$a = a_0 + \mathbf{a} \tag{1}$$

where $\mathbf{a} = a_1 e_1 + a_2 e_2 + a_3 e_3 \in \boldsymbol{G}_3^{1,2}$ is a *complex vector* (vector + bivector), and $a_0, a_1, a_2, a_3 \in \boldsymbol{G}_3^{0,3}$ are *complex scalars* (scalars + pseudoscalars or trivectors). Note that the complex scalars make up the *center* of the algebra $\boldsymbol{G}_3$ in that they commute with all of the elements of $\boldsymbol{G}_3$.

The geometric algebra $\boldsymbol{G}_3$ has a structure that is closely related to the Gibbs-Heaviside vector algebra (see Sobczyk [8]). Let $a = a_0 + \mathbf{a}$ and $b = b_0 + \mathbf{b}$ be elements of $\boldsymbol{G}_3$. The geometric product

$$ab = (a_0 + \mathbf{a})(b_0 + \mathbf{b}) = (a_0 b_0 + \mathbf{a} \circ \mathbf{b}) + (b_0 \mathbf{a} + a_0 \mathbf{b} + \mathbf{a} \otimes \mathbf{b}) \tag{2}$$

where the *complex inner product* $\mathbf{a} \circ \mathbf{b} := a_1 b_1 + a_2 b_2 + a_3 b_3$, and the *complex vector product*

$$\mathbf{a} \otimes \mathbf{b} := i \det \begin{pmatrix} e_1 & e_2 & e_3 \\ a_1 & a_2 & a_3 \\ b_1 & b_2 & b_3 \end{pmatrix} = i\mathbf{a} \times \mathbf{b}$$

where $\mathbf{a} \times \mathbf{b}$ is the ordinary Gibbs-Heaviside cross product when the vectors $\mathbf{a}$ and $\mathbf{b}$ are real. For real vectors $\mathbf{a}$ and $\mathbf{b}$, $i\mathbf{a} \times \mathbf{b} := i(\mathbf{a} \times \mathbf{b})$ has the geometric interpretation of the bivector normal to the vector $\mathbf{a} \times \mathbf{b}$. Note also that we have the *complex triple product*

$$\mathbf{a} \circ \mathbf{b} \otimes \mathbf{c} := i \det \begin{pmatrix} a_1 & a_2 & a_3 \\ b_1 & b_2 & b_3 \\ c_1 & c_2 & c_3 \end{pmatrix}.$$

The geometric product $\mathbf{ab}$ of complex vectors $\mathbf{a}, \mathbf{b} \in \boldsymbol{G}_3^{1,2}$ can be expressed in the form

$$\mathbf{ab} = \frac{1}{2}(\mathbf{ab} + \mathbf{ba}) + \frac{1}{2}(\mathbf{ab} - \mathbf{ba}) = \mathbf{a} \circ \mathbf{b} + \mathbf{a} \otimes \mathbf{b}$$

where $\mathbf{a} \circ \mathbf{b} = \frac{1}{2}(\mathbf{ab} + \mathbf{ba})$ and $\mathbf{a} \otimes \mathbf{b} = \frac{1}{2}(\mathbf{ab} - \mathbf{ba})$, and has no analogy in the Gibbs-Heaviside vector algebra. To better understand the complex vector products on $\boldsymbol{G}_3$, we calculate $\mathbf{abc}$ for the three complex vectors $\mathbf{a}, \mathbf{b}, \mathbf{c} \in \boldsymbol{G}_3^{1,2}$,

$$\mathbf{a}(\mathbf{bc}) = \mathbf{a}(\mathbf{b} \circ \mathbf{c} + \mathbf{b} \otimes \mathbf{c}) = (\mathbf{b} \circ \mathbf{c})\mathbf{a} + \mathbf{a} \circ (\mathbf{b} \otimes \mathbf{c}) + \mathbf{a} \otimes (\mathbf{b} \otimes \mathbf{c}).$$

The last triple product can be expanded as $\mathbf{a} \otimes (\mathbf{b} \otimes \mathbf{c}) = (\mathbf{a} \circ \mathbf{b})\mathbf{c} - (\mathbf{a} \circ \mathbf{c})\mathbf{b}$.

Let $a \in \boldsymbol{G}_3$. By a^* we mean the *inversion* obtained from a by replacing $\mathbf{v}$ by $-\mathbf{v}$ for all real vectors $\mathbf{v}$ contained in a. For example, if $a = s + \mathbf{v} + \mathbf{b} + t$ for real scalar s, real vector $\mathbf{v}$, bivector $\mathbf{b}$ and trivector t , then $a^* = s - \mathbf{v} + \mathbf{b} - t$. The *reversion* $a^\dagger$ of a is defined by reversing the order of the geometric product of all real vectors in a. For $a \in \boldsymbol{G}_3$ given above, $a^\dagger = s + \mathbf{v} - \mathbf{b} - t$. Finally, the *conjugation* $\bar{a}$ is defined by $\bar{a} = (a^*)^\dagger$. For the given a, $\bar{a} = s - \mathbf{v} - \mathbf{b} + t$. We see that, unlike the usual complex conjugation, for $b = b_0 + \mathbf{b} \in \boldsymbol{G}_3$, $\bar{b} = b_0 - \mathbf{b}$. Clearly, the various multivector parts of a are all expressible in terms of $a, a^*, a^\dagger, \bar{a}$.

A general complex element $a = a_0 + \mathbf{a} \in \boldsymbol{G}_3$ will have an inverse iff

$$\det a := a\bar{a} = a_0^2 - \mathbf{a}^2 \neq 0, \tag{3}$$

and it is natural to define the determinant function $\det a$ of $a \in \boldsymbol{G}_3$ in terms of this quantity.

3   The Matrix Geometric Algebra $M(2, \boldsymbol{G}_3)$

The matrix algebra $M(2, \boldsymbol{G}_3)$ consists of all matrices of the form $[g] = \begin{pmatrix} a & b \\ c & d \end{pmatrix}$ where $a, b, c, d \in \boldsymbol{G}_3$. Addition and multiplication of matrices in $M(2, \boldsymbol{G}_3)$ is the usual matrix addition and multiplication, but attention should be paid to the order of the elements in the product (since, in general, $ab \neq ba$ in $\boldsymbol{G}_3$). We say that $M(2, \boldsymbol{G}_3)$ is the hybrid 2×2 matrix geometric algebra over $\boldsymbol{G}_3$. Matrices have been studied over the complex quaternions in (Tian [10]), and in other works.

Let us now see how a general element $[g] \in M(2, \boldsymbol{G}_3)$ represents an element g in the larger geometric algebra $\boldsymbol{G}_{4,1}$. We consider the geometric algebra $\boldsymbol{G}_{4,1}$ to be the geometric algebra $\boldsymbol{G}_3$ *extended* by two additional orthonormal basis vectors σ, γ satisfying $\sigma^2 = 1 = -\gamma^2$. The geometric algebra $\boldsymbol{G}_3$ is thus a subalgebra of the larger algebra $\boldsymbol{G}_{4,1} = gen\{e_1, e_2, e_3, \sigma, \gamma\}$.

In order to represent an element $g \in \boldsymbol{G}_{4,1}$ as a matrix $[g] \in M(2, \boldsymbol{G}_3)$, we introduce the special elements

$$u = \sigma\gamma, \quad u_\pm = \frac{1}{2}(1 \pm u) \tag{4}$$

and note that

$$\sigma u = -u\sigma, \quad \sigma u_\pm = u_\mp \sigma.$$

Because $u^2 = 1$, it easily follows that u_+ and u_- are mutually annihilating idempotents which partition 1,

$$(1 \quad \sigma) u_+ \begin{pmatrix} 1 \\ \sigma \end{pmatrix} = u_+ + u_- = 1. \tag{5}$$

Noting that $ua = au$ and $\sigma a = a^* \sigma$ for any element $a \in \boldsymbol{G}_3$, we can express any $g \in \boldsymbol{G}_{4,1}$ in the form

$$g = au_+ + bu_+\sigma + c^* u_-\sigma + d^* u_-, \tag{6}$$

for $a, b, c, d \in \boldsymbol{G}_3$, (see Pozo and Sobczyk [9]). Now using (4), (5) and (6), we easily find the matrix form $[g] \in M(2, \boldsymbol{G}_3)$,

$$g = (1 \quad \sigma) u_+ \begin{pmatrix} 1 \\ \sigma \end{pmatrix} g (1 \quad \sigma) u_+ \begin{pmatrix} 1 \\ \sigma \end{pmatrix}$$

$$= (1 \quad \sigma) u_+ \begin{pmatrix} g & g\sigma \\ \sigma g & \sigma g \sigma \end{pmatrix} u_+ \begin{pmatrix} 1 \\ \sigma \end{pmatrix} = (1 \quad \sigma) u_+ [g] \begin{pmatrix} 1 \\ \sigma \end{pmatrix},$$

where

$$[g] := \begin{pmatrix} a & b \\ c & d \end{pmatrix}.$$

Extending the operations of inversion g^*, reversion $g^\dagger$, and conjugation $\bar{g}$ to the larger algebra $\boldsymbol{G}_{4,1}$, we find

$$g^* = a^* u_+ - b^* u_+\sigma - cu_-\sigma + du_-,$$
$$g^\dagger = \bar{d}u_+ + \bar{b}u_+\sigma + c^\dagger u_-\sigma + a^\dagger u_-,$$

and

$$\bar{g} = d^\dagger u_+ - b^\dagger u_+\sigma - \bar{c}u_-\sigma + \bar{a}u_-,$$

from which follows the corresponding operations in the Hybrid matrix algebra

$$[g]^\dagger := [g^\dagger] = \begin{pmatrix} \bar{d} & \bar{b} \\ \bar{c} & \bar{a} \end{pmatrix}, \quad [g]^* := [g^*] = \begin{pmatrix} a^* & -b^* \\ -c^* & d^* \end{pmatrix},$$

and

$$\overline{[g]} := [\bar{g}] = \begin{pmatrix} d^\dagger & -b^\dagger \\ -c^\dagger & a^\dagger \end{pmatrix}.$$

Our objective now is to determine the condition on the elements in $\boldsymbol{G}_3$ which will guarantee that a given hybrid element $[g]$ will have an inverse $[g]^{-1}$. First, we extend the determinant function to a function on the hybrid algebra by requiring that it satisfies the following important properties: For $a, b, c, d \in \boldsymbol{G}_3$,

$$\det \begin{pmatrix} a & b \\ 0 & d \end{pmatrix} = a\bar{a}d\bar{d}, \quad \det \begin{pmatrix} ea & eb \\ c & d \end{pmatrix} = e\bar{e} \det \begin{pmatrix} a & b \\ c & d \end{pmatrix}, \text{ and}$$

$$\det \begin{pmatrix} a & b \\ ea+c & eb+d \end{pmatrix} = \det \begin{pmatrix} a & b \\ c & d \end{pmatrix}.$$

These properties agree with the usual definition of the determinant function, a consequence of what is known as the *generalized algorithm of Gauss*, see Gantmacher [3– p.45].

We can now prove the following theorem about hybrid determinants.

Theorem: $\det \begin{pmatrix} a & b \\ c & d \end{pmatrix} = a\bar{a}d\bar{d} + b\bar{b}c\bar{c} - (\bar{a}b\bar{d}c + \bar{c}d\bar{b}a).$

Proof. For convenience, let $\alpha = a\bar{a}$.

$$\det \begin{pmatrix} a & b \\ c & d \end{pmatrix} = \frac{1}{\alpha} \det \begin{pmatrix} \bar{a}a & \bar{a}b \\ c & d \end{pmatrix} = \alpha^{-3} \det \begin{pmatrix} \bar{a}a & \bar{a}b \\ \bar{a}ac & \bar{a}ad \end{pmatrix}$$

$$= \alpha^{-3} \det \begin{pmatrix} \bar{a}a & \bar{a}b \\ 0 & \bar{a}ad - c\bar{a}b \end{pmatrix} = \frac{1}{\alpha}[(a\bar{a}d - c\bar{a}b)(a\bar{a}\bar{d} - \bar{b}a\bar{c})]$$

$$= a\bar{a}d\bar{d} + b\bar{b}c\bar{c} - (\bar{a}b\bar{d}c + \bar{c}d\bar{b}a).$$

In the proof of the above theorem, we assumed that the element $a \in \boldsymbol{G}_3$ was invertible, which is not always the case. Never-the-less, the theorem remains valid in these cases, as can be argued by standard continuity requirements.

The non-vanishing of the determinant function $\det[g] \neq 0$ is the condition for the invertibility of a general element $g \in \boldsymbol{G}_{4,1}$ where $[g] = \begin{pmatrix} a & b \\ c & d \end{pmatrix}$. An expression for the inverse of the Hybrid 2-matrix $\begin{pmatrix} a & b \\ c & d \end{pmatrix}$ can be derived by the usual trick of applying the generalized Gauss algorithm to the augmented matrix

$$\begin{pmatrix} a & b & 1 & 0 \\ c & d & 0 & 1 \end{pmatrix}$$

to get the result

$$\left(\begin{pmatrix} 1 & 0 \\ 0 & 1 \end{pmatrix} \begin{pmatrix} a & b \\ c & d \end{pmatrix}^{-1} \right).$$

The computations are a little tricky because of the lack of general commutativity of the elements of $\boldsymbol{G}_3$. The final result is

$$\begin{pmatrix} a & b \\ c & d \end{pmatrix}^{-1} = \frac{1}{\det \begin{pmatrix} a & b \\ c & d \end{pmatrix}} \begin{pmatrix} \bar{a}d\bar{d} - \bar{c}d\bar{b} & b\bar{b}\bar{c} - \bar{a}b\bar{d} \\ \bar{b}c\bar{c} - \bar{d}c\bar{a} & a\bar{a}\bar{d} - \bar{b}a\bar{c} \end{pmatrix},$$

which is related to a similar result which can be found in a web page by Roweis, [7].

4   The Geometric Algebra $\boldsymbol{G}_{4,1}$

Using (6), the matrix representation of the basis elements of $\boldsymbol{G}_{4,1}$ are easily calculated:

$$[e_k] = \begin{pmatrix} e_k & 0 \\ 0 & -e_k \end{pmatrix}, \quad [\sigma] = \begin{pmatrix} 0 & 1 \\ 1 & 0 \end{pmatrix}, \quad [\gamma] = \begin{pmatrix} 0 & -1 \\ 1 & 0 \end{pmatrix}.$$

It is also interesting to note that

$$[u] = [\sigma\gamma] = \begin{pmatrix} 1 & 0 \\ 0 & -1 \end{pmatrix}, \quad [u_+] = \begin{pmatrix} 1 & 0 \\ 0 & 0 \end{pmatrix}, \quad [u_-] = \begin{pmatrix} 0 & 0 \\ 0 & 1 \end{pmatrix}.$$

The matrix representation of any other element can be easily calculated by taking sums of products of the matrix representations of the vector basis elements. For example, defining the null vectors $e = \frac{1}{2}(\sigma + \gamma)$ and $\bar{e} = \sigma - \gamma$, we find that

$$[e] = \begin{pmatrix} 0 & 0 \\ 1 & 0 \end{pmatrix}, \quad [\bar{e}] = \begin{pmatrix} 0 & 2 \\ 0 & 0 \end{pmatrix}.$$

Calculating the various conjugation operations on a general element

$$g = s + v + b + t + f + p \in \boldsymbol{G}_{4,1},$$

where $s = \langle g \rangle_0, v = \langle g \rangle_1, b = \langle g \rangle_2, t = \langle g \rangle_3, f = \langle g \rangle_4, p = \langle g \rangle_5$, gives

$$g^* = s - v + b - t + f - p, \quad g^\dagger = s + v - b - t + f + p,$$

and

$$\bar{g} = s - v - b + t + f - p,$$

which we use to find

$$\frac{1}{4}[g + g^* + g^\dagger + \bar{g}] = s + f, \quad \frac{1}{4}[g - g^* + g^\dagger - \bar{g}] = v + p,$$

and

$$\frac{1}{4}[g + g^* - g^\dagger - \bar{g}] = b, \quad \frac{1}{4}[g - g^* - g^\dagger + \bar{g}] = t.$$

We see that for a general element in $\boldsymbol{G}_{4,1}$, the conjugation operations cannot by themselves distinguish between a real scalar and a 4-vector, or a vector and a pseudoscalar.

A general (real) vector $x \in \mathbb{R}^{4,1}$ can be written in the form

$$x = \mathbf{x} + \alpha e + \frac{1}{2}\beta\bar{e}, \tag{7}$$

where $\mathbf{x} \in \mathbb{R}^3$ and $\alpha, \beta \in \mathbb{R}$. A general *complex* vector $x \in \boldsymbol{G}_{4,1}^{1+4}$ has the form

$$x = \mathbf{x} + iu\mathbf{y} + (\alpha_1 + iu\alpha_2)e + \frac{1}{2}(\beta_1 + iu\beta_2)\bar{e}$$

The matrix representation of the real x is

$$[x] = \begin{pmatrix} \mathbf{x} & \beta \\ \alpha & -\mathbf{x} \end{pmatrix},$$

and for the complex x,

$$[x] = \begin{pmatrix} \mathbf{x} + i\mathbf{y} & \beta_1 + i\beta_2 \\ \alpha_1 + i\alpha_2 & -\mathbf{x} - i\mathbf{y} \end{pmatrix},$$

as is easily verified. The matrix representation of the complex $x \in \boldsymbol{G}_{4,1}^{1+4}$ is exactly the same as for the real case, except that the real vector $\mathbf{x} \in \mathbb{R}^3$ is replaced by the complex vector $\mathbf{x} + i\mathbf{y} \in \boldsymbol{G}_3^{1+2}$ and the real scalars $\alpha, \beta \in \mathbb{R}$ are replaced by the complex scalars $\alpha_1 + i\alpha_2, \beta_1 + i\beta_2 \in \boldsymbol{G}_3^{0+3}$. The determinant of $[x]$ for both the real and complex x is

$$\det [x] = \mathbf{x}^4 + 2\alpha\beta\mathbf{x}^2 + \alpha^2\beta^2 = (\mathbf{x}^2 + \alpha\beta)^2.$$

The group of all invertible elements of $\boldsymbol{G}_{4,1}$, which is algebraically isomorphic to the general linear group $M_{4\times4}(C)$, is denoted by $\boldsymbol{G}_{4,1}^*$ (see Maks [5– p.18]). The *Lipschitz subgroup* $\Gamma_{4,1}$ of $\boldsymbol{G}_{4,1}^*$ consists of those elements in $\boldsymbol{G}_{4,1}^*$ for which $gx\bar{g} \in \mathbb{R}^{4,1}$ for all $x \in \mathbb{R}^{4,1}$ and is known to be generated by the product of invertible vectors $x \in \mathbb{R}^{4,1}$, see Lounesto [4– p. 220], and Porteous [6– p. 168]. For an invertible element $g \in \Gamma_{4,1}$, it follows that both $gg^\dagger \in \mathbb{R}^* := R - \{0\}$ and $g\bar{g} = \pm gg^\dagger \in \mathbb{R}^*$. This is the basis for the definition of the *pseudo-determinant* of $[g] = \begin{pmatrix} a & b \\ c & d \end{pmatrix}$, $pdet \begin{pmatrix} a & b \\ c & d \end{pmatrix} := ad^\dagger - bc^\dagger \in \mathbb{R}^*$. We see that

$$[g\bar{g}] = \begin{pmatrix} a & b \\ c & d \end{pmatrix} \overline{\begin{pmatrix} a & b \\ c & d \end{pmatrix}} = \begin{pmatrix} a & b \\ c & d \end{pmatrix} \begin{pmatrix} d^\dagger & -b^\dagger \\ -c^\dagger & a^\dagger \end{pmatrix}$$

$$= \begin{pmatrix} ad^\dagger - bc^\dagger & 0 \\ 0 & da^\dagger - cb^\dagger \end{pmatrix} = (ad^\dagger - bc^\dagger) \begin{pmatrix} 1 & 0 \\ 0 & 1 \end{pmatrix},$$

and

$$[gg^\dagger] = \begin{pmatrix} a & b \\ c & d \end{pmatrix} \begin{pmatrix} a & b \\ c & d \end{pmatrix}^\dagger = \begin{pmatrix} a & b \\ c & d \end{pmatrix} \begin{pmatrix} \bar{d} & \bar{b} \\ \bar{c} & \bar{a} \end{pmatrix}$$

$$= \begin{pmatrix} a\bar{d} + b\bar{c} & 0 \\ 0 & d\bar{a} + c\bar{b} \end{pmatrix} = (a\bar{d} + b\bar{c}) \begin{pmatrix} 1 & 0 \\ 0 & 1 \end{pmatrix}$$

from which it follows that

$$\det[gg^\dagger] = (a\bar{d} + b\bar{c})^2 = \det[g\bar{g}] = (ad^\dagger - bc^\dagger)^2.$$

The vanishing of the off diagonal terms

$$cd^\dagger - dc^\dagger = ab^\dagger - ba^\dagger = c\bar{d} + d\bar{c} = a\bar{b} + b\bar{a} = 0 \tag{8}$$

further implies that $a_0 b_0 = \mathbf{a} \circ \mathbf{b}$ and $c_0 d_0 = \mathbf{c} \circ \mathbf{d}$.

It is interesting, and perhaps new, to go through the complexification of the above arguments, see Porteous [6– p. 49, p. 168], which also exactly spells out the relationship of the definition of the pseudodeterminant function to the standard determinant function.

Definition: By a *complex vector* $[x] \in M_{2\times2}$ we mean any element of the form $[x] = \begin{pmatrix} \mathbf{x} & \beta \\ \alpha & -\mathbf{x} \end{pmatrix}$ for $\mathbf{x} \in \boldsymbol{G}_3^{1+2}$ and $\alpha, \beta \in \boldsymbol{G}_3^{0,3}$.

The *complex Lipschitz subgroup* $\Gamma_{4,1}^c$ of $\boldsymbol{G}_{4,1}^*$ consists of those elements of $g \in \boldsymbol{G}_{4,1}^*$ for which $gxg^\dagger \in \boldsymbol{G}_{4,1}^{1+4}$ for all $x \in \boldsymbol{G}_{4,1}^{1+4}$.

Letting $[g] = \begin{pmatrix} a & b \\ c & d \end{pmatrix}$, we find that

$$[g][x][g^\dagger] = \begin{pmatrix} a x \bar{d} + \alpha b \bar{d} + \beta a \bar{c} - b x \bar{c} & a x \bar{b} + \alpha b \bar{b} + \beta a \bar{a} - b x \bar{a} \\ c x \bar{d} + \alpha d \bar{d} + \beta c \bar{c} - d x \bar{c} & c x \bar{b} + \alpha d \bar{b} + \beta c \bar{a} - d x \bar{a} \end{pmatrix}.$$

Examining the complex products, we find that

$$< a x \bar{d} - b x \bar{c} >_{0+3} = \mathbf{x} \circ [b_0 \mathbf{c} - a_0 \mathbf{d} + d_0 \mathbf{a} - c_0 \mathbf{b} + \mathbf{a} \otimes \mathbf{d} - \mathbf{b} \otimes \mathbf{c}] = 0 \qquad (9)$$

for all $\mathbf{x}$, or equivalently, $< \bar{a} d - \bar{b} c >_{1+2} = 0$. We also have

$$< \alpha b \bar{d} + \beta a \bar{c} >_{0+3} = 0$$

for all $\alpha, \beta \in \boldsymbol{G}_3^{0+3}$, or equivalently, $b \bar{d} = -d \bar{b}$ and $a \bar{c} = -c \bar{a}$.

Using the relations (8) and (9), we can now directly relate the pseudodeterminant function to the ordinary determinant function for elements $g \in \Gamma_{4,1}^c$. We find that

$$\det[g] = \det \begin{pmatrix} a & b \\ b & c \end{pmatrix} = (\bar{d} a \bar{a} - \bar{b} a \bar{c}) \frac{1}{a \bar{a}} (a \bar{a} d - c \bar{a} b)$$

$$= (\bar{d} a \bar{a} + \bar{b} c \bar{a}) \frac{1}{a \bar{a}} (a \bar{a} d + a \bar{c} b) = (\bar{d} a + \bar{b} c)(\bar{a} d + \bar{c} b)$$

$$= (\bar{a} d + \bar{c} b)^2 = (\mathrm{pdet}[g])^2,$$

since $\bar{a} d + \bar{c} b = < \bar{a} d + \bar{c} b >_{0+3}$. Whereas it is clear that $\Gamma_{4,1} \subset \Gamma_{4,1}^c$, it is not obvious that the condition (9) implies that $g \in \Gamma_{4,1}^c$ is a product of invertible complex vectors in $\boldsymbol{G}_{4,1}^{1+4}$.

5   3-Dimensional Horosphere

We now consider special representations of the points $\mathbf{x} \in \mathbb{R}^3$ in the larger pseudoeuclidean space $\mathbb{R}^{4,1}$. We have already seen in (7) that each real vector $x \in \mathbb{R}^{4,1}$ has the form $x = \mathbf{x} + \alpha e + \frac{1}{2} \beta \bar{e}$, where $\alpha, \beta \in \mathbb{R}$. If $\alpha = 1$ and $\beta = 0$, we say that $x_h = \mathbf{x} + e \in \mathbb{R}^{4,1}$ represents $\mathbf{x} \in \mathbb{R}^3$ in the *affine plane*

$$\mathcal{A}_e(\mathbb{R}^3) := \{x = \mathbf{x} + e | \ \mathbf{x} \in \mathbb{R}^3\}.$$

Properties of the affine plane $\mathcal{A}_e(\mathbb{R}^3)$ have been studied in Bayro and Sobczyk [1– p. 35, p. 263]. The *horosphere* $H(\mathbb{R}^3)$ is a *non linear* representation of the points in Euclidean space $\mathbb{R}^3$, defined by the condition that $x_c := x_h + \beta \bar{e} = \mathbf{x} + e + \beta \bar{e}$ is a null vector for all $\mathbf{x} \in \mathbb{R}^3$. We see that

$$x_c^2 = x_h^2 + 2 \beta \bar{e} \cdot x_h = \mathbf{x}^2 + 2 \beta = 0$$

or $\beta = -\frac{1}{2} \mathbf{x}^2$. Thus,

$$H(\mathbb{R}^3) := \{x_c = \mathbf{x} - \frac{\mathbf{x}^2}{2} \bar{e} + e | \ \mathbf{x} \in \mathbb{R}^3\}.$$

Note that the horosphere can equally well be considered to be the horosphere of homogeneous points, since $x_c = \frac{\alpha x_c}{\bar{e} \cdot (\alpha x_c)}$ for all $x_c \in H(R^3)$ and $\alpha \in I\!\!R^*$.

Calculating $[x_c]$, we find

$$[x_c] = [x_h] - [\frac{\mathbf{x}^2}{2}\bar{e}] = \begin{pmatrix} \mathbf{x} & -\mathbf{x}^2 \\ 1 & -\mathbf{x} \end{pmatrix} = \begin{pmatrix} \mathbf{x} \\ 1 \end{pmatrix} (1 \quad -\mathbf{x}).$$

The fact that $[x_c]$ can be written as the product of a column matrix and a row matrix suggest that we may equally well represent points in the horosphere by the *column h-twistor* $[x_c]_t$, defined by

$$[x_c]_t = \begin{pmatrix} \mathbf{x} \\ 1 \end{pmatrix}.$$

More generally, by the space of *column h-twistors* $\mathcal{T}_{\mathbf{G}_3}$ of $\mathbf{G}_3$ we mean

$$\mathcal{T}_{\mathbf{G}_3} := \{[w]_t = \begin{pmatrix} a \\ b \end{pmatrix} \mid a, b \in \mathbf{G}_3\}. \tag{10}$$

Given a column h-twistors $[w]_t = \begin{pmatrix} a \\ b \end{pmatrix}$, it is natural to define a corresponding *conjugate row h-twistor* by $[w]_t^\dagger := (\bar{b} \quad \bar{a})$, and also an *h-twistor inner product* by

$$< [w_1]_t, [w_2] >_t := [w_1]_t^\dagger [w_2]_t = \bar{b_1} a_2 + \bar{a_1} b_2 \in \mathbf{G}_3^{1+3},$$

where $[w_1]_t = \begin{pmatrix} a_1 \\ b_1 \end{pmatrix}$ and $[w_2]_t = \begin{pmatrix} a_2 \\ b_2 \end{pmatrix}$ are h-twistors.

With these definitions in tow, we can now express any point x_c on the horosphere by

$$[x_c] = [x_c]_t [x_c]_t^\dagger. \tag{11}$$

Actually, we have gained much more. Let us say that two h-twistor are *equivalent*, $[w_1]_t \equiv [w_2]_t$ if $[w_1]_t [w_1]_t^\dagger = [w_2]_t [w_2]_t^\dagger$, and that they are *projectively equivalent* if $[w_1]_t [w_1]_t^\dagger = \alpha [w_2]_t [w_2]_t^\dagger$ for $\alpha \in I\!\!R^*$. It follows that $[x_c]_t$ and $[x_c h]_t$ are projectively equivalent for all $h \in \mathbf{G}_3$ such that $h\bar{h} \in I\!\!R^*$. Thus, points on the horosphere need only be defined up to an invertible multivector $h \in \mathbf{G}_3$. The concept of an h-twistor cuts calculations on the horosphere in half. For example, for any $g \in \mathbf{G}_{4,1}$ with $[g] = \begin{pmatrix} a & b \\ c & d \end{pmatrix}$

$$[g x_c g^\dagger] = [g][x_c][g]^\dagger = [g][x_c]_t ([g][x_c]_t)^\dagger.$$

Reflections in $I\!\!R^3$, when represented on the horosphere, have the homogeneous form

$$S_{\mathbf{a}}(x_c) := \mathbf{a} x_c \mathbf{a} = \mathbf{a} \mathbf{x} \mathbf{a} - \mathbf{a}^2 (e - \frac{\mathbf{x}^2}{2}\bar{e}).$$

In terms of the h-twistor representation, we have

$$[S_{\mathbf{a}}(x_c)] = ([\mathbf{a}][x_c]_t)([\mathbf{a}][x_c]_t)^\dagger = \begin{pmatrix} \mathbf{a} & 0 \\ 0 & -\mathbf{a} \end{pmatrix} \begin{pmatrix} \mathbf{x} \\ 1 \end{pmatrix} \begin{pmatrix} \mathbf{x} \\ 1 \end{pmatrix}^\dagger \begin{pmatrix} \mathbf{a} & 0 \\ 0 & -\mathbf{a} \end{pmatrix}^\dagger$$

$$= \begin{pmatrix} \mathbf{ax} \\ -\mathbf{a} \end{pmatrix} (\mathbf{a} \quad \mathbf{xa}) = \begin{pmatrix} \mathbf{axa} & \mathbf{a}^2\mathbf{x}^2 \\ -\mathbf{a}^2 & -\mathbf{axa} \end{pmatrix}.$$

Rotations are the composition of two reflections. We find that

$$S_{\mathbf{b}}S_{\mathbf{a}}(x_c) = \mathbf{ba}x_c(\mathbf{ba})^\dagger = \mathbf{ba}x_c\mathbf{ab}.$$

In terms of the h-twistor construction, we find

$$[S_{\mathbf{c}}S_{\mathbf{a}}(x_c)] = ([\mathbf{ba}][x_c]_t)([\mathbf{ba}][x_c]_t)^\dagger = \begin{pmatrix} \mathbf{ba} & 0 \\ 0 & \mathbf{ba} \end{pmatrix} \begin{pmatrix} \mathbf{x} \\ 1 \end{pmatrix} \begin{pmatrix} \mathbf{x} \\ 1 \end{pmatrix}^\dagger \begin{pmatrix} \mathbf{ba} & 0 \\ 0 & \mathbf{ba} \end{pmatrix}^\dagger$$

$$= \begin{pmatrix} \mathbf{bax} \\ \mathbf{ba} \end{pmatrix} (\mathbf{ab} \quad -\mathbf{xab}) = \begin{pmatrix} \mathbf{baxab} & -\mathbf{a}^2\mathbf{b}^2\mathbf{x}^2 \\ \mathbf{a}^2\mathbf{b}^2 & -\mathbf{baxab} \end{pmatrix}.$$

We can also represent *translations* in the horosphere. For $\mathbf{a} \in \mathbb{R}^3$,

$$T_{\mathbf{a}}(x_c) := (1 + \frac{\mathbf{a}\bar{e}}{2})x_c(1 - \frac{\mathbf{a}\bar{e}}{2}).$$

In terms of the h-twistor construction,

$$[T_{\mathbf{a}}(x_c)] = [1 + \frac{\mathbf{a}\bar{e}}{2}][x_c]_t([1 + \frac{\mathbf{a}\bar{e}}{2}][x_c]_t)^\dagger$$

$$= \begin{pmatrix} 1 & \mathbf{a} \\ 0 & 1 \end{pmatrix} \begin{pmatrix} \mathbf{x} \\ 1 \end{pmatrix} \begin{pmatrix} \mathbf{x} \\ 1 \end{pmatrix}^\dagger \begin{pmatrix} 1 & \mathbf{a} \\ 0 & 1 \end{pmatrix}^\dagger = \begin{pmatrix} \mathbf{x} + \mathbf{a} \\ 1 \end{pmatrix} (1 \quad -\mathbf{x} - \mathbf{a})$$

$$= \begin{pmatrix} \mathbf{x} + \mathbf{a} & -(\mathbf{x} + \mathbf{a})^2 \\ 1 & -\mathbf{x} - \mathbf{a} \end{pmatrix}.$$

For a general element $g \in G_{4,1}^*$, with $[g] = \begin{pmatrix} a & b \\ c & d \end{pmatrix}$, the h-twistor transformation

$$\begin{pmatrix} a & b \\ c & d \end{pmatrix} \begin{pmatrix} \mathbf{x} \\ 1 \end{pmatrix} = \begin{pmatrix} a\mathbf{x} + b \\ c\mathbf{x} + d \end{pmatrix},$$

leads to the general linear fraction *Möbius transformation* or conformal transformation

$$f(\mathbf{x}) = (a\mathbf{x} + b)(c\mathbf{x} + d)^{-1},$$

because of the projective equivalence of the h-twistors

$$\begin{pmatrix} a\mathbf{x} + b \\ c\mathbf{x} + d \end{pmatrix} \quad \text{and} \quad \begin{pmatrix} (a\mathbf{x} + b)(c\mathbf{x} + d)^{-1} \\ 1 \end{pmatrix}$$

at all points where $(c\mathbf{x} + d)^{-1}$ is defined, see Pozo and Sobczyk [9], Lounesto [4, p. 244], and Porteous [6, p. 245].

To check that any linear fractional transformation $f(\mathbf{x})$ does indeed define a conformal transformation, we set $f = f(\mathbf{x})$ and calculate the differential of the equation $f(\mathbf{x})(c\mathbf{x} + d) = (a\mathbf{x} + b)$, getting $df(c\mathbf{x} + d) + fc\,d\mathbf{x} = a\,d\mathbf{x}$ or $df = (a - fc)d\mathbf{x}(c\mathbf{x} + d)^{-1}$. Continuing the calculation,

$$df = \frac{[a(c\mathbf{x} + b)(\bar{d} - \mathbf{x}\bar{c}) - (a\mathbf{x} + b)(\bar{d} - \mathbf{x}\bar{c})c]d\mathbf{x}(\bar{d} - \mathbf{x}\bar{c})}{(c\mathbf{x} + b)^2(\bar{d} - \mathbf{x}\bar{c})^2}.$$

We now work to simplify the first part of the numerator, in particular, our objective is to move $\mathbf{x}$ to the right with the help of the relations (8) and (9),

$$a(c\mathbf{x} + b)(\bar{d} - \mathbf{x}\bar{c}) - (a\mathbf{x} + b)(\bar{d} - \mathbf{x}\bar{c})c$$

$$= ac\mathbf{x}\bar{d} - ad\mathbf{x}\bar{c} + bc\bar{c}c - a\mathbf{x}\bar{d}c + ad\bar{d} - ac\bar{c}\mathbf{x}^2 - b\bar{d}c + a\mathbf{x}^2 c\bar{c}$$

$$= ac\mathbf{x}\bar{d} - ad\mathbf{x}\bar{c} + bc\bar{c}c - a\mathbf{x}\bar{d}c + d(\bar{d}a + \bar{b}c)$$

$$= ac\mathbf{x}\bar{d} + (ad\mathbf{x} + b\mathbf{x}c)\bar{c} - a\mathbf{x}\bar{d}c + d\,\mathrm{pdet}(g)$$

$$= \cdots = -a\bar{c}d\mathbf{x} + b\bar{c}c\mathbf{x} + d\,\mathrm{pdet}(g) = \mathrm{pdet}(g)(c\mathbf{x} + d).$$

Putting everything back together gives the result

$$df = \frac{\mathrm{pdet}(g)(c\mathbf{x} + d)d\mathbf{x}(\bar{d} - \mathbf{x}\bar{c})}{(c\mathbf{x} + d)^2(\bar{d} - \mathbf{x}\bar{c})^2} = \mathrm{pdet}(g)(\bar{d} - \mathbf{x}\bar{c})^{-1}d\mathbf{x}(c\mathbf{x} + d)^{-1}.$$

Squaring both sides gives the differential relation

$$(df)^2 = \frac{\det g}{(c\mathbf{x} + d)^2(\mathbf{x}\bar{c} - \bar{d})^2}(d\mathbf{x})^2,$$

which shows that $\mathbf{y} = f(\mathbf{x})$ is conformal at all points at which $\det(c\mathbf{x} + d) \neq 0$. (Note in the calculations above care must be taken to always distinguish between the *differential* $d\mathbf{x}$ and the *geometric product* $d\mathbf{x}$ of d and $\mathbf{x}$.)

6   Conversions

In this section, we give formulas for the representation of elements in $\boldsymbol{G}_3$ as 2×2 matrices of complex scalars, and the representation of elements in $\boldsymbol{G}_{4,1}$ as 4×4 complex matrices.

We have already seen in (6) that $g \in \boldsymbol{G}_{4,1}$ can be written in the form

$$g = au_+ + bu_+\sigma + c^*u_-\sigma + d^*u_- = (1 \quad \sigma)u_+ \begin{pmatrix} a & b \\ c & d \end{pmatrix}\begin{pmatrix} 1 \\ \sigma \end{pmatrix}. \qquad (12)$$

Note now that $a = a_0 + a_1 e_1 + a_2 e_2 + a_3 e_3 \in \boldsymbol{G}_3$, where each $a_\mu \in \boldsymbol{G}_3^{0+3}$, can also be expressed in the matrix form

$$a = (1 \quad e_1)v_+ \begin{pmatrix} a_0 + a_3 & a_1 - ia_2 \\ a_1 + ia_2 & a_0 - a_3 \end{pmatrix}\begin{pmatrix} 1 \\ e_1 \end{pmatrix},$$

where $v_\pm = \frac{1}{2}(1 \pm e_3)$ and $i = e_1 e_2 e_3$. We can put these last two expressions together, by using the Kronecker product of matrices, to find that (where $u = \sigma \gamma$)

$$g = $$

$$(1 \quad \sigma \quad e_1 \quad \sigma e_1) u_+ v_+ \begin{pmatrix} a_0 + a_3 & a_1 - ia_2 & b_0 + b_3 & b_1 - ib_2 \\ a_1 + ia_2 & a_0 - a_3 & b_1 + ib_2 & b_0 - b_3 \\ c_0 + c_3 & c_1 - ic_2 & d_0 + d_3 & d_1 - id_2 \\ c_1 + ic_2 & c_0 - c_3 & d_1 + id_2 & d_0 - d_3 \end{pmatrix} \begin{pmatrix} 1 \\ \sigma \\ e_1 \\ e_1 \sigma \end{pmatrix}.$$

The matrix representation (12) can be directly solved for each of the elements $a, b, c, d \in \mathbf{G}_3$. Noting that $au_+ = u_+ g u_+$ and $au_- = u_- \sigma g^* \sigma u_-$, and similar expressions for the other components, we have

$$a = au_+ + au_- = u_+ g u_+ + u_- \sigma g^* \sigma u_-,$$

$$b = bu_+ + bu_- = u_+ g \sigma u_+ - u_- \sigma g^* u_-,$$

$$c = cu_- + cu_- = u_+ \sigma g u_+ - u_- g^* \sigma u_-,$$

and

$$d = du_+ + du_- = u_+ \sigma g \sigma u_+ + u_- g^* u_-,$$

or

$$\begin{pmatrix} a & b \\ c & d \end{pmatrix} = u_+ \begin{pmatrix} g & g\sigma \\ \sigma g & \sigma g\sigma \end{pmatrix} u_+ + u_- \begin{pmatrix} \sigma g^* \sigma & -\sigma g^* \\ -g^* \sigma & g^* \end{pmatrix} u_-.$$

7   Mathematica Implementation of $M(2, \mathbf{G}_3)$

The algebra described above encompasses multiple common subspaces currently in use by the geometric algebra community. It is therefore appropriate to develop a symbolic package that would allow the manipulation of elements in $\mathbf{G}_3$ and $M(2, \mathbf{G}_3)$. A useful package should include tools to convert between $\mathbf{G}_{4,1}$ and $M(2, \mathbf{G}_3)$, and output expressions in both spaces in a format consistent with standard mathematical notation. This is accomplished through the formatting facilities provided by the Mathematica system. Although many important details are missing from this exposition, we hope the examples motivate the reader to try out the package and provide feedback. The package has been developed to operate at a symbolic level, and provides comprehensive simplification and manipulation tools. As much as possible, we avoid decomposing expressions into their coordinate components.

In order to provide a maximum amount of flexibility, we have constructed a system based on a low level representation for paravectors and paramatrices. A paravector is defined as a scalar and a vector pair $a = a_0 + \mathbf{a}$. Input of paravectors takes several forms: pV[a], pVv[a], and pVs[a], and a useful shortcut pV[a, b, c] to set multiple paravectors. The first form defines the paravector as a symbolic value. The second defines it as a scalar/vector pair, and the third defines the paravector in terms of its four components. A paramatrix similarly has similar input mechanisms, $\mathsf{pM}[a]$, $\mathsf{pMs}[a]$, and $\mathsf{pMs}[a, b, c, d]$, where in the last case, the arguments of pM are paravectors. The basic operations on elements of $\boldsymbol{G}_3$ are the oWedge (outer product), oDot (inner product), and oGeom (geometric product) operators defined earlier in the text. Note however, that we have modified the oDot operator to act on general elements of $\boldsymbol{G}_3$ according to $a \odot b := a_v \odot b_v + a_0 b_v + a_v b_0$. Commutativity of the geometric product, invariance of the triple product with respect to argument cycling, and other properties (such as $a \otimes a = 0$) are built into the system to ease the simplification task.

The operator $\mathsf{expandAll}$ provides the means for generic simplification. All products are maximally expanded, consistent with the chosen representation for either the paravectors or paramatrices. For example,

$$\mathrm{In}[10] := \quad aa = \mathsf{pVv}[a];\, \mathsf{oGeom}[aa, aa]//\mathsf{expandAll}$$
$$\mathrm{Out}[10] = \quad \mathbf{a}_v^2 + 2\mathbf{a}_v a_0 + a_0^2 \tag{13}$$

In its internal implementation, $\mathsf{expandAll}$ successively applies the rules $\mathsf{flattenGeomRules}$, $\mathsf{geomRules}$, $\mathsf{expandRules}$, $\mathsf{expandpVRules}$, $\mathsf{expandDotRules}$, $\mathsf{expandWedgeRules}$, $\mathsf{conjugationRules}$, $\mathsf{inversionRules}$, $\mathsf{reversionRules}$, $\mathsf{tripleRules}$, $\mathsf{orderRules}$, and $\mathsf{ExpandAll}$ until there is no further change. The definitions of these rules can be found in the source files. The above rules apply to paramatrices as well, and act on a component by component basis.

The operations of conjugation, inversion, and reversion have been implemented via the operators of the same name, along with appropriate simplification rules. Two such rules are that each of these operators, when squared, is the identity operator, and that a reversion applied to an inversion is a conjugation.

One of the important simplification strategies is to recognize and extract scalar quantities from complex expressions. For example, the geometric product $\mathsf{oGeom}[a, \mathsf{conjugation}[a]]$ is known to be a scalar:

$$\mathrm{In}[14] := \quad \{aa, bb, cc\} = \mathsf{pV}[a, b, c];\, \mathsf{oGeom}[ba, aa, \mathsf{conjugation}[aa], cc]$$
$$\mathrm{Out}[14] = \quad \mathsf{scalar}[a\bar{a}]\mathsf{oGeom}[b, c]$$

The addition of the scalar operator was necessary to clearly differentiate the geometric multiplication from regular multiplication, both represented without an explicit operator. In early experiments, we also found that not isolating scalar components within an explicit operator could lead to ambiguities and errors when working with expressions involving conjugation. Two very useful operators are $\mathsf{scalarPartF}$ and $\mathsf{vectorPartF}$. When applied to a general paravector expression, they extract its scalar and vector parts respectively. If this extraction is not explicitly possible, the desired operation is indicated by enclosing the expression in angular brackets with a subscript s or v.

An interesting feature of our package is the ability to check potential identities. Because there is no general theory of simplification, different equivalent mathematical expressions may simplify in different ways. As a consequence, it is not always straightforward to prove whether a proposed identity is or is not valid.

We make available the operator `convertToComponents`, which simply replaces every paravector to component form. For example, if $aa = \mathsf{pVv}[a]$,

$$\mathsf{In}[13] := \mathsf{convertToComponents}[aa]$$
$$\mathsf{Out}[13] = LL[a_0, a_1, a_2, a_3] \tag{14}$$

The arguments of LL are the four components of the paravector pV.

Consider the trace operator applied to n paravectors p_1 through p_n. It is defined as the scalar part of the geometric product of its arguments. Alternatively, $\mathsf{dtrace}[expr\_] := (expr + \mathsf{conjugation}[expr])/2$ It is easy to prove that the trace acting on n arguments in $\boldsymbol{G}_3$ is invariant with respect to cyclic shifting of its arguments. With the above definitions, consider two *supposedly* equivalent forms for the trace of the geometric product of six *pure* vectors $a_v, b_v, c_v, d_v, e_v, f_v$:

$$((a_v \otimes b_v) \odot f_v)((c_v \otimes d_v) \otimes e_v) - ((a_v \otimes b_v) \odot e_v)((c_v \otimes d_v) \odot f_v)$$
$$+((a_v \odot f_v)(b_v \odot e_v) - (a_v \odot e_v)(b_v \odot f_v))(c_v \odot d_v)$$
$$+(a_v \odot b_v)((c_v \odot f_v)(d_v \odot e_v) - (c_v \odot e_v)(d_v \odot f_v))$$
$$+((a_v \odot d_v)(b_v \odot c_v) - (a_v \odot c_v)(b_v \odot d_v) + (a_v \odot b_v)(c_v \odot d_v))(e_v \odot f_v)$$

and

$$(a_v \odot f_v)(b_v \odot c_v)(d_v \odot e_v) - (a_v \odot c_v)(b_v \odot f_v)(d_v \odot e_v)$$
$$+(a_v \odot b_v)(c_v \odot f_v)(d_v \odot e_v) - (a_v \odot e_v)(b_v \odot c_v)((d_v \odot f_v)$$
$$+(a_v \odot c_v)(b_v \odot e_v)(d_v \odot f_v) - (a_v \odot b_v)(c_v \odot e_v)(d_v \odot f_v)$$
$$+(a_v \odot d_v)(b_v \odot c_v)(e_v \odot f_v) - (a_v \odot c_v)(b_v \odot d_v)(e_v \odot f_v)$$
$$+(a_v \odot b_v)(c_v \odot d_v)(e_v \odot f_v) + \mathsf{triple}[a_v, b_v, c_v]\mathsf{triple}[d_v, e_v, f_v]$$

Subtraction and simplification produces an even longer expression. Applying the `convertToComponents` operator to the difference of the above two expressions gives $LL[0, 0, 0, 0]$, which indicates that the two expressions are in fact equal.

As an application of our symbolic package, we compute an explicit formula for the determinant of a 3×3 matrix $A = \{a_{ij}\}$ of elements in $\boldsymbol{G}_3$ in terms of its elements and their conjugates. We wrote a function to implement a version of Gaussian elimination to accomplish the task. Following the rules prescribed in the paper,

$$\begin{pmatrix} a_{11} & a_{12} & a_{13} \\ a_{21} & a_{22} & a_{23} \\ a_{31} & a_{32} & a_{33} \end{pmatrix} = \det(a_{11})^{-1} \begin{pmatrix} \overline{a_{11}}a_{11} & \overline{a_{11}}a_{12} & \overline{a_{11}}a_{13} \\ a_{21} & a_{22} & a_{23} \\ a_{31} & a_{32} & a_{33} \end{pmatrix}$$
$$= \det(a_{11})^{-3} \begin{pmatrix} a_{22}\det(a_{11}) - a_{21}\overline{a_{11}}a_{12} & a_{23}\det(a_{11}) - a_{21}\overline{a_{11}}a_{13} \\ a_{32}\det(a_{11}) - a_{31}\overline{a_{12}}a_{12} & a_{33}\det(a_{11}) - a_{21}\overline{a_{12}}a_{13} \end{pmatrix} \tag{15}$$

where $\det(a_{11}) = a_{11}\overline{a_{11}}$. We have built in a formula for the determinant of a general element of $M(2, \boldsymbol{G}_3)$ through the `determinant` operator. For example,

$$\mathsf{In}[17] := \mathsf{determinant}[\mathsf{pMs}[b]] \tag{16}$$
$$\mathsf{Out}[17] = a_{11}\overline{a_{11}}a_{22}\overline{a_{22}} + a_{12}\overline{a_{12}}a_{21}\overline{a_{21}} - a_{11}\overline{a_{21}}a_{22}\overline{a_{12}} - a_{12}\overline{a_{22}}a_{21}\overline{a_{11}}$$

Clearly, using (17) to compute (15) would be very tedious. Instead, we let Mathematica do the work for us. A direct evaluation of the above determinant generates

$$D_{13}\,D_{22}\,D_{31} + D_{12}\,D_{23}\,D_{31} + D_{13}\,D_{21}\,D_{32}$$
$$+\,D_{11}\,D_{23}\,D_{32} + D_{12}\,D_{21}\,D_{33} + D_{11}\,D_{22}\,D_{33}$$
$$-\,D_{11}\,\mathrm{Tr}(c_{22},\overline{c}_{32},c_{33},\overline{c}_{23}) - D_{21}\,\mathrm{Tr}(c_{12},\overline{c}_{32},c_{33},\overline{c}_{13}) - D_{31}\,\mathrm{Tr}(c_{22},\overline{c}_{12},c_{13},\overline{c}_{23})$$
$$+\,\frac{4\mathrm{Tr}(\overline{c_{11}}c_{12}\overline{c_{32}}c_{31}\,\mathrm{Tr}\overline{c_{11}}c_{13}\overline{c_{23}}c_{21})}{D_{11}} + \frac{4\mathrm{Tr}(\overline{c_{11}}c_{12}\overline{c_{22}}c_{21}\,\mathrm{Tr}\overline{c_{11}}c_{13}\overline{c_{33}}c_{31})}{D_{11}}$$
$$-\,\frac{2\mathrm{Tr}(\overline{c_{11}}c_{12}\overline{c_{32}}c_{31}\overline{c_{11}}c_{13}\overline{c_{23}}c_{21})}{D_{11}} - \frac{2\mathrm{Tr}(\overline{c_{11}}c_{13}\overline{c_{33}}c_{31}\overline{c_{11}}c_{12}\overline{c_{22}}c_{21})}{D_{11}}$$
$$-\,D_{12}\,\mathrm{Tr}(\overline{c}_{21},c_{23},\overline{c}_{33},c_{31}) - D_{22}\,\mathrm{Tr}(\overline{c}_{11},c_{13},\overline{c}_{33},c_{31}) - D_{32}\,\mathrm{Tr}(\overline{c}_{11},c_{13},\overline{c}_{23},c_{21})$$
$$-\,D_{13}\,\mathrm{Tr}(\overline{c}_{21},c_{22},\overline{c}_{32},c_{31}) - D_{23}\,\mathrm{Tr}(\overline{c}_{11},c_{12},\overline{c}_{32},c_{31}) - D_{33}\,\mathrm{Tr}(\overline{c}_{11},c_{12},\overline{c}_{22},c_{21})$$
$$+\,\mathrm{Tr}(\overline{c}_{11},c_{12},\overline{c}_{32},c_{33},\overline{c}_{23},c_{21}) + \mathrm{Tr}(\overline{c}_{11},c_{13},\overline{c}_{33},c_{32},\overline{c}_{22},c_{21})$$
$$+\,\mathrm{Tr}(c_{12},\overline{c}_{32},c_{31},\overline{c}_{21},c_{23},\overline{c}_{13}) + \mathrm{Tr}(c_{13},\overline{c}_{33},c_{31},\overline{c}_{21},c_{22},\overline{c}_{12})$$
$$+\,\mathrm{Tr}(c_{22},\overline{c}_{12},c_{11},\overline{c}_{31},c_{33},\overline{c}_{23}) + \mathrm{Tr}(c_{22},\overline{c}_{32},c_{31},\overline{c}_{11},c_{13},\overline{c}_{23}), \tag{17}$$

where $D_{ij} = a_{ij}\overline{a_{ij}}$. In order to further simplify this expression (i.e., remove the denominator, which we know should not be there), we created a rule that recognizes that the trace is invariant with respect to the cyclic permutation of its arguments. Furthermore, $\mathrm{Tr}(a)\mathrm{Tr}(b) = \frac{1}{2}[\mathrm{Tr}(a,b) + \mathrm{Tr}(\overline{a},b)]$ and $\mathrm{Tr}(a\overline{a}b) = a\overline{a}\mathrm{Tr}b$. Taking these rules into account, we finally find the following for the determinant of a 3×3 matrix of elements in $M(2, \boldsymbol{G}_3)$:

$$\det = D_{13}\,D_{22}\,D_{31} + D_{12}\,D_{23}\,D_{31} + D_{13}\,D_{21}\,D_{32}$$
$$+\,D_{11}\,D_{23}\,D_{32} + D_{12}\,D_{21}\,D_{33} + D_{11}\,D_{22}\,D_{33}$$
$$-\,D_{11}\,\mathrm{Tr}(c_{22},\overline{c}_{32},c_{33},\overline{c}_{23}) - D_{21}\,\mathrm{Tr}(c_{12},\overline{c}_{32},c_{33},\overline{c}_{13}) - D_{31}\,\mathrm{Tr}(c_{22},\overline{c}_{12},c_{13},\overline{c}_{23})$$
$$-\,D_{12}\,\mathrm{Tr}(\overline{c}_{21},c_{23},\overline{c}_{33},c_{31}) - D_{22}\,\mathrm{Tr}(\overline{c}_{11},c_{13},\overline{c}_{33},c_{31}) - D_{32}\,\mathrm{Tr}(\overline{c}_{11},c_{13},\overline{c}_{23},c_{21})$$
$$-\,D_{13}\,\mathrm{Tr}(\overline{c}_{21},c_{22},\overline{c}_{32},c_{31}) - D_{23}\,\mathrm{Tr}(\overline{c}_{11},c_{12},\overline{c}_{32},c_{31}) - D_{33}\,\mathrm{Tr}(\overline{c}_{11},c_{12},\overline{c}_{22},c_{21})$$
$$+\,\mathrm{Tr}(\overline{c}_{11},c_{12},\overline{c}_{32},c_{33},\overline{c}_{23},c_{21}) + \mathrm{Tr}(\overline{c}_{11},c_{13},\overline{c}_{33},c_{32},\overline{c}_{22},c_{21})$$
$$+\,\mathrm{Tr}(c_{12},\overline{c}_{32},c_{31},\overline{c}_{21},c_{23},\overline{c}_{13}) + \mathrm{Tr}(c_{13},\overline{c}_{33},c_{31},\overline{c}_{21},c_{22},\overline{c}_{12})$$
$$+\,\mathrm{Tr}(c_{22},\overline{c}_{12},c_{11},\overline{c}_{31},c_{33},\overline{c}_{23}) + \mathrm{Tr}(c_{22},\overline{c}_{32},c_{31},\overline{c}_{11},c_{13},\overline{c}_{23}) \tag{18}$$

To the authors' knowledge, this explicit form has not been presented previously. While the structure of this expression is clear, an analytical derivation is still out of reach. Documentation, along with the package source code and instructions for use can be found online ([2]).

Acknowledgements

The first author thanks the School of Research (DIP) at the Universidad de Las Americas - Puebla for support of this work, and is a member of the National Research Community of Mexico (SNI) No. 14587. He also thanks the organizers for their gracious hospitality in China. The second author performed this work while on sabbatical at the Mathematics Department at the Universidad de Las Americas, Puebla.

References

1. Bayro Corrochano, E., Sobczyk, G.: Geometric Algebra with Applications in Science and Engineering. Birkhäuser, Boston 2001
2. Erlebacher, G. and Sobczyk, G.: New Geometric Algebra Symbolic Package, http://www.csit.fsu.edu/~erlebach/geometric_algebra/gl3
3. Gantmacher F. R.: The Theory of Matrices, Vol. 1. Chelsea Publishing Company, New York 1960
4. Lounesto, P.: Clifford Algebras and Spinors. Cambridge University Press, Cambridge 2001
5. Maks, J. G.: Modulo (1,1) Periodicity of Clifford Algebras and Generalized Anti Möbius Transformations. Doctoral Thesis, Delft Technical University, Netherlands 1989
6. Porteous, I. R.: Clifford Algebras and the Classical Groups. Cambridge University Press, Cambridge 1995
7. http://www.cs.toronto.edu/~roweis/notes/matrixid.pdf
8. Sobczyk, G.: Spacetime Vector Analysis. Physics Letters A: **84A** (1981) 45–48
9. Pozo, J. M., Sobczyk, G.: Geometric Algebra in Linear Algebra and Geometry. Acta Applicandae Mathematicae **71** (2002) 207–244
10. Tian, Y.: Matrix Theory over the Complex Quaternion Algebra. arXiv:math.L/0004005 v1 1 Apr 2000

Intrinsic Differential Geometry with Geometric Calculus⋆

Hongbo Li, Lina Cao, Nanbin Cao, and Weikun Sun

Mathematics Mechanization Key Laboratory,
Academy of Mathematics and Systems Science,
Chinese Academy of Sciences,
Beijing 100080, P. R. China

Abstract. Setting up a symbolic algebraic system is the first step in mathematics mechanization of any branch of mathematics. In this paper, we establish a compact symbolic algebraic framework for local geometric computing in intrinsic differential geometry, by choosing only the Lie derivative and the covariant derivative as basic local differential operators. In this framework, not only geometric entities such as the curvature and torsion of an affine connection have elegant representations, but their involved local geometric computing can be simplified.

Keywords: Intrinsic differential geometry, Clifford algebra, Mathematics mechanization, Symbolic geometric computing.

1 Introduction

Mathematics mechanization focuses on solving mathematical problems with symbolic computation techniques, particularly on mathematical reasoning by algebraic manipulation of mathematical symbols. In differential geometry, the mechanization viz. mechanical theorem proving, is initiated by [7] using local coordinate representation. On the other hand, in modern differential geometry the dominant algebraic framework is moving frames and differential forms [1], which are independent of local coordinates. For mechanical theorem proving in spatial surface theory, [3], [4], [5] proposed to use differential forms and moving frames as basic algebraic tools. It appears that differential geometry benefits from both local coordinates and global invariants [6].

For *extrinsic differential geometry*, [2] proposed to use Clifford algebra and vector derivative viz. Dirac operator as basic algebraic tools. This framework combines local and global representations, and conforms with the classical framework of vector analysis. The current paper intends to extend this representation to *intrinsic differential geometry*.

In intrinsic differential geometry, there is already an algebraic system abundant in coordinate-free operators:

⋆ Supported partially by NKBRSF 2004CB318001 and NSFC 10471143.

H. Li, P. J. Olver and G. Sommer (Eds.): IWMM-GIAE 2004, LNCS 3519, pp. 207–216, 2005.

Algebraic operators: tensor product $\otimes$, exterior product $\wedge$, interior product i_x, contraction $\mathcal{C}$, the pairing $\rfloor$ between a vector space and its dual, Riemannian metric tensor g, etc.

Differential operators: exterior differentiation d, Lie derivative $\mathcal{L}_x$, the bracket $[x, y]$ of tangent vector fields x, y, covariant differentiation D_x, curvature tensor R, torsion tensor T, etc.

In our opinion, the existing algebraic system contains a lot of redundancy, which may complicate its mechanization. The first task should be to reduce the number of basic operators in the above list, in order to gain both compact representation and effective symbolic manipulation for both local and global geometric computing. This task is fulfilled in this paper.

We propose a compact algebraic framework for local geometric computing in intrinsic differential geometry. We select the following operators as basic ones:

Basic algebraic operators: tensor product $\otimes$, exterior product $\wedge$, interior product ".".

Basic local differential operators: Lie derivative $\mathcal{L}_a$, vector derivative ∂, covariant differentiations δ_a and δ.

Basic global differential operators: Lie derivative $\mathcal{L}_x$, exterior differentiation d, covariant differentiation δ.

This framework is suitable for studying intrinsic properties of general vector bundles, affine connections, Riemannian vector bundles, etc, and can be extended to include spin structure and integration. Some typical benefits of this framework include:

1. Exterior differentiation d is equal to the composition of the covariant differentiation δ_a (or δ in the case of a torsion-free affine connection) and the wedge product.
2. The curvature and torsion tensors can be represented elegantly by covariant differentiation. By these representations, the famous Bianchi identities are given "transparent" algebraic representations.
3. Some geometric theorems can be given simplified proofs, e.g., Schur's Lemma in Riemannian geometry.

This paper is organized as follows. In Section 2 we introduce some terminology and notations. In Section 3 we construct exterior differentiation locally by Lie derivative. In Section 4 we represent curvature tensor, torsion tensor and Bianchi identities locally by covariant differentiation. In Section 5 we represent the curvature tensor of Riemannian manifold explicitly as a 2-tensor of the tangent bivector bundle, and use it to give a simple proof of Schur's Lemma.

2   Some Terminology and Notations

We use the same symbol "." to denote the pairing between a vector space and its dual space, the interior product of a tangent vector field and a differential form, and the inner product induced by a pseudo-Riemannian metric tensor. The explanation is that if no metric tensor occurs, then the dot symbol denotes the interior product which includes the pairing naturally; if there is a fixed pseudo-Riemannian metric tensor, then the tangent space and cotangent space are identified, so are the pairing and the inner product induced by the metric. The latter identification is further justified by the fact that in Riemannian geometry, we use only the Levi-Civita covariant differentiation, which preserves both the inner product and the pairing.

In this paper, we use the following extension [2] of the pairing "$\cdot$" from $\mathcal{V}$ and $\mathcal{V}^*$ to their Grassmann spaces $\Lambda(\mathcal{V})$ and $\Lambda(\mathcal{V}^*)$: for vectors $x_i \in \mathcal{V}$ and $y^j \in \mathcal{V}^*$,

$$(x_1 \wedge \cdots \wedge x_r) \cdot (y^1 \wedge \cdots \wedge y^s) = \begin{cases} (x_1 \cdot (x_2 \cdot (\cdots (x_r \cdot (y^1 \wedge \cdots \wedge y^s) \cdots), & \text{if } r \le s; \\ (\cdots (x_1 \wedge \cdots \wedge x_r) \cdot y^1) \cdot y^2) \cdots) \cdot y^s), & \text{if } r > s. \end{cases}$$
$$(2.1)$$

The topic "vector derivative" is explored in detail in [2]. Let $\mathcal{V}$ and $\mathcal{V}^*$ be an nD vector space and its dual space, Let $\{a_i \mid i = 1 \ldots n\}$ and $\{a_j\}$ be a pair of dual bases of $\mathcal{V}$ and $\mathcal{V}^*$ respectively. The *vector derivative* at a point $x \in \mathcal{V}$ is the following $\mathcal{V}$-valued first-order differential operator:

$$\partial_x = \sum_{i=1}^{n} a^i \left. \frac{\partial}{\partial a_i} \right|_x . \tag{2.2}$$

Here $\partial/\partial a_i|_x$ denotes the *directional derivative* at point x. Obviously, $\partial = \partial_x$ is independent of the basis $\{a_i\}$ chosen, and for any vector a,

$$a \cdot \partial = \partial/\partial a, \quad a = \sum_{i=1}^{n} (a \cdot a^i) a_i. \tag{2.3}$$

Henceforth we consider an nD differentiable manifold M. The tangent and cotangent bundles of M are denoted by TM and T^*M respectively. The set of smooth tangent and cotangent vector fields on M are denoted by $\Gamma(TM)$ and $\Gamma(T^*M)$ respectively.

The *Lie derivatives* of smooth scalar field f, tangent vector field y, and cotangent vector field ω with respect to a smooth tangent vector field x, are defined by

$$\begin{aligned} \mathcal{L}_x f &= x(f), \\ \mathcal{L}_x y &= [x, y], \\ y \cdot \mathcal{L}_x(\omega) &= \mathcal{L}_x(y \cdot \omega) - \omega \cdot \mathcal{L}_x y. \end{aligned} \tag{2.4}$$

Here $[x, y]$ is the *bracket* defined by

$$[x, y](f) = x(y(f)) - y(x(f)), \ \forall \ f \in C^\infty(M). \tag{2.5}$$

Below we introduce two notations. Let π be a differential operator taking values in $\Lambda(\Gamma(T^*M))$. Then in $\pi \wedge u \wedge v$, where $u, v \in \Lambda(\Gamma(T^*M))$, the differentiation is carried out to the right hand side of π, i.e., to both u and v. If we want the differentiation be made only to u, we use $\pi \wedge \dot{u} \wedge v$ to denote this, else if we want it be made only to v, we use $\pi \wedge \breve{u} \wedge v$ to denote this.

If u itself is a differential operator taking values in $\Lambda(\Gamma(T^*M))$, then it has two parts: the $\Lambda(\Gamma(T^*M))$-part and the scalar-valued differential operator part. If the differentiation in π is made only to the $\Lambda(\Gamma(T^*M))$-part of u, we use $\pi \wedge \dot{u} \wedge v$ to denote this. If the differentiation is not made to the $\Lambda(\Gamma(T^*M))$-part of u, we use $\pi \wedge \breve{u} \wedge v$ to denote this. The scalar-valued differentiation parts $s(\pi)$, $s(u)$ of π, u always act on v in the manner $s(\pi)s(u)v$. If v is also a $\Lambda(\Gamma(T^*M))$-valued differential operator, and the differentiation in π is made only to the $\Lambda(\Gamma(T^*M))$-part of v, we use $\pi \wedge u \wedge \dot{v}$ or $\pi \wedge \breve{u} \wedge \dot{v}$ to denote this, depending on how u is differentiated.

3   Exterior Differentiation

Definition 1. A *local natural basis* $\{a_i\}$ of $\Gamma(TM)$ at $x \in M$ is a set of n smooth tangent vector fields defined in a neighborhood N of x, such that there exists a local coordinate system $\{u_i\}$ of N with the property

$$a_i(f) = \frac{\partial f}{\partial u_i}, \ \forall \ f \in C^\infty(N). \tag{3.6}$$

Let $\{a_i\}$ be a local basis of $\Gamma(TM)$, and let $\{a^i\}$ be its dual basis in $\Gamma(T^*M)$. Define

$$d^a = \sum_{i=1}^n a^i \wedge \mathcal{L}_{a_i} = \sum_{i=1}^n a^i \mathcal{L}_{a_i} \wedge . \tag{3.7}$$

Lemma 1. (1) Let $\{a_i\}$ be a local basis of $\Gamma(TM)$. Then it is natural if only if $[a_i, a_j] = 0$ for any $1 \le i < j \le n$.
(2) $(d^a)^2 = 0$ if and only if $\{a_i\}$ is natural.

Proof. We only prove the second part. By the definition of d^a, to prove $(d^a)^2 = 0$ we only need to prove $(d^a)^2 f = 0$ for any $f \in C^\infty(M)$.

$$\begin{aligned}
(d^a)^2 f &= \sum_{i,j=1}^n a^i \wedge \mathcal{L}_{a_i}(a^j a_j(f)) \\
&= \sum_{i,j=1}^n (a^i \wedge a^j a_i(a_j(f))) + a_j(f) a^i \wedge \mathcal{L}_{a_i}(a^j) \\
&= \sum_{i<j} [a_i, a_j](f) a^i \wedge a^j - \sum_{i,j,k=1}^n a_j(f) a^i \wedge ([a_i, a_k] \cdot a^j) a^k \\
&= \sum_{i<j} [a_i, a_j](f) a^i \wedge a^j - \sum_{i,k=1}^n [a_i, a_k](f) a^i \wedge a^k \\
&= -\sum_{i<j} [a_i, a_j](f) a^i \wedge a^j.
\end{aligned}$$

Thus $(d^a)^2 f = 0$ if and only if $[a_i, a_j](f) = 0$, i.e., $\{a_i\}$ is natural.

Proposition 1. d^a is independent of the local natural basis $\{a_i\}$. It is defined globally on M, called the *exterior differentiation operator* of M, denoted by d.

Proof. It is easy to prove that $d^a|_{C^\infty(M)}$ is independent of a_i. So we only need to prove that $d^a|_{\Gamma(T^*M)}$ is independent of a_i. The latter is true because

$$d^a(fd^a(g)) = d^a f \wedge d^a g + f(d^a)^2 g = d^a f \wedge d^a g$$

is independent of $\{a_i\}$, for any $f, g \in C^\infty(M)$.

For local calculus, we can further decompose d into two parts: the wedge product "$\wedge$", and the $\Gamma(T^*M)$-valued local differential operator

$$\delta_a = \sum_{i=1}^n a^i \mathcal{L}_{a_i}. \tag{3.8}$$

This operator depends on the local basis $\{a_i\}$. The benefit of introducing δ_a can be seen from the following local expression of $[x, y]$.

Proposition 2. (1) For any local natural basis $\{a_i\}$,

$$\delta_a \wedge \delta_a = 0. \tag{3.9}$$

(2) For any smooth local tangent vector fields x, y,

$$[x, y] = x \cdot \delta_a y - y \cdot \delta_a x. \tag{3.10}$$

Proof. We only prove the second part. For any $f \in C^\infty(M)$,

$$\begin{aligned}
[x, y](f) &= x \cdot \delta_a(y \cdot \delta_a f) - y \cdot \delta_a(x \cdot \delta_a f) \\
&= (x \cdot \delta_a y) \cdot \delta_a f + y \cdot (x \cdot \delta_a \delta_a f) - (y \cdot \delta_a x) \cdot \delta_a f - x \cdot (y \cdot \delta_a \delta_a f) \\
&= (x \cdot \delta_a y - y \cdot \delta_a x)(f) - (x \wedge y) \cdot (\delta_a \wedge \delta_a) f \\
&= (x \cdot \delta_a y - y \cdot \delta_a x)(f).
\end{aligned}$$

4   Connection on a Vector Bundle

A *connection* of a vector bundle E over M is denoted by D. Its *covariant derivative* with respect to a vector field a is denoted by D_a. As in the case of exterior differentiation, we can decompose D into two parts: the tensor product "$\otimes$", and the $\Gamma(T^*M)$-valued local differential operator

$$\delta = \sum_{i=1}^n a^i D_{a_i}, \tag{4.11}$$

where $\{a_i\}$ is a local basis of $\Gamma(TM)$. The difference is that here the operator δ is independent of the basis $\{a_i\}$ chosen, called the *covariant differentiation operator*. We have

$$D = \delta \otimes, \qquad D_a = a \cdot \delta. \tag{4.12}$$

The first benefit of introducing δ is that we can add it up with any differential form ω. The sum is a new connection, and the new covariant derivative is $a \cdot (\delta + \omega)$. The second benefit is its intrinsic characterization of the curvature tensor, and in the case of affine connection, also the torsion tensor.

The scalar-valued operator $(x \wedge y) \cdot (\delta \wedge \delta)$ represents the difference between the two compositions $D_x D_{\check{y}}$ and $D_y D_{\check{x}}$ of the differentiations D_x and D_y. So $\delta \wedge \delta$ measures the asymmetry of two successive covariant derivatives. In this section, we decompose this asymmetry into two parts: the curvature part and the torsion part. This characterization clarifies the intrinsic relationship between an affine connection and its curvature and torsion tensors.

Definition 2. The differential operator

$$R(x, y) = D_x D_y - D_y D_x - D_{[x,y]}, \quad \forall x, y \in \Gamma(TM) \tag{4.13}$$

is called the *curvature operator* of the connection. The following (3,1)-tensor is called the *curvature tensor* of the connection:

$$R(x, y, z, \omega) = \omega \cdot ((D_x D_y - D_y D_x - D_{[x,y]})z), \quad \forall x, y, z \in \Gamma(TM), \omega \in \Gamma(T^*M). \tag{4.14}$$

We use the same symbol R with different number of arguments to denote both the curvature operator and the curvature tensor.

Proposition 3. The curvature operator has the following local representation with respect to a local natural basis $\{a_i\}$:

$$R(x, y) = -(x \wedge y) \cdot (\delta \wedge \check{\delta}). \tag{4.15}$$

Proof. By $R(a_i, a_j) = a_i \cdot \delta(a_j \cdot \delta) - a_j \cdot \delta(a_i \cdot \delta)$ and

$$\delta \wedge \check{\delta} = -\sum_{i=1}^{n} a^i \wedge \delta(a_i \cdot \delta) = \sum_{i,j=1}^{n} (a^i \wedge a^j) a_i \cdot \delta(a_j \cdot \delta), \tag{4.16}$$

we get $\delta \wedge \check{\delta} = \sum_{i<j} (a^i \wedge a^j) R(a_i, a_j)$. From this and the fact that $R(x, y)$ is a function of $x \wedge y$, we obtain (4.15).

Corollary 1. $\delta \wedge \check{\delta}$ is independent of $\{a_i\}$ as long as the local basis is natural. Furthermore, $\delta \wedge \check{\delta}\big|_{C^\infty(M)} = 0$.

Definition 3. The *torsion tensor* of an *affine connection* D is defined by

$$T(x, y) = D_x y - D_y x - [x, y], \quad \forall x, y \in \Gamma(TM). \tag{4.17}$$

The *dual connection* and *dual covariant differentiation* of D and δ are defined by their actions on a local natural basis $\{a_i\}$:

$$\overline{D}_{a_i}(a_j) = D_{a_j} a_i, \qquad a_i \cdot \overline{\delta} a_j = a_j \cdot \delta a_i. \tag{4.18}$$

In fact, for any tangent vector fields x, y,

$$\overline{D}_x(y) = y \cdot \delta x + [x, y],$$
$$T(x, y) = x \cdot (\delta - \bar{\delta}y). \tag{4.19}$$

Proposition 4. For any tangent vector fields x, y and cotangent vector field ω,

$$T(x, y) \cdot \delta = (x \wedge y) \cdot (\delta \wedge \dot{\delta}),$$
$$T(x, y) \cdot \omega = (x \wedge y) \cdot (\delta \wedge \omega - d\omega). \tag{4.20}$$

In particular, $\delta \wedge \dot{\delta}$ is independent of $\{a_i\}$ as long as the local basis is natural.

Corollary 2. 1. $d = \delta \wedge$ if and only if δ is torsion-free.
2. If δ is torsion-free, then

$$\overbrace{\delta \wedge \cdots \wedge \delta}^{r} = \sum_{i_1, \cdots, i_r = 1}^{n} a^{i_1} \wedge \cdots \wedge a^{i_r} a_{i_1} \cdot \delta \cdots a_{i_r} \cdot \delta. \tag{4.21}$$

Proposition 5. For a local natural basis $\{a_i\}$, the curvature and torsion of the connection $\delta_a = \sum_{i=1}^{n} a^i \mathcal{L}_{a_i}$ are both zero. It is called a *local natural affine connection*.

Theorem 1. [Commutation Formula]

$$(x \wedge y) \cdot (\delta \wedge \delta) = T(x, y) \cdot \delta - R(x, y). \tag{4.22}$$

Proof. In a local natural basis $\{a_i\}$, the conclusion follows (4.15), (4.20) and the Lebniz rule $\delta \wedge \delta = \delta \wedge \dot{\delta} + \delta \wedge \check{\delta}$. If $\{a_i\}$ is not natural, then

$$- R = \delta \wedge \check{\delta} + d\delta, \quad T \cdot \delta = \delta \wedge \dot{\delta} - d\delta. \tag{4.23}$$

The relation (4.22) still holds.

When $T = 0$, then $R = \delta \wedge \delta$. This representation crystalizes the famous Bianchi identities as follows.

Proposition 6. [Bianchi] Let δ be torsion-free, R be its curvature. Let $\{a_i\}$ be a local natural basis of $\Gamma(TM)$. The components of R and DR with respect to $\{a_i\}$ and its dual basis $\{a^i\}$ are denoted by $R^j_{ikl}, R^j_{ikl,h}$ respectively. Then

$$R^i_{jkl} + R^i_{klj} + R^i_{ljk} = 0$$
$$R^i_{jkl,h} + R^i_{jlh,k} + R^i_{jhk,l} = 0 \tag{4.24}$$

Proof. We have $-R^i_{jkl} = a^i \cdot ((a_k \wedge a_l) \cdot (\delta \wedge \dot{\delta})a_j)$, and the relation

$$(a_k \wedge a_l) \cdot (\delta \wedge \dot{\delta})(a^i \cdot a_j) = a^i \cdot ((a_k \wedge a_l) \cdot (\delta \wedge \dot{\delta})a_j) + a_j \cdot ((a_k \wedge a_l) \cdot (\delta \wedge \dot{\delta})a^i) = 0.$$

Then

$$R^i_{jkl} + R^i_{klj} + R^i_{ljk} = a_j \cdot ((a_k \wedge a_l) \cdot (\delta \wedge \dot{\delta})a^i) + a_k \cdot ((a_l \wedge a_j) \cdot (\delta \wedge \dot{\delta})a^i)$$
$$+ a_l \cdot ((a_j \wedge a_k) \cdot (\delta \wedge \dot{\delta})a^i)$$
$$= (a_j \wedge a_k \wedge a_l) \cdot (\delta \wedge \dot{\delta} \wedge a^i).$$

The first Bianchi identity is equivalent to

$$\delta \wedge \breve{\delta} \wedge a^i = 0 \tag{4.25}$$

when $T = 0$ and $\{a_i\}$ is natural. It follows $d^2 = 0$.

From

$$R^i_{jkl,h} = (a_h \cdot \delta R)(a_k, a_l, a_j, a^i) = (a_k \wedge a_l) \cdot (a_h \cdot \delta R)(a_j, a^i),$$

we get

$$R^i_{jkl,h} + R^i_{jlh,k} + R^i_{jhk,l} = (a_k \wedge a_l \wedge a_h) \cdot (\delta \wedge \dot{R})(a_j, a^i).$$

The second Bianchi identity is equivalent to

$$\delta \wedge \dot{R} = \delta \wedge (\delta \wedge \breve{\delta})^{\cdot} = 0 \tag{4.26}$$

When $T = 0$ and $\{a_i\}$ is natural. It follows $\delta \wedge \dot{\delta} = 0$ and

$$\delta \wedge (\delta \wedge \breve{\delta})^{\cdot} = \delta \wedge \dot{\delta} \wedge (\breve{\delta})^{\vee} + \delta \wedge \breve{\delta} \wedge (\breve{\delta})^{\cdot}.$$

5   Riemannian Geometry

The *Levi-Civita connection* δ of a generalized Riemannian manifold is the unique torsion-free connection preserving the metric tensor. Its curvature tensor R induces a new differential operator, denoted by R':

$$R'(z, \omega)(x, y) = R(x, y, z, \omega). \tag{5.27}$$

Since $R'(z, \omega)$ is a $C^\infty(M)$-linear function on $\Lambda^2(\Gamma(TM))$, there exists a unique $\Omega(z, \omega) \in \Lambda^2(\Gamma(T^*(M)))$ such that $R'(z, \omega)(x, y) = (x \wedge y) \cdot \Omega(z, \omega)$. Below we compute Ω.

By (4.15),

$$\Omega(z, \omega) = -\delta \wedge \breve{\delta}(\dot{z} \cdot \omega) = \delta \wedge \breve{\delta}(\dot{\omega} \cdot z) = z \cdot (\delta \wedge \breve{\delta} \wedge \omega) - (z \cdot (\delta \wedge \breve{\delta})) \wedge \omega,$$

so for any local natural basis $\{a_i\}$,

$$\begin{aligned}
\Omega(a_i, a_j) &= -(a_i \cdot (\delta \wedge \breve{\delta})) \wedge a_j \\
&= -\sum_{k=1}^{n} a^k \wedge (a_i \cdot \delta a_k \cdot \delta a_j - a_k \cdot \delta a_i \cdot \delta a_j) \\
&= -\sum_{k=1}^{n} a^k \wedge (a_i \cdot \delta a_j \cdot \delta a_k) + \delta \wedge (a_i \cdot \delta a_j).
\end{aligned}$$

Since $\Omega(a_i, a_j) = -\Omega(a_j, a_i)$ and $a_i \cdot \delta a_j = a_j \cdot \delta a_i$,

$$\Omega(a_i, a_j) = -\frac{1}{2} \sum_{k=1}^{n} a^k \wedge R(a_i, a_j) a_k. \tag{5.28}$$

A direct corollary of (5.28) is that $R = R'$, i.e.,

$$R(a_k, a_l)(a_i, a_j) = R'(a_i, a_j)(a_k, a_l) = (a_k \wedge a_l) \cdot \Omega(a_i, a_j) = R(a_i, a_j)(a_k, a_l).$$

Because of this, we can identify the curvature tensor with the following symmetric 2-tensor in the space $\Lambda^2(\Gamma(TM))$, still denoted by R:

$$R(x \wedge y, z \wedge \omega) = R(x, y)(z, \omega) = R(x, y, z, \omega). \tag{5.29}$$

Proposition 7. For any local basis $\{a_i\}$ which is not necessarily natural,

$$R = -\frac{1}{2}\sum_{k=1}^{n}(a^k \wedge \delta) \otimes (\check{\delta} \wedge a_k) = -\frac{1}{2}(\delta \wedge \check{\delta}) \otimes (\sum_{k=1}^{n} \dot{a}_k \wedge a^k) = -\frac{1}{2}(\delta \wedge \check{\delta}) \otimes (\sum_{k=1}^{n} \dot{a}^k \wedge a_k).$$

Let B_2 be a 2-blade of T_pM, $p \in M$. Then

$$K(B_2) = R(B_2^{-1}, B_2) \tag{5.30}$$

is called the *sectional curvature* at point p.

Proposition 8. At every point p of M, R is determined by K.

Proof. Let ∂ be the vector derivative in the vector space $\Lambda^2(T_pM)$. For any $A_2 \in \Lambda^2(\Gamma(TM))$, by the multilinearity of R, we have

$$A_2 \cdot \partial(R(B_2^{-1}, B_2)) = 2\frac{R(A_2, B_2)B_2 \cdot B_2 - R(B_2, B_2)A_2 \cdot B_2}{(B_2 \cdot B_2)^2}, \tag{5.31}$$

so

$$R(A_2, B_2) = K A_2 \cdot B_2 + \frac{1}{2}(B_2 \cdot B_2)(A_2 \cdot \partial K). \tag{5.32}$$

If at $p \in M$, $K(B_2)$ is independent of B_2, we say M is *isotropic* at p. If M is isotropic at every point and $K = K_p$ is independent of p, we call M a *constant-curvature space*.

Proposition 9. [F. Schur] If M is an nD connected Riemannian manifold ($n > 2$) and is isotropic at every point, then it is a constant-curvature space.

Proof. We only need to prove that $K = K_p$ is independent of p. First, since K_p is constant at $\Lambda^2(T_pM)$, i.e., $A_2 \cdot \partial K_p = 0$ for any $A_2 \in \Lambda(\Gamma(T_pM))$, by (5.32), we have $R(A_2, B_2) = K_p A_2 \cdot B_2$ for any $B_2 \in \Lambda(\Gamma(T_pM))$, so

$$R(B_2) = K B_2. \tag{5.33}$$

Below we compute $\delta \wedge R(B_2)$ using the relation $\delta \wedge \dot{R} = 0$:

$$\delta \wedge R(B_2) = \delta \wedge \dot{R}(B_2) + \delta \wedge \check{R}(\dot{B}_2) = K\delta \wedge B_2. \tag{5.34}$$

On the other hand, by differentiating the right-hand side of (5.33) directly, we get

$$\delta \wedge R(B_2) = \delta \wedge (KB_2) = (\delta \dot{K}) \wedge B_2 + K\delta \wedge B_2. \tag{5.35}$$

So for any $B_2 \in \Lambda^2(\Gamma(T_pM))$, $(\delta \dot{K}) \wedge B_2 = 0$. Since the dimension of $\Lambda^2(\Gamma(T_pM))$ is greater than one, it must be that $\delta K = 0$ for any p, i.e., K is constant over M.

6   Conclusion

In this paper, we set up a compact symbolic algebraic system for local geometric computing in intrinsic differential geometry. In this system, the exterior differentiation, the curvature tensor of a connection, the torsion tensor of an affine connection, the Bianchi identities of a torsion-free affine connection, and the curvature tensor of a Levi-Civita connection are given elegant representations, which can be used to simplify the involved geometric computing, e.g. in the proof of Schur's Lemma. The idea behind the system is the use of vector derivative and Clifford algebra. Our investigation shows that this is a promising research direction for both mathematics mechanization and differential geometry.

References

1. S. S. Chern, W.-H. Chen and K.-S. Lan (1999): *Lectures on Differential Geometry*, World Scientific, Singapore.
2. D. Hestenes and G. Sobczyk (1984): *Clifford Algebra to Geometric Calculus*, D. Reidel, Dordrecht, Boston.
3. H. Li (1995). Automated reasoning with differential forms. In: *Proc. ASCM'95*, Scientists Inc., Tokyo, pp. 29–32.
4. H. Li (1997). On mechanical theorem proving in differential geometry – local theory of surfaces. *Science in China* A, 40(4): 350–356.
5. H. Li (2000). Mechanical theorem proving in differential geometry. In: *Mathematics Mechanization and Applications*, X.-S. Gao and D. Wang (eds.), Academic Press, London, pp. 147-174.
6. T. J. Willmore (1993): *Riemannian Geometry*, Clarendon Press, Oxford, New York.
7. W.-T. Wu (1979): On the mechanization of theorem-proving in elementary differential geometry. *Scientia Sinica* (Math Suppl. I): 94-102.

On Miquel's Five-Circle Theorem*

Hongbo Li, Ronghua Xu, and Ning Zhang

Mathematics Mechanization Key Laboratory,
Academy of Mathematics and Systems Science,
Chinese Academy of Sciences,
Beijing 100080, P. R. China

Abstract. Miquel's Five-Circle Theorem is difficult to prove algebraically. In this paper, the details of the first algebraic proof of this theorem is provided. The proof is based on conformal geometric algebra and its accompanying invariant algebra called null bracket algebra, and is the outcome of the powerful computational techniques of null bracket algebra embodying the novel idea **breefs**.

Keywords: Miquel's Theorem, Mathematics mechanization, Conformal geometric algebra, Null bracket algebra, "breefs".

1   Introduction

Miquel's Five-Circle Theorem is among a sequence of wonderful theorems in plane geometry bearing his name. It states that by starting from a pentagon in the plane, one can construct a 5-star whose vertices are the intersections of the non-neighboring edges of the pentagon. If five circles are drawn such that each circle circumscribes a triangular corner of the star, then the neighboring circles intersect at five new points which are cocircular.

The sequence of Miquel's Theorems is as follows: Let there be an n-gon in the plane. When $n = 3$, the three vertices of a triangle are on a unique circle, which can be taken as the unique circle determined by the three edges of the triangle, called the *Miquel 3-circle*. When $n = 4$, the 4 edges of a quadrilateral form 4 distinct 3-tuples of edges, each determining a Miquel 3-circle, and Miquel's 4-Circle Theorem says that the 4 Miquel 3-circles pass through a common point (i.e., are concurrent), called the *Miquel 4-point*. Miquel's Five Circle Theorem says that the 5 Miquel 4-points of a pentagon are cocircular, and the unique circle is the *Miquel 5-circle* of the pentagon. In general, Miquel's $2m$-Circle Theorem says that the $2m$ Miquel $(2m - 1)$-circles of an $(2m)$-gon are concurrent, and Miquel's $(2m + 1)$-Circle Theorem says that the $2m + 1$ Miquel $(2m)$-points of an $(2m + 1)$-gon are cocircular.

The first proof of Miquel's n-Circle Theorem is given by Clifford in the 19th century in a purely geometric manner. By now, an algebraic (analytic) proof has not been found yet. While for $n = 4$ an algebraic proof is easy, for $n = 5$ it is

* Supported partially by NKBRSF 2004CB318001 and NSFC 10471143.

H. Li, P. J. Olver and G. Sommer (Eds.): IWMM-GIAE 2004, LNCS 3519, pp. 217–228, 2005.

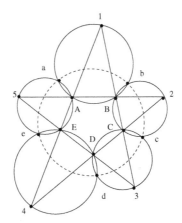

Fig. 1. Miquel's Five-Circle Theorem

extremely difficult. In fact, the first algebraic proof for $n = 5$ was found in 2001 [4]. In mechanical geometric theorem proving, the first machine proof for $n = 5$ was produced in 1994 [1], which is a purely geometric one.

Below let us analyze how to prove Miquel's 5-Circle Theorem algebraically. The geometric configuration has linear construction, i.e., the constrained objects can be constructed sequentially as unique geometric intersections, whose algebraic representations can be given explicitly by polynomial expressions. The following is a typical linear construction:

Free points in the plane: A, B, C, D, E.

Intersections:

$$1 = EA \cap BC, \quad 2 = AB \cap CD, \quad 3 = BC \cap DE, \quad 4 = CD \cap EA,$$
$$5 = DE \cap AB, \quad a = AE5 \cap AB1, \quad b = BA1 \cap BC2, \quad c = CB2 \cap CD3,$$
$$d = DC3 \cap DE4, \quad e = ED4 \cap EA5.$$

Conclusion: a, b, c, d, e are cocircular.

Here $1 = EA \cap BC$ denotes that point 1 is the intersection of lines EA and BC, and $a = AE5 \cap AB1$ denotes the second intersection of circles $AE5$ and $AB1$ other than A if they intersect, or A itself if the two circles are tangent to each other.

By symmetry, we only need to prove the cocircularity of a, b, c, d. Since the construction is linear, we only need to substitute the explicit expressions of a, b, c, d in terms of the free points A, B, C, D, E into the algebraic equality f representing the cocircularity of a, b, c, d, and then expand the result to get zero.

Indeed, the proof is strategically very easy: it is composed of a first procedure of substitution (elimination) and a second procedure of simplification. However, both procedures are too difficult to handle by either coordinates and any other classical invariants, because extremely complicated symbolic computations are involved.

In 2001, a new powerful algebraic tool for geometric computation, called *conformal geometric algebra*, was proposed [3]. Its accompanying invariant algebra called *null bracket algebra* was proposed at the same time [4]. The two algebras can provide elegant simplifications in symbolic computation of geometric problems. The idea leading to such simplifications is recently summarized as "**breefs**" (**b**racket-oriented **r**epresentation, **e**limi- nation and **e**xpansion for **f**actored and **s**hortest result) [6], [7], [8], [9], [10], [11].

In this paper, we are going to show the detailed procedure of the algebraic proof of Miquel's 5-Circle Theorem [4], which has never been shown before. We will show how the idea **breefs** is used in the proof, and how the simplifications are achieved by using conformal geometric algebra and its accompanying invariant algebra. Due to the limit of space we only cite the necessary theorems and formulas in both algebras needed in the proof. For details we refer to [3], [5], [9].

2   Conformal Geometric Algebra

We omit the introduction of *geometric algebra*, and recommend [2] for a clear reading.

Conformal geometric algebra is the geometric algebra set on the so-called "conformal model" of classical geometry: To study nD Euclidean (or hyperbolic, or spherical) geometry, we elevate the geometric space into the null cone of an $(n + 2)$D Minkowski space $\mathcal{M}$. Let $e_1, \ldots, e_n$ be an orthonormal basis of $\mathcal{R}^n$, let $\mathbf{e}, \mathbf{e}_0$ be a standard null basis of the Minkowski plane orthogonal to $\mathcal{R}^n$, i.e., $\mathbf{e}^2 = \mathbf{e}_0^2 = 0$ and $\mathbf{e} \cdot \mathbf{e}_0 = -1$, then a point $a \in \mathcal{R}^n$ corresponds to the following null vector:

$$\mathbf{a} = \mathbf{e}_0 + a + \frac{a^2}{2}\mathbf{e}. \tag{2.1}$$

By setting $a = 0$ we see that the origin of $\mathcal{R}^n$ corresponds to vector $\mathbf{e}_0$. No point corresponds to $\mathbf{e}$: it represents the *point at infinity*. Vector $\mathbf{a}$ satisfies $\mathbf{a} \cdot \mathbf{e} = -1$. It is the *inhomogeneous* representation of point a. The *homogeneous* representation is any null vector collinear with $\mathbf{a}$, and is more convenient for symbolic computation. On the other hand, the inhomogeneous representation is more convenient for geometric interpretation. For example, in the inhomogeneous representation,

$$\mathbf{a}_1 \cdot \mathbf{a}_2 = -\frac{d(a_1, a_2)^2}{2}. \tag{2.2}$$

i.e., the squared distance of two points becomes the inner product of the corresponding null vectors. The circles and spheres of various dimensions correspond to Minkowski subspaces, and can be computed homogeneously. The conformal transformations are realized by orthogonal transformations in $\mathcal{M}$ and can be computed by spin representations.

In conformal geometric algebra, geometric objects and relations often have universal and compact algebraic representations. For example, when 1 is the point at infinity, then $123 \cap 12'3'$ is exactly the intersection of two lines $23, 2'3'$. In fact, when any of the points is replaced by the points at infinity, the corresponding circle becomes a straight line, but the algebraic representation is still the same. The algebraic representation of $123 \cap 12'3'$, denoted by $\mathbf{123} \cap \mathbf{12'3'}$, has two different forms:

$\mathbf{123} \cap \mathbf{12'3'}$

$$= 2 \cdot 3[122'3'][132'3']1 - \frac{1}{2}[12312'3'][132'3']2 + \frac{1}{2}[12312'3'][122'3']3 \quad (2.3)$$

$$= 2' \cdot 3'[1232'][1233']1 + \frac{1}{2}[12312'3'][1233']2' - \frac{1}{2}[12312'3'][1232']3'.$$

Obviously, $\mathbf{123} \cap \mathbf{12'3'} = \mathbf{132} \cap \mathbf{12'3'} = \mathbf{123} \cap \mathbf{13'2'} = \mathbf{12'3'} \cap \mathbf{123}$. Let $2'', 3''$ be points on circle $12'3'$. Then

$$\mathbf{1} \cdot (\mathbf{123} \cap \mathbf{12'3'}) = [12312'3']^2, \quad \mathbf{2} \cdot (\mathbf{123} \cap \mathbf{12'3'}) = (\mathbf{1} \cdot \mathbf{3})(\mathbf{2} \cdot \mathbf{3})[122'3']^2. \quad (2.4)$$

3   Null Bracket Algebra

Null bracket algebra is the accompanying invariant algebra of conformal geometric algebra when all geometric constructions are based on points. However, it can be defined and employed independent of conformal geometric algebra. The following is a self-contained definition of this algebra.

Definition 1. *Let $\mathcal{K}$ be a field of characteristic $\neq 2$. Let $n \leq m$ be two positive integers. The nD null bracket algebra generated by symbols $\mathbf{a}_1, \ldots, \mathbf{a}_m$, is the quotient of the polynomial ring over $\mathcal{K}$ with indeterminates $\langle \mathbf{a}_{i_1} \ldots \mathbf{a}_{i_{2p}} \rangle$ and $[\mathbf{a}_{j_1} \ldots \mathbf{a}_{j_{n+2q-2}}]$ for $p, q \geq 1$ and $1 \leq i_1, \ldots, i_{2p}, j_1, \ldots, j_{n+2q-2} \leq m$, modulo the ideal generated by the following 8 types of elements:*
   B1. $[\mathbf{a}_{i_1} \ldots \mathbf{a}_{i_n}]$ *if $i_j = i_k$ for some $j \neq k$.*
   B2. $[\mathbf{a}_{i_1} \ldots \mathbf{a}_{i_n}] - \text{sign}(\sigma)[\mathbf{a}_{i_{\sigma(1)}} \ldots \mathbf{a}_{i_{\sigma(n)}}]$ *for a permutation σ of $1, \ldots, n$.*
   B3. $\langle \mathbf{a}_i \mathbf{a}_j \rangle - \langle \mathbf{a}_j \mathbf{a}_i \rangle$ *for $i \neq j$.*
   N. $\langle \mathbf{a}_i \mathbf{a}_i \rangle$ *for any i.*
   GP1. $\displaystyle\sum_{k=1}^{n+1} (-1)^{k+1} \langle \mathbf{a}_j \mathbf{a}_{i_k} \rangle [\mathbf{a}_{i_1} \ldots \check{\mathbf{a}}_{i_k} \ldots \mathbf{a}_{i_{n+1}}].$
   GP2. $[\mathbf{a}_{i_1} \ldots \mathbf{a}_{i_n}][\mathbf{a}_{j_1} \ldots \mathbf{a}_{j_n}] - \det(\langle \mathbf{a}_{i_k} \mathbf{a}_{j_l} \rangle)_{k,l=1..n}.$
   AB. $\langle \mathbf{a}_{i_1} \cdots \mathbf{a}_{i_{2l}} \rangle - \displaystyle\sum_{j=2}^{2l} (-1)^j \langle \mathbf{a}_{i_1} \mathbf{a}_{i_j} \rangle \langle \mathbf{a}_{i_2} \cdots \check{x}_{i_j} \cdots \mathbf{a}_{i_{2l}} \rangle.$
   SB. $[\mathbf{a}_{i_1} \cdots \mathbf{a}_{i_{n+2l}}] - \sum_\sigma \text{sign}(\sigma, \check{\sigma}) \langle x_{\sigma(1)} \cdots x_{\sigma(2l)} \rangle \times [x_{\check{\sigma}(1)} \cdots x_{\check{\sigma}(n)}]$ *for all partitions $\sigma, \check{\sigma}$ of $1, 2, \ldots, n+2l$ into two subsequences of length $2l$ and n respectively.*

In practice we usually use notations $\mathbf{a} \cdot \mathbf{a}_j$ and $\mathbf{a}_i^2$ instead of $\langle \mathbf{a}_i \mathbf{a}_j \rangle$ and $\langle \mathbf{a}_i \mathbf{a}_j \rangle$.
When $n = 4$, the following are some typical brackets and their geometric meanings:

1.

$$\langle \mathbf{e} \mathbf{a}_1 \mathbf{a}_2 \mathbf{a}_3 \rangle = (a_3 - a_2) \cdot (a_1 - a_2) = |a_1 a_2||a_2 a_3| \cos \angle (a_2 a_3, a_2 a_1);$$
$$[\mathbf{e} \mathbf{a}_1 \mathbf{a}_2 \mathbf{a}_3] = \quad 2 S_{a_1 a_2 a_3} \quad = |a_1 a_2||a_2 a_3| \sin \angle (a_2 a_3, a_2 a_1). \quad (3.1)$$

Here $S_{a_1a_2a_3}$ is the signed area of triangle $a_1a_2a_3$, and $\angle(a_2a_3, a_2a_1)$ is the oriented angle from vector a_2a_3 to vector a_2a_1. In particular, a_1, a_2, a_3 are collinear if and only if $[\mathbf{ea_1a_2a_3}] = 0$.

2.

$$\langle \mathbf{a_1a_2a_3a_4} \rangle = -8\frac{\rho_{a_1a_2a_3}\rho_{a_1a_3a_4}S_{a_1a_2a_3}S_{a_1a_3a_4}}{|a_1a_3|^2} \cos \angle(N^{a_1}_{a_1a_2a_3}, N^{a_1}_{a_1a_3a_4})$$

$$[\mathbf{a_1a_3a_2a_4}] = -8\frac{\rho_{a_1a_2a_3}\rho_{a_1a_3a_4}S_{a_1a_2a_3}S_{a_1a_3a_4}}{|a_1a_3|^2} \sin \angle(N^{a_1}_{a_1a_2a_3}, N^{a_1}_{a_1a_3a_4})$$

$$= (\rho^2_{a_1a_2a_3} - |o_{a_1a_2a_3}a_4|^2)S_{a_1a_2a_3}.$$

$$(3.2)$$

Here $o_{a_1a_2a_3}$ is the center of circle $a_1a_2a_3$; $\rho_{a_1a_2a_3}, \rho_{a_1a_3a_4}$ are the radii of circles $a_1a_2a_3$, $a_1a_3a_4$ respectively, and $N^{a_1}_{a_1a_2a_3}, N^{a_1}_{a_1a_3a_4}$ are the outward normal directions of circles $a_1a_2a_3$, $a_1a_3a_4$ at point a_1 respectively. In particular, points a_1, a_2, a_3, a_4 are cocircular if and only if $[\mathbf{a_1a_2a_3a_4}] = 0$.

3.

$$[\mathbf{ea_2a_3ea_4a_5}] = -4S_{a_2a_4a_3a_5} = 2|a_2a_3||a_4a_5| \sin \angle(a_4a_5, a_2a_3),$$

$$\langle \mathbf{ea_2a_3ea_4a_5} \rangle = 2(a_3 - a_2) \cdot (a_5 - a_4) = 2|a_2a_3||a_4a_5| \cos \angle(a_4a_5, a_2a_3).$$

$$(3.3)$$

Here $S_{a_2a_4a_3a_5}$ is the signed area of quadrilateral $a_2a_4a_3a_5$. In particular, $[\mathbf{ea_2a_3ea_4a_5}] = 0$ if and only if $a_2a_3 \parallel a_4a_5$; $\langle \mathbf{ea_2a_3ea_4a_5} \rangle = 0$ if and only if $a_2a_3 \perp a_4a_5$.

In this paper, we often use the following shorthand notation:

$$[\mathbf{a_1a_2; a_3a_4}] = -\frac{1}{2}[\mathbf{ea_1a_2ea_3a_4}] = \mathbf{e} \cdot a_1[\mathbf{ea_2a_3a_4}] - \mathbf{e} \cdot a_2[\mathbf{ea_1a_3a_4}]. \quad (3.4)$$

4.

$$\langle \mathbf{a_1a_2a_3a_1a_4a_5} \rangle = 8\rho_{a_1a_2a_3}\rho_{a_1a_4a_5}S_{a_1a_2a_3}S_{a_1a_4a_5} \cos \angle(N^{a_1}_{a_1a_2a_3}, N^{a_1}_{a_1a_4a_5}),$$

$$[\mathbf{a_1a_2a_3a_1a_4a_5}] = 8\rho_{a_1a_2a_3}\rho_{a_1a_4a_5}S_{a_1a_2a_3}S_{a_1a_4a_5} \sin \angle(N^{a_1}_{a_1a_2a_3}, N^{a_1}_{a_1a_4a_5})$$

$$= 16S_{a_1a_2a_3}S_{a_1a_4a_5}S_{a_1o_{123}o_{145}}.$$

$$(3.5)$$

Here o_{123}, o_{145} are the centers of circles $a_1a_2a_3, a_1a_4a_5$ respectively. In particular, $[\mathbf{a_1a_2a_3a_1a_4a_5}] = 0$ if and only if circles $a_1a_2a_3$, $a_1a_4a_5$ are tangent to each other; $\langle \mathbf{a_1a_2a_3a_1a_4a_5} \rangle = 0$ if and only if they are perpendicular.

5.

$$[\mathbf{a_1; a_2a_3; a_4a_5}] = \mathbf{e} \cdot a_4 \, \mathbf{e} \cdot a_5[\mathbf{ea_1a_2a_3}] - \mathbf{e} \cdot a_2 \, \mathbf{e} \cdot a_3[\mathbf{ea_1a_4a_5}] + \mathbf{e} \cdot a_2 \, \mathbf{e} \cdot a_4 \, [\mathbf{ea_1a_3a_5}]$$

is twice the signed area of pentagon $a_1a_2a_3a_5a_4$. Furthermore,

$$[\mathbf{a_1; a_2a_3; a_4a_5}] = -[\mathbf{a_1; a_4a_5; a_2a_3}];$$
$$[\mathbf{a_1; a_2a_3; a_5a_4}] = [\mathbf{a_2; a_3a_4; a_1a_5}] = \cdots = [\mathbf{a_5; a_1a_2; a_4a_3}].$$

$$(3.6)$$

The following are some typical formulas for algebraic computation in 4D null bracket algebra:

− **Fundamental formulas:**

$$\mathbf{a}_1\mathbf{a}_2\cdots\mathbf{a}_k\mathbf{a}_1 = 2\sum_{i=2}^{k}(-1)^i\mathbf{a}_1\cdot\mathbf{a}_i\,(\mathbf{a}_2\cdots\breve{\mathbf{a}}_i\cdots\mathbf{a}_k\mathbf{a}_1).\qquad(3.7)$$

In particular, $\mathbf{a}_1\mathbf{a}_2\mathbf{a}_3\mathbf{a}_1 = -\mathbf{a}_1\mathbf{a}_3\mathbf{a}_2\mathbf{a}_1$.

− **Expansion formulas:**

$$\frac{1}{2}[\mathbf{a}_1\mathbf{a}_2\mathbf{a}_3\mathbf{a}_4\mathbf{a}_1\mathbf{a}_5\cdots\mathbf{a}_{2l+5}]$$
$$= \langle\mathbf{a}_1\mathbf{a}_2\mathbf{a}_3\mathbf{a}_4\rangle[\mathbf{a}_1\mathbf{a}_5\cdots\mathbf{a}_{2l+5}] + [\mathbf{a}_1\mathbf{a}_2\mathbf{a}_3\mathbf{a}_4]\langle\mathbf{a}_1\mathbf{a}_5\cdots\mathbf{a}_{2l+5}\rangle,$$
$$\frac{1}{2}\langle\mathbf{a}_1\mathbf{a}_2\mathbf{a}_3\mathbf{a}_4\mathbf{a}_1\mathbf{a}_5\cdots\mathbf{a}_{2l+5}\rangle$$
$$= \langle\mathbf{a}_1\mathbf{a}_2\mathbf{a}_3\mathbf{a}_4\rangle\langle\mathbf{a}_1\mathbf{a}_5\cdots\mathbf{a}_{2l+5}\rangle - [\mathbf{a}_1\mathbf{a}_2\mathbf{a}_3\mathbf{a}_4][\mathbf{a}_1\mathbf{a}_5\cdots\mathbf{a}_{2l+5}],$$
$$\frac{1}{2}[\mathbf{a}_1\mathbf{a}_2\mathbf{a}_3\mathbf{a}_1\mathbf{a}_4\mathbf{a}_5\mathbf{a}_1\mathbf{a}_6\cdots\mathbf{a}_{2l+6}]$$
$$= [\mathbf{a}_1\mathbf{a}_2\mathbf{a}_3\mathbf{a}_1\mathbf{a}_4\mathbf{a}_5]\langle\mathbf{a}_1\mathbf{a}_6\cdots\mathbf{a}_{2l+6}\rangle + \langle\mathbf{a}_1\mathbf{a}_2\mathbf{a}_3\mathbf{a}_1\mathbf{a}_4\mathbf{a}_5\rangle[\mathbf{a}_1\mathbf{a}_6\cdots\mathbf{a}_{2l+6}],$$
$$\frac{1}{2}\langle\mathbf{a}_1\mathbf{a}_2\mathbf{a}_3\mathbf{a}_1\mathbf{a}_4\mathbf{a}_5\mathbf{a}_1\mathbf{a}_6\cdots\mathbf{a}_{2l+6}\rangle$$
$$= \langle\mathbf{a}_1\mathbf{a}_2\mathbf{a}_3\mathbf{a}_1\mathbf{a}_4\mathbf{a}_5\rangle\langle\mathbf{a}_1\mathbf{a}_6\cdots\mathbf{a}_{2l+6}\rangle - [\mathbf{a}_1\mathbf{a}_2\mathbf{a}_3\mathbf{a}_1\mathbf{a}_4\mathbf{a}_5][\mathbf{a}_1\mathbf{a}_6\cdots\mathbf{a}_{2l+6}].$$
$$(3.8)$$

− **Distribution formulas:**

$$\frac{1}{2}[\mathbf{a}_1\mathbf{a}_2\mathbf{a}_4\mathbf{a}_5][\mathbf{a}_3\mathbf{a}_1\mathbf{a}_2\mathbf{a}_3\mathbf{a}_4\mathbf{a}_5] = \mathbf{a}_1\cdot\mathbf{a}_2\,[\mathbf{a}_1\mathbf{a}_3\mathbf{a}_4\mathbf{a}_5][\mathbf{a}_2\mathbf{a}_3\mathbf{a}_4\mathbf{a}_5]$$
$$-\mathbf{a}_4\cdot\mathbf{a}_5\,[\mathbf{a}_1\mathbf{a}_2\mathbf{a}_3\mathbf{a}_4][\mathbf{a}_1\mathbf{a}_2\mathbf{a}_3\mathbf{a}_5],$$
$$\frac{1}{2}[\mathbf{a}_1\mathbf{a}_2\mathbf{a}_4\mathbf{a}_5]\langle\mathbf{a}_3\mathbf{a}_1\mathbf{a}_2\mathbf{a}_3\mathbf{a}_4\mathbf{a}_5\rangle = \mathbf{a}_1\cdot\mathbf{a}_2\,[\mathbf{a}_1\mathbf{a}_3\mathbf{a}_4\mathbf{a}_5]\langle\mathbf{a}_2\mathbf{a}_3\mathbf{a}_4\mathbf{a}_5\rangle$$
$$+\mathbf{a}_4\cdot\mathbf{a}_5\,\langle\mathbf{a}_1\mathbf{a}_2\mathbf{a}_3\mathbf{a}_4\rangle[\mathbf{a}_1\mathbf{a}_2\mathbf{a}_3\mathbf{a}_5].$$
$$(3.9)$$

− **Factorization formulas:**
  Group 1.

$$\mathbf{a}_2\cdot\mathbf{a}_3[\mathbf{a}_1\mathbf{a}_2\mathbf{a}_5\mathbf{a}_6][\mathbf{a}_3\mathbf{a}_4\mathbf{a}_5\mathbf{a}_6] + \mathbf{a}_5\cdot\mathbf{a}_6[\mathbf{a}_1\mathbf{a}_2\mathbf{a}_3\mathbf{a}_6][\mathbf{a}_2\mathbf{a}_3\mathbf{a}_4\mathbf{a}_5]$$
$$= -\frac{1}{2}[\mathbf{a}_1\mathbf{a}_2\mathbf{a}_3\mathbf{a}_4\mathbf{a}_5\mathbf{a}_6][\mathbf{a}_2\mathbf{a}_3\mathbf{a}_5\mathbf{a}_6],$$
$$\mathbf{a}_2\cdot\mathbf{a}_3[\mathbf{a}_1\mathbf{a}_2\mathbf{a}_5\mathbf{a}_6]\langle\mathbf{a}_3\mathbf{a}_4\mathbf{a}_5\mathbf{a}_6\rangle - \mathbf{a}_5\cdot\mathbf{a}_6[\mathbf{a}_1\mathbf{a}_2\mathbf{a}_3\mathbf{a}_6]\langle\mathbf{a}_2\mathbf{a}_3\mathbf{a}_4\mathbf{a}_5\rangle$$
$$= -\frac{1}{2}\langle\mathbf{a}_1\mathbf{a}_2\mathbf{a}_3\mathbf{a}_4\mathbf{a}_5\mathbf{a}_6\rangle[\mathbf{a}_2\mathbf{a}_3\mathbf{a}_5\mathbf{a}_6].$$
$$(3.10)$$

  Group 2.

$$\langle\mathbf{a}_1\mathbf{a}_2\mathbf{a}_3\mathbf{a}_4\rangle[\mathbf{a}_1\mathbf{a}_2\mathbf{a}_3\mathbf{a}_1\mathbf{a}_5\mathbf{a}_6] - \langle\mathbf{a}_1\mathbf{a}_2\mathbf{a}_3\mathbf{a}_1\mathbf{a}_5\mathbf{a}_6\rangle[\mathbf{a}_1\mathbf{a}_2\mathbf{a}_3\mathbf{a}_4]$$
$$= -2\mathbf{a}_1\cdot\mathbf{a}_2\,\mathbf{a}_2\cdot\mathbf{a}_3[\mathbf{a}_1\mathbf{a}_3\mathbf{a}_4\mathbf{a}_1\mathbf{a}_5\mathbf{a}_6],$$
$$\langle\mathbf{a}_1\mathbf{a}_2\mathbf{a}_3\mathbf{a}_4\rangle\langle\mathbf{a}_1\mathbf{a}_2\mathbf{a}_3\mathbf{a}_1\mathbf{a}_5\mathbf{a}_6\rangle + [\mathbf{a}_1\mathbf{a}_2\mathbf{a}_3\mathbf{a}_1\mathbf{a}_5\mathbf{a}_6][\mathbf{a}_1\mathbf{a}_2\mathbf{a}_3\mathbf{a}_4]$$
$$= -2\mathbf{a}_1\cdot\mathbf{a}_2\,\mathbf{a}_2\cdot\mathbf{a}_3\langle\mathbf{a}_1\mathbf{a}_3\mathbf{a}_4\mathbf{a}_1\mathbf{a}_5\mathbf{a}_6\rangle.$$
$$(3.11)$$

Group 3.

$$a_2 \cdot a_3[a_1a_2a_5a_6][a_1a_2a_5a_1a_3a_4] - a_2 \cdot a_5[a_1a_2a_3a_4][a_1a_2a_3a_1a_5a_6]$$

$$= \frac{1}{2}[a_1a_2a_3a_4a_1a_2a_5a_6][a_1a_2a_3a_5],$$

$$a_2 \cdot a_3\langle a_1a_2a_5a_6\rangle[a_1a_2a_5a_1a_3a_4] - a_2 \cdot a_5[a_1a_2a_3a_4]\langle a_1a_2a_3a_1a_5a_6\rangle$$

$$= \frac{1}{2}\langle a_1a_2a_3a_4a_1a_2a_5a_6\rangle[a_1a_2a_3a_5].$$

$$(3.12)$$

4 The Five-Circle Theorem: Elimination

We are now equipped with all necessary algebraic tools to prove the 5-circle theorem. In order to simplify the proof, we need to make some symmetry analysis of the geometric configuration of the theorem before the formal start of the proof.

- The conclusion that a, b, c, d are cocircular can be represented by $[\mathbf{abcd}] = 0$. We need to compute the expressions of $\mathbf{a, b, c, d}$ by the 5 free points $\mathbf{A, B, C, D, E}$. Since $[\mathbf{abcd}] = [(\mathbf{a} \wedge \mathbf{b}) \wedge (\mathbf{c} \wedge \mathbf{d})]$, we need to compute $\mathbf{a} \wedge \mathbf{b}$ and $\mathbf{c} \wedge \mathbf{d}$.
- If $\mathbf{a} = f(\mathbf{A}, 1, 5, \mathbf{B}, \mathbf{E}) = g(\mathbf{A}, \mathbf{C}, \mathbf{D}, \mathbf{B}, \mathbf{E})$, then since points a, b are symmetric with respect to line $1D$ combinatorially, it must be that

$$\mathbf{b} = f(\mathbf{B}, 1, 2, \mathbf{A}, \mathbf{C}) = g(\mathbf{B}, \mathbf{E}, \mathbf{D}, \mathbf{A}, \mathbf{C}), \qquad (4.1)$$

  i.e., $\mathbf{b}$ can be derived from $\mathbf{a}$ by the interchanges $\mathbf{A} \longleftrightarrow \mathbf{B}$ and $\mathbf{C} \longleftrightarrow \mathbf{E}$.
- Since (d, c) and (a, b) are symmetric with respect to line $2E$ combinatorially, if we have obtained $\mathbf{a} \wedge \mathbf{b} = f(1, 2, 5, \mathbf{A}, \mathbf{B}, \mathbf{C}, \mathbf{D}, \mathbf{E}) = g(\mathbf{A}, \mathbf{B}, \mathbf{C}, \mathbf{D}, \mathbf{E})$, then we can get directly

$$\mathbf{c} \wedge \mathbf{d} = -f(3, 2, 4, \mathbf{D}, \mathbf{C}, \mathbf{B}, \mathbf{A}, \mathbf{E}) = -g(\mathbf{D}, \mathbf{C}, \mathbf{B}, \mathbf{A}, \mathbf{E}). \qquad (4.2)$$

Thus, the first stage of the proof, i.e., elimination, should contain the following steps: (1) compute $\mathbf{a}$, (2) compute $\mathbf{b}$ by interchanges of symbols, then compute $a \wedge b$, (3) compute $\mathbf{c} \cdot \mathbf{d}$ by interchanges of symbols, then compute $[\mathbf{abcd}]$. In this stage, computation means the elimination of $1, \cdots, 5$.

Step 1. Compute $a = AB1 \cap AE5$: We express $\mathbf{a}$ as a linear combination of $\mathbf{A}, \mathbf{B}, 1$ instead of $\mathbf{A}, \mathbf{E}, 5$ in order to employ the symmetry between $\mathbf{a}$ and $\mathbf{b}$.

$$\mathbf{a} = \mathbf{B} \cdot 1 \, [\mathbf{AE5B}][\mathbf{AE51}] \, \mathbf{A} - \frac{1}{2}[\mathbf{AE51}][\mathbf{AB1AE5}] \, \mathbf{B} + \frac{1}{2}[\mathbf{AE5B}][\mathbf{AB1AE5}] \, 1.$$

$$(4.3)$$

To eliminate $1, 5$ from the brackets, in $\mathbf{A} \cdot 1$ we use $1 = 1(\mathbf{e}, \mathbf{A}, \mathbf{E})$ and (2.4), and in $\mathbf{B} \cdot 1$ we use $1 = 1(\mathbf{e}, \mathbf{B}, \mathbf{C})$. The results are two bracket monomials. In $[\mathbf{AE5B}]$ we use $5 = 5(\mathbf{e}, \mathbf{A}, \mathbf{B})$ so that $\mathbf{A}, \mathbf{B}$ in the bracket annihilate the same vector symbols in the representation of **5**, leading to a bracket monomial result. Similarly, in [**AE51**] we use $1 = 1(e, A, E)$, and either of $5 = 5(e, A, B)$ and $5 = 5(e, D, E)$. We get

$$[\mathbf{AE51}] = -\mathbf{A} \cdot \mathbf{E}\,[\mathbf{eABC}][\mathbf{eABE}][\mathbf{eADE}][\mathbf{eBCE}][\mathbf{AB;DE}],$$

$$[\mathbf{AE5B}] = -\mathbf{A} \cdot \mathbf{B}\,[\mathbf{eABE}][\mathbf{eADE}][\mathbf{eBDE}],$$

$$\mathbf{A} \cdot 1 \;\;= \mathbf{e} \cdot \mathbf{E}\,\mathbf{A} \cdot \mathbf{E}\,[\mathbf{eABC}]^2,$$

$$\mathbf{B} \cdot 1 \;\;= \mathbf{e} \cdot \mathbf{C}\,\mathbf{B} \cdot \mathbf{C}\,[\mathbf{eABE}]^2,$$

and

$$-\frac{1}{2}[\mathbf{AB1AE5}]$$

$$= \mathbf{A} \cdot \mathbf{B}[\mathbf{AE51}] - \mathbf{A} \cdot 1[\mathbf{AE5B}]$$

$$= \mathbf{A} \cdot \mathbf{B}\,\mathbf{A} \cdot \mathbf{E}\,[\mathbf{eABC}][\mathbf{eABE}][\mathbf{eADE}]$$
$$(-[\mathbf{eBCE}][\mathbf{AB;DE}] + \mathbf{e} \cdot \mathbf{E}\,[\mathbf{eABC}][\mathbf{eBDE}])$$

$$= \mathbf{A} \cdot \mathbf{B}\,\mathbf{A} \cdot \mathbf{E}\,[\mathbf{eABC}][\mathbf{eABE}][\mathbf{eADE}]$$
$$(\mathbf{e} \cdot \mathbf{D}[\mathbf{eABE}][\mathbf{eBCE}] - \mathbf{e} \cdot \mathbf{E}[\mathbf{eABD}][\mathbf{eBCE}] + \mathbf{e} \cdot \mathbf{E}\,[\mathbf{eABC}][\mathbf{eBDE}])$$

$$= \mathbf{A} \cdot \mathbf{B}\,\mathbf{A} \cdot \mathbf{E}\,[\mathbf{eABC}][\mathbf{eABE}][\mathbf{eADE}]$$
$$(\mathbf{e} \cdot \mathbf{D}[\mathbf{eABE}][\mathbf{eBCE}] - \mathbf{e} \cdot \mathbf{E}[\mathbf{eABE}][\mathbf{eBCD}])$$

$$= \mathbf{A} \cdot \mathbf{B}\,\mathbf{A} \cdot \mathbf{E}\,[\mathbf{eABC}][\mathbf{eABE}]^2[\mathbf{eADE}][\mathbf{BC;DE}].$$

The next to the last step is based on a *Grassmann-Plücker relation* [10]:

$$-\mathbf{e} \cdot \mathbf{E}[\mathbf{eABD}][\mathbf{eBCE}] + \mathbf{e} \cdot \mathbf{E}\,[\mathbf{eABC}][\mathbf{eBDE}] = -\mathbf{e} \cdot \mathbf{E}\,[\mathbf{eABE}][\mathbf{eBCD}]. \quad (4.4)$$

After removing 8 common bracket factors $\mathbf{A} \cdot \mathbf{B}\,\mathbf{A} \cdot \mathbf{E}\,[\mathbf{eABC}][\mathbf{eABE}]^3[\mathbf{eADE}]$, we get

$$\mathbf{a} = \mathbf{e} \cdot \mathbf{C}\,\mathbf{B} \cdot \mathbf{C}\,[\mathbf{eABE}][\mathbf{eBCE}][\mathbf{eBDE}][\mathbf{AB;DE}]\,\mathbf{A}$$
$$+ \mathbf{A} \cdot \mathbf{E}\,[\mathbf{eABC}][\mathbf{eBCE}][\mathbf{AB;DE}][\mathbf{BC;DE}]\,\mathbf{B} \qquad (4.5)$$
$$- \mathbf{A} \cdot \mathbf{B}\,[\mathbf{eBDE}][\mathbf{BC;DE}]\,1.$$

Step 2. Compute $b = AB1 \cap BC2$ and $\mathbf{a} \wedge \mathbf{b}$:

$$\mathbf{b} = \qquad \mathbf{B} \cdot \mathbf{C}\,[\mathbf{eABE}][\mathbf{eACE}][\mathbf{AB;CD}][\mathbf{EA;CD}]\,\mathbf{A}$$
$$- \mathbf{e} \cdot \mathbf{E}\,\mathbf{A} \cdot \mathbf{E}\,[\mathbf{eABC}][\mathbf{eACD}][\mathbf{eACE}][\mathbf{AB;CD}]\,\mathbf{B} \qquad (4.6)$$
$$+ \mathbf{A} \cdot \mathbf{B}\,[\mathbf{eACD}][\mathbf{EA;CD}]\,1$$

and $\mathbf{a} \wedge \mathbf{b} = \lambda_{AB}\mathbf{A} \wedge \mathbf{B} + \lambda_{A1}\mathbf{A} \wedge 1 + \lambda_{B1}\mathbf{B} \wedge 1$, where

$$\lambda_{AB} = -\mathbf{A} \cdot \mathbf{E}\,\mathbf{B} \cdot \mathbf{C}\,[\mathbf{eABC}][\mathbf{eABE}][\mathbf{eACE}][\mathbf{eBCE}][\mathbf{eCDE}][\mathbf{AB;CD}]$$
$$[\mathbf{AB;DE}][\mathbf{D;AC;BE}],$$

$$\lambda_{A1} = -\mathbf{A} \cdot \mathbf{B}\,\mathbf{B} \cdot \mathbf{C}\,[\mathbf{eABC}][\mathbf{eABE}][\mathbf{eBDE}][\mathbf{eCDE}][\mathbf{EA;CD}][\mathbf{A;CE;DB}],$$

$$\lambda_{B1} = \mathbf{A} \cdot \mathbf{B}\,\mathbf{A} \cdot \mathbf{E}\,[\mathbf{eABC}][\mathbf{eABE}][\mathbf{eACD}][\mathbf{eCDE}][\mathbf{BC;DE}][\mathbf{B;DA;EC}].$$

$$(4.7)$$

After removing 4 common factors $[\mathbf{eABC}][\mathbf{eABE}][\mathbf{eCDE}][\mathbf{D}; \mathbf{AC}; \mathbf{BE}]$, we get

$$
\begin{aligned}
\mathbf{a} \wedge \mathbf{b} = &-\mathbf{A} \cdot \mathbf{E}\,\mathbf{B} \cdot \mathbf{C}\,[\mathbf{eACE}][\mathbf{eBCE}][\mathbf{AB}; \mathbf{CD}][\mathbf{AB}; \mathbf{DE}]\,\mathbf{A} \wedge \mathbf{B} \\
&+\mathbf{A} \cdot \mathbf{B}\,\mathbf{B} \cdot \mathbf{C}\,[\mathbf{eBDE}][\mathbf{EA}; \mathbf{CD}]\,\mathbf{A} \wedge \mathbf{1} \\
&+\mathbf{A} \cdot \mathbf{B}\,\mathbf{A} \cdot \mathbf{E}\,[\mathbf{eACD}][\mathbf{BC}; \mathbf{DE}]\,\mathbf{B} \wedge \mathbf{1}.
\end{aligned}
\tag{4.8}
$$

Step 3. Compute $\mathbf{c} \wedge \mathbf{d}$ and$[\mathbf{abcd}]$:

$$
\begin{aligned}
\mathbf{c} \wedge \mathbf{d} = &\ \lambda_{CD}\mathbf{C} \wedge \mathbf{D} + \lambda_{C3}\mathbf{C} \wedge \mathbf{3} + \lambda_{D3}\mathbf{D} \wedge \mathbf{3} \\
= &\ \ \mathbf{B} \cdot \mathbf{C}\,\mathbf{D} \cdot \mathbf{E}\,[\mathbf{eBCE}][\mathbf{eBDE}][\mathbf{AB}; \mathbf{CD}][\mathbf{EA}; \mathbf{CD}]\,\mathbf{C} \wedge \mathbf{D} \\
&+\mathbf{C} \cdot \mathbf{D}\,\mathbf{D} \cdot \mathbf{E}\,[\mathbf{eABD}][\mathbf{EA}; \mathbf{BC}]\,\mathbf{C} \wedge \mathbf{3} \\
&+\mathbf{B} \cdot \mathbf{C}\,\mathbf{C} \cdot \mathbf{D}\,[\mathbf{eACE}][\mathbf{AB}; \mathbf{DE}]\,\mathbf{D} \wedge \mathbf{3}
\end{aligned}
\tag{4.9}
$$

and

$$
\begin{aligned}
[\mathbf{abcd}] = &\ \lambda_{AB}\lambda_{CD}[\mathbf{ABCD}] + \lambda_{AB}\lambda_{C3}[\mathbf{ABC3}] + \lambda_{AB}\lambda_{D3}[\mathbf{ABD3}] \\
&+\lambda_{A1}\lambda_{CD}[\mathbf{A1CD}] + \lambda_{A1}\lambda_{C3}[\mathbf{A1C3}] + \lambda_{A1}\lambda_{D3}[\mathbf{A1D3}] \\
&+\lambda_{B1}\lambda_{CD}[\mathbf{B1CD}] + \lambda_{B1}\lambda_{C3}[\mathbf{B1C3}] + \lambda_{B1}\lambda_{D3}[\mathbf{B1D3}].
\end{aligned}
\tag{4.10}
$$

Here $\lambda_{AB}, \lambda_{A1}, \lambda_{B1}$ denote the coefficients of $\mathbf{A} \wedge \mathbf{B}, \mathbf{A} \wedge \mathbf{1}, \mathbf{B} \wedge \mathbf{1}$ in (4.8).

After eliminating $\mathbf{1}, \mathbf{3}$ from (4.10), using the first of the distribution formulas (3.9) to get rid of the square brackets not involving $\mathbf{e}$, then removing 5 common bracket factors $(\mathbf{B} \cdot \mathbf{C})^2\,[\mathbf{eACE}][\mathbf{eBCE}][\mathbf{eBDE}]$, we get an expression of 14 terms:

$[\mathbf{abcd}]$

$$
\begin{aligned}
= &\ \mathbf{A} \cdot \mathbf{E}\,\mathbf{C} \cdot \mathbf{D}\,\mathbf{D} \cdot \mathbf{E}[\mathbf{AB}; \mathbf{CD}][\mathbf{AB}; \mathbf{DE}][\mathbf{EA}; \mathbf{CD}][\mathbf{eBCE}][\mathbf{eABC}][\mathbf{eABD}] \\
&- \mathbf{A} \cdot \mathbf{B}\,\mathbf{A} \cdot \mathbf{E}\,\mathbf{D} \cdot \mathbf{E}[\mathbf{AB}; \mathbf{CD}][\mathbf{AB}; \mathbf{DE}][\mathbf{EA}; \mathbf{CD}][\mathbf{eACD}][\mathbf{eBCD}][\mathbf{eBCE}] \\
&- \mathbf{A} \cdot \mathbf{E}\,\mathbf{C} \cdot \mathbf{D}\,\mathbf{D} \cdot \mathbf{E}[\mathbf{AB}; \mathbf{CD}][\mathbf{AB}; \mathbf{DE}][\mathbf{EA}; \mathbf{BC}][\mathbf{eABC}][\mathbf{eABD}][\mathbf{eCDE}] \\
&- \mathbf{A} \cdot \mathbf{E}\,\mathbf{B} \cdot \mathbf{C}\,\mathbf{C} \cdot \mathbf{D}[\mathbf{AB}; \mathbf{CD}][\mathbf{AB}; \mathbf{DE}]^2[\mathbf{eABD}][\mathbf{eACE}][\mathbf{eCDE}] \\
&+ \mathbf{A} \cdot \mathbf{E}\,(\mathbf{C} \cdot \mathbf{D})^2[\mathbf{AB}; \mathbf{DE}]^2[\mathbf{BC}; \mathbf{DE}][\mathbf{eABC}][\mathbf{eABD}][\mathbf{eACE}] \\
&- \mathbf{A} \cdot \mathbf{B}\,\mathbf{A} \cdot \mathbf{E}\,\mathbf{C} \cdot \mathbf{D}[\mathbf{AB}; \mathbf{DE}]^2[\mathbf{BC}; \mathbf{DE}][\mathbf{eACD}][\mathbf{eACE}][\mathbf{eBCD}] \\
&+ \mathbf{A} \cdot \mathbf{B}\,\mathbf{B} \cdot \mathbf{C}\,\mathbf{D} \cdot \mathbf{E}[\mathbf{AB}; \mathbf{CD}][\mathbf{EA}; \mathbf{CD}]^2[\mathbf{eABE}][\mathbf{eACD}][\mathbf{eBDE}] \\
&+ \mathbf{A} \cdot \mathbf{B}\,\mathbf{C} \cdot \mathbf{D}\,\mathbf{D} \cdot \mathbf{E}[\mathbf{EA}; \mathbf{BC}][\mathbf{EA}; \mathbf{CD}]^2[\mathbf{eABC}][\mathbf{eABD}][\mathbf{eBDE}] \\
&- (\mathbf{A} \cdot \mathbf{B})^2\,\mathbf{D} \cdot \mathbf{E}[\mathbf{EA}; \mathbf{BC}][\mathbf{EA}; \mathbf{CD}]^2[\mathbf{eACD}][\mathbf{eBCD}][\mathbf{eBDE}] \\
&+ \mathbf{A} \cdot \mathbf{B}\,\mathbf{A} \cdot \mathbf{E}\,\mathbf{C} \cdot \mathbf{D}[\mathbf{AB}; \mathbf{DE}][\mathbf{EA}; \mathbf{CD}][\mathbf{BC}; \mathbf{DE}][\mathbf{eABC}][\mathbf{eADE}][\mathbf{eBCD}] \\
&- \mathbf{A} \cdot \mathbf{B}\,\mathbf{C} \cdot \mathbf{D}\,\mathbf{D} \cdot \mathbf{E}[\mathbf{AB}; \mathbf{DE}][\mathbf{EA}; \mathbf{BC}][\mathbf{EA}; \mathbf{CD}][\mathbf{eABC}][\mathbf{eADE}][\mathbf{eBCD}] \\
&+ \mathbf{A} \cdot \mathbf{B}\,\mathbf{A} \cdot \mathbf{E}\,\mathbf{D} \cdot \mathbf{E}[\mathbf{AB}; \mathbf{CD}][\mathbf{EA}; \mathbf{CD}][\mathbf{BC}; \mathbf{DE}][\mathbf{eABE}][\mathbf{eACD}][\mathbf{eBCD}] \\
&+ \mathbf{e} \cdot \mathbf{B}\,\mathbf{A} \cdot \mathbf{B}\,\mathbf{C} \cdot \mathbf{D}\,\mathbf{D} \cdot \mathbf{E}[\mathbf{EA}; \mathbf{BC}][\mathbf{EA}; \mathbf{CD}][\mathbf{eABC}][\mathbf{eABD}][\mathbf{eADE}][\mathbf{eCDE}] \\
&- \mathbf{e} \cdot \mathbf{C}\,\mathbf{A} \cdot \mathbf{B}\,\mathbf{A} \cdot \mathbf{E}\,\mathbf{C} \cdot \mathbf{D}[\mathbf{AB}; \mathbf{DE}][\mathbf{BC}; \mathbf{DE}][\mathbf{eABE}][\mathbf{eACD}][\mathbf{eADE}][\mathbf{eBCD}].
\end{aligned}
\tag{4.11}
$$

5   The Five-Circle Theorem: Simplification

On a Pentium III/500MHz, setting $A = (0,0)$, $B = (1,0)$, we get zero from (4.11) in 0.315 seconds. This is pretty satisfactory for a machine proof. However, we shall show that by means of the term reduction and factorization techniques in null bracket algebra, we can get zero from (4.11) in two steps of term collection without resorting to either coordinate representation or computer program.

Step 4. Grouping terms on the right side of (4.11) according to the inner products not involving **e**, we can reduce the number of terms to 6 by factorization within each group of terms:

(1) $\mathbf{A} \cdot \mathbf{E}\,\mathbf{C} \cdot \mathbf{D}\,\mathbf{D} \cdot \mathbf{E}[\mathbf{AB};\mathbf{CD}][\mathbf{AB};\mathbf{DE}][\mathbf{EA};\mathbf{CD}][\mathbf{eBCE}][\mathbf{eABC}][\mathbf{eABD}]$
$\quad -\mathbf{A} \cdot \mathbf{E}\,\mathbf{C} \cdot \mathbf{D}\,\mathbf{D} \cdot \mathbf{E}[\mathbf{AB};\mathbf{CD}][\mathbf{AB};\mathbf{DE}][\mathbf{EA};\mathbf{BC}][\mathbf{eABC}][\mathbf{eABD}][\mathbf{eCDE}]$

$= \; \mathbf{e} \cdot \mathbf{E}\,\mathbf{A} \cdot \mathbf{E}\,\mathbf{C} \cdot \mathbf{D}\,\mathbf{D} \cdot \mathbf{E}[\mathbf{AB};\mathbf{CD}][\mathbf{AB};\mathbf{DE}][\mathbf{eABC}][\mathbf{eABD}][\mathbf{eACE}][\mathbf{eBCD}];$

(2) $-\mathbf{A} \cdot \mathbf{B}\,\mathbf{A} \cdot \mathbf{E}\,\mathbf{D} \cdot \mathbf{E}[\mathbf{AB};\mathbf{CD}][\mathbf{AB};\mathbf{DE}][\mathbf{EA};\mathbf{CD}][\mathbf{eACD}][\mathbf{eBCD}][\mathbf{eBCE}]$
$\quad +\mathbf{A} \cdot \mathbf{B}\,\mathbf{A} \cdot \mathbf{E}\,\mathbf{D} \cdot \mathbf{E}[\mathbf{AB};\mathbf{CD}][\mathbf{EA};\mathbf{CD}][\mathbf{BC};\mathbf{DE}][\mathbf{eABE}][\mathbf{eACD}][\mathbf{eBCD}]$

$= -\mathbf{e} \cdot \mathbf{E}\,\mathbf{A} \cdot \mathbf{B}\,\mathbf{A} \cdot \mathbf{E}\,\mathbf{D} \cdot \mathbf{E}[\mathbf{AB};\mathbf{CD}][\mathbf{EA};\mathbf{CD}][\mathbf{eABC}][\mathbf{eACD}][\mathbf{eBCD}][\mathbf{eBDE}];$

(3) $-\mathbf{A} \cdot \mathbf{B}\,\mathbf{A} \cdot \mathbf{E}\,\mathbf{C} \cdot \mathbf{D}[\mathbf{AB};\mathbf{DE}][\mathbf{BC};\mathbf{DE}][\mathbf{AB};\mathbf{DE}][\mathbf{eACD}][\mathbf{eACE}][\mathbf{eBCD}]$
$\quad +\mathbf{A} \cdot \mathbf{B}\,\mathbf{A} \cdot \mathbf{E}\,\mathbf{C} \cdot \mathbf{D}[\mathbf{AB};\mathbf{DE}][\mathbf{EA};\mathbf{CD}][\mathbf{BC};\mathbf{DE}][\mathbf{eABC}][\mathbf{eADE}][\mathbf{eBCD}]$
$\quad -\mathbf{e} \cdot \mathbf{C}\,\mathbf{A} \cdot \mathbf{B}\,\mathbf{A} \cdot \mathbf{E}\,\mathbf{C} \cdot \mathbf{D}[\mathbf{AB};\mathbf{DE}][\mathbf{BC};\mathbf{DE}][\mathbf{eABE}][\mathbf{eACD}][\mathbf{eADE}][\mathbf{eBCD}]$

$= \; \mathbf{A} \cdot \mathbf{B}\,\mathbf{A} \cdot \mathbf{E}\,\mathbf{C} \cdot \mathbf{D}[\mathbf{AB};\mathbf{DE}][\mathbf{BC};\mathbf{DE}][\mathbf{eBCD}]$
$\quad\quad ([\mathbf{AB};\mathbf{CD}][\mathbf{eACE}][\mathbf{eADE}] - [\mathbf{AB};\mathbf{DE}][\mathbf{eACD}][\mathbf{eACE}])$

$= -\mathbf{e} \cdot \mathbf{A}\,\mathbf{A} \cdot \mathbf{B}\,\mathbf{A} \cdot \mathbf{E}\,\mathbf{C} \cdot \mathbf{D}[\mathbf{AB};\mathbf{DE}][\mathbf{BC};\mathbf{DE}][\mathbf{eABD}][\mathbf{eACE}][\mathbf{eBCD}][\mathbf{eCDE}];$

(4) $\mathbf{A} \cdot \mathbf{B}\,\mathbf{C} \cdot \mathbf{D}\,\mathbf{D} \cdot \mathbf{E}[\mathbf{EA};\mathbf{BC}][\mathbf{EA};\mathbf{CD}][\mathbf{EA};\mathbf{CD}][\mathbf{eABC}][\mathbf{eABD}][\mathbf{eBDE}]$
$\quad -\mathbf{A} \cdot \mathbf{B}\,\mathbf{C} \cdot \mathbf{D}\,\mathbf{D} \cdot \mathbf{E}[\mathbf{AB};\mathbf{DE}][\mathbf{EA};\mathbf{BC}][\mathbf{EA};\mathbf{CD}][\mathbf{eABC}][\mathbf{eADE}][\mathbf{eBCD}]$
$\quad +\mathbf{e} \cdot \mathbf{B}\,\mathbf{A} \cdot \mathbf{B}\,\mathbf{C} \cdot \mathbf{D}\,\mathbf{D} \cdot \mathbf{E}[\mathbf{EA};\mathbf{BC}][\mathbf{EA};\mathbf{CD}][\mathbf{eABC}][\mathbf{eABD}][\mathbf{eADE}][\mathbf{eCDE}]$

$= \; \mathbf{A} \cdot \mathbf{B}\,\mathbf{C} \cdot \mathbf{D}\,\mathbf{D} \cdot \mathbf{E}[\mathbf{EA};\mathbf{BC}][\mathbf{EA};\mathbf{CD}][\mathbf{eABC}]$
$\quad\quad (-[\mathbf{AB};\mathbf{CD}][\mathbf{eADE}][\mathbf{eBDE}] + [\mathbf{EA};\mathbf{CD}][\mathbf{eABD}][\mathbf{eBDE}])$

$= \; \mathbf{e} \cdot \mathbf{D}\,\mathbf{A} \cdot \mathbf{B}\,\mathbf{C} \cdot \mathbf{D}\,\mathbf{D} \cdot \mathbf{E}[\mathbf{EA};\mathbf{BC}][\mathbf{EA};\mathbf{CD}][\mathbf{eABC}][\mathbf{eABE}][\mathbf{eACD}][\mathbf{eBDE}];$

(5) $-\mathbf{A} \cdot \mathbf{E}\,\mathbf{B} \cdot \mathbf{C}\,\mathbf{C} \cdot \mathbf{D}[\mathbf{AB};\mathbf{CD}][\mathbf{AB};\mathbf{DE}]^2[\mathbf{eABD}][\mathbf{eACE}][\mathbf{eCDE}]$
$\quad +\mathbf{A} \cdot \mathbf{E}\,\mathbf{C} \cdot \mathbf{D}\,\mathbf{C} \cdot \mathbf{D}[\mathbf{AB};\mathbf{DE}]^2[\mathbf{BC};\mathbf{DE}][\mathbf{eABC}][\mathbf{eABD}][\mathbf{eACE}]$

$= \; -\dfrac{1}{4}\mathbf{A} \cdot \mathbf{E}\,\mathbf{C} \cdot \mathbf{D}[\mathbf{AB};\mathbf{DE}]^2[\mathbf{eCBAeCDE}][\mathbf{eABD}][\mathbf{eACE}][\mathbf{eBCD}];$

(6) $\mathbf{A} \cdot \mathbf{B}\,\mathbf{B} \cdot \mathbf{C}\,\mathbf{D} \cdot \mathbf{E}[\mathbf{AB};\mathbf{CD}][\mathbf{EA};\mathbf{CD}]^2[\mathbf{eABE}][\mathbf{eACD}][\mathbf{eBDE}]$
$\quad -\mathbf{A} \cdot \mathbf{B}\,\mathbf{A} \cdot \mathbf{B}\,\mathbf{D} \cdot \mathbf{E}[\mathbf{EA};\mathbf{BC}][\mathbf{EA};\mathbf{CD}]^2[\mathbf{eACD}][\mathbf{eBCD}][\mathbf{eBDE}]$

$= \; -\dfrac{1}{4}\mathbf{A} \cdot \mathbf{B}\,\mathbf{D} \cdot \mathbf{E}[\mathbf{EA};\mathbf{CD}]^2[\mathbf{eCBAeCDE}][\mathbf{eABC}][\mathbf{eACD}][\mathbf{eBDE}].$

In factorization (5) and (6), we have used the first of formulas (3.12). The result after the term reduction is

$[abcd]$

$$
\begin{aligned}
= \; & e \cdot \mathbf{E\,A} \cdot \mathbf{E\,C} \cdot \mathbf{D\,D} \cdot \mathbf{E}[AB;CD][AB;DE][eABC][eABD][eACE][eBCD] \\
& - e \cdot \mathbf{E\,A} \cdot \mathbf{B\,A} \cdot \mathbf{E\,D} \cdot \mathbf{E}[AB;CD][EA;CD][eABC][eACD][eBCD][eBDE] \\
& - e \cdot \mathbf{A\,A} \cdot \mathbf{B\,A} \cdot \mathbf{E\,C} \cdot \mathbf{D}[AB;DE][BC;DE][eABD][eACE][eBCD][eCDE] \\
& + e \cdot \mathbf{D\,A} \cdot \mathbf{B\,C} \cdot \mathbf{D\,D} \cdot \mathbf{E}[EA;BC][EA;CD][eABC][eABE][eACD][eBDE] \\
& - \frac{1}{4} \mathbf{A} \cdot \mathbf{E\,C} \cdot \mathbf{D}[AB;DE]^2[eCBAeCDE][eABD][eACE][eBCD] \\
& - \frac{1}{4} \mathbf{A} \cdot \mathbf{B\,D} \cdot \mathbf{E}[EA;CD]^2[eCBAeCDE][eABC][eACD][eBDE].
\end{aligned}
$$

$$(5.12)$$

Step 5. Grouping terms on the right side of (5.12) according to the inner products not involving **e** and doing factorization within each group of terms, using the first of formulas (3.8) and (3.11), we get zero from each group, thus finishing the proof of the theorem.

(a)
$$
\begin{aligned}
& e \cdot \mathbf{E}\underline{\mathbf{A}} \cdot \underline{\mathbf{E\,C}} \cdot \underline{\mathbf{D}}\mathbf{D} \cdot \mathbf{E}[AB;CD][AB;DE][eABC][eABD][\underline{eACE}][eBCD] \\
& -e \cdot \mathbf{A\,A} \cdot \mathbf{B}\underline{\mathbf{A}} \cdot \underline{\mathbf{E\,C}} \cdot \mathbf{D}[AB;DE]\underline{[BC;DE]}[eABD][\underline{eACE}][\underline{eBCD}][eCDE] \\
& -\frac{1}{4}\mathbf{A} \cdot \underline{\mathbf{E\,C}} \cdot \mathbf{D}[AB;DE][AB;DE][eCBAeCDE][eABD][eACE][eBCD] \\[4pt]
= \; & \frac{1}{2}\mathbf{A} \cdot \mathbf{E\,C} \cdot \mathbf{D}[AB;DE][eABD][eACE][eBCD] \\
& \quad (2\,e \cdot \mathbf{E\,D} \cdot \mathbf{E}[AB;CD][eABC] + \langle eCDE\rangle[AB;DE][eABC] \\
& \quad -2\,e \cdot \mathbf{A\,A} \cdot \mathbf{B}[BC;DE][eCDE] - \langle eABC\rangle[AB;DE][eCDE]) \\[4pt]
= \; & 0;
\end{aligned}
$$

(b)
$$
\begin{aligned}
& -e \cdot \mathbf{E}\underline{\mathbf{A}} \cdot \underline{\mathbf{B}}\mathbf{A} \cdot \underline{\mathbf{E\,D}} \cdot \mathbf{E}[AB;CD]\underline{[EA;CD]}[eABC][eACD][eBCD][eBDE] \\
& +e \cdot \mathbf{D}\underline{\mathbf{A}} \cdot \underline{\mathbf{B}}\mathbf{C} \cdot \mathbf{D}\underline{\mathbf{D}} \cdot \mathbf{E}[EA;BC]\underline{[EA;CD]}[eABC][eABE][eACD][eBDE] \\
& -\frac{1}{4}\mathbf{A} \cdot \mathbf{B\,D} \cdot \mathbf{E}[EA;CD][EA;CD][eCBAeCDE][eABC][eACD][eBDE] \\[4pt]
= \; & \frac{1}{2}\mathbf{A} \cdot \mathbf{B\,D} \cdot \mathbf{E}[EA;CD][eABC][eACD][eBDE] \\
& \quad (-2\,e \cdot \mathbf{E\,A} \cdot \mathbf{E}[AB;CD][eBCD] - \langle eEAB\rangle[EA;CD][eBCD] \\
& \quad +2\,e \cdot \mathbf{D\,C} \cdot \mathbf{D}[EA;BC][eABE] + \langle eBCD\rangle[EA;CD][eABE]) \\[4pt]
= \; & 0.
\end{aligned}
$$

Step 6. Nondegeneracy conditions: The above algebraic proof is valid under the assumption that ABC, BCD, CDE, DEA, EAB are triplets of non-collinear points, i.e., the pentagon is nondegenerate.

6   Conclusion

In this paper, we present the details of an algebraic proof of Miquel's 5-circle theorem. The proof shows that tremendous simplification by using conformal geometric algebra and null bracket algebra, in that the expressions can be more easily factored, and the number of terms can be more easily reduced. The major idea behind the simplification is **breefs**, which in this example is:

- Eliminate constrained points within each bracket.
- Choose suitable elimination rules to make the number of terms after the elimination as small as possible.
- Group terms by inner products, preferably those not involving **e**, before doing factorization.

References

1. S. Chou, X. Gao and J. Zhang. *Machine Proof in Geometry*, World Scientific, Singapore, 1994.
2. D. Hestenes, G. Sobczyk. *Clifford Algebra to Geometric Calculus*, Kluwer, Dordrecht, 1984.
3. H. Li, D. Hestenes, A. Rockwood. Generalized Homogeneous Coordinates for Computational Geometry. In: *Geometric Computing with Clifford Algebras*, G. Sommer (ed.), pp. 27–60, Springer, Heidelberg, 2001.
4. H. Li. Automated Theorem Proving in the Homogeneous Model with Clifford Bracket Algebra. In: *Applications of Geometric Algebra in Computer Science and Engineering*, L. Dorst et al. (eds.), pp. 69–78, Birkhauser, Boston, 2002.
5. H. Li. Clifford Expansions and Summations. *MM Research Preprints* **21**: 112-154, 2002.
6. H. Li. Clifford Algebras and Geometric Computation. In: *Geometric Computation*, F. Chen and D. Wang (eds.), World Scientific, Singapore, pp. 221-247, 2004.
7. H. Li. Automated Geometric Theorem Proving, Clifford Bracket Algebra and Clifford Expansions. In: *Trends in Mathematics: Advances in Analysis and Geometry*, Birkhauser, Basel, pp. 345-363, 2004.
8. H. Li. Algebraic Representation, Elimination and Expansion in Automated Geometric Theorem Proving. In: *Automated Deduction in Geometry*, F. Winkler (ed.), Springer, Berlin, Heidelberg, pp. 106-123, 2004.
9. H. Li. Symbolic Computation in the Homogeneous Geometric Model with Clifford Algebra. In: *Proc. ISSAC 2004*, J. Gutierrez (ed.), ACM Press, pp. 221-228, 2004.
10. H. Li, Y. Wu. Automated Short Proof Generation in Projective Geometry with Cayley and Bracket Algebras I. Incidence Geometry. *J. Symb. Comput.* **36**(5): 717-762, 2003.
11. H. Li, Y. Wu. Automated Short Proof Generation in Projective Geometry with Cayley and Bracket Algebras II. Conic Geometry. *J. Symb. Comput.* **36**(5): 763-809, 2003.
12. B. Sturmfels. *Algorithms in Invariant Theory*. Springer, Wien, 1993.

On Averaging in Clifford Groups[*]

Sven Buchholz and Gerald Sommer

Cognitive Systems Group, University of Kiel, Germany

Abstract. Averaging measured data is an important issue in computer vision and robotics. Integrating the pose of an object measured with multiple cameras into a single mean pose is one such example. In many applications data does not belong to a vector space. Instead, data often belongs to a non-linear group manifold as it is the case for orientation data and the group of three-dimensional rotations $SO(3)$. Averaging on the manifold requires the utilization of the associated Riemannian metric resulting in a rather complicated task. Therefore the Euclidean mean with best orthogonal projection is often used as an approximation. In $SO(3)$ this can be done by rotation matrices or quaternions. Clifford algebra as a generalization of quaternions allows a general treatment of such approximated averaging for all classical groups. Results for the two-dimensional Lorentz group $SO(1,2)$ and the related groups $SL(2,\mathbb{R})$ and $SU(1,1)$ are presented. The advantage of the proposed Clifford framework lies in its compactness and easiness of use.

1 Introduction

Averaging measured data is one of the most frequently arising problems in many different applications. For example, integrating the pose of an object measured with multiple cameras into a single mean pose is a standard task in computer vision. Feature–based registration of images would be another such example. The original motivation for this paper has been the following. In a neural network where every neuron represents a geometric transformation, say three–dimensional rotation, one has to average over several neurons in order to adjust the network topology to new presented data.

Surely, averaging data belonging to some vector space is rather trivial. For a set of points $\{x\}_i^n$ one only has to calculate the barycentre

$$\mathcal{A} = \frac{1}{n}\sum_{i=1}^{n} x_i \,.\tag{1}$$

In $\mathbb{R}^d$ this also minimizes the sum of the squared distances to the given points. Because of that variational property, (1) is then also called the arithmetic mean. Distance here refers of course to the usual Euclidean metric $d_E(\cdot,\cdot)$ yielding

---

[*] This work has been supported by DFG Grant So-320/2-3.

H. Li, P. J. Olver and G. Sommer (Eds.): IWMM-GIAE 2004, LNCS 3519, pp. 229–238, 2005.

$$\mathcal{A} = \arg\min_{x \in \mathbb{R}^d} \sum_{i=1}^{n} d_E(x, x_i)^2 . \tag{2}$$

All of the examples given above, however, involve data which does not belong to a vector space. Rather, the data are elements of a group, like SO(3) as for the case of three–dimensional rotations. In fact one therefore has to deal with non–linear manifolds having different geometrical structure than that of "flat" vector spaces. So let M be a matrix group. Then the Riemannian distance between two group elements is given by

$$d_R(G_1, G_2) = \frac{1}{\sqrt{2}} \| \log(G_1^T, G_2) \| , \tag{3}$$

where log refers to matrix logarithm and the usual Frobenius norm is applied (see e.g. [5]). The Riemannian metric (3) measures the length of the shortest geodesic connecting G_1 and G_2. Since every group acts transitively on itself there is a closed form solution to that. However, the shortest geodesic may not be unique. The Riemannian mean associated with (3) now reads

$$\mathcal{R} = \arg\min_{G \in M} \sum_{i=1}^{n} \| \log(G_i^T, G) \|^2 . \tag{4}$$

For many important groups solving (4) analytically is not possible. Recently, in [11] it was proven that there is no closed form solution of (4) for SO(3). Therein it was also demonstrated that Riemannian averaging is already a very hard problem for one–parameter subgroups of SO(3).

As an alternative to computing the Riemannian mean, approximative embedding techniques are well established (see e.g. [4]). The basic idea is to embed the data into a larger vector space in which operations are then performed, and project the result back onto the manifold. Technically, one thereby performs constrained optimization and usually uses orthogonal projection. Both aspects are treaded extensively in [3]. The natural embedding for a matrix group is of course the covering general linear group GL(n), which in turn is also a vector space. Hence the Frobenius norm induces the following metric on GL(n)

$$d_F(G_1, G_2) = \| G_1 - G_2 \| . \tag{5}$$

Associated with (5) is the mean

$$\mathcal{E} = \arg\min_{G \in M} \sum_{i=1}^{n} \| G_i - G) \|^2 , \tag{6}$$

which will be termed Euclidean mean from now on. Note that without back–projection $\mathcal{E}$ does not have to be an element of M.

Whenever the group M is a differentiable manifold, i.e. a Lie group, there is a further alternative for an approximation of the Riemannian mean $\mathcal{R}$ (4). Roughly speaking, the Riemannian distance in the Lie group can be approximated by the

Euclidean distance in the corresponding Lie algebra. This is done by virtue of the famous Baker–Campbell–Hausdorff formula [8]. The whole method is very common in robotics [13]. A recent example for its use for motion estimation is [6].

This paper, however, concentrates on the study of Clifford groups for approximating the Riemannian mean $\mathcal{R}$. It is well known, that any three–dimensional rotation can be represented by a rotation matrix or a unit quaternion (among other possible representations like Euler angles). Unit quaternions do form a group which acts by a two–sided operation. That way a different Riemannian mean approximation results than the one induced by rotation matrices. Unit quaternions are a particularly example of a Clifford group. In fact all classical groups do have a covering Clifford group. Hence the Clifford algebra framework offers a general alternative for averaging on such groups.

The remainder of this paper is organized as follows. In section 2 we briefly present basic facts about Clifford groups. This is then followed in section 3 by reviewing what is known about averaging on SO(3) using both rotations matrices and unit quaternions. Additionally, an experimental comparison of the two approaches for noisy data is presented. Averaging on the two–dimensional Lorenz group SO(1,2) and its covering Clifford group $\Gamma_{1,2}$ is discussed in section 4. Therein a closer look on the related groups SL(2,$\mathbb{R}$) and SU(1,1) is also provided. The paper finishes with some concluding remarks.

2 Clifford Groups

Associated with every Clifford algebra there is the so–called Clifford group (or Lipschitz group) formed by all of its invertible elements. A Clifford algebra can be constructed from a quadratic space. Here we are particularly interested in real quadratic spaces $\mathbb{R}^{p,q}$, meaning $\mathbb{R}^{p+q}$ equipped with a quadratic form Q of signature (p,q).

Definition 1 ([10]). *An associative algebra over $\mathbb{R}$ with unity 1 is the Clifford algebra $\mathcal{C}_{p,q}$ of $\mathbb{R}^{p,q}$ if it contains $\mathbb{R}^{p+q}$ and $1 \cdot \mathbb{R} = \mathbb{R}$ as distinct subspaces so that*

(a) $\mathbf{x}^2 = Q(x)$ *for any* $x \in \mathbb{R}^{p+q}$
(b) $\mathbb{R}^{p+q}$ *generates* $\mathcal{C}_{p,q}$
(c) $\mathcal{C}_{p,q}$ *is not generated by any proper subspace of* $\mathbb{R}^{p+q}$.

Let $\{e_1, e_2, \ldots, e_n\}$ be an orthonormal basis of $\mathbb{R}^{p,q}$. Then the following relations hold

$$e_i^2 = 1, \ 1 \leq i \leq p, \quad e_i^2 = -1, \ p < i \leq n \quad e_i e_j = -e_j e_i, \ i < j. \tag{7}$$

That way an algebra of dimension 2^n is generated (putting $e_0 = 1$). The canonical basis of a Clifford algebra is therefore formed by

$$B_\mu = e_{j_1} e_{j_2} \cdots e_{j_r}, \ 1 \leq j_1 < \ldots j_r \leq p+q. \tag{8}$$

Those basis vectors which consist of an even number of factors do form a subalgebra. This subalgebra $\mathcal{C}_{p,q}^+$ is called the even part of $\mathcal{C}_{p,q}$.

For Clifford algebras the following isomorphisms hold

$$\mathcal{C}_{0,0} \cong \mathbb{R} \tag{9a}$$

$$\mathcal{C}_{0,1} \cong \mathbb{C} \tag{9b}$$

$$\mathcal{C}_{0,2} \cong \mathbb{H}, \tag{9c}$$

with $\mathbb{C}$ denoting complex numbers and $\mathbb{H}$ denoting quaternions as usual. The embedding used in Definition 1 (a) above can be made more explicit as follows. Define

$$\alpha : \mathbb{R}^{p+q} \to \mathcal{C}_{p,q}, x \mapsto \mathbf{x} = \sum_{i=1}^{n} x_i e_i \tag{10}$$

and identify $\mathbb{R}^{p+q}$ with its image under that mapping. The elements $\alpha(\mathbb{R}^{p+q})$ are termed vectors again. All invertible vectors $(\mathbf{x}^2 \neq 0)$ already generate the Clifford group. For a more revealing characterization the following two mappings are required. Inversion, which is an automorphism, is defined by $\hat{\mathbf{x}} = -\mathbf{x}$ and $\widehat{ab} = \hat{a}\hat{b}$. Reversion is an anti–automorphism defined by $\tilde{\mathbf{x}} = \mathbf{x}$ and $\widetilde{ab} = \tilde{b}\tilde{a}$. Using (8) these mappings become

$$\widehat{B_\mu} = (-1)^r B_\mu \quad \widetilde{B_\mu} = (-1)^{\frac{r(r-1)}{2}} B_\mu . \tag{11}$$

Definition 2. *The Clifford group associated with the Clifford algebra $\mathcal{C}_{p,q}$ is defined by*

$$\Gamma_{p,q} = \{s \in \mathcal{C}_{p,q} \mid \forall x \in \mathbb{R}^{p,q}, s\mathbf{x}\hat{s}^{-1} \in \mathbb{R}^{p,q}\}.$$

Hence the Clifford group is determined by its two–sided action on vectors. Furthermore the map $\mathbf{x} \mapsto s\mathbf{x}\hat{s}^{-1}$ is an orthogonal automorphism of $\mathbb{R}^{p,q}$ [12].

Normalizing the Clifford group $\Gamma_{p,q}$ yields

$$\mathrm{Pin}(\mathrm{p},\mathrm{q}) = \{s \in \Gamma_{p,q} \mid s\tilde{s} = \pm 1\}. \tag{12}$$

The group $\mathrm{Pin}(\mathrm{p},\mathrm{q})$ is a two–fold covering of the orthogonal group $\mathrm{O}(\mathrm{p},\mathrm{q})$. Further subgroups of $\mathrm{Pin}(\mathrm{p},\mathrm{q})$ are

$$\mathrm{Spin}(\mathrm{p},\mathrm{q}) = \mathrm{Pin}(\mathrm{p},\mathrm{q}) \cap \mathcal{C}_{p,q}^+ \tag{13}$$

and

$$\mathrm{Spin}^+(\mathrm{p},\mathrm{q}) = \{s \in \mathrm{Spin}(\mathrm{p},\mathrm{q}) \mid s\tilde{s} = 1\}. \tag{14}$$

Both groups are again two–fold covers of their classical counterparts. The whole situation can be summarized as

$$\mathrm{Pin}(\mathrm{p},\mathrm{q})\backslash\{\pm 1\} \cong \mathrm{O}(\mathrm{p},\mathrm{q}) \tag{15a}$$
$$\mathrm{Spin}(\mathrm{p},\mathrm{q})\backslash\{\pm 1\} \cong \mathrm{SO}(\mathrm{p},\mathrm{q}) \tag{15b}$$
$$\mathrm{Spin}^+(\mathrm{p},\mathrm{q})\backslash\{\pm 1\} \cong \mathrm{SO}^+(\mathrm{p},\mathrm{q}). \tag{15c}$$

Thereby $SO^+(p, q)$ is formed by those elements which are connected with the identity. This does not carry over to the covering groups, i.e. $Spin^+(p, q)$ does not have to be connected [10]. In case of $q = 0$ one has $SO^+(p, 0) = SO(p, 0)$ and $Spin^+(p, 0) = Spin(p, 0)$ (analogously for $q = 0$). Further on we simply write $SO(p)$ and $Spin(p)$ then. Finally note that every Lie group can be represented as spin group [2]. Hence averaging using Lie methods is also possible inside the Clifford framework.

3   Averaging on Spin(3) and SO(3)

The already mentioned group of unit quaternions is isomorphic to the three–sphere S^3. The latter, in turn, being isomorphic to the group $Spin(3)$. Additionally $\mathcal{C}_{0,3}^+ \cong \mathcal{C}_{0,2} \cong \mathbb{H}$ yields, and therefore averaging rotations using $Spin(3)$ is the same as if using quaternions. To remain consistent everything in the following will be denoted in terms of $Spin(3)$.

Before actually turning to the problem of averaging rotations, a more theoretical remark may be in order. The group $Spin(3)$ can also be used for averaging on the three-sphere, which of course can not be done by using rotation matrices. On the other hand no disadvantage results from the fact that $Spin(3)$ is a two–fold cover of $SO(3)$. Formally, some care has to be taken due to the existence of antipodal points ($s, -s \in Spin(3)$ induce the same rotation). A consistent set of group elements, however, can always be chosen easily if necessary.

As for representing rotations itself, the Euclidean mean can be defined both in terms of $SO(3)$

$$\mathcal{E}_{SO(3)} = \arg\min_{R \in SO(3)} \sum_{i=1}^{n} \|R_i - R)\|^2 \qquad (16)$$

and $Spin(3)$

$$\mathcal{E}_{Spin(3)} = \arg\min_{s \in Spin(3)} \sum_{i=1}^{n} \|s_i - s)\|^2. \qquad (17)$$

For the latter note that every Clifford algebra is of course also a real vector space. The following was derived (using quaternions) in [7] for the Euclidean mean on $Spin(3)$

$$\mathcal{E}_{Spin(3)} = \arg\max_{s \in Spin(3)} \sum_{i=1}^{n} ss_i = \arg\max_{s \in Spin(3)} s \sum_{i=1}^{n} s_i = \frac{\sum_{i=1}^{n} s_i}{\|\sum_{i=1}^{n} s_i\|^2} \qquad (18)$$

which is the ordinary arithmetic mean with normalization. Solving for the matrix mean (16) is a special case of the famous Procrustes problem yielding

$$\mathcal{E}_{SO(3)} = \arg\max_{R \in SO(3)} \sum_{i=1}^{n} tr(R^T R_i) = \arg\max_{R \in SO(3)} tr(R^T \sum_{i=1}^{n} R_i)$$

$$= \arg\max_{R \in SO(3)} tr\left(R^T \frac{\sum_{i=1}^{n} R_i}{n}\right), \qquad (19)$$

which is the orthogonal projection of the arithmetic mean onto SO(3). The actual solution can then be obtained by using Singular Value Decomposition (SVD). This, although not too complicated, is somehow more costly than simple normalization as in (18). Moreover the two discussed approximation methods for the Riemannian mean are indeed based on different linearizations.

A rotation is represented in the algebra $\mathcal{C}_{0,2}$ as

$$\cos(\frac{\phi}{2})e_0 + \sin(\frac{\phi}{2})(xe_1 + ye_2 + ze_1e_2),\qquad(20)$$

with (x, y, z) being the rotation axis and θ being the angle of rotation. Since every Clifford group operates by a two–sided action (see Definition 2 again) the approximation (18) is based on half the angle θ. Contrary, every matrix group acts by ordinary matrix multiplication and the approximation (16) is therefore based on the whole angle θ. An experimental comparison of the two methods have been already provided in [7] using a Gaussian sampling for the angle and a uniform one for the axis. Both methods have been reported as equally very good. In our opinion a Gaussian sampling for both parameter types seems to be at least as realistic.

For demonstration a little experiment on synthetic data has been carried out. The sample size is set to 20, the mean is set to $[1, 2, 3, 4]/\|[1, 2, 3, 4]\|$ with standard deviation of 0.2 (first setup) and 0.5 (second setup). Each setup is re-

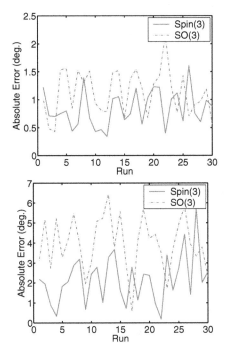

Fig. 1. Abbreviation of the approximated mean from the Riemannian mean for standard deviation of 0.2 (top) and 0.5 (bottom). See text for details

peated 30 times. The true Riemannian mean is computed by using non–linear optimization from `MATLAB`. The angle of the rotation transforming the approximated mean into the Riemannian mean is used as error measure. The obtained results are shown in Fig. 1.

The average error in the small standard deviation setup is 0.83 deg. for Spin(3) averaging versus 1.13 deg. for SO(3) averaging. Both values are very good and the difference is rather insignificant. For the larger standard deviation setup the value for Spin(3) is 2.17 deg. compared to 3.98 deg. for SO(3). Here the difference is obviously noticeable. All in all in our chosen setup averaging using Spin(3) seems therefore preferable.

Approximating the Riemannian mean using the aforementioned projected Euclidean means works quite nice for three–dimensional rotations in reasonable setups (taken into consideration both [7] and the above experiment). One prerequisite for any approximation to make sense is of course that the entity being approximated makes sense itself. The Riemannian metric (shortest geodesic) for three–dimensional rotation does. For example, it is bi–invariant [11]. For the Special Euclidean group SE(3), of which SO(3) is a subgroup, however, no bi–invariant metric does exist [1].

4   Averaging on Spin(1,2) and SO(1, 2)

When not familiar with Clifford algebra everything about quaternions seems to be quite exceptional at first sight. As we have seen this is not true. All Clifford groups do operate in the same way by a two–sided action yielding an orthogonal automorphism. Hence there is also a general treatment of averaging on other orthogonal (sub–)groups in terms of Clifford algebra. Quaternions being just one example. Another such example will be studied in this section. In oder to simplify notations we will use the canonical ordered basis derivable from (8) do denote elements of a Clifford algebra. That is we just write $(a, b, c, d) \in \mathcal{C}_{0,2}$ instead of $ae_0 + be_1 + ce_2 + de_1e_2$, for example.

In the following we want to study the two–dimensional Lorentz group SO(1,2). As shown in [9] this group comes into play whenever measurements with respect to motion are (realistically) considered as taking their own time. The group SO(1,2) has time dimension (t) and spatial dimension (x, y) leaving invariant the scalar product

$$< (t, x, y), (t', x'y') >= tt' - xx' - yy' . \tag{21}$$

More precisely, it is formed by those 3×3 matrices with determinant one which preserve (21). Geometrically, everything about SO(1,2) is related to cones. An important example being the future cone $\{(t, x, y) \mid t^2 - x^2 - y^2 \geq 0, t \geq 0\}$.

The group SO(1,2) has the two well known covering groups SU(1,1) and SL(2, $\mathbb{R}$), which are defined by

$$\left\{ \begin{pmatrix} y_1 & y_2 \\ \overline{y_2} & \overline{y_1} \end{pmatrix} \mid y_1, y_2 \in \mathbb{C}, |y_1|^2 - |y_2|^2 = 1 \right\}, \tag{22}$$

and

$$\left\{ \begin{pmatrix} a & b \\ c & d \end{pmatrix} \mid a, b, c, d \in \mathbb{R}, \, ad - bc = 1 \right\}, \tag{23}$$

respectively. Using the Euclidean mean as approximation for the Riemannian mean requires to project the arithmetic mean onto the manifold. This, yet, is a rather complicated problem for SO(1,2) since the orthogonality condition is now $RR^T = diag(1, -1, -1)$. Of course things are easier using SL(2, $\mathbb{R}$) instead. In that case, however, there is no need to use a matrix group at all since the following relations hold

$$\mathrm{SL}(2, \mathbb{R}) \cong \mathrm{SU}(1, 1) \cong \mathrm{Spin}(1, 2), \tag{24}$$

the latter being a two–fold cover of SO(1,2) by definition. Hence as abstract groups all groups in question are isomorphic. Moreover, the different representations as elements of the Clifford algebra $\mathcal{C}_{1,2}$ do only differ by permutation. That can be easily checked using the fact $\mathcal{C}_{1,2} \cong \mathbb{C}(2)$, where $\mathbb{C}(2)$ denotes the space of all complex 2×2 matrices. A matrix $\begin{pmatrix} a & b \\ c & d \end{pmatrix} \in \mathrm{SL}(2, \mathbb{R})$ is represented in $\mathcal{C}_{1,2}$ by

$$\frac{1}{2}(a + d, b + c, 0, c - b, 0, a - d, 0, 0). \tag{25}$$

Setting $y_1 = \frac{1}{2}((a + d) + i(b - c))$ and $y_2 = \frac{1}{2}((b + c) - i(d - a))$ yields the corresponding SU(1,1) matrix, which in turn is represented by

$$\frac{1}{2}(a + d, b + c, b + c, -(c - b), 0, d - a, 0, 0). \tag{26}$$

Hence all groups have essentially the same representation in the Clifford algebra $\mathcal{C}_{1,2}$. Furthermore the Euclidean means

$$\mathcal{E}_{\mathrm{Spin}(1,2)} = \arg \min_{s \in \mathrm{Spin}(1,2)} \sum_{i=1}^{n} \|s_i - s)\|^2 \tag{27a}$$

$$\mathcal{E}_{\mathrm{SL}(2,\mathbb{R})} = \arg \min_{R \in \mathrm{SL}(2,\mathbb{R})} \sum_{i=1}^{n} \|R_i - R)\|^2 \tag{27b}$$

$$\mathcal{E}_{\mathrm{SU}(1,1)} = \arg \min_{U \in \mathrm{SU}(1,1)} \sum_{i=1}^{n} \|U_i - U)\|^2 \tag{27c}$$

are then also identical and can be computed all inside the algebra $\mathcal{C}_{1,2}$ just by simple normalization

$$\frac{\sum_{i=1}^{n} m_i}{\|\sum_{i=1}^{n} m_i\|^2}, \tag{28}$$

with $m_i = s_i$, $m_i = R_i$, or $m_i = U_i$ accordingly to the cases in (27). Moreover, everything could also been carried out in the algebra $\mathcal{C}_{3,0}$, which is isomorphic to $\mathcal{C}_{1,2}$. For example,

$$\frac{1}{2}(a + d, 0, b + c, d - a, 0, 0, b - c, 0) \tag{29}$$

corresponds to (25). In the following we will only consider the group Spin(1,2). In oder to evaluate the quality of approximation by (28) the Riemannian mean has to be studied first a little bit closer. Formally, (4) does apply again. So we are rather looking for a parameterization of SO(1,2). One such parameterization is the Cartan decomposition SO(1,2)= KAK having factors

$$K = \begin{pmatrix} 1 & 0 & 0 \\ 0 & \cos\phi & -\sin\phi \\ 0 & \sin\phi & \cos\phi \end{pmatrix} \quad \text{and} \quad A = \begin{pmatrix} \cosh\psi & 0 & \sinh\psi \\ 0 & 1 & 0 \\ \sinh\psi & 0 & \cosh\psi \end{pmatrix}. \quad (30)$$

Again an experiment on synthetic data has been performed to compare the approximated Euclidean mean on Spin(1,2) with the true Riemannian mean. Both angles arising from the Cartan decomposition (30) have been sampled using a Gaussian distribution. In the first experiment we used $\phi = 30$ deg. and $\psi = 20$ deg. as mean values and a standard deviation of 2 deg. in both cases for a sample of size 30. The obtained results are separately reported for both angles in Fig. 2.

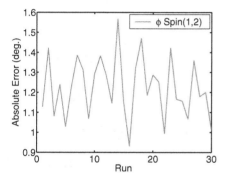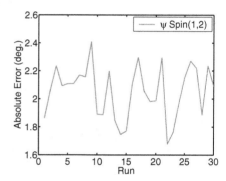

Fig. 2. Abbreviation of the approximated mean from the Riemannian mean for standard deviation of 2 deg. for ϕ (left) and ψ (right). See text for details

The results are quite good. The average error is 1.27 deg. for ϕ and 2.03 for ψ. Using a standard deviation of both 5 deg. resulted in average errors of 2.56 and 3.87, respectively. As before the error was measured as angles of the transformation needed to carry over the approximated mean into the true Riemannian one. The latter was computed using non–linear optimization from `Matlab` again. From the obtained results averaging in Spin(1,2) seems to be a useful approximation method for practical applications.

5   Conclusions

In this paper we studied averaging in Clifford groups. More precisely, the approximation of the Riemannian means by Euclidean means of such groups have been discussed. The Clifford algebra framework allows a general and elegant treatment of averaging problems. The particular case of three–dimensional rotations has been reviewed comparing averaging in SO(3) (rotation matrices) with averaging in Spin(3) (unit quaternions). In the chosen setup the latter performed slightly better. More important, the Euclidean mean is always easy to compute for a Clifford group, namely by just performing normalization in the associated algebras. This was further demonstrated on $SO(1,2)$, where it has been also demonstrated how related groups can be handled in the same manner. The obtained results suggest that Clifford algebra is a useful and flexible tool for averaging. Future work will be on testing the proposed methods for particular neural networks in practical applications. Also a comparison with common Lie algebra averaging seems to be interesting. Studying the influence of embeddings like the conformal model of Clifford algebra for averaging in SE(3) might also be worthwhile.

References

1. H. Bruyninckx. Some Invariance Problems in Robotics. Technical Report PMA 91R4, Kathollieke Universiteit Leuven, 1991.
2. C. Doran, D. Hestenes, F. Sommen, and N. Van Acker. Lie Groups as Spin Groups. *J. Math. Phys.*, 34:3642–3669, 1993.
3. A. Edelman, T. A. Arias, and S. T. Smith. The Geometry of Algorithms with Orthogonality Constraints. *SIAM J. Matrix Anal. Appl.*, 20:303–353, 1999.
4. N. I. Fisher, T. Lewis, and B. J. Embleton. *Analysis of Spherical Data*. Cambridge University Press, 1987.
5. G. H. Golub. *Matrix Computations*. The Johns Hopkins University Press, 1990.
6. V. M. Govindu. Lie–algebraic Averaging for Globally Consistent Motion Estimation. *Proc. of the CVPR*, 1:684–691, 2004.
7. C. Gramkow. On Averaging Rotations. *J. Math. Imaging and Vision*, 15:7–16, 2001.
8. A. Iserles, H. Z. Munthe-Kaas, S. P. Nørsett, and A. Zanna. Lie–group Methods. *Acta Numer.*, 9:215–365, 2000.
9. R. Lenz and G. Granlund. If i had a fisheye i would not need so(1,n), or is hyperbolic geometry useful in image processing? *Proc. of the SSAB Symposium, Uppsala, Sweden*, pages 49–52, 1998.
10. P. Lounesto. *Clifford Algebras and Spinors*. Cambridge University Press, 1997.
11. M. Moakher. Means and Averaging in the Group of Rotations. *SIAM J. Matrix Anal. Appl.*, 24:1–16, 2002.
12. I. R. Porteous. *Clifford Algebras and the Classical Groups*. Cambridge University Press, Cambridge, 1995.
13. J. M. Selig. *Geometrical Methods in Robotics*. Springer–Verlag, 1996.

Combinatorics and Representation Theory of Lie Superalgebras over Letterplace Superalgebras

Andrea Brini, Francesco Regonati, and Antonio Teolis

Dipartimento di Matematica, University of Bologna,
Piazza di Porta San Donato 5, 40126 Italy
brini@dm.unibo.it

Abstract. We state three combinatorial lemmas on Young tableaux, and show their role in the proof of the triangularity theorem about the action of Young-Capelli symmetrizers on symmetrized bitableaux. As an application, we describe in detail the way to specialize general results to the representation theory of the symmetric group and to classical invariant theory.

1   Introduction

The theory of letterplace superalgebras, regarded as bimodules with respect to the action of a pair of general linear Lie superalgebras, is a fairly general one, and encompasses a variety of classical theories such as the ordinary representation theory of the symmetric group, the representation theory of general linear and symmetric groups over tensor spaces, as well as a substantial part of classical algebraic invariant theory, to name but a few.

The purpose of the present work is twofold. On the one hand, we prove that the general theory can be founded, by the systematic use of the superalgebraic version of Capelli's method of virtual variables, on a handful of combinatorial lemmas on Young supertableaux; this fact should be regarded as an ultimate by-product of the van der Waerden-von Neumann argument in the representation theory of the symmetric group. On the other hand, we describe in detail the way to specialize the general theory to two classical settings: the representation theory of the symmetric group and the invariant theory of algebraic forms over vector variables.

The paper is organized as follows. In section 2, we provide a survey of the basics of the representation theory of general linear Lie superalgebras acting on letterplace superalgebras. This theory gets its effectiveness from the superalgebraic version of Capelli's method of virtual variables. For an exposition of the method of virtual variables, we refer the reader to our recent work [3].

We envision section 3 as the methodological core of the present contribution; we state three combinatorial lemmas (whose proofs will be published elsewhere) from which we derive in a rather immediate way the cornerstone of the general theory, namely, the *triangularity theorem* about the action of Young-Capelli symmetrizer operators on standard symmetrized bitableaux.

H. Li, P. J. Olver and G. Sommer (Eds.): IWMM-GIAE 2004, LNCS 3519, pp. 239–257, 2005.

By way of application, in section 4 we derive from the general theory a three pages self-contained treatment of the whole theory of ordinary representations of the symmetric group, up to the Young natural form of irreducible matrix representations; in section 5, we prove a canonical form of the beautiful Capelli's polar expansion formula and briefly discuss its relevance in classical algebraic invariant theory.

The authors thank Stefano Sarti for his valuable comments on the subject of section 4.

2   The Letterplace Algebra as a Bimodule $pl(L) \cdot Super[L|P] \cdot pl(P)$

2.1   Letterplace Superalgebras

A *signed set* is a set $\mathcal{A}$ endowed with a *signature* map $|\ | : \mathcal{A} \to \mathbb{Z}_2$; the sets $\mathcal{A}^+ = \{a \in \mathcal{A}; |a| = 0\}$ and $\mathcal{A}^- = \{a \in \mathcal{A}; |a| = 1\}$ are called the subsets of *positive* and *negative* symbols, respectively.

In the following, we consider a pair of signed alphabets $X = X^- \cup X^+$ and $Y = Y^- \cup Y^+$ (X^-, X^+, Y^-, Y^+ countable sets), that we call the *letter alphabet* and the *place alphabet*, respectively. The *letterplace alphabet*

$$[X|Y] = \{(x|y); \ x \in X, \ y \in Y\}$$

inherits a signature by setting $|(x|y)| = |x| + |y| \in \mathbb{Z}_2$.

In the following, $\mathbb{K}$ will denote a field, $char(\mathbb{K}) = 0$.

The *letterplace* $\mathbb{K}-$*superalgebra* $Super[X|Y]$ is the quotient algebra of the free associative $\mathbb{K}-$algebra with 1 generated by the letterplace alphabet $[X|Y]$ modulo the bilateral ideal generated by the elements of the form:

$$(x|y)(z|t) - (-1)^{(|x|+|y|)(|z|+|t|)}(z|t)(x|y), \qquad x, z \in X, \ y, t \in Y.$$

Remarks

1. $Super[X|Y]$ is a commutative superalgebra (i.e., it is $\mathbb{Z}_2-$graded and super-symmetric);
2. $Super[X|Y]$ is an $\mathbb{N}-$graded algebra

$$Super[X|Y] = \bigoplus_{n \in \mathbb{N}} Super_n[X|Y]$$

$$Super_n[X|Y] = \langle (x_{i_1}|y_{i_1})(x_{i_2}|y_{i_2})\dots(x_{i_n}|y_{i_n}), \ x_{i_h} \in X, \ y_{j_k} \in Y\rangle_{\mathbb{K}};$$

3. the $\mathbb{Z}_2-$gradation and the $\mathbb{N}-$gradation of $Super[X|Y]$ are *coherent*, that is

$$(Super[X|Y])^i = \bigoplus_{n \in \mathbb{N}} (Super_n[X|Y])^i, \qquad i \in \mathbb{Z}_2.$$

2.2   Superpolarization Operators

Let $x', x \in X$. The *superpolarization* $\mathcal{D}_{x',x}$ *of the letter* x *to the letter* x' is the unique *left* superderivation of $\mathbb{Z}_2$−grade $|x'| + |x|$, that is a linear operator $\mathcal{D}_{x',x} : Super[X|Y] \to Super[X|Y]$ such that

$$\mathcal{D}_{x',x}(AB) = \mathcal{D}_{x',x}(A)B + (-1)^{(|x'|+|x|)|A|} A \mathcal{D}_{x',x}(B),$$

for all monomials $A, B \in Super[X|Y]$, defined by the conditions

$$\mathcal{D}_{x',x}(z|t) = \delta_{x,z}(x'|t),$$

for every $(z|t) \in [X|Y]$.

Here and in the following the Greek letter δ will denote the Kronecker symbol.

Let $y', y \in Y$. The *superpolarization* $_{y,y'}\mathcal{D}$ *of the place* y *to the place* y' is the unique *right* superderivation of $\mathbb{Z}_2$−grade $|y'| + |y|$, that is a linear operator (mirror notation) $Super[X|Y] \leftarrow Super[X|Y] : {}_{y,y'}\mathcal{D}$ such that

$$(-1)^{(|y|+|y'|)|B|}(A) \; _{y,y'}\mathcal{D}B + A(B) \; _{y,y'}\mathcal{D} = (AB) \; _{y,y'}\mathcal{D},$$

for all monomials $A, B \in Super[X|Y]$, defined by the conditions

$$(z|t) \; _{y,y'}\mathcal{D} = \delta_{t,y}(z|y'),$$

for every $(z|t) \in [X|Y]$.

We hardly need to recall that every letter-polarization operator *commutes* with every place-polarization operator.

Let $L = L^+ \cup L^- \subset X$ and $P = P^+ \cup P^- \subset Y$ be *finite* subsets of the "universal" letter and place alphabets X and Y, respectively. The elements $x \in L$ ($y \in P$) are called *proper* letters (*proper* places), and the elements $x \in X \setminus L$ ($y \in Y \setminus P$) are called *virtual* letters (places). The signed subset $[L|P] = \{(x|y); \; x \in L, \; y \in P\} \subset [X|Y]$ is called a *proper letterplace alphabet*.

The $\mathbb{Z}_2$−graded subalgebra $Super[L|P] \subset Super[X|Y]$ generated by the set of proper letterplace variables $[L|P]$ is called the *proper letterplace superalgebra* generated by the proper alphabets L and P.

2.3   General Linear Lie Superalgebras, Representation Theory and Polarization Operators

Consider the (finite dimensional) *general linear Lie superalgebras* $pl(L)$ and $pl(P)$ associated to the finite signed subsets L and P, respectively, and their $\mathbb{Z}_2$−homogeneous standard bases $\{E_{x'x}; \; x, x' \in L\}$ and $\{E_{y'y}; \; y, y' \in P\}$ (see e.g. [3], [13]).

The (even) mappings

$$E_{x'x} \mapsto \mathcal{D}_{x',x}, \quad x, x' \in L, \qquad E_{y'y} \mapsto {}_{y,y'}\mathcal{D}, \quad y, y' \in P$$

induce Lie superalgebra actions of $pl(L)$ and $pl(P)$ over any $\mathbb{N}$−*homogeneous component* $Super_n[L|P]$ of the proper letterplace algebra.

Furthermore, by the commutation property, $Super_n[L|P]$ is a bimodule over the universal enveloping algebras $\mathcal{U}(pl(L))$ and $\mathcal{U}(pl(P))$.

In the following, we will denote by $\mathcal{B}_n$, $_n\mathcal{B}$ the (finite dimensional) subalgebras of $End_\mathbb{K}(Super_n[L|P])$ induced by the actions of $\mathcal{U}(pl(L))$ and $\mathcal{U}(pl(P))$, respectively. The operator algebras $\mathcal{B}_n$, $_n\mathcal{B}$ are therefore the algebras generated by the proper letter and place polarization operators (restricted to $Super_n[L|P]$), respectively.

2.4   Young Tableux on a Signed Set: Basic Definitions

Assume that a finite signed linearly ordered set $\mathcal{A} = \mathcal{A}^+ \cup \mathcal{A}^-$ is given. The symbol $Tab(\mathcal{A})$ will denote the set of all Young tableaux on $\mathcal{A}$. The *shape* of a tableau $S \in Tab(\mathcal{A})$ is the *partition* $\lambda = (\lambda_1 \geq \lambda_2 \geq \ldots)$ whose parts λ_i are the lengths of the rows of the tableau S and is denoted by the symbol $sh(S)$; we write $sh(S) = \lambda \vdash n$ to mean that the shape $sh(S)$ is a partition of the integer n. The *content* of a Young tableau T is the set of the symbols that appear in T, together with their multiplicities.

A Young tableau $S \in Tab(\mathcal{A})$ is called *(super)standard* when each row of S is non-decreasing, with no repeated negative symbols, and each column of S is non-decreasing, with no repeated positive symbols. The set of all the standard tableaux over $\mathcal{A}$ is denoted by $Stab(\mathcal{A})$. Consistently, we denote by $Stab_\lambda(\mathcal{A})$ the set of all the standard tableaux S over $\mathcal{A}$ such that $sh(S) = \lambda$.

Let $H(\mathcal{A}) = \{\lambda = (\lambda_1 \geq \lambda_2 \geq \ldots); \quad \lambda_{r+1} < s+1\}$, where r denotes the cardinality of the set $\mathcal{A}^+$ and s denotes the cardinality of the set $\mathcal{A}^-$. We recall that the set of standard tableaux of shape λ over $\mathcal{A}$ is non empty if and only if $\lambda \in H(\mathcal{A})$.

A tableau C is said to be of *coDeruyts* type whenever any two symbols in the same row of C are equal, while any two symbols in the same column of C are distinct. A tableau D is said to be of *Deruyts* type whenever any two symbols in the same column of D are equal, while any two symbols in the same row of D are distinct. In the following, the symbol C will denote a coDeruyts tableau filled with *virtual* positive symbols, and the symbol D will denote a Deruyts tableau filled with *virtual* negative symbols. In the formulas below, the shapes of the tableaux C and D, and the fact that the virtual symbols were letter or place symbols should be easily inferred from the context.

2.5   Symmtrized Bitableaux and Young-Capelli Symmetrizers

Given a Young tableau $S \in Tab(\mathcal{A})$, by reading its entries from left to right and from top to bottom, we get a word $w(S)$ on $\mathcal{A}$, called the *row word* of S.

Let $T \in Tab(X)$, $U \in Tab(Y)$, with $sh(T) = sh(U) \vdash n$. The *bitableau monomial* $\underline{TU}$ is defined as follows:

$$\underline{TU} = (t_1|u_1)(t_2|u_2)\ldots(t_n|u_n) \in Super_n[X|Y],$$

where $t_1 \ldots t_n = w(T)$ and $u_1 \ldots u_n = w(U)$.

Let $S', S \in Tab(X)$, with $sh(S') = sh(S) \vdash n$. The *letter polarization monomial* of the tableau S to the tableau S' is defined to be the $\mathbb{K}$−linear operator

$$S'S = \mathcal{D}_{x_1' x_1} \mathcal{D}_{x_2' x_2} \dots \mathcal{D}_{x_n' x_n} \in End_{\mathbb{K}}[Super_n[X|Y]],$$

where $x_1' \dots x_n' = w(S')$ and $x_1 \dots x_n = w(S)$.

Let $V, V' \in Tab(Y)$, with $sh(V) = sh(V') \vdash n$. The *place polarization monomial* of the tableau V to the tableau V' is defined to be the $\mathbb{K}$−linear operator

$$VV' =_{y_1 y_1'} \mathcal{D} \ _{y_2 y_2'} \mathcal{D} \ \dots \ _{y_n y_n'} \mathcal{D} \in End_{\mathbb{K}}[Super_n[X|Y]],$$

where $y_1 \dots y_n = w(V)$ and $y_1' \dots y_n' = w(V')$.

Definition 1. *(Right Symmetrized Bitableaux) For every $\lambda \vdash n$ and every $T \in Tab(L)$, $U \in Tab(P)$, with $sh(T) = \lambda = sh(U)$, we define the right symmetrized bitableau $(T \boxed{\|} \, U) \in Super_n[L|P]$ by setting*

$$(T \boxed{\|} \, U) = TC \ \underline{CD} \ DU$$

where C is any virtual tableaux of coDeruyts type, D is any virtual tableaux of Deruyts type, all of shape λ.

A quite useful (see section 3), but not trivial, fact is that symmetrized bitableaux admit different equivalent definitions.

Proposition 1. *For every $\lambda \vdash n$ and every $T \in Tab(L)$, $U \in Tab(P)$, with $sh(T) = \lambda = sh(U)$, we have*

$$(T \boxed{\|} \, U) = \underline{TC_1} \ C_1 D \ DU = TC \ \underline{CD} \ DU = TC \ CD_1 \ \underline{D_1 U}$$

where C, C_1 are any virtual tableaux of coDeruyts type, D, D_1 are any virtual tableaux of Deruyts type, all of shape λ.

The symmetrized bitableau $(T \boxed{\|} \, U)$ is supersymmetric in the rows of T, i.e. any transposition of two letters t' and t in the same row of T gives rise to a sign change $(-1)^{|t'||t|}$, and dual supersymmetric in the columns of U, i.e. any transposition of two places u' and u in the same column of U gives rise to a sign change $(-1)^{(|u'|+1)(|u|+1)}$. A symmetrized bitableau is called *standard* when both its letter tableau and its place tableau are standard.

Theorem 1. *(The Gordan–Capelli Basis [3]) The set of standard right symmetrized bitableaux $(T \boxed{\|} \, U)$, with $sh(T) = sh(U) \vdash n$, is a $\mathbb{K}$−linear basis of $Super_n[L|P]$.*

Given an element f in $Super_n[L|P]$, it can be expanded it into a linear combination of the elements of the Gordan-Capelli basis:

$$f = \sum_{\lambda \vdash n} \sum_{\substack{T \in Stab_\lambda(L) \\ U \in Stab_\lambda(P)}} c_{T,U}^f \, (\boxed{T} \, | U), \quad c_{T,U}^f \in \mathbb{K}.$$

Such an expansion is called the *Gordan-Capelli expansion* of f.

Definition 2. *(Right Young-Capelli symmetrizers) Let* $\lambda \vdash n$ *and let* $S', S \in Tab(L)$, *with* $sh(S') = \lambda = sh(S)$. *The product of letter bitableau polarization monomials* $S'C\ CD\ DS$ *defines, by restriction, a linear operator*

$$\gamma_n(S', \boxed{S}) = S'C\ CD\ DS\ \in End_{\mathbb{K}}[Super_n[L|P]],$$

which is independent from the choice of the virtual tableau C *of coDeruyts type and of the virtual tableau* D *of Deruyts type. The operator* $\gamma_n(S', \boxed{S})$ *is called a (right) Young–Capelli symmetrizer.*

By the metatheoretic significance of the method of virtual variables (see, e.g. [3]), the crucial fact is that $\gamma_n(S', \boxed{S})$ belongs to the subalgebra $\mathcal{B}_n$, the algebra generated by proper letter polarizations. In plain words, even though the operator $\gamma_n(S', \boxed{S})$ is defined by using virtual variables, it admits presentations involving only superpolarizations between proper letters.

A right Young–Capelli symmetrizer is called *standard* when both its tableaux are standard.

We define a *partial* order on the set of all standard tableaux over L whose shapes are partitions of a given integer n : for every standard tableau S, we consider the sequence $S^{(p)}$ $p = 1, 2, \ldots$, of the subtableaux obtained from S by considering only the first p symbols of the alphabet, and consider the family $sh(S^{(p)})$, $p = 1, 2, \ldots$, of the corresponding shapes. Since the alphabet is assumed to be finite, this sequence is finite and its last term is $sh(S)$. For standard tableaux S, T we set

$$S \leq T \ \Leftrightarrow \ sh(S^{(p)}) \leq sh(T^{(p)}), \ p = 1, 2, \ldots,$$

where $\leq$ stands for the *dominance order* on partitions. We recall that the dominance order on the set of all the partitions of a given positive integer n is defined as follows: $\lambda = (\lambda_1 \geq \lambda_2 \geq \ldots) \leq \mu = (\mu_1 \geq \mu_2 \geq \ldots)$ if and only if

$$\lambda_1 + \cdots + \lambda_i \leq \mu_1 + \cdots + \mu_i,$$

for every $i = 1, 2, \ldots$

Theorem 2. *(Triangularity Theorem, [3]) The action of standard Young–Capelli symmetrizers on standard symmetrized bitableaux is given by*

$$\gamma_n(S', \boxed{S})(T\|\boxed{U}) = \begin{cases} h_\lambda \theta_{ST}^D (S'\|\boxed{U}), & sh(S) = sh(T) = \lambda \\ 0 & otherwise \end{cases},$$

where θ_{ST}^D *are integers,* $\theta_{ST}^D = 0$ *unless* $S \geq T$, $\theta_{ST}^D \neq 0$ *for* $S = T$, *and* h_λ *is a non zero integer.*

We claim that the integer h_λ admits a deep combinatorial interpretation: h_λ is the product of the hook lengths of the shape λ (see e.g. [12], [10]).

Given any linear extension of the partial order defined above, the matrix $[\theta_{S,T}^D]$ is lower triangular with nonzero integral diagonal entries; the matrix

$$[\varrho_{S,T}^D] = [\theta_{S,T}^D]^{-1}$$

is called the *Rutherford matrix*.

For every $S', S \in Stab(L)$, with $sh(S') = sh(S) = \lambda \vdash n$, we define the *orthonormal generator* $Y_n(S', \boxed{S}) \in \mathcal{B}_n$ by setting

$$Y_n(S', \boxed{S}) = \frac{1}{h_\lambda} \sum_{T \in Stab(L)} \varrho_{ST}^D \, \gamma_n(S', \boxed{T}).$$

From Theorem 2 and the definitions above, the next result immediately follows.

Theorem 3. *The action of the orthonormal generators on the standard symmetrized bitableaux is given by*

$$Y_n(S', \boxed{S})(T\boxed{U}) = \delta_{S,T}(S'\boxed{U});$$

therefore, the orthonormal generators $Y_n(S', \boxed{S})$ *form a* $\mathbb{K}-$*linear basis of the algebra* $\mathcal{B}_n$.

2.6   Complete Decomposition Theorems

In the following, given a subset E of a $\mathbb{K}-$ vector space, the symbol

$$\langle E \rangle_\mathbb{K}$$

will denote the $\mathbb{K}-$ vector subspace spanned by the subset E.

As matter of fact, the following results are corollaries of Theorem 3.

Theorem 4. *We have the following complete decomposition of* $Super_n[L|P]$ *as a module with respect to the action of the general linear Lie superalgebra* $pl(L)$:

$$Super_n[L|P] = \bigoplus_{\substack{\lambda \in H(L) \cap H(P) \\ \lambda \vdash n}} \bigoplus_{\substack{T \in Stab(P) \\ sh(T) = \lambda}} \langle (S\boxed{T}), \ S \in Stab(L) \rangle_\mathbb{K},$$

where the outer sum indicates the isotypic decomposition of the semisimple module, and the inner sum describes a complete decomposition of each isotypic component into irreducible submodules .

Given a standard tableau $T \in Stab(P)$, we recall that the *Schur module* $\mathcal{S}_T$ is the $pl(L)-$module defined as follows:

$$\mathcal{S}_T = \langle (S\boxed{T}), \ S \in Tab(L) \rangle_\mathbb{K}.$$

By using the straightening formula of Grosshans–Rota–Stein [9], it can be proved that the set $\{(S\boxed{T}), \ S \in Stab(L)\}$ is a $\mathbb{K}-$linear basis of $\mathcal{S}_T$.

Therefore, the decomposition formula of Theorem 4 can be rewritten in the more compact form:

$$Super_n[L|P] = \bigoplus_{\substack{\lambda \in H(L) \cap H(P) \\ \lambda \vdash n}} \bigoplus_{\substack{T \in Stab(P) \\ sh(T) = \lambda}} \mathcal{S}_T.$$

Theorem 5. *We have the following complete decomposition of the operator algebra* $\mathcal{B}_n$ *generated by the letter polarization operators acting on* $Super_n[L|P]$:

$$\mathcal{B}_n = \bigoplus_{\substack{\lambda \in H(L) \cap H(P) \\ \lambda \vdash n}} \bigoplus_{\substack{S \in Stab(L) \\ sh(S) = \lambda}} \langle Y_n(S', \boxed{S}), \quad S' \in Stab(L) \rangle_{\mathbb{K}},$$

where the outer sum indicates the isotypic components of the semisimple algebra, and the inner sum describes a complete decomposition of each simple subalgebra into minimal left ideals.

A significant and extremely useful (see, e.g., sect.5) alternative ("dual") version of the theory above can be obtained just by interchanging the roles of the virtual tableaux D and C of Deruyts and coDeruyts type; this is an instance of the *Schur-Weyl duality.* Specifically, we define the *left symmetrized bitableau* $(\boxed{T}|U) \in Super_n[L|P]$ by setting $(\boxed{T}|U) = TD\ \underline{DC}\ CU$ and, consistently, the *left Young–Capelli symmetrizers* $\gamma_n(\boxed{S'}, S) \in \mathcal{B}_n$ by setting $\gamma_n(\boxed{S'}, S) = S'D\ DC\ CS.$

We have a "dual" version of the triangularity theorem:

$$\gamma_n(\boxed{S'}, S)(\boxed{T}|U) = \begin{cases} h_\lambda \theta^C_{ST}(\boxed{S'}|U), & sh(S) = sh(T) = \lambda \\ 0 & otherwise \end{cases},$$

where $\theta^C_{ST} = \theta^D_{TS}.$

Therefore, writing $[\varrho^C_{S,T}]$ for the matrix $[\theta^C_{S,T}]^{-1}$, we set

$$Y_n(\boxed{S'}, S) = \frac{1}{h_\lambda} \sum_{T \in Stab(L)} \varrho^C_{ST}\ \gamma_n(\boxed{S'}, S).$$

and, obviously, we get

$$Y_n(\boxed{S'}, S)((\boxed{T}|U) = \delta_{S,T}((\boxed{S'}|U).$$

In closing this section, we submit that a completely parallel theory holds for the operator algebra $_n\mathcal{B}$ generated by the proper place polarization operators acting on $Super_n[L|P]$. As a consequence, we immediately get:

Theorem 6. *(The Double Commutator Theorem) The subalgebras* $\mathcal{B}_n$ *and* $_n\mathcal{B}$ *of* $End_{\mathbb{K}}(Super_n[L|P])$ *are the centralizer of each other.*

3   Combinatorial Foundations

In the following, we will use the dominance (partial) order for shapes and the (partial) order defined in the previous section for standard tableaux.

Lemma 1. *(A Gale-Ryser type Lemma) Let* $\underline{AB}$ *be a letterplace monomial with all letters negative and all places positive, let* D *be a Deruyts tableau of the same content of* A, *and* C *be a coDeruyts tableau of the same content of* B. *We have*

1. *if $sh(D) \not\trianglerighteq sh(C)$, then $\underline{AB} = 0$;*
2. *if $sh(D) = sh(C)$, then $\underline{AB} = \eta \ \underline{DC}$, where $\underline{DC} \neq 0$ and $\eta \in \{0, \pm 1\}$.*

Let DS be a polarization monomial and $\underline{TC}$ a letterplace monomial, and assume that $sh(D) = sh(S)$ and $sh(T) = sh(C)$ are partitions of the same integer. We consider the action $DS \ \underline{TC}$ of DS on $\underline{TC}$; note that the result is 0 unless S and T have the same content. If S and T have the same content, then $DS \ \underline{TC} = \sum_i A_i B_i$, where the content of each A_i is the same as the content of D and the content of each B_i is the same as the content of C.

By Lemma 1, if $sh(S) \not\trianglerighteq sh(T)$, then each summand $\underline{A_i B_i}$ equals zero, and, therefore,

$$DS \ \underline{TC} = 0.$$

Again by Lemma 1, if $sh(S) = sh(T)$, then each summand $\underline{A_i B_i}$ equals $\eta_i \underline{DC}$, with $\eta_i \in \{0, \pm 1\}$, and, therefore,

$$DS \ \underline{TC} = \theta^D_{ST} \ \underline{DC},$$

where θ^D_{ST} is an integer coefficient depending on S and T. These coefficients are called *symmetry transition coefficients.*

Lemma 2. *(Triangularity Lemma) For every standard letter-tableaux S, T, with $sh(S) = sh(T)$, we have*

$$\theta^D_{ST} = \begin{cases} = 0 \ if \ S \not\trianglerighteq T \\ \neq 0 \ if \ S = T \end{cases}$$

Moreover, each diagonal coefficient θ^D_{SS} is, up to a sign, the product of the factorials of the multiplicities of positive symbols in each row and of negative symbols in each column of the tableau S.

We claim that, for every choice of coDeruyts tablaux C, C_1, C_2 and Deruyts tableau D, D_1, D_2 of the same shape λ, the following identities hold:

$$\underline{CD_2} \ D_2C_1 \ C_1D = CD_1 \ \underline{D_1C_1} \ C_1D = CD_1 \ D_1C_2 \ \underline{C_2D} = h_\lambda \underline{CD},$$

where h_λ is an integer coefficient.

Lemma 3. *For every shape λ, the coefficient h_λ is a positive integer.*

Proof of Theorem 2. First of all, by definition, we have

$$\gamma_n(S', \boxed{S})(T\boxed{U}) = S'C_1 \ (C_1D_1 \ (D_1S \ \underline{TC_2} \ ) C_2D_2) \ D_2U$$

We note that, by Lemma 1, $D_1S \ \underline{TC_2} \neq 0$ implies that $sh(D_1) \geq sh(C_2)$ and $C_1D_1 \ D_1S \ \underline{TC_2} \ C_2D_2 \neq 0$ implies that $sh(C_1) \leq sh(D_2)$. Thus, the whole expression is nonzero only if S and T have the same shape, say $sh(S) = sh(T) = \lambda$. Under this condition, again by Lemma 1, we have

$$\gamma_n(S', \boxed{S})(T\boxed{U}) = S'C_1 \ C_1D_1 \ (D_1S \ \underline{TC_2}) \ C_2D_2 \ D_2U$$
$$= \theta^D_{ST}S'C_1 \ (C_1D_1 \ \underline{D_1C_2} \ C_2D_2) \ D_2U$$
$$= h_\lambda\theta^D_{ST}S'C_1 \ \underline{C_1D_2} \ D_2U$$
$$= h_\lambda\theta^D_{ST}(S'\boxed{U}).$$

The coefficients θ_{ST}^D and h_λ satisfy the triangularity and nondegeneracy conditions by Lemma 2 and 3.

The previous Lemmas have several non trivial consequences, obtained by passing from letterplace identities to operator identities. For example, the identity $DS\ \underline{TC} = \theta_{ST}^D\underline{DC}$ has the operator analog $DS\ TC = \theta_{ST}^D DC$, when the action of the operators DS, TC, DC is restricted to the subspace $Super_n[X|Y]$.

Furthermore, the symmetry transition coefficients θ_{ST}^C arise, in a natural way, just by interchanging the roles of C and D. As a matter of fact, they are defined by the relations $CS\ \underline{TD} = \theta_{ST}^C\ \underline{CD}$, and satisfy the identity $\theta_{ST}^C = \theta_{TS}^D$. In the sequel, the qualifiers D and C in the θ's will be dropped, when clear by the context.

4   The Natural Form of Irreducible Matrix Representations

4.1   The General Case

Let L, P be arbitrary finite signed sets, and m a fixed positive integer. In the sequel, given a partition $\lambda \vdash m$, we will write $S_1, S_2, \ldots, S_{f_\lambda}$ to mean the list of all standard tableaux of shape λ over L with respect to a linear order which is a linear extension of the partial order defined in section 2. In the following, we write θ_{ij}^λ in place of $\theta_{S_iS_j}$, and ϱ_{ij}^λ in place of $\varrho_{S_iS_j}$.

Given an operator $G \in \mathcal{B}_m = \oplus_{\lambda \vdash m}\mathcal{B}_\lambda$, we will denote by G_λ its component in the simple subalgebra $\mathcal{B}_\lambda$ (Theorem 5).

We have

$$Y_m(S_h, \boxed{S_h})\ G\ Y_m(S_k, \boxed{S_k}) = d_{hk}^\lambda(G)Y_m(S_h, \boxed{S_k}),$$

where the coefficients $d_{hk}^\lambda(G)$ are precisely the coefficients that appear in the expansion of G_λ with respect to the basis $\{Y_m(S_h, \boxed{S_k});\ h, k = 1, 2, \ldots f_\lambda\}$ of $\mathcal{B}_\lambda$.

On the other hand, given any standard tableau T of shape λ over P, we have

$$G(S_k\boxed{T}) = G_\lambda(S_k\boxed{T}) = \sum_h c_{hk}^T(G)(S_h\boxed{T}), \qquad c_{hk}^T(G) \in \mathbb{K}.$$

Clearly, we have:

Proposition 2. *For every tableau T, standard over P, with $sh(T) = \lambda$,*

$$d_{hk}^\lambda(G_\lambda) = d_{hk}^\lambda(G) = c_{hk}^T(G) = c_{hk}^T(G_\lambda),$$

for every $G \in \mathcal{B}_m$, and for every $h, k = 1, 2, \ldots f_\lambda$.

In the following, we write c_{hk}^λ in place of c_{hk}^T.

Note that, by setting

$$DS_h\ G\ \underline{S_jC} = \theta_{hj}^\lambda(G)\ \underline{DC},$$

we have:

$$c_{ij}^\lambda(G) \; (S_i\boxed{\boxed{T}}) = Y_m(S_i, \boxed{S_i}) \; G \; (S_j\boxed{\boxed{T}})$$

$$= \frac{1}{h_\lambda} \; S_i C \; CD \sum_h \varrho_{ih}^\lambda \left(DS_h \; G \; \underline{S_j C} \right) CD \; DT$$

$$= \frac{1}{h_\lambda} \; S_i C \; CD \sum_h \varrho_{ih}^\lambda \; \theta_{hj}^\lambda(G) \; \underline{DC} \; CD \; DT$$

$$= \sum_h \varrho_{ih}^\lambda \; \theta_{hj}^\lambda(G) \; \frac{1}{h_\lambda} \; S_i C \; CD \; \underline{DC} \; CD \; DT$$

$$= \sum_h \varrho_{ih}^\lambda \; \theta_{hj}^\lambda(G) \; S_i C \; \underline{CD} \; DT$$

$$= \sum_h \varrho_{ih}^\lambda \; \theta_{hj}^\lambda(G) \; (S_i\boxed{\boxed{T}}).$$

Since $[\theta_{ij}^\lambda] = [\theta_{ij}^\lambda(I)]$, where I denotes the identity in $\mathcal{B}_m$, and $[\varrho_{ij}^\lambda] = [\theta_{ij}^\lambda(I)]^{-1}$, then we get a compact expression for the coefficients $c_{ij}^\lambda(G)$ in terms of the ordinary product of the matrices $\Theta^\lambda(I)^{-1} = [\theta_{ij}^\lambda(I)]^{-1}$ and $\Theta^\lambda(G) = [\theta_{hj}^\lambda(G)]$.

Theorem 7. $C^\lambda(G) = \Theta^\lambda(I)^{-1} \times \Theta^\lambda(G).$

For every $\lambda \vdash m$ such that $\lambda \in H(L) \cap H(P)$ and every standard place tableau T of shape λ, the module structure $\mathcal{U}(pl(L)) \cdot \mathcal{S}_T$, induces a surjective algebra morphism

$$\nu_T : \; \mathcal{U}(pl(L)) \rightarrow End_{\mathbb{K}}[\mathcal{S}_T];$$

by choosing the basis of the standard symmetrized bitableaux $(S_i\boxed{T})$ in the irreducible Schur module $\mathcal{S}_T$, the morphism ν_T induces an irreducible matrix representation

$$\overline{\nu}_T : \; \mathcal{U}(pl(L)) \rightarrow M_{f_\lambda}$$

where, for every $\mathcal{G} \in \mathcal{U}(pl(L))$,

$$\overline{\nu}_T(\mathcal{G}) = [c_{hk}^\lambda(\nu_T(\mathcal{G}))].$$

4.2   The Symmetric Group $\mathbf{S}_n$

In this subsection, we fix the linearly ordered alphabets $L = L^- = \underline{n} = P^- = P$, where $\underline{n} = \{1, 2, \ldots, n\}$. Consider the subspace

$$Super_n^M[\underline{n}|\underline{n}] \subset Super_n[\underline{n}|\underline{n}]$$

spanned by all the monomials (in commutative letterplace variables) of the form

$$(\tau(1)|1)(\tau(2)|2) \cdots (\tau(n)|n), \quad \tau \in \mathbf{S}_n.$$

The space $Super_n^M[\underline{n}|\underline{n}]$ is a $\mathbb{K}[\mathbf{S}_n]$−module with respect to the action

$$\sigma \cdot ((\tau(1)|1)(\tau(2)|2) \cdots (\tau(n)|n)) = (\sigma\tau(1)|1)(\sigma\tau(2)|2) \cdots (\sigma\tau(n)|n), \quad \forall \sigma \in \mathbf{S}_n.$$

The map

$$F : (\tau(1)|1)(\tau(2)|2) \cdots (\tau(n)|n) \mapsto \tau$$

induces a $\mathbb{K}[\mathbf{S}_n]$−module isomorphism

$$F : Super_n^M [\underline{n}|\underline{n}] \rightarrow \mathbb{K}[\mathbf{S}_n],$$

where the group algebra $\mathbb{K}[\mathbf{S}_n]$ is regarded as a *regular left* $\mathbb{K}[\mathbf{S}_n]$−*module*.

A tableau S over the set $\underline{n}$ is said to be *multilinear* whenever each symbol appears just once in S. In the sequel, given a partition $\lambda \vdash n$, we will write $S_1, S_2, \ldots, S_{f_\lambda^M}$ to mean the set of all multilinear standard tableaux of shape λ over $\underline{n}$ endowed with a linear order which is a linear extension of the partial order defined in section 2.

By specializing Theorem 1, the set of right symmetrized bitableaux $(S_i \boxed{S_j})$, $i, j = 1, 2, \ldots, f_\lambda^M$, is a $\mathbb{K}$−linear basis of $Super_n^M [\underline{n}|\underline{n}]$.

Proposition 3. *Let S_i, S_j be standard multilinear tableaux of shape $\lambda \vdash n$. Then*

$$F((S_i \boxed{S_j})) = \pi_{ij} (\sum_{\substack{\tau \,\in\, R(S_j) \\ \xi \,\in\, C(S_j)}} (-1)^{|\tau|} \tau \xi),$$

where the subgroups $R(S_j)$ and $C(S_j)$ of $\mathbf{S}_n$ are the row-stabilizer and the column-stabilizer of S_j, respectively, and $\pi_{ij} \in \mathbf{S}_n$ is the (unique) permutation such that

$$\pi_{ij}(S_j) = S_i.$$

In the classical notation, $F((S_i \boxed{S_j})) = \pi_{ij} e_j = e_{ij}$, where e_j denotes the *Young symmetrizer* associated to the tableau S_j (see, e.g. [10]).

The module structure

$$\mathbb{K}[\mathbf{S}_n] \cdot Super_n^M [\underline{n}|\underline{n}]$$

induces a representation

$$\rho^M : \mathbb{K}[\mathbf{S}_n] \rightarrow End_\mathbb{K}[Super_n^M [\underline{n}|\underline{n}]].$$

We define a semisimple subalgebra of $End_\mathbb{K}[Super_n^M [\underline{n}|\underline{n}]]$ by setting

$$\mathcal{B}_n^M = \bigoplus_{\lambda \vdash n} \bigoplus_{j=1,2,\ldots,f_\lambda^M} \langle Y(S_i \boxed{S_j}), \; i = 1, 2, \ldots, f_\lambda^M \rangle_\mathbb{K} = \bigoplus_{\lambda \vdash n} \mathcal{B}_\lambda^M,$$

where each simple component $\mathcal{B}_\lambda^M$ is isomorphic to $M_{f_\lambda^M}$, the full $\mathbb{K}$−algebra of square matrices of order f_λ^M.

Theorem 8. *We have*

$$\rho^M[\mathbb{K}[\mathbf{S}_n]] = \mathcal{B}_n^M,$$

and, therefore, ρ^M induces a $\mathbb{K}-$algebra isomorphism $\mathbb{K}[\mathbf{S}_n] \cong \mathcal{B}_n^M$. Moreover,

$$\rho^M(e_{ij}) = \gamma_n(S_i, \boxed{S_j}).$$

The proof of the theorem above follows from the proof of Prop. 3 of [3] in combination with the Robinson-Schensted correspondence.

By combining Theorem 4 with Theorem 8, we get immediately

Theorem 9. *(Complete decomposition of $\mathbb{K}[\mathbf{S}_n]$ as a left regular module)*

$$\mathbb{K}[\mathbf{S}_n] = F[Super_n^M[\underline{n}|\underline{n}]]$$

$$= F\left[\bigoplus_{\lambda \vdash n} \bigoplus_{j=1,2,\ldots,f_\lambda^M} \langle(S_i|\boxed{S_j}),\ i=1,2,\ldots,f_\lambda^M\rangle_{\mathbb{K}}\right]$$

$$= \bigoplus_{\lambda \vdash n} \bigoplus_{j=1,2,\ldots,f_\lambda^M} F\left[\langle(S_i|\boxed{S_j}),\ i=1,2,\ldots,f_\lambda^M\rangle_{\mathbb{K}}\right]$$

$$= \bigoplus_{\lambda \vdash n} \bigoplus_{j=1,2,\ldots,f_\lambda^M} \langle \pi_{ij}e_j;\ i=1,2,\ldots,f_\lambda^M\rangle_{\mathbb{K}}$$

$$= \bigoplus_{\lambda \vdash n} \bigoplus_{j=1,2,\ldots,f_\lambda^M} \mathbb{K}[\mathbf{S}_n]e_j,$$

where the outer sum is the isotypic decomposition, and, given λ,

$$\langle \pi_{ij}e_j;\ i=1,2,\ldots,f_\lambda^M\rangle_{\mathbb{K}} = \mathbb{K}[\mathbf{S}_n]e_j, \quad j=1,2,\ldots,f_\lambda^M$$

are minimal left ideals of $\mathbb{K}[\mathbf{S}_n]$.

The irreducible $\mathbb{K}[\mathbf{S}_n]-$module

$$\mathcal{S}_{S_j}^M = \langle(S_i|\boxed{S_j}),\ i=1,2,\ldots,f_\lambda^M\rangle_{\mathbb{K}} = F^{-1}[\mathbb{K}[\mathbf{S}_n] \cdot e_j]$$

is the *Specht module* (of the first kind) associated to the multilinear standard tableau S_j of shape λ (see e.g. [5, 12]).

For every $\lambda \vdash n$ and every multilinear standard place tableau S_j, the module structure $\mathbb{K}[\mathbf{S}_n] \cdot \mathcal{S}_{S_j}^M$ induces an surjective algebra morphism

$$\nu_{\lambda,j}^M:\ \mathbb{K}[\mathbf{S}_n] \to End_{\mathbb{K}}[\mathcal{S}_{S_j}^M].$$

By choosing the basis of the $(S_i|\boxed{S_j})$ in $\mathcal{S}_{S_j}^M$, the morphism $\nu_{\lambda,j}^M$ induces an irreducible matrix representation

$$\overline{\nu}_{\lambda,j}^M:\ \mathbb{K}[\mathbf{S}_n] \to M_{f_\lambda^M}$$

where, for every $\sigma \in \mathbf{S}_n$,

$$\overline{\nu}_{\lambda,j}^{M}(\sigma) = [c_{hk}^{\lambda}(\nu_{\lambda,j}^{M}(\sigma))].$$

For every $\lambda \vdash n$, the module structure $\mathbb{K}[\mathbf{S}_n] \cdot Super_n^M[\underline{n}|\underline{n}]$ induces an surjective algebra morphism

$$\rho_\lambda^M : \mathbb{K}[\mathbf{S}_n] \to \mathcal{B}_\lambda^M;$$

by choosing the basis of the $Y(S_i \boxed{S_j})$, in $\mathcal{B}_\lambda^M$, the morphism ρ_λ^M induces an irreducible matrix representation

$$\overline{\rho}_\lambda^M : \mathbb{K}[\mathbf{S}_n] \to M_{f_\lambda^M}$$

where, for every $\sigma \in \mathbf{S}_n$,

$$\overline{\rho}_\lambda^M(\sigma) = [d_{hk}^\lambda(\rho_\lambda^M(\sigma))].$$

By Proposition 1, we have

Proposition 4. *The irreducible representations $\overline{\rho}_\lambda^M$ and $\overline{\nu}_{\lambda,j}^M$ are equal.*

In closing this section, we specialize Theorem 7 to the multilinear case, therefore obtaining a simple combinatorial interpretation of the coefficients $c_{hk}^\lambda(\nu_{\lambda,j}^M(\sigma)) = d_{hk}^\lambda(\rho_\lambda^M(\sigma))$. We have:

$$C^{\lambda M}(\sigma) = \Theta^{\lambda M}(I)^{-1} \times \Theta^{\lambda M}(\sigma),$$

where

$$\theta_{ij}^{\lambda M}(\sigma) = \theta_{S_i S_j}(\sigma) = \theta_{S_i \ \sigma S_j}.$$

In the multilinear case, the symmetry transition coefficients θ_{ST}

$$DS\ \underline{TC} = \theta_{ST}\ \underline{DC}$$

admit a simple combinatorial description; specifically

$$\theta_{ST} = \begin{cases} (-1)^{|\tau|} & if \quad \exists\ (!)\ \sigma \in C(S),\ \tau \in R(T) :\ \sigma S = \tau T \\ 0 & otherwise \end{cases},$$

where the subgroups $R(T)$ and $C(S)$ of $\mathbf{S}_n$ are the row-stabilizer and the column-stabilizer of T, S, respectively. Hence, the matrix $\Theta^{\lambda M}(I)^{-1}$ is the same as the transition matrix from the (normalized) generalized Young symmetrizers $\frac{1}{h_\lambda} e_{ij}$ to the *Young natural units* γ_{ij} in $\mathbb{K}[\mathbf{S}_n]$, (see e.g. [10, 11]). As a matter of fact, we have

$$\rho^M(\gamma_{ij}) = \rho^M\left(\sum_h \varrho_{ih}^\lambda \frac{1}{h_\lambda}\ e_{hj}\right) = \sum_h \varrho_{ih}^\lambda \frac{1}{h_\lambda}\ \gamma(S_h, \boxed{S_j}) = Y(S_i, \boxed{S_j}).$$

Therefore, $\overline{\rho}_\lambda^M = \overline{\nu}_{\lambda,j}^M$ is indeed the *Young natural form* of the irreducible matrix representations of $\mathbf{S}_n$, for every $\lambda \vdash n$ (see,e.g. [8]).

5   Capelli's Polar Expansion Formula

As a further by-product of the present approach, we state and prove a *canonical* version of Capelli's *polar expansion formula*. This expansion formula appeared in Capelli's book ([4], pagg.124-153), but not in a canonical form. It is one of the deepest achievements of classical invariant theory, and leads to the so-called *reduction principle* (see, e.g. [14], [7]).

Informally speaking, the polar expansion formula says that any homogeneous polynomial function f in n vector variables in dimension d may be expressed as a linear combination of polarized *determinantal polynomials* involving only the first d vector variables; furthermore, these determinantal polynomials may be obtained, in turn, by polarizing the original polynomial function f.

Since the polarization process is an invariantive process, the study of invariant homogeneous polynomial functions in n vector variables in dimension d is reduced to the study of invariant homogeneous polynomial functions in $d - 1$ vector variables (see Corollary 1).

5.1   Preliminaries

Let $L = \{x_1, x_2, \ldots, x_n\}$ and $P = \{1, 2, 3, \ldots, d\}$ be two *negatively* signed proper $\mathbb{Z}_2$-graded alphabets, that is $|x_i| = |j| = 1$, for every $i = 1, 2, \ldots, n$ and $j = 1, 2, \ldots, d$.

The letterplace superalgebra is a commutative algebra

$$Super[L|P] = \mathbb{K}[(x_i|j)]_{i=1,\ldots,n;j=1,\ldots,d}$$

and, therefore, if $char(\mathbb{K}) = 0$, it can be regarded as the algebra $\mathbb{K}[V^{\oplus n}]$ of *polynomial functions over the direct sum* of n copies of a vector space V, $dim(V) = d$. Actually, given a basis of the dual space V^*, we can read the letterplace variable $(x_i|j)$ as the j-*th coordinate function over the i-th copy of* V in the direct sum $V^{\oplus n}$, for every $i = 1, 2, \ldots, n$, $j = 1, 2, \ldots, d$.

The *contragradient action* of the general linear Lie algebra $gl(d)$ on any homogeneous component

$$Super_m[L|P] = \mathbb{K}_m[(x_i|j)] \cong \mathbb{K}_m[V^{\oplus n}]$$

is described by the algebra of P-place polarizations; more specifically, the action of the elementary matrix E_{hk} is given by $-_{hk}D$, the opposite of the polarization of the place h with respect to the place k.

Since, as it is well known, the operator algebra induced by the (contragradient) action of the enveloping algebra $U(gl(d))$ equals the operator algebra induced by the (contragradient) action of the group algebra $\mathbb{K}[GL(d)]$ of the general linear group $GL(d)$, this algebra is the algebra generated by the place-polarization operators, and the algebra of $GL(d)$−equivariant endomorphisms is generated by the letter polarization operators (see Theorem 6).

It is worth to recall a basic definition.

Definition 3. *(Determinantal bitableaux [1], [6]) For every $T \in Tab(L)$, $U \in Tab(P)$, with $sh(T) = \lambda = sh(U)$, the determinantal bitableaux of the pair (T, U) is the element*

$$(T|U) \in \mathbb{K}_[(x_i|j)]$$

defined by setting

$$(T|U) = \underline{TC_2}\ C_2 U = (\lambda!)^{-1} TC_1\ \underline{C_1 C_2}\ C_2 U = TC_1\ \underline{C_1 U}$$

where C_1, C_2 are virtual tableaux of coDeruyts type of shape $\lambda = (\lambda_1, \lambda_2, \ldots, \lambda_p)$, and $\lambda! = \lambda_1! \lambda_2! \ldots, \lambda_p!$.

Even though the virtual definition given above plays a crucial role in our proof of Capelli's polar expansion formula, determinantal bitableaux admit also a direct combinatorial description.

If $\lambda = (t)$ is a partition with one part, and $T = x_{i_1}\ x_{i_2}\ \cdots\ x_{i_t}$, $U = j_1\ j_2\ \cdots\ j_t$, then

$$(T|U) = (x_{i_1}\ x_{i_2}\ \cdots\ x_{i_t}\ |\ j_1\ j_2\ \cdots\ j_t) = (-1)^{\binom{t}{2}} det[(x_{i_r}|j_s)]_{r,s=1,\ldots,t}$$

where $det[(x_{i_r}|j_s)]$ is the formal determinant of the matrix $[(x_{i_r}|j_s)]$.

In general, if $T_1, T_2, \ldots, T_p$ are the rows of the tableau T and $U_1, U_2, \ldots, U_p$ are the rows of the tableau U, then the determinantal bitableau $(T|U)$ equals, up to a sign, the product $(T_1|U_1)(T_2|U_2)\cdots(T_p|U_p)$.

5.2   Capelli's Polar Expansion Formula

In the sequel, for every $\lambda \vdash m$, with $\lambda_1 \leq min(n, d)$, we will denote by

$$X_\lambda = \begin{matrix} x_1\ x_2\ \cdots & & x_{\lambda_1} \\ x_1\ x_2\ \cdots & & x_{\lambda_2} \\ \vdots \\ x_1\ x_2\ \cdots\ x_{\lambda_p} \end{matrix}$$

the Deruyts tableau of shape λ filled with the first λ_1 proper letters (vectors); when no ambiguity will arise, we will write X in place of X_λ.

We recall that, for every standard letter tableau S of shape $\lambda \vdash m$, we have an *idempotent* in the algebra of polarizations on $\mathbb{K}_m[(x_i|j)]$:

$$Y_m(\boxed{S}, S) = SD\ DC\ \frac{1}{h_\lambda} \sum_{A \in Stab_\lambda(L)} \varrho_{S,A}\ CA\ ,$$

where h_λ is the product of the hook lengths of λ and the ϱ_{SA}'s are elements of the Rutherford matrix for the shape λ. By the final remarks of section 3, we can write

$$Y_m(\boxed{S}, S) = \frac{1}{c_\lambda}\ SD\ DX\ \frac{1}{h_\lambda}\ XC \sum_A \varrho_{S,A}\ CA\ ,$$

where c_λ is the product of the factorials of the lengths of the columns of λ; then, by setting

$$\mathcal{P}_{SX} = \frac{1}{c_\lambda} \, SD \, DX, \qquad \mathcal{P}'_{XS} = \frac{1}{h_\lambda} \, XC \sum_A \varrho_{S,A} \, CA \, ,$$

we have

$$Y_m(\boxed{S}, S) = \mathcal{P}_{SX} \mathcal{P}'_{XS}.$$

The operators $\mathcal{P}_{SX}$ and $\mathcal{P}'_{XS}$ implement the transition from determinantal bitableaux to symmetrized bitableaux, and viceversa.

Theorem 10. *For every standard tableaux S, T, U whose shapes are partitions of the integer m, we have*

1. $\mathcal{P}_{SX}\,(X|T) = (\boxed{S}|T)$;

2. $\mathcal{P}'_{XS} = \frac{1}{c_\lambda} \, Y_m(\boxed{X}, S)$;

3. $\mathcal{P}'_{XS}\,(\boxed{U}|T) = \delta_{S,U}(X|T)$;

Proof.

1. $\mathcal{P}_{SX}\,(X|T) = \frac{1}{c_\lambda}\,SD\,DX\,\underline{XC}\,CT = SD\,\underline{DC}\,CT = (\boxed{S}|T)$;

2. $\frac{1}{h_\lambda}\,XC\sum_A \varrho_{S,A}\,CA\; = \frac{1}{c_\lambda}\frac{1}{h_\lambda}\,XD\,DC\sum_A \varrho_{S,A}\,CA\; = \frac{1}{c_\lambda}\,Y_m(\boxed{X}, S)$;

3. $\mathcal{P}'_{XS}\,(\boxed{U}|T) = \frac{1}{c_\lambda}\,Y_m(\boxed{X}, S)(\boxed{U}|T) = \frac{1}{c_\lambda}\,\delta_{SU}(\boxed{X}|T) = \delta_{SU}(X|T)$. $\square$

Theorem 11. *(Capelli's Polar Expansion Theorem [4]) Let $f \in \mathbb{K}_m[(x_i|j)]$; then f can be written in the form*

$$f = \sum_{\substack{\lambda \vdash m \\ \lambda_1 \leq d}} \sum_{S \in Stab_\lambda(L)} \mathcal{P}_{SX_\lambda}\,\mathcal{F}_{X_\lambda S},$$

where X_λ is the Deruyts tableau of shape λ, filled with the first $\lambda_1 \leq d$ letters, and

1. $\mathcal{F}_{X_\lambda S} = \sum_T c^f_{S,T}\,(X_\lambda|T), \quad c^f_{S,T} \in \mathbb{K}$;

2. $\mathcal{F}_{X_\lambda S} = \mathcal{P}'_{X_\lambda S}(f)$.

Proof. Let

$$f = \sum_{\substack{\lambda \vdash m \\ \lambda_1 \leq d}} \sum_{\substack{S \in Stab_\lambda(L) \\ T \in Stab_\lambda(P)}} c^f_{S,T}\,(\boxed{S}|T)$$

be the Gordan-Capelli expansion of f (see subsection 2.5).

From assertion 1 of Theorem 10 it follows that

$$f = \sum_{\substack{\lambda \vdash m \\ \lambda_1 \leq d}} \sum_{\substack{S \in Stab_\lambda(L) \\ T \in Stab_\lambda(P)}} c_{S,T}^f \left(\boxed{S}|T\right)$$

$$= \sum_\lambda \sum_{S,T} c_{S,T}^f \, \mathcal{P}_{SX_\lambda} \left(X_\lambda|T\right)$$

$$= \sum_\lambda \sum_S \mathcal{P}_{SX_\lambda} \sum_T c_{S,T}^f \left(X_\lambda|T\right) = \sum_\lambda \sum_S \mathcal{P}_{SX_\lambda} \, \mathcal{F}_{X_\lambda S}.$$

On the other hand, from assertion 3 of Theorem 10 it follows that

$$\mathcal{P}'_{X_\lambda S}(f) = \mathcal{P}'_{X_\lambda S} \sum_{\substack{\mu \vdash m \\ \mu_1 \leq d}} \sum_{\substack{U \in Stab_\mu(L) \\ T \in Stab_\mu(P)}} c_{U,T}^f \left(\boxed{U}|T\right)$$

$$= \sum_\mu \sum_{U,T} c_{U,T}^f \, \mathcal{P}'_{X_\lambda S}\left(\boxed{U}|T\right) = \sum_T c_{S,T}^f \left(X_\lambda|T\right)$$

$$= \mathcal{F}_{X_\lambda S}.$$

□

Note that every $\mathcal{F}_{X_\lambda S}$ can be written in the form

$$\mathcal{F}_{X_\lambda S} = [x_1 x_2 \ldots x_d]^{q_\lambda} \varphi_S(x_1, x_2, \ldots, x_{d-1}),$$

where $q_\lambda = \#\{i; \ \lambda_i = d\}$. Therefore, if $\mathcal{F}_{X_\lambda S}$ is a relative G−invariant, then $\varphi_S(x_1 x_2 \ldots x_{d-1})$ is also G−invariant, for every subgroup G of $GL(d)$.

Corollary 1. *(The reduction principle of classical invariant theory) Let G be a subgroup of $GL(d)$, and let $f = f(x_1, x_2, \ldots, x_n)$ be a relative G−invariant m−homogeneous polynomial function in n vector variables, $n \geq d$. Then*

$$f = \sum_{\substack{\lambda \vdash m \\ \lambda_1 \leq d}} \sum_{S \in Stab_\lambda(L)} \mathcal{P}_{SX_\lambda} \left([x_1 x_2 \ldots x_d]^{q_\lambda} \varphi_S(x_1, x_2, \ldots, x_{d-1})\right),$$

where the φ_S's are relative G−invariants involving only the first $d - 1$ vector variables $x_1, x_2, \ldots, x_{d-1}$.

For example, since the $GL(d)$−invariant polynomial functions involving at most $d - 1$ vector variables are just the constant functions and a polarization of a bracket yields a new bracket, then the following well-known result immediately follows.

Corollary 2. *(The first fundamental theorem for vector $GL(d)$−invariants) Let $f = f(x_1, x_2, \ldots, x_n)$ be a relative $GL(d)$−invariant homogeneous polynomial function on n vector variables, $n \geq d$. Then f can be written as a homogeneous polynomial in the brackets*

$$[x_{i_1} x_{i_2} \ldots x_{i_d}]$$

filled with the variables $x_1, x_2, \ldots, x_n$.

References

1. Brini, A., Palareti, A., Teolis, A.: Gordan-Capelli Series in Superalgebras, *Proc. Natl. Acad. Sci. USA* **85** (1988) 1330-1333
2. Brini, A., Teolis, A.: Young-Capelli Symmetrizers in Superalgebras, *Proc. Natl. Acad. Sci. USA* **86** (1989) 775-778
3. Brini, A., Regonati, F., Teolis, A.: The Method of Virtual Variables and Representations of Lie Superalgebras, in *Clifford Algebras – Applications to Mathematics, Physics, and Engeneeering.* Ablamowicz, R., ed., Birkhauser Boston (2004) 245–263
4. Capelli, A.: *Lezioni sulla teoria delle forme algebriche,* Pellerano, Napoli (1902)
5. Clausen, M.: Letter Place Algebras and a Characteristic-Free Approach to the Representation Theory of the General Linear and Symmetric Groups, I, *Adv. Math.* **33** (1979) 161–191
6. Doubilet, P., Rota, G.-C., Stein, J,: On the foundation of combinatorial theory IX. Combinatorial methods in invarint theory, *Studies in Appl.Math.* **53** (1974) 185-216
7. Fulton, W., Harris, J.: *Representation theory. A first course,* Graduate Texts in Mathematics, vol. 129, Springer-Verlag, New York etc. (1991)
8. Garsia, A., Mc Larnan, T.J.: Relations between Young's natural and the Kazhdan-Lusztig representation of S_n, *Adv. Math.* **69** (1988) 32-92
9. Grosshans, F., Rota, G.-C., Stein, J.: *Invariant Theory and Superalgebras,* Am. Math. Soc., Providence, RI (1987)
10. James, G., Kerber, A.: *The Representation Theory of the Symmetric Group,* Encyclopedia of Mathematics and Its Applications, vol. 16, Addison–Wesley, Reading, MA (1981)
11. Rutherford, D. E.: *Substitutional analysis,* Edinburgh University Publications: Science and Mathematics, Edinburgh-London (1948)
12. Sagan, B.E.: *The Symmetric Group. Representations, Combinatorial Algorithms, and Symmetric Functions*, Springer Graduate Texts in Mathematics, vol. 203, Springer Verlag (2001)
13. Scheunert, M.: *The Theory of Lie Superalgebras: An Introduction,* Lecture Notes in Mathematics, vol. 716, Springer Verlag, New York (1979)
14. Weyl, H.: *The Classical Groups,* Princeton Univ. Press, Princeton, NY (1946)

Applications of Geometric Algebra in Robot Vision

Gerald Sommer

Cognitive Systems Group, University of Kiel,
Christian-Albrechts-Platz 4, 24118 Kiel, Germany
gs@ks.informatik.uni-kiel.de

Abstract. In this tutorial paper we will report on our experience in
the use of geometric algebra (GA) in robot vision. The results could be
reached in a long term research programme on modelling the perception-
action cycle within geometric algebra. We will pick up three important
applications from image processing, pattern recognition and computer
vision. By presenting the problems and their solutions from an engineering point of view, the intention is to stimulate other applications of GA.

1   Introduction

In this paper we want to present a survey on applications of geometric algebra (GA) in robot vision. We will restrict our scope to results contributed by
the Kiel Cognitive Systems Group in the last few years. For more details and
for getting a wider view on this topic a visit of the publications on the website http://www.ks.informatik.uni-kiel.de is recommended. We will take on an
engineer's viewpoint to give some impression of the need of such complex mathematical framework as geometric algebra for designing robot vision systems more
easily. In fact, we are using GA as a mathematical language for modelling. This
includes the task related shaping of that language itself.

The aim of robot vision is to make robots seeing, i.e. to endow robots with
visual capabilities comparable to those of human. While contemporary applications of robots are restricted to industrial artificial environments, the hope is
that future generations of robots are able to cope with real world conditions. The
term robot vision indicates a concentration of research activities onto modelling
of useful visual architectures. In the framework of behaviour-based robotics the
coupling of (visual) perception and action is the key paradigm of system design.
A behaviour is represented in the so-called perception-action cycle. Of practical
importance are the projections of the perception-action cycle: "vision for action"
means controlling actions by vision and "action for vision" means controlling the
gaze (or body) for making vision more easier.

Robot vision is emerging from several contributing disciplines which are as
different as image processing, computer vision, pattern recognition and robotics.
Each of these have their own scientific history and are using different mathematical languages. The demanding task of system design is aiming at a unique

H. Li, P. J. Olver and G. Sommer (Eds.): IWMM-GIAE 2004, LNCS 3519, pp. 258–277, 2005.

framework of modelling. With respect to the quest of a useful cognitive architecture the perception-action cycle may be the right representation. With respect to unifying the mathematical language geometric algebra is hopefully the framework of merging the above mentioned disciplines in a coherent system.

Such system has to be an embodiment of the geometry and the stochastic nature of the external world. This should be understood in a dynamic sense which enables internal processes converging at reasonable interpretations of the world and useful actions in the environment.

While merging the different disciplines is one main motivation of using geometric algebra in robot vision, the other one is to overcome several serious limitations of modelling within the disciplines themselves. These limitations result from the dominant use of vector algebra. A vector space is a completely unstructured algebraic framework endowed with a simple product, the scalar product, which only can destroy information originally represented in the pair of vectors by mapping them to a scalar.

Imagine, for instance, that a cognitive system as a human could reason on the world or could act in the world only by its decomposition into sets of points and having no other operations at hand than the scalar product. This in fact is an impossible scenario. The phenomena of the world we can cope with are of global nature at the end in comparison to the local point-like entities we have in vector space. They are phenomena of higher order in the language of vector algebra. Hence, most of the basic disciplines of robot vision are getting stuck in non-linearities because of the complexity of the problem at hand, which are non-tractable in real-time applications.

In fact, only the tight relations of GA to geometric modelling [12] have been considered, yet. We are just starting with the fusion of stochastic modelling and GA. The benefit we derive from using GA in geometric modelling is rooted in the rich algebraic structure of the 2^n-dimensional linear space $\mathbb{R}_{p,q,r}$ resulting from the vector space $\mathbb{R}^{p,q,r}$. The indexes p, q, r with $p + q + r = n$ mark its signature. By choosing a certain signature the decision for certain geometries adequate to the problem at hand can be made. On the other hand, the blade structure of a GA represents higher-order relations between vectors which, once computed, can be further processed in a linear manner. Hence, multi-dimensional or higher-order phenomena can be reduced to one-dimensional or linear representations. This has not only impact on modelling of geometric problems but also of stochastic ones. Therefore, there is need of using GA in context of higher-order statistics too.

In the following sections we want to show the advantages of using GA in three different areas of robot vision. These are signal analysis, pattern recognition using the paradigm of neural computing, and computer vision. In each case we will emphasize the inherent limitations of modelling in the classical scheme and the benefits we gain from using GA instead. In the conclusion section we will give a summary of all these advantages in a more general way.

2   Analysis of Multi-dimensional Signals

Image analysis as a fundamental part of robot vision does not mean to deliver complete descriptions of scenes or to recognize certain objects. This is in fact the subject of computer vision. Instead, the aim of image analysis is deriving a set of rich features from visual data for further processing and analysis. In (linear) signal theory we are designing (linear shift invariant) operators which should extract certain useful features from signals. The Fourier transform plays a major role because representations of both operators and images in spectral domain are useful for interpretation.

In this section we want to give an overview on our endavours of overcoming a not well-known representation gap in Fourier domain in the case of multi-dimensional signals. In fact, the well-known complex-valued Fourier transform cannot explicitly represent multi-dimensional phase concepts. The Fourier transform is a global integral transform and amplitude and phase are global spectral representations. In practice more important is the incompleteness of local spectral representations in the multi-dimensional case. This is a representation problem of the Hilbert transform which delivers the holomorphic complement of a one-dimensional real-valued signal. Therefore, the aim of this section is showing generalizations of both Fourier and Hilbert transform from the one-dimensional to the multi-dimensional case which are derived from embeddings into geometric algebra.

In that respect two different concepts of the term dimension of a function have to be distinguished. The first one is the global embedding dimension, which is related to the Fourier transform. The second one is the (local) intrinsic dimension, which is related to the Hilbert transform and which means the degrees of freedom that locally define the function.

2.1   Global Spectral Representations

It is well-known that the complex-valued Fourier transform, $F^c \in \mathbb{C}$,

$$F^c(\mathbf{u}) = \mathcal{F}^c \{f(\mathbf{x})\} = \int f(\mathbf{x}) \exp(-j2\pi \mathbf{u} \cdot \mathbf{x}) \, d^N \mathbf{x} \qquad (1)$$

enables computing the spectral representations amplitude $A^c(\mathbf{u})$ and phase $\Phi^c(\mathbf{u})$,

$$A^c(\mathbf{u}) = |F^c(\mathbf{u})| \qquad (2a)$$

$$\Phi^c(\mathbf{u}) = \arg\left(F^c(\mathbf{u})\right), \qquad (2b)$$

of the real-valued N–dimensional (ND) function $f(\mathbf{x})$, $\mathbf{x} \in \mathbb{R}^N$. Although it is commonsense that the phase function is representing the structural information of the function, there is no way in the complex domain to explicitly access to ND structural information for $N > 1$. This problem is related to the impossibility of linearly representing symmetries of dimension $N > 1$ in the complex domain.

In fact, the Fourier transform in the 1D case can be seen as a way to compute the integral over all local symmetry decompositions of the function $f(x)$ into an even and an odd component according to

$$f(x) = f_e(x) + f_o(x) \tag{3}$$

$$F^c(u) = F_e^c(u) + F_o^c(u). \tag{4}$$

This results from the Euler decomposition of the Fourier basis functions

$$Q^c(u, x) = \exp(-j2\pi ux) = \cos(2\pi ux) - j\sin(2\pi ux). \tag{5}$$

In the case of 2D complex Fourier transform the basis functions can also be separately decomposed,

$$Q^c(\mathbf{u}, \mathbf{x}) = \exp(-j2\pi \mathbf{ux}) = \exp(-j2\pi ux)\exp(-j2\pi vy). \tag{6}$$

From this results a signal decomposition into products of symmetries with respect to coordinate axes,

$$F^c(\mathbf{u}) = F_{ee}^c(\mathbf{u}) + F_{oo}^c(\mathbf{u}) + F_{oe}^c(\mathbf{u}) + F_{eo}^c(\mathbf{u}). \tag{7}$$

But this concept of line symmetry is partially covered in the complex domain because of the limited degrees of freedom. If $F_R^c(\mathbf{u})$ and $F_I^c(\mathbf{u})$ are the real and imaginary components of the spectrum, respectively, $F^c(\mathbf{u}) = F_R^c(\mathbf{u}) + jF_I^c(\mathbf{u})$, then

$$F_R^c(\mathbf{u}) = F_{ee}^c(\mathbf{u}) + F_{oo}^c(\mathbf{u}) \ , \ F_I^c(\mathbf{u}) = F_{eo}^c(\mathbf{u}) + F_{oe}^c(\mathbf{u}). \tag{8}$$

If we consider the Hartley transform [10] as a real-valued Fourier transform, $F^r(u)$, then also in the 1D case the components $F_e^r(u)$ and $F_o^r(u)$ are totally covered in $F^r \in \mathbb{R}$. This observation supports the following statement [6]. Given a real valued function $f(\mathbf{x})$, $\mathbf{x} \in \mathbb{R}^N$, all its global decompositions into even and odd symmetry components with respect to coordinate axes are given by the Clifford valued Fourier transform $F^{Cl} \in \mathbb{R}_{2^N}$,

$$F^{Cl}(\mathbf{u}) = \mathcal{F}^{Cl}\{f(\mathbf{x})\} = \int_{\mathbb{R}^N} f(\mathbf{x}) \prod_{k=1}^{N} \exp(-i_k 2\pi u_k x_k) \, d^N\mathbf{x}. \tag{9}$$

In the 2D case the Fourier transform has to be quaternionic, $F^q \in \mathbb{H}$,

$$F^q(\mathbf{u}) = \mathcal{F}^q\{f(\mathbf{x})\} = \int_{\mathbb{R}^2} f(\mathbf{x}) \exp(-i2\pi ux)\exp(-j2\pi vy) \, d^2\mathbf{x} \tag{10}$$

with

$$Q^q(\mathbf{u}, \mathbf{x}) = \exp(-i2\pi ux)\exp(-j2\pi vy). \tag{11}$$

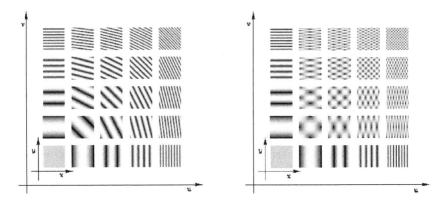

Fig. 1. Left: basis functions of the complex 2D Fourier transform. Right: basis functions of the quaternionic 2D Fourier transform. Only the even/even-even part is shown each

Because

$$F^q(\mathbf{u}) = F_R^q(\mathbf{u}) + iF_I^q(\mathbf{u}) + jF_J^q(\mathbf{u}) + kF_K^q(\mathbf{u}) \tag{12a}$$

$$F^q(\mathbf{u}) = F_{ee}^q(\mathbf{u}) + F_{oe}^q(\mathbf{u}) + F_{eo}^q(\mathbf{u}) + F_{oo}^q(\mathbf{u}), \tag{12b}$$

all symmetry combinations are explicitely accessible by one real and three imaginary parts. As figure 1 visualizes, the basis functions according to (11) are intrinsically two-dimensional in contrast to those of (6) and, thus, they can represent 2D structural information. This can also be seen in the polar representation of the quaternionic Fourier transform [6], where ϕ and θ represent each a 1D phase, ψ is a 2D phase.

$$F^q(\mathbf{u}) = |F^q(\mathbf{u})| \exp\left(i\phi(\mathbf{u})\right) \exp\left(k\psi(\mathbf{u})\right) \exp\left(j\theta(\mathbf{u})\right). \tag{13}$$

2.2   Local Spectral Representations

We call functions of local energy, $e(x)$, (or amplitude, $a(x) = \sqrt{e(x)}$), and local phase, $\phi(x)$, as local spectral representations:

$$e(x) = f^2(x) + f_H^2(x) \tag{14a}$$

$$\phi(x) = \arg\left(f_A(x)\right), \tag{14b}$$

which are derived from a complex-valued extension, $f_A(x)$, of a real 1D function $f(x)$ by an operator $\mathcal{A}$,

$$f_A(x) = \mathcal{A}\left\{f(x)\right\} = f(x) + jf_H(x). \tag{15}$$

In signal theory $f_A(x)$ is called analytic signal and $f_H(x)$ is derived from $f(x)$ by a local phase shift of $-\frac{\pi}{2}$, which is gained by the Hilbert transform, $\mathcal{H}$, of $f(x)$,

$$f_H(x) = \mathcal{H}\{f(x)\}. \tag{16}$$

In the complex analysis this corrresponds to computing the holomorphic complement of $f(x)$ by solving the Laplace equation. In the complex Fourier domain the Hilbert transform reads

$$H(u) = -j\frac{u}{|u|}, \tag{17}$$

thus, the operator of the analytic signal is given by the frequency transfer function

$$A(u) = 1 + \frac{u}{|u|}. \tag{18}$$

The importance of the analytic signal in signal analysis results from the evaluation of the local spectral representations, equations (14a) and (14b): If the local energy exceeds some threshold at x, then this indicates an interesting location. The evaluation of the local phase enables a qualitative signal interpretation with respect to a mapping of the signal to a basis system of even and odd symmetry according to equation (3). In image processing we can interprete lines as even symmetric structures and edges as odd symmetric ones, both of intrinsic dimension one. But in images we have as an additional unknown the orientation of lines or edges. Hence, the operators $\mathcal{A}$ or $\mathcal{H}$, respectively, should be rotation invariant. Although the analytic signal is also used in image processing since one decade, only by embedding into GA, this problem could be solved.

Simply to take over the strategy described in section 2.1 from global to local domain is not successful. The quaternionic analytic signal [7], which is derived from a quaternionic Hilbert transform, delivers no rotation invariant local energy. Hence, the applied concept of line symmetry cannot succeed in formulating isotropic operators. Instead, an isotropic generalization of the Hilbert transform in the case of a multidimensional embedding of a 1D function can be found by considering point symmetry in the image plane. This generalization is known in Clifford analysis [3] as Riesz transform.

The Riesz transform of a real ND function is an isotropic generalization of the Hilbert transform for $N > 1$ in an $(N+1)$D space. In Clifford analysis, i.e. an extension of the complex analysis of 1D functions to higher dimensions, a real ND function is considered as a harmonic potential field, $\mathbf{f}(\mathbf{x})$, in the geometric algebra $\mathbb{R}_{N+1}$. The Clifford valued extension of $\mathbf{f}(\mathbf{x})$ is called monogenic extension and is located in the hyperplane where the $(N+1)$th component, say s, vanishes. It can be computed by solving the $(N+1)$D Laplace equation of the corresponding harmonic potential as a boundary value problem of the second kind for $s = 0$ (see [8]).

Let us consider the case of a real 2D signal, $f(\mathbf{x})$, represented as $\mathbf{e}_3$-valued vector field $\mathbf{f}(\mathbf{x}) = f(\mathbf{x})\mathbf{e}_3$ in the geometric algebra $\mathbb{R}_3$. Then in the plane $s = 0$ a monogenic signal $\mathbf{f}_M$ exists [9],

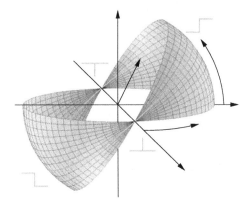

Fig. 2. The phase decomposition of a monogenic signal expressing symmetries of a local i1D structure embedded in $\mathbb{R}_3$. The great circle is passing $|\boldsymbol{f}_M|e_3$ and $\boldsymbol{f}_M$ defines the orientation of a complex plane in $\mathbb{R}^3$

$$\mathbf{f}_M(\mathbf{x}) = \mathbf{f}(\mathbf{x}) + \mathbf{f}_R(\mathbf{x}), \tag{19}$$

where $\mathbf{f}_R(\mathbf{x})$ is the vector of the Riesz transformed signal. In our considered case the monogenic signal is a generalization of the analytic signal, equation (15), for embedding an intrinsically 1D function into a 2D signal domain. Similarly the Riesz transform with the impulse response

$$\mathbf{h}_R(\mathbf{x}) = \frac{\mathbf{x}e_3}{2\pi|\mathbf{x}|^3} = -\frac{x}{2\pi|\mathbf{x}|^3}e_{31} + \frac{y}{2\pi|\mathbf{x}|^3}e_{23} \tag{20}$$

and the frequency transfer function

$$\mathbf{H}_R(\mathbf{u}) = \frac{\mathbf{u}}{|\mathbf{u}|}\mathbf{I}_2^{-1} \quad , \ \mathbf{I}_2 = \mathbf{e}_{12} \tag{21}$$

generalizes all properties known from the Hilbert transform to 2D.

The local energy, derived from the monogenic signal is rotation invariant. Furthermore, the phase generalizes to a phase vector, whose one component is the normal phase angle and the second component represents the orientation angle of the intrinsic 1D structure in the image plane. Hence, the set of local spectral features is augmented by a feature of geometric information, that is the orientation. In figure 2 can be seen how the complex plane of the local spectral representations rotates in the augmented 3D space. This results in an elegant scheme of analyzing intrinsically 1D (i1D) structures in images.

So far, we only considered the solution of the Laplace equation for $s = 0$. The extension to $s > 0$ results in a set of operators, called Poisson kernels and conjugated Poisson kernels, respectively, which not only form a Riesz triple with quadrature phase relation as the components in equation (19), but transform the monogenic signal into a representation which is called scale-space representation. In fact, the s-component is a scale-space coordinate. A scale-space is spanned by $(\mathbf{x}, s)$. This is a further important result because for nearly 50 years only the Gaussian scale-space, which results as a solution of the heat equation, has been considered as scale-space for image processing. One advantage of the Poisson scale-space in comparison to the Gaussian scale-space is its alibity to naturally embed the complete local spectral analysis into a scale-space theory. Recently we could demonstrate the superiority of using the monogenic scale space for signal reconstruction from phase in comparison to the use of the orientation selective phase of an analytic signal. Recently we extended the generalization of the monogenic signal to 3D images. The 3D monogenic signal represents a phase vector with three components. In the context of image sequence analysis our first application is computing the optical flow. There one of the orientation angles represents the velocity of motion.

Regrettably, this nice theory derived from the Clifford analysis framework is relevant only for i1D signals, yet. The extension to i2D structures in images like curved lines/edges, crosses or any other more general structure in a linear way will be matter of future research. Nevertheless, in [8] a so-called structure multivector is proposed from which a rich set of features can be derived for a special model of an i2D structure. That is built from two perpendicularly crossing lines/edges. The linear operator model uses spherical harmonics up to the order three. The local image structure is represented by a set of five independent features which are main orientation, two local i1D amplitudes and two local i1D phases. That local filtering scheme enables classification between i1D and i2D structures. But a linear approach to filtering a more flexible model is not at hand, yet. Instead, we recently used a tensor representation of the monogenic signal resulting from two crossing lines/edges with flexible angles for decomposing that mixed signal into both single contributing monogenic signals. The idea is based on an eigenvalue decomposition of the local signal tensor. But the used tensor model is a non-linear one.

3   Knowledge Based Neural Computing

Neural nets are computational tools for learning certain competences of a robot vision system within the perception-action cycle. A net of artificial neurons as primitive processing units can learn for instance classification, function approximation, prediction or control. Artificial neural nets are determined in their functionality by the kind of neurons used, their topological arrangement and communication, and the weights of the connections. By embedding neural computing into the framework of GA, any prior algebraic knowledge can be used for increasing the performance of the neural net. The main benefit we get from this approach is constraining the decision space and, thus, preventing the curse of dimensionality. From this can follow faster convergence of learning and increased generalization capability. For instance, if the task is learning the parameters of a transformation group from noisy data, then noise does not follow the algebraic constraints of the transformation group and will not be learned.

In subsection 3.1 we will consider such type of problems under orthogonal constraints. Another advantage of embedding neural computing into geometric algebras is related to learning of functions. Neural nets composed by neurons of perceptron type - this is the type of neurons we will consider here - are able to learn nonlinear functions. If the nonlinear problem at hand is transformed in an algebraic way to a linear one, then learning becomes much easier and with less ressources. We will consider such problems in the case of manifold learning in subsection 3.1 and in the case of learning non-linear decision boundaries in subsection 3.3. In subsection 3.2 we will show the learning of the cross-ratio with a Möbius transform by using only one spinor Clifford neuron in $\mathbb{R}_{1,2}$.

3.1   The Spinor Clifford Neuron

In this subsection we will consider the embedding of perceptron neurons into a chosen geometric algebra. The output, y, of such neuron for a given input vector $\mathbf{x}$ and a weight vector $\mathbf{w}$, $\mathbf{x}, \mathbf{w} \in \mathbb{R}^n$, is given by

$$y = g\left(f(\mathbf{x}; \mathbf{w})\right). \tag{22}$$

The nonlinear function $g : \mathbb{R} \longrightarrow \mathbb{R}$ is called activation function. It shall be omitted in the moment, say, by setting it to the identity. More important for our concern is the propagation function $f : \mathbb{R}^n \longrightarrow \mathbb{R}$,

$$f(\mathbf{x}) = \sum_{i=1}^n w_i x_i + \theta, \tag{23}$$

where $\theta \in \mathbb{R}$ is a bias (threshold). Obviously, one single neuron can learn a linear function, represented by equation (23), as linear regression by minimizing the sum of squared errors over all samples $(\mathbf{x}^j, r^j)$ from a sufficiently composed sample set (universe). We assume a supervised training scheme where $r^j \in \mathbb{R}$ is the requested output of the neuron. By arranging a certain number of neurons in a single layer fully connected to the input vector x, we will get a single layer perceptron network (SLP). A SLP obviously enables task sharing in a parallel fashion and, hence, can perform a multi-linear regression. If $\mathbf{x}, \mathbf{y} \in \mathbb{R}^n$, then

$$\mathbf{y} = W\mathbf{x} \tag{24}$$

is representing a linear transformation by the weight matrix W.

After introducing the computational principles of a real-valued neuron, we will extend now our neuron model by embedding it into the GA $\mathbb{R}_{p,q}, p + q = n$, see [5]. Hence, for $\mathbf{x}, \mathbf{w}, \theta \in \mathbb{R}_{p,q}$ the propagation function $\mathbf{f} : \mathbb{R}_{p,q} \longrightarrow \mathbb{R}_{p,q}$ reads

$$\mathbf{f}(\mathbf{x}) = \mathbf{wx} + \theta, \tag{25}$$

where $\mathbf{wx}$ is the geometric product instead of the scalar product of equation (23). A neuron with a propagation function according to equation (25) is called a Clifford neuron. The superiority of equation (25) over equation (23) follows from explicitly introducing a certain algebraic model as constraint by choosing $\mathbb{R}_{p,q}$ for learning the weight matrix of equation (24) instead of additionally learning the required constraints. It is obviously more advantageous to use a model than to perform its simulation. This leads in addition to a reduction of the required resources (neurons, connections). By explicitly introducing algebraic constraints, statistical learning will become a simpler task.

We will further extend our model. If the weight matrix W in equation (24) is representing an orthogonal transformation, we can introduce the constraint of an orthogonal transformation group. This is done by embedding the propagation function now into $\mathbb{R}_{p,q}^+$. Instead of equation (25) we get for $\mathbf{f}: \mathbb{R}_{p,q}^+ \longrightarrow \mathbb{R}_{p,q}^+$:

$$\mathbf{f}(\mathbf{x}) = \mathbf{w}\mathbf{x}\tilde{\mathbf{w}} + \theta. \tag{26}$$

We will call a neuron with such propagation function a spinor Clifford neuron because the spinor product of the input vector with one weight vector is computed. The representation of an orthogonal transformation by a spinor [12] has several computational advantages in comparison to a matrix representation which will not be explicitly discussed here.

3.2   Learning the Cross-Ratio with Möbius Transformation

In this subsection we will describe the application of the concept of spinor Clifford neurons to a really hard problem which cannot be learned in practice by real neurons. We will see that only one single spinor neuron in the algebra $\mathbb{R}_{1,2}$ is sufficient, although some effort will be needed.

The cross-ratio is an important projective invariant. It is defined in the following way. Let A, B, C, D be four collinear points with the coordinates $\mathbf{a}, \mathbf{b}, \mathbf{c}$ and $\mathbf{d}$ in the real projective plane $\mathbb{R}_{2,1}$. Then their collinearity may be expressed by $\mathbf{c} = \alpha\mathbf{a} + \beta\mathbf{b}$ and $\mathbf{d} = \gamma\mathbf{a} + \delta\mathbf{b}$ for suitable numbers $\alpha, \beta, \gamma, \delta, \in \mathbb{R}$. The cross-ratio, $[\mathbf{a}, \mathbf{b}, \mathbf{c}, \mathbf{d}]$, is then defined by the ratio of the ratios β/α and δ/γ,

$$[\mathbf{a}, \mathbf{b}, \mathbf{c}, \mathbf{d}] = \frac{\beta\gamma}{\alpha\delta}. \tag{27}$$

To compute the cross-ratio in the real projective plane is the standard way if projective transformations mapping lines to lines are of interest. This will be also our concern. But instead of using real numbers $\alpha, \beta, \gamma, \delta$, we will use complex numbers $\mathbf{z}, \mathbf{q}, \mathbf{r}, \mathbf{s} \in \mathbb{R}_{0,1}$ of the complex plane. Then there exists a known theorem [14] which states: The cross-ratio $[\mathbf{z}, \mathbf{q}, \mathbf{r}, \mathbf{s}]$ is the image of $\mathbf{z}$ under that Möbius transformation that maps $\mathbf{q}$ to 0, $\mathbf{r}$ to 1 and $\mathbf{s}$ to ∞, respectively. The Möbius transformation $m(\mathbf{z}) \in M(0,1)$ of a complex number $\mathbf{z} \in \mathbb{R}_{0,1}$ is the fractional, bilinear transformation

$$m(\mathbf{z}) = \frac{a\mathbf{z} + b}{c\mathbf{z} + d} \tag{28}$$

with $\mathbf{a}, \mathbf{b}, \mathbf{c}, \mathbf{d} \in \mathbb{R}_{0,1}$ and $\mathbf{ad} - \mathbf{bc} \neq 0$. A Möbius transformation is a conformal transformation of the extended complex plane. It is uniquely determined by three points of the complex plane. The Möbius group $M(p, q)$ of the geometric algebra $\mathbb{R}_{p,q}$ is covered by the orthogonal group $O(p+1, q+1)$ of $\mathbb{R}_{p+1,q+1}$ [11]. Hence,

Table 1. Transformation parameters

Parameter	Value	Learned parameters
a	0.2 + 0.5 i	0.20019 + 0.50010 i
b	0.11 - 0.16 i	0.11001 - 0.15992 i
c	- 0.2 + i	- 0.20079 + 0.99939 i
d	- 0.08 - 0.64 i	- 0.07930 - 0.64007 i

Table 2. Cross-ratios of test points (rounded)

z	Value	Clifford neuron output
2 + 3i	0.3578 - 0.3357 i	0.3577 - 0.3364 i
4 - 7i	0.4838 - 0.3044 i	0.4838 - 0.3051 i
0.3 + 0.1 i	0.0077 - 0.6884 i	- 0.0082 + 0.6884 i

by embedding our problem, which is formulated in $\mathbb{R}_{0,1}$, into the algebra $\mathbb{R}_{1,2}$, we can interpret Möbius transformations as orthogonal ones. Then we are able to learn those transformations and, thus, also the cross-ratio, by one single spinor neuron in $\mathbb{R}_{1,2}$. This has been published in [4, 5]. Details of the implementation will therefore be omitted here. Instead, in tables 1 and 2 the results are shown for the following components of the cross-ratio: $q = 0.2 + 0.3i, r = 0.4 - 0.7i$ and $s = 0.6 - 0.2i$. In table 1 we compare the learned parameters of the Möbius transformation with the expected ones. The results are quiet acceptable. In table 2 we see the learned cross-ratios in comparison to the true ones for several test points.

As a whole, the presented results are a convincing demonstration of the proposed model of a Clifford spinor neuron for linear learning of orthogonal transformations in the chosen algebraic domain which are non-linear transformations in the Euclidean domain.

3.3   The Hypersphere Neuron

In subsection 3.1 we neglected the activation function in describing our algebraic neuron models. This is possible if a kind of function learning is the task at hand. In the case of a classification task the non-linear activation function is mandatory to complete the perceptron to a non-linear computing unit which in the trained state represents a linear decision boundary in $\mathbb{R}^n$. The decision boundary enables solving a 2-class decision problem by dividing the sample space in two half-spaces. Regrettably, the higher the dimension n of the feature vectors, the less likely a 2-class problem can be treated with a linear hyperplane of $\mathbb{R}^n$. Therefore, parallel (SLP) or sequential (MLP - multilayer perceptron) compositions of neurons are necessary for modelling non-linear decision boundaries.

One special case is a hypersphere decision boundary. This can be only very inefficiently modelled by using perceptrons. But spheres are the geometric basis entities of conformal geometry. That is, in conformal geometry we can operate

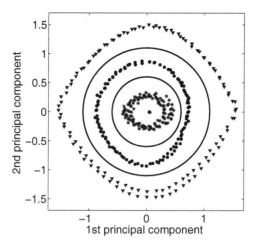

Fig. 3. The decision boundaries and the first two principal components of three toy objects rotated by $360°$ in steps of one degree

on hyperspheres in a linear manner just as operating on points in Euclidean geometry (points are the basis entities of Euclidean geometry).

In the last few years the conformal geometric algebra (GA) $\mathbb{R}_{p+1,q+1}$ [13] became an important embedding framework of different problems relevant to robot vision because of several attractive properties. We will come back to that point in section 4. As mentioned above, the conformal group $C(p,q)$, which is highly non-linear in $\mathbb{R}^{p,q}$ transforms to a representation which is isomorphic to the orthogonal group $O(p+1,q+1)$ in $\mathbb{R}_{p+1,q+1}$. Second, a subspace of the conformal space $\mathbb{R}^{n+1,1}$ which is called horosphere, N_e^n, is a non-Euclidean model of the Euclidean space $\mathbb{R}^n$ with the remarkable property of metrical equivalence. This is the property we want to exploit for constructing the hypersphere neuron. Here we need in fact only the vector space $\mathbb{R}^{n+1,1} = \langle \mathbb{R}_{n+1,1} \rangle_1$ of the CGA.

Any point $\mathbf{x} \in \mathbb{R}^n$ transforms to a point $\underline{\mathbf{x}} \in \mathbb{R}^{n+1,1}$ with $\underline{\mathbf{x}}^2 = 0$. That is, points are represented as null vectors on the horosphere. Because hyperspheres are the basis entities of $\mathbb{R}^{n+1,1}$, a point $\underline{\mathbf{x}}$ can be seen as a degenerate hypersphere $\underline{\mathbf{s}}$ with radius zero.

Let be $\mathbf{x}, \mathbf{y} \in \mathbb{R}^n$ two points with Euclidean distance $d(\mathbf{x},\mathbf{y}) = \|\mathbf{x} - \mathbf{y}\|$ and let be $\underline{\mathbf{x}}, \underline{\mathbf{y}} \in N_e^n \subset \mathbb{R}^{n+1,1}$ with distance $d(\underline{\mathbf{x}},\underline{\mathbf{y}}) = \underline{\mathbf{x}} \cdot \underline{\mathbf{y}}$ (scalar product), then

$$d(\underline{\mathbf{x}},\underline{\mathbf{y}}) = -\frac{1}{2}d^2(\mathbf{x},\mathbf{y}). \tag{29}$$

Therefore, the distance of a point $\underline{\mathbf{x}}$ to a hypersphere $\underline{\mathbf{s}}$ will be

$$d(\underline{\mathbf{x}},\underline{\mathbf{s}}) : \begin{cases} > 0 & \text{if} \quad \underline{\mathbf{x}} \text{ is outside } \underline{\mathbf{s}} \\ = 0 & \text{if} \quad \underline{\mathbf{x}} \text{ is on } \underline{\mathbf{s}} \\ < 0 & \text{if} \quad \underline{\mathbf{x}} \text{ is inside } \underline{\mathbf{s}}. \end{cases} \tag{30}$$

That distance measure is used for designing the propagation function of a hypersphere neuron [1]. If the parameters of the hypersphere are interpreted as the weights of a perceptron, then by embedding any data points into $\mathbb{R}^{n+1,1}$, the decision boundary will be a hypersphere. In fact, the hypersphere neuron subsumes the classical perceptron because a hypersphere with infinite radius is a hyperplane.

To simplify the implementation we take advantage of the equivalence of the scalar products in $\mathbb{R}^{n+1,1}$ and in $\mathbb{R}^{n+2}$. This enables a data coding which makes the hypersphere neuron to a perceptron with a second bias component. For an example see figure (3).

4   Pose Estimation of Free-Form Objects

As a rather complex example for using GA in robot vision we will report on estimating the pose of an object which is moving in 3D space. In that example computer vision and robotics meet very obviously. As pose we denote position and orientation. Hence, pose estimation means estimating the parameters of the special Euclidean transformation in space by observations of the rigid body motion in the image plane. Pose estimation is a basic task of robot vision which can be used in several respects as part of more complex tasks as visual tracking, homing or navigation. Although there is plenty of solutions over the years, only CGA [13] gives a framework which is adequate to the problem at hand [15].

The diversity of approaches known so far results from the hidden complexity of the problem. Estimation of the parameters of the special Euclidean transformation, which is composed by rotation and translation, has to be done in a projective scenario. But the coupling of projective and special Euclidean transformations, both with non-linear representations in the Euclidean space, results in a loss of the group properties. Another problem is how to model the object which serves as reference. In most cases a set of point features is used. But this does not enable to exploit internal constraints contained in higher order entities as lines, planes, circles, etc. Finally, it can be distinguished between 2D and 3D representations of either model data and measurement data.

To be more specific, the task of pose estimation is the following: Given a set of geometric features represented in an object centered frame and given the projections of these features onto the image plane of a camera, then, for a certain parametrized projection model, determine the parameters of the rigid body motion between two instances of an object centered frame by observations in a camera centered frame.

In our scenario we are assuming a 2D-3D approach, i.e. having 2D measurement data from a calibrated full perspective monocular camera and 3D model data. As model we are taking either a set of geometric primitives as points, lines, planes, circles or spheres, or more complex descriptions as a kinematic chain (coupled set of piecewise rigid parts) or free-form curves/surfaces given as CAGD-model. From the image features we are projectively reconstructing lines or planes in space. Now the task is to move the reference model in such a way that the spatial distance of the considered object features to the projection rays/planes becomes zero. This is an optimization task which is done by a gradient descent method on the spatial distance as error measure. Hence, we prevent minimizing any error measure directly on the manifold.

4.1   Pose Estimation in Conformal Geometric Algebra

We cannot introduce here the conformal geometric algebra (CGA) [13] because of limited space. There is plenty of good introductions, see e.g. the PhD thesis [15] with respect to an extended description of pose estimation in that framework.

We are using the CGA $\mathbb{R}_{4,1}$ of the Euclidean space $\mathbb{R}^3$. If $\mathbb{R}^3$ is spanned by the unit vectors $\mathbf{e}_i$, $i = 1, 2, 3$, then $\mathbb{R}_{4,1}$ is the algebra of the augmented space $\mathbb{R}^{4,1} = \mathbb{R}^{3,0} \oplus \mathbb{R}^{1,1}$ with the additional basis vectors $\mathbf{e}_o$ (representing the point at origin) and $\mathbf{e}$ (representing the point at infinity). The basis $\{\mathbf{e}_o, \mathbf{e}\}$ constitutes a null basis. From this follows that the representation of geometric entities by null vectors is the characteristic feature of that special conformal model of the Euclidean space.

The advantages we can profit from within our problem are the following. First, $\mathbb{R}_{4,1}$ constitutes a unique framework for affine, projective and Euclidean geometry. If $\mathbb{R}_{3,0}$ is the geometric algebra of the 3D Euclidean space and $\mathbb{R}_{3,1}$ is that one of the 3D projective space, then we have the following relations between these linear spaces:

$$\mathbb{R}_{3,0} \subset \mathbb{R}_{3,1} \subset \mathbb{R}_{4,1}.$$

Because the special Euclidean transformation is a special affine transformation, we can handle either kinematic, projective or metric aspects of our problem in the same algebraic frame. Higher efficiency is reached by simply transforming the representations of the geometric entities to the respective partial space.

Second, the special Euclidean group $SE(3)$ is a subgroup of the conformal group $C(3)$, which is an orthogonal group in $\mathbb{R}_{4,1}^+$. Hence, the members of $SE(3)$, represented in $\mathbb{R}_{4,1}$, which we call motors, $\mathbf{M}$, are spinors representing a general rotation. Any entity $\underline{\mathbf{u}} \in \mathbb{R}_{4,1}$ will be transformed by the spinor product,

$$\underline{\mathbf{u}}' = \mathbf{M}\underline{\mathbf{u}}\widetilde{\mathbf{M}}. \tag{31}$$

Let be $\mathbf{R} \in \mathbb{R}_{4,1}^+$ a rotor as representation of pure rotation and let be $\mathbf{T} \in \mathbb{R}_{4,1}^+$ another rotor, called translator, representing translation in $\mathbb{R}^3$, then any $g \in SE(3)$ is given by the motor

$$\mathbf{M} = \mathbf{T}\mathbf{R}. \tag{32}$$

Motors have some nice properties. They concatenate multiplicatively. If e.g. $\mathbf{M} = \mathbf{M}_2\mathbf{M}_1$, then

$$\underline{\mathbf{u}}'' = \mathbf{M}\underline{\mathbf{u}}\widetilde{\mathbf{M}} = \mathbf{M}_2\underline{\mathbf{u}}'\widetilde{\mathbf{M}}_2 = \mathbf{M}_2\mathbf{M}_1\underline{\mathbf{u}}\widetilde{\mathbf{M}}_1\widetilde{\mathbf{M}}_2. \tag{33}$$

Furthermore, motors are linear operations (as used also in context of spinor Clifford neurons). That is, we can exploit the outermorphism [11], which is the preservation of the outer product under linear transformation.

This leads directly to the third advantage of CGA. The incidence algebra of projective geometry generalizes in CGA. If $\underline{s}_1, \underline{s}_2 \in \langle \mathbb{R}_{4,1} \rangle_1$ are two spheres, then their outer product is a circle, $\underline{z} \in \langle \mathbb{R}_{4,1} \rangle_2$,

$$\underline{z} = \underline{s}_1 \wedge \underline{s}_2. \tag{34}$$

If the circle is undergoing a rigid motion, then

$$\underline{z}' = \mathbf{M}\underline{z}\widetilde{\mathbf{M}} = \mathbf{M}(\underline{s}_1 \wedge \underline{s}_2)\widetilde{\mathbf{M}} = \mathbf{M}\underline{s}_1\widetilde{\mathbf{M}} \wedge \mathbf{M}\underline{s}_2\widetilde{\mathbf{M}} = \underline{s}'_1 \wedge \underline{s}'_2. \tag{35}$$

In very contrast to the (homogeneously extended) Euclidean space, we can handle the rigid body motion not only as linear transformation of points but of any entities derived from points or spheres. These entities are no longer only set concepts as subspaces in a vector space but algebraic entities. This enables also to handle any free-form object as one algebraic entity which can be used for gradient based pose estimation in the same manner as points. That is a real cognitive approach to the pose problem.

Finally, we can consider the special Euclidean group $SE(3)$ as Lie group and are able to estimate the parameters of the generating operator of the group member. This approach leads to linearization and very fast iterative estimation of the parameters. Although this approach is also applicable to points in an Euclidean framework, in CGA it is applicable to any geometric entity built from points or spheres.

The generating operator of a motor is called twist, $\Psi \in se(3)$, represented in $\mathbb{R}_{4,1}$. The model of the motor as a Lie group member changes in comparison to equation (32) to

$$\mathbf{M} = \mathbf{T}\mathbf{R}\widetilde{\mathbf{T}}. \tag{36}$$

This equation, which expresses a general rotation as Lie group member in $\mathbb{R}_{4,1}$, can be easily interpreted. The general rotation is performed as normal rotation, $\mathbf{R}$, in the rotation plane $\mathbf{l} \in \langle \mathbb{R}_3 \rangle_2$, which passes the origin, by an angle $\theta \in \mathbb{R}$, after correcting the displacement $\mathbf{t} \in \langle \mathbb{R}_3 \rangle_1$ of the rotation axis $\mathbf{l}^* \in \langle \mathbb{R}_3 \rangle_1$ from the origin with the help of the translator $\widetilde{\mathbf{T}}$. Finally, the displacement has to be reconstructed by the translator $\mathbf{T}$. From this follows equation (36) in terms of the parameters of the rigid body motion in $\mathbb{R}_3$,

$$\mathbf{M} = \exp\left(\frac{\mathbf{e}\mathbf{t}}{2}\right) \exp\left(-\frac{\theta}{2}\mathbf{l}\right) \exp\left(-\frac{\mathbf{e}\mathbf{t}}{2}\right). \tag{37}$$

This factorized representation can be compactly written as

$$\mathbf{M} = \exp\left(-\frac{\theta}{2}\Psi\right). \tag{38}$$

Here

$$\Psi = \mathbf{l} + \mathbf{e}(\mathbf{t} \cdot \mathbf{l}) \tag{39}$$

is the twist of rigid body motion. Geometrically interpreted is the twist representing the line $\underline{\mathbf{l}} \in \langle \mathbb{R}_{4,1} \rangle_2$ around which the rotation is performed in $\mathbb{R}_{4,1}$.

Now we return to our problem of pose estimation. The above mentioned optimization problem is nothing else than a problem of minimizing the spatial distance of e.g. a projection ray to a feature on the silhouette of the reference model. If this distance is zero, then a certain geometric constraint is fulfilled. These constraints are either collinearity or coplanarity for points, lines and planes, or tangentiality in case of circles or spheres.

For instance, the collinearity of a point $\underline{\mathbf{x}} \in \langle \mathbb{R}_{4,1} \rangle_1$, after being transformed by the motor $\mathbf{M}$, with a projection line $\underline{\mathbf{l}}_x$ is written as their vanishing commutator product,

$$(\mathbf{M}\underline{\mathbf{x}}\widetilde{\mathbf{M}}) \times \underline{\mathbf{l}}_x = 0. \tag{40}$$

This equation has to be written in more details to see its coupling with the observation of a point x in the image plane of the camera:

$$\lambda \left((\mathbf{M}\underline{\mathbf{x}}\widetilde{\mathbf{M}}) \times (\mathbf{e} \wedge (\mathbf{O} \wedge x)) \right) \cdot \mathbf{e}_+ = 0. \tag{41}$$

This in fact is the complete pose estimation problem written as a symbolic dense but nevertheless algebraic correct equation. The outer product $\mathbf{O} \wedge x$ of the image point x with the optical center $\mathbf{O}$ of the camera results in the projection ray representation in projective space $\mathbb{R}_{3,1}$. By wedging it with $\mathbf{e}$, we transform the projection ray to the conformal space $\mathbb{R}_{4,1}$, hence, $\underline{\mathbf{l}}_x = \mathbf{e} \wedge \mathbf{O} \wedge x$. Now the collinearity constraint can be computed in $\mathbb{R}_{4,1}$. Finally, to assign the constraint an Euclidean distance zero, the expression has to be transformed to the Euclidean space $\langle \mathbb{R}_3 \rangle_1$ by computing the inner product with the basis vector $\mathbf{e}_+$ which is derived from the null basis and by scaling with $\lambda \in \mathbb{R}$. In reality the commutator product will take on a minimum as a result of the optimization process.

4.2   Twist Representations of Free-Form Objects

So far we only considered the pose estimation problem on the base of modelling a rigid object by a set of different order geometric entities. In robotics exists another important object model which is piecewise rigid. A kinematic chain is the formalization of several rigid bodies coupled by either revolute or prismatic joints. If pose estimation is applied to observatons of e.g. a robot arm for controlling grasping movements, this is the adequate model. Each of the parts of an arm is performing movements in mutual dependence. Hence, the motion of the j-th joint causes also motions of the preceding ones. If $\underline{\mathbf{u}}_j \in \mathbb{R}_{4,1}$ is a geometric entity attached to the j-th segment, e.g. a fingertip, its net displacement caused by a motor $\mathbf{M}$, represented in the (fixed) base coordinate system, is given by

$$\underline{\mathbf{u}}'_j = \mathbf{M}(\mathbf{M}_1...\mathbf{M}_j\underline{\mathbf{u}}_j\widetilde{\mathbf{M}}_j...\widetilde{\mathbf{M}}_1)\widetilde{\mathbf{M}}. \tag{42}$$

Here, the motors $\mathbf{M}_i$ are representing the constrained motion of the i-th joint. While in the homogeneous space $\mathbb{R}^4$ this equation is limited to points as moving entities and in the framework of the motor algebra [2] lines and planes can be considered in addition, there is no such restriction in CGA.

The idea of the kinematic chain can be further generalized to a generator of free-form shape. We will consider the trajectory caused by the motion of the entity $\underline{u}$ in space as the orbit of a multi-parameter Lie group of $SE(3)$. This multi-parameter Lie group represents a possibly very complex general rotation in $\mathbb{R}_{4,1}$ which is generated by a set of nested motors, contributing each with its own constrained elementary general rotation. If the entity $\underline{u}$ would be a point, then the orbit can be a space curve, a surface or a volume. But $\underline{u}$ can also be of higher order (e.g. a tea pot). Hence, the generated orbit may be of any complexity.

This kinematic model of shape in CGA [16] has not been known before, although there is a long history of relating kinematics and geometry, e.g. with respect to algebraic curves, or of the geometric interpretation of Lie groups.

The mentioned generalization of the model of a kinematic chain is twofold. First, the rotation axes of the motors must not be positioned at the periphery of the shape but may be (virtually) located anywhere. Second, several different oriented axes may be fixed at the same location.

The principle of constructing higher order 3D curves or surfaces as orbit of a point which is controlled by only a few coupled twists as generators is simple. Let be $\underline{x}_z$ an arbitrary point on a circle $\underline{z}$, that is $\underline{x}_z \cdot \underline{z} = 0$, and let be $\mathbf{M}(\underline{l}; \theta)$ the corresponding motor, then for all $\theta \in [0, ..., 2\pi]$

$$\underline{x}_z^{(1)} = \mathbf{M}\underline{x}_z\widetilde{\mathbf{M}} \qquad (43)$$

generates the circle $\underline{z}$. If the motor is representing a pure translator, then any arbitrary oriented line $\underline{l}$ will be generated. Such primitive curves will be called 3D-1twist curves.

By coupling a second twist to the first one, either a 3D-2twist curve or surface will be generated. In fact, already this simple system has a lot of degrees of freedom which enable generating quite different figures [15]. This can be seen by the following equation:

$$\underline{x}_c^{(2)} = \mathbf{M}_2\mathbf{M}_1\underline{x}_c\widetilde{\mathbf{M}_1}\widetilde{\mathbf{M}_2} \qquad (44)$$

with

$$\mathbf{M}_i(\underline{l}_i; \lambda_i, \theta_i) = \exp\left(-\frac{\lambda_i\theta_i}{2}\Psi_i\right), \qquad (45)$$

Here $\lambda_i \in \mathbb{R}$ is the ratio of the angular frequency, $\theta_i \in [\alpha_1, ..., \alpha_2]$ is the angular segment which is covered by θ_i and finally Ψ_i defines the position and orientation of the general rotation axis $\underline{l}_i$ and the type of motion, that is rotation-like, translation-like, or a mixed form.

Finally, the twist model of a closed shape is equivalent to the well-known Fourier representation, we started with this survey. Since the Fourier series expansion of a closed 3D curve $C(\phi)$ can be interpreted as the action of an infinite

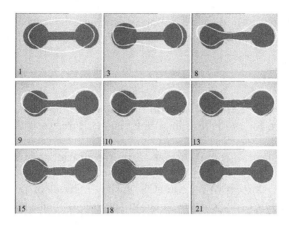

Fig. 4. Regularization of iterative pose estimation by Fourier representation (numbers indicate the iteration steps)

set of rotors, $\mathbf{R}_k^\phi$, fixed at the origin of $\langle\mathbb{R}_3\rangle_1$, onto phase vectors $\mathbf{p}_k$, there is no need of using CGA. A planar closed curve is given by

$$C(\phi) = \lim_{N\to\infty} \sum_{k=-N}^{N} \mathbf{p}_k \exp\left(\frac{2\pi k\phi}{T}\mathbf{l}\right) = \lim_{N\to\infty} \sum_{k=-N}^{N} \mathbf{R}_k^\phi \mathbf{p}_k \widetilde{\mathbf{R}}_k^\phi. \tag{46}$$

The phase vectors $\mathbf{p}_k$ are the Fourier series coefficients. Instead of the imaginary unit of the harmonics, the unit bivector $\mathbf{l} \in \mathbb{R}_3, \mathbf{l}^2 = -1$, is used which defines the rotation plane of the phase vectors. Based on the assumption that any closed 3D curve can be reconstructed from three orthogonal projections, a spatial curve is represented by

$$C(\phi) = \lim_{N\to\infty} \sum_{m=1}^{3} \sum_{k=-N}^{N} \mathbf{p}_k^m \exp\left(\frac{2\pi k\phi}{T}\mathbf{l}_m\right). \tag{47}$$

This scheme can be extended to free-form surfaces as well and has been used for pose estimation in the presented framework. In that respect the inverse Fourier transform of discretized curves/surfaces was applied. The necessary transformation of the Euclidean Fourier representation, $C_E(\phi)$, to a conformal one, $C_C(\phi)$, can simply be done by the following operation:

$$C_C(\phi) = \mathbf{e} \wedge (C_E(\phi) + \mathbf{e}_-), \tag{48}$$

where $\mathbf{e}_-$ stands for the homogeneous coordinate of the projective space $\mathbb{R}_{3,1}$ and wedging with $\mathbf{e}$ again realizes the transformation from $\mathbb{R}_{3,1}$ to $\mathbb{R}_{4,1}$.

One advantage of using the Fourier interpretation of the twist approach in pose estimation is the possibility of regularizing the optimization in the case of non-convex objects. This is demonstrated in figure (4). During the iteration process, which needs for that object only a few milliseconds, successively more Fourier coefficients are used. Hence, getting stuck in local minima is prevented.

The presented methods used in pose estimation clearly demonstrate the strength of the applied algebraic model of conformal geometric algebra.

5   Conclusions

We have demonstrated the application of geometric algebra as universal mathematical language for modelling in robot vision. Here we will summarize some general conclusions.

First, GA supports generalizations of representations. In section 4 we used the stratification of spaces of different geometry. This enables simple switching between different aspects of a geometric entity. Instead of points, in a certain GA as CGA, higher order entities take on the role of basis entities from which object concepts with algebraic properties can be constructed. We have shown that kinematics, shape theory and Fourier theory are unified in the framework of CGA. Besides, in section 2 we could handle multi-dimensional functions in a linear signal theory.

Second, the transformation of non-linear problems to linear ones is another important advantage. This could be demonstrated in all presented application fields. In learning theory this enables designing algebraically constrained knowledge based neural nets.

Third, GA is a mathematical language which supports symbolic dense formulations with algebraic meaning. We are able to lift up representations where besides the above mentioned qualitative aspects also reduced computational complexity results. This is important in real-time critical applications as robot vision.

We can state that the progress we made in robot vision could only be possible on the base of using geometric algebra and this enforces its use also in other application fields.

Acknowledgements

We acknowledge the European Community, Deutsche Forschungsgemeinschaft and Studienstiftung des deutschen Volkes for supporting the research which is reported in this paper. Especially contributions of Thomas Bülow, Michael Felsberg, Sven Buchholz, Vladimir Banarer, Bodo Rosenhahn, Martin Krause, Di Zang and Christian Perwass contributed to the reported results.

References

1. V. Banarer, C. Perwass, and G. Sommer. The hypersphere neuron. In *Proc. 11th European Symposium on Artificial Neural Networks, ESANN 2003, Bruges*, pages 469–474. d-side publications, Evere, Belgium, 2003.
2. E. Bayro-Corrochano, K. Daniilidis, and G. Sommer. Motor algebra for 3d kinematics: The case of hand-eye calibration. *Journal of Mathematical Imaging and Vision*, 13:79–100, 2000.

3. F. Brackx, R. Delanghe, and F. Sommen. *Clifford Analysis*. Pitman Advanced Publ. Program, Boston, 1982.
4. S. Buchholz and G. Sommer. Learning geometric transformations with Clifford neurons. In G. Sommer and Y. Zeevi, editors, *2nd International Workshop on Algebraic Frames for the Perception-Action Cycle, AFPAC 2000, Kiel*, volume 1888 of *LNCS*, pages 144–153. Springer-Verlag, 2000.
5. S. Buchholz and G. Sommer. Introduction to neural computation in Clifford algebra. In G. Sommer, editor, *Geometric Computing with Clifford Algebra*, pages 291–314. Springer-Verlag, Heidelberg, 2001.
6. T. Bülow. Hypercomplex spectral signal representations for the processing and analysis of images. Technical Report Number 9903, Christian-Albrechts-Universität zu Kiel, Institut für Informatik und Praktische Mathematik, August 1999.
7. T. Bülow and G. Sommer. Hypercomplex signals - a novel extension of the analytic signal to the multidimensional case. *IEEE Transactions on Signal Processing*, 49(11):2844–2852, 2001.
8. M. Felsberg. Low-level image processing with the structure multivector. Technical Report Number 0203, Christian-Albrechts-Universität zu Kiel, Institut für Informatik und Praktische Mathematik, März 2002.
9. M. Felsberg and G. Sommer. The monogenic signal. *IEEE Transactions on Signal Processing*, 49(12):3136–3144, 2001.
10. R.V.L. Hartley. A more symmetrical Fourier analysis applied to transmission problems. *Proc. IRE*, 30:144–150, 1942.
11. D. Hestenes. The design of linear algebra and geometry. *Acta Appl. Math.*, 23:65–93, 1991.
12. D. Hestenes, H. Li, and A. Rockwood. New algebraic tools for classical geometry. In G. Sommer, editor, *Geometric Computing with Clifford Algebras*, pages 3–23. Springer-Verlag, Heidelberg, 2001.
13. H. Li, D. Hestenes, and A. Rockwood. Generalized homogeneous coordinates for computational geometry. In G. Sommer, editor, *Geometric Computing with Clifford Algebras*, pages 27–59. Springer-Verlag, Heidelberg, 2001.
14. T. Needham. *Visual Complex Analysis*. Clarendon Press, Oxford, 1997.
15. B. Rosenhahn. Pose estimation revisited. Technical Report Number 0308, Christian-Albrechts-Universität zu Kiel, Institut für Informatik und Praktische Mathematik, September 2003.
16. B. Rosenhahn, C. Perwass, and G. Sommer. Free-form pose estimation by using twist representations. *Algorithmica*, 38:91–113, 2004.

Twists - An Operational Representation of Shape[*]

Gerald Sommer[1], Bodo Rosenhahn[2], and Christian Perwass[1]

[1] Institut für Informatik und Praktische Mathematik,
Christian-Albrechts-Universität zu Kiel,
Christian-Albrechts-Platz 4, 24118 Kiel, Germany
[2] Centre for Image Technology and Robotics,
University of Auckland,
Tamaki Campus, Auckland, New Zealand

Abstract. We give a contribution to the representation problem of free-form curves and surfaces. Our proposal is an operational or kinematic approach based on the Lie group $SE(3)$. While in Euclidean space the modelling of shape as an orbit of a point under the action of $SE(3)$ is limited, we are embedding our problem into the conformal geometric algebra $\mathbb{R}_{4,1}$ of the Euclidean space $\mathbb{R}^3$. This embedding results in a number of advantages which makes the proposed method a universal and flexible one with respect to applications. It makes possible the robust and fast estimation of the pose of 3D objects from incomplete and noisy image data. Especially advantagous is the equivalence of the proposed shape model to that of the Fourier representations.

1 Introduction

Shape is a geometric concept of the appearance of an object, a data set or a function which "can be defined as the total of all information that is invariant under translations, rotations, and isotropic rescalings. Thus two objects can be said to have the same shape if they are similar in the sense of Euclidean geometry." This quotation from [26] is very general because of the consideration of scale invariance. By leaving out that property, we can define the shape of an object as that geometric concept that is invariant under the special Euclidean group, e.g. $SE(3)$ if we consider 3D shape in Euclidean space $\mathbb{R}^3$. Furthermore, we allow our objects to change their shape in a well-defined manner under the action of some external forces which may include also a re-normalization of size.

The literature on shape modelling and applications is vast. Examples are visualization and animation in computer graphics or shape and motion recognition in computer vision. The central problem for the usefulness of a model in either field is the chosen representation of shape.

---

[*] This work has been partially supported (G.S. and C.P.) by EC Grant IST-2001-3422 (VISATEC), by DFG Grant RO 2497/1-1 (B.R.), and by DFG Graduiertenkolleg No. 357 (B.R. and C.P.).

Here we present a new approach to the modelling of free-form shape of curves and surfaces which has some features that make it especially attractive for computer vision and computer graphics. In our applications of pose estimation of 3D objects we could easily handle incomplete and noisy image data for numerically stable estimations with nearly video real-time capability.

That new representation results from the fusion of two concepts:

1) Free-form curves and surfaces are modelled as the orbit of a point under the action of the Lie group $SE(3)$, caused by a set of coupled infinitesimal generators of the group, called twists [18].
2) These object models are embedded in the conformal geometric algebra (CGA) of the Euclidean space $\mathbb{R}^3$ [16], that is $\mathbb{R}_{4,1}$. Only in conformal geometry does the (above mentioned) modelling of shape unfold a rich set of useful features.

The concept of fusing a local with a global algebraic framework has been proposed already in [27]. But only the pioneering work of Hestenes, Li and Rockwood [16] made it feasible to consider the Lie algebra $se(3)$, the space of tangents to an object, embedded in $\mathbb{R}_{4,1}$, as the source of our shape model instead of using $se(3)$ in $\mathbb{R}^3$.

The tight relations of geometry and kinematics are known to the mathematicians for centuries, see e.g. [4]. But in contrast to most applications in mechanical engineering, we are not restricted in our approach by physically feasible motions nor will we get problems in generating spatial curves or surfaces.

By embedding our design method into CGA, both primitive geometric entities such as points or objects on the one hand and actions on the other hand will have algebraic representations in one single framework. Furthermore, objects are defined by actions, and also actions can take on the role of operands.

Our proposed kinematic definition of shape uses infinitesimal actions to generate global patterns of low intrinsic dimension. This phenomen corresponds to the interpretation of the special Euclidean group in CGA, $SE(3)$, as a Lie group, where an element $g \in SE(3)$ performs a transformation of an entity $\underline{u} \in \mathbb{R}_{4,1}$,

$$\underline{u}' = \underline{u}(\theta) = g\left\{\underline{u}(0)\right\} \tag{1}$$

with respect to the parameter θ of g. Any special $g \in SE(3)$ that represents a general rotation in CGA corresponds to a Lie group operator $M \in \mathbb{R}_{4,1}^+$ which is called a motor and which is applied by the bilinear spinor product

$$\underline{u}' = M\underline{u}\widetilde{M}, \tag{2}$$

where $\widetilde{M}$ is the reverse of M. This product indicates that M is an orthogonal operator. If g is an element of the Lie group $SE(3)$, its infinitesimal generator, ξ, is defined in the corresponding Lie algebra, that is $\xi \in se(3)$. That Lie algebra element of the rigid body motion is geometrically interpreted as the rotation axis $\underline{l}$ in conformal space. Then the motor M results from the exponential map of the generator $\underline{l}$ of the group element, which is called a twist:

$$M = \exp\left(-\frac{\theta}{2}\underline{l}\right). \tag{3}$$

While θ is the rotation angle as the parameter of the motor, its generator is defined by the five degrees of freedom of a line $\underline{l}$ in space.

In our approach, the motor M is the effective operator which causes arbitrarily complex object shape. This operator may result from the multiplicative coupling of a set of primitive motors $\{M_i | i = n, ..., 1\}$,

$$M = M_n M_{n-1} ... M_2 M_1. \tag{4}$$

Each of these motors M_i represents a circular motion of a point around its own axis.

Based on that approach, rather complex free-form objects can be designed which behave as algebraic entities. That means, they can be transformed by motors in a covariant and linear way. To handle complete objects in that way as unique entities makes sense from both a cognitive and a numeric point of view.

The conformal geometric algebra $\mathbb{R}_{4,1}$ makes this possible. This is due to two essential facts. First, the representation of the special Euclidean group $SE(3)$ in $\mathbb{R}_{4,1}$ as a subgroup of the conformal group $C(3)$ is isomorphic to the special orthogonal group $SO^+(4,1)$. Hence, rigid body motion can be performed as rotation in CGA and therefore has a covariant representation. Second, the basic geometric entity of the conformal geometric algebra of the Euclidean space is the sphere. All geometric entities derived by incidence operations from the sphere can be transformed in CGA by an element $g \in SE(3)$, that is a motor $M \in \mathbb{R}_{4,1}^+$, in the same linear way, just as a point in the homogeneous Euclidean space $\mathbb{R}^4$. Because there exists a dual representation of a sphere (and of all derived entities) in CGA, which considers points as the basic geometric entity of the Euclidean space in the conformal space, all the known concepts from Euclidean space can be transformed to the conformal one.

Finally, we can take advantage of the stratification of spaces by CGA. Since the seminal paper [5] the purposive use of stratified geometries became an important design principle of vision systems. This means that an observer in dependence of its possibilities and needs can have access to different geometries as projective, affine or metric ones. So far this could hardly be realized. In CGA we have quite another situation.

The CGA $\mathbb{R}_{4,1}$ is a linear space of dimension 32. This mighty space represents not only conformal geometry but also affine geometry. Note that the special Euclidean group is a special affine group. Because $\mathbb{R}_{4,1}$ is derived from the Euclidean space $\mathbb{R}^3$, it encloses also Euclidean geometry, which is represented by the geometric algebra $\mathbb{R}_{3,0}$. In addition, the projective geometric algebra $\mathbb{R}_{3,1}$ is enclosed in $\mathbb{R}_{4,1}$. Thus, we have the stratification of the geometric algebras $\mathbb{R}_{3,0} \subset \mathbb{R}_{3,1} \subset \mathbb{R}_{4,1}$. This enables to consider metric (Euclidean), projective and kinematic (affine) problems in one single algebraic framework.

2 Geometric Entities and Motion in Conformal Geometric Algebra

After giving a bird's eye view on construction of the conformal geometric algebra, its features and geometric entities, will present the possibilities of representing the rigid body motion in CGA.

2.1 CGA of the Euclidean Space

A geometric algebra (GA) $\mathbb{R}_{p,q,r}$ is a linear space of dimension 2^n, $n = p + q + r$, which results from a vector space $\mathbb{R}^{p,q,r}$. We call (p, q, r) the signature of the vector space of dimension n. This indicates that there are $p/q/r$ unit vectors e_i which square to $+1/-1/0$, respectively. While $n = p$ in case of the Euclidean space $\mathbb{R}^3$, $\mathbb{R}^{p,q,r}$ indicates a vector space with a metric different than the Euclidean one. In the case of $r \neq 0$ there is a degenerate metric. We will omit the signature indexes from right if the interpretation is unique, as in the case of $\mathbb{R}^3$. The basic product of a GA is the associative and anticommutative geometric product, indicated by juxtaposition of the operands. There can be used a lot of other product forms in CA too, as the outer product ($\wedge$), the inner product ($\cdot$), the commutator product ($\times$) and the anticommutator product $\overline{\times}$. The space $\mathbb{R}_{p,q,r}$ is spanned by a set of 2^n linear subspaces of different grade called blades.

The conformal geometry of Euclidean and non-Euclidean spaces is known for a long time [32] without giving strong impact on the modelling in engineering with the exception of electrical engineering. There are different representations of the conformal geometry. Most disseminated is a complex formulation [19]. Based on an idea in [13], in [16] and in two other papers of the same authors in [28], the conformal geometries of the Euclidean, spherical and hyperbolic spaces have been worked out in the framework of GA.

The basic approach is that a conformal geometric algebra (CGA) $\mathbb{R}_{p+1,q+1}$ is built from a pseudo-Euclidean space $\mathbb{R}^{p+1,q+1}$. If we start with an Euclidean space $\mathbb{R}^n$, the construction $\mathbb{R}^{n+1,1} = \mathbb{R}^n \oplus \mathbb{R}^{1,1}$, $\oplus$ being the direct sum, uses a plane with Minkowski signature for augmenting the basis of $\mathbb{R}^n$ by the additional basis vectors $\{e_+, e_-\}$ with $e_+^2 = 1$ and $e_-^2 = -1$. Because that model can be interpreted as a homogeneous stereographic projection of all points $x \in \mathbb{R}^n$ to points $\underline{x} \in \mathbb{R}^{n+1,1}$, this space is called the homogeneous model of $\mathbb{R}^n$. Furthermore, by replacing the basis $\{e_+, e_-\}$ with the basis $\{e, e_0\}$, the homogeneous stereographic representation will become a representation of null vectors. This is caused by the properties $e^2 = e_0^2 = 0$ and $e \cdot e_0 = -1$. The relation between the null basis $\{e, e_0\}$ and the basis $\{e_+, e_-\}$ is given by

$$e := (e_- + e_+) \text{ and } e_0 := \frac{1}{2}(e_- - e_+). \tag{5}$$

Any point $x \in \mathbb{R}^n$ transforms to a point $\underline{x} \in \mathbb{R}^{n+1,1}$ according to

$$\underline{x} = x + \frac{1}{2}x^2 e + e_0 \tag{6}$$

with $\underline{x}^2 = 0$.

In fact, any point $\boldsymbol{x} \in \mathbb{R}^{n+1,1}$ is lying on an n-dimensional subspace $N_e^n \subset \mathbb{R}^{n+1,1}$, called horosphere [16]. The horosphere is a non-Euclidean model of the Euclidean space $\mathbb{R}^n$.

It must be mentioned that the basis vectors $\boldsymbol{e}$ and $\boldsymbol{e}_0$ have a geometric interpretation. In fact, $\boldsymbol{e}$ corresponds the north pole and $\boldsymbol{e}_0$ corresponds the south pole of the hypersphere of the stereographic projection, embedded in $\mathbb{R}^{n+1,1}$. Thus, $\boldsymbol{e}$ is representing the points at infinity and $\boldsymbol{e}_0$ is representing the origin of $\mathbb{R}^n$ in the space $\mathbb{R}^{n+1,1}$.

By setting apart these two points from all others of the $\mathbb{R}^n$ makes $\mathbb{R}^{n+1,1}$ a homogeneous space in the sense that each $\boldsymbol{x} \in \mathbb{R}^{n+1,1}$ is a homogeneous null vector without having reference to the origin. This enables coordinate-free computing to a large extent. Hence, $\underline{\boldsymbol{x}} \in N_e^n$ constitutes an equivalence class $\{\lambda\underline{\boldsymbol{x}}, \lambda \in \mathbb{R}\}$ on the horosphere. The reduction of that equivalence class to a unique entity with metrical equivalence to the point $\boldsymbol{x} \in \mathbb{R}^n$ needs a normalization.

The CGA $\mathbb{R}_{4,1}$, derived from the Euclidean space $\mathbb{R}^3$, offers 32 blades as basis of that linear space. This rich structure enables one to represent low order geometric entities in a hierarchy of grades. These entities can be derived as solutions of either the IPNS or the OPNS depending on what we assume as the basis geometric entity of the conformal space, see [21]. So far we only considered the mapping of an Euclidean point $\boldsymbol{x} \in \mathbb{R}^3$ to a point $\underline{\boldsymbol{x}} \in N_e^3 \subset \mathbb{R}^{4,1}$. But the null vectors on the horosphere are only a special subset of all the vectors of $\mathbb{R}^{4,1}$. All the vectors of $\mathbb{R}^{4,1}$ are representing spheres as the basic entities of the conformal space. A sphere $\underline{\boldsymbol{s}} \in \mathbb{R}^{4,1}$ is defined by its center position, $\boldsymbol{c} \in \mathbb{R}^3$, and its radius $\rho \in \mathbb{R}$ according to

$$\underline{\boldsymbol{s}} = \boldsymbol{c} + \frac{1}{2}(\boldsymbol{c} - \rho)^2 \boldsymbol{e} + \boldsymbol{e}_0. \tag{7}$$

And because $\underline{\boldsymbol{s}}^2 = \rho^2 > 0$, it must be a non-null vector. A point $\underline{\boldsymbol{x}} \in N_e^3$ can be considered as a degenerate sphere of radius zero and a plane $\boldsymbol{p} \in \mathbb{R}^{4,1}$ can be interpreted as a sphere of infinite radius. Hence, spheres $\underline{\boldsymbol{s}}$, points $\underline{\boldsymbol{x}}$ and planes $\boldsymbol{p}$ are entities of grade 1. By taking the outer product of spheres $\underline{\boldsymbol{s}}_i$, other entities of higher grade can be constructed.

So we get a circle $\underline{\boldsymbol{z}}$, a point pair $\underline{\boldsymbol{q}}$ and a point $\underline{\boldsymbol{y}}$ as entities of grade 2, 3 and 4, respectively, which exist outside the null cone in $\mathbb{R}^{4,1}$,

$$\underline{\boldsymbol{z}} = \underline{\boldsymbol{s}}_1 \wedge \underline{\boldsymbol{s}}_2 \tag{8}$$

$$\underline{\boldsymbol{q}} = \underline{\boldsymbol{s}}_1 \wedge \underline{\boldsymbol{s}}_2 \wedge \underline{\boldsymbol{s}}_3 \tag{9}$$

$$\underline{\boldsymbol{y}} = \underline{\boldsymbol{s}}_1 \wedge \underline{\boldsymbol{s}}_2 \wedge \underline{\boldsymbol{s}}_3 \wedge \underline{\boldsymbol{s}}_4 \tag{10}$$

as solutions of the IPNS. If we consider the OPNS on the other hand, we are starting with points $\underline{\boldsymbol{x}}_i \in N_e^3$ and can proceed similarly to define a point pair $\underline{\boldsymbol{Q}}$, a circle $\underline{\boldsymbol{Z}}$ and a sphere $\underline{\boldsymbol{S}}$ as entities of grade 2, 3 and 4 derived from points $\underline{\boldsymbol{x}}_i$ on the null cone of $\mathbb{R}_{4,1}$ according to

$$\underline{\boldsymbol{Q}} = \underline{\boldsymbol{x}}_1 \wedge \underline{\boldsymbol{x}}_2 \tag{11}$$

$$\underline{\boldsymbol{Z}} = \underline{\boldsymbol{x}}_1 \wedge \underline{\boldsymbol{x}}_2 \wedge \underline{\boldsymbol{x}}_3 \tag{12}$$

$$\underline{\boldsymbol{S}} = \underline{\boldsymbol{x}}_1 \wedge \underline{\boldsymbol{x}}_2 \wedge \underline{\boldsymbol{x}}_3 \wedge \underline{\boldsymbol{x}}_4. \tag{13}$$

These sets of entities are obviously related by the duality $\underline{\boldsymbol{u}}^* = \boldsymbol{U}$.

In OPNS, for lines $\boldsymbol{L}$ and planes $\boldsymbol{P}$, we have the definitions

$$\underline{\boldsymbol{L}} = e \wedge \underline{\boldsymbol{x}}_1 \wedge \underline{\boldsymbol{x}}_2 \tag{14}$$

$$\underline{\boldsymbol{P}} = e \wedge \underline{\boldsymbol{x}}_1 \wedge \underline{\boldsymbol{x}}_2 \wedge \underline{\boldsymbol{x}}_3 \tag{15}$$

and in IPNS we get the lines $\underline{l}$ and the planes $\underline{p}$ as entities of grade 2 and 1 as the dual of $\underline{\boldsymbol{L}}$ and $\underline{\boldsymbol{P}}$, respectively. Finally,

$$\underline{\boldsymbol{X}} = e \wedge \underline{\boldsymbol{x}}$$

is called the affine representation of a point [16]. This representation of a point is used if the interplay of the projective with the conformal representation is of interest in applications as in [24]. The same is with the line $\underline{\boldsymbol{L}}$ and the plane $\underline{\boldsymbol{P}}$.

Let us come back to the stratification of spaces mentioned in Section 1. Let be $\boldsymbol{x} \in \mathbb{R}^n$ a point of the Euclidean space, $\boldsymbol{X} \in \mathbb{R}^{n,1}$ a point of the projective space and $\underline{\boldsymbol{X}} \in \mathbb{R}^{n+1,1}$ a point of the conformal space. Then the operations which transform the representation between the spaces are for $\mathbb{R}_3 \longrightarrow \mathbb{R}_{3,1} \longrightarrow \mathbb{R}_{4,1}$

$$\underline{\boldsymbol{X}} = e \wedge \boldsymbol{X} = e \wedge (\boldsymbol{x} + e_-), \tag{16}$$

and for $\mathbb{R}_{4,1} \longrightarrow \mathbb{R}_{3,1} \longrightarrow \mathbb{R}_3$

$$\boldsymbol{x} = -\frac{\boldsymbol{X}}{\boldsymbol{X} \cdot e_-} = \frac{((e_+ \cdot \underline{\boldsymbol{X}}) \wedge e_-) \cdot e_-}{(e_+ \cdot \underline{\boldsymbol{X}}) \cdot e_-}. \tag{17}$$

2.2   The Special Euclidean Group in CGA

A geometry is defined by its basic entity, the geometric transformation group which is acting in a linear and covariant manner on all the entities which are constructed from the basic entity by incidence operations, and the resulting invariances with respect to that group. The search for such a geometry was motivated in Section 1. Next we want to specify the required features of the special Euclidean group in CGA.

To make a geometry a proper one, we have to require that any action $\mathcal{A}$ of that group on an entity, say $\boldsymbol{u}$, is grade preserving, or in other words structure preserving. This makes it necessary that the operator $\boldsymbol{A}$ applies as versor product [22]

$$\mathcal{A}\{\boldsymbol{u}\} = \boldsymbol{A}\boldsymbol{u}\boldsymbol{A}^{-1}. \tag{18}$$

This means that the entity $\boldsymbol{u}$ should transform covariantly [15], [3]. If $\boldsymbol{u}$ is composed by e.g. two representants $\boldsymbol{u}_1$ and $\boldsymbol{u}_2$ of the basis entities of the geometry, then $\boldsymbol{u}$ should transform according to

$$\mathcal{A}\{\boldsymbol{u}\} = \mathcal{A}\{\boldsymbol{u}_1 \circ \boldsymbol{u}_2\} = (\boldsymbol{A}\boldsymbol{u}_1\boldsymbol{A}^{-1}) \circ (\boldsymbol{A}\boldsymbol{u}_2\boldsymbol{A}^{-1}) = \boldsymbol{A}\boldsymbol{u}\boldsymbol{A}^{-1}. \tag{19}$$

The invariants of the conformal group $C(3)$ in $\mathbb{R}^3$ are angles. The conformal group $C(3)$ is mighty [19], but other than (18) and (19) it is nonlinear and transforms not covariantly in $\mathbb{R}^3$. Besides, in $\mathbb{R}^3$ there exist no entities other than points which could be transformed.

As we have shown in Section 2.1, in $\mathbb{R}_{4,1}$ the situation is quite different because all the geometric entities derived there can be seen also as algebraic entities in the sense of Section 1. Not only the elements of the null cone transform covariantly but also those of the dual space of $\mathbb{R}_{4,1}$. Furthermore, the representation of the conformal group $C(3)$ in $\mathbb{R}_{4,1}$ has the required properties of (18) and (19), see [16] and [15]. All vectors with positive signature in $\mathbb{R}_{4,1}$, that is a sphere, a plane as well as the components inversion and reflection of $C(3)$ compose a multiplicative group. That is called the versor representation of $C(3)$. This group is isomorphic to the Lorentz group of $\mathbb{R}_{4,1}$. The subgroup, which is composed by products of an even number of these vectors, is the spin group $Spin^+(4,1)$, that is the spin representation of $O^+(4,1)$. To that group belong the subgroups of rotation, translation, dilatation, and transversion of $C(3)$. They are applied as a spinor $\boldsymbol{S}$, $\boldsymbol{S} \in \mathbb{R}_{4,1}^+$ and $\boldsymbol{S}\tilde{\boldsymbol{S}} = |\boldsymbol{S}|^2$. A rotor $\boldsymbol{R}$, $\boldsymbol{R} \in \langle\mathbb{R}_{4,1}\rangle_2$ and $\boldsymbol{R}\boldsymbol{R}^2 = 1$, is a special spinor. Rotation and translation are represented in $\mathbb{R}_{4,1}$ as rotors.

The special Euclidean group $SE(3)$ is defined by $SE(3) = SO(3) \oplus \mathbb{R}^3$. Therefore, the rigid body motion $g = (R, t)$, $g \in SE(3)$ of a point $\boldsymbol{x} \in \mathbb{R}^3$ writes in Euclidean space

$$\boldsymbol{x}' = g\{\boldsymbol{x}\} = R\boldsymbol{x} + \boldsymbol{t}. \tag{20}$$

Here R is a rotation matrix and $\boldsymbol{t}$ is a translation vector. Because $SE(3) \subset C(3)$, in our choice of a special rigid body motion the representation of $SE(3)$ in CGA is isomorphic to the special orthogonal group, $SO^+(4,1)$. Hence, such $g \in SE(3)$, which does not represent the full screw, is represented as rotation in $\mathbb{R}_{4,1}$. This rotation is a general one, that is the rotation axis in $\mathbb{R}^3$ is shifted out of the origin by the translation vector $\boldsymbol{t}$.

That transformation $g \in SE(3)$ is represented in CGA by a special rotor $\boldsymbol{M}$ called a motor, $\boldsymbol{M} \in \langle\mathbb{R}_{4,1}\rangle_2$. The motor may be written

$$\boldsymbol{M} = \exp\left(-\frac{\theta}{2}\underline{\boldsymbol{l}}\right), \tag{21}$$

where $\theta \in \mathbb{R}$ is the rotation angle and $\underline{\boldsymbol{l}} \in \langle\mathbb{R}_{4,1}\rangle_2$ is indicating the line of the general rotation. To specify $\underline{\boldsymbol{l}}$ by the rotation and translation in $\mathbb{R}^3$, the motor has to be decomposed into its rotation and translation components. The normal rotation in CGA is given by the rotor

$$\boldsymbol{R} = \exp\left(-\frac{\theta}{2}l\right) \tag{22}$$

with $l \in \langle\mathbb{R}_3\rangle_2$ indicating the rotation plane which passes the origin. The translation in CGA is given by a special rotor, called a translator,

$$\boldsymbol{T} = \exp\left(\frac{et}{2}\right) \tag{23}$$

with $t \in \langle \mathbb{R}_3 \rangle_1$ as the translation vector. Because rotors constitute a multiplicative group, a naive formulation of the coupling of R and T would be

$$M = TR. \tag{24}$$

But if we interpret the rotor R as that entity of $\mathbb{R}_{4,1}$ which should be transformed by translation in a covariant manner, a better choice is

$$M = TR\widetilde{T}. \tag{25}$$

We call this special motor representation the twist representation. Its exponential form is given by

$$M = \exp\left(\frac{1}{2}et\right) \exp\left(-\frac{\theta}{2}l\right) \exp\left(-\frac{1}{2}et\right). \tag{26}$$

This equation expresses the shift of the rotation axis l^* in the plane l by the vector t to perform the normal rotation and finally shifting back the axis.

Because $SE(3)$ is a Lie group, the line $\underline{l} \in \langle \mathbb{R}_{4,1} \rangle_2$ is the representation of the infinitesimal generator of M, $\xi \in se(3)$. We call the generator representation a twist because it represents rigid body motion as general rotation. It is parameterized by the position and orientation of $\underline{l}$ which are the Plücker coordinates, represented by the rotation plane l and the inner product $(t \cdot l)$, [24],

$$\underline{l} = l + e(t \cdot l). \tag{27}$$

The most general formulation of the rigid body motion is the screw motion [23]. It is formulated in CGA as

$$M_s = T_s TR\widetilde{T} \tag{28}$$

with the pitch translator

$$T_s = \exp\left(\frac{d}{2}el^*\right), \tag{29}$$

where $l^* \in \langle \mathbb{R}_3 \rangle_1$ is the screw axis as the dual of l and $t_s = dl^*$ is a translation vector parallel to that axis and of length d. If we formulate M_s as

$$M_s = \exp\left(-\frac{\theta}{2}(l + em)\right) \tag{30}$$

with the vector $m \in \langle \mathbb{R}_3 \rangle_1$

$$m = t \cdot l - \frac{d}{\theta}l^*, \tag{31}$$

then all special cases of the rigid body motion, represented in the CGA $\mathbb{R}_{4,1}$ can be derived from (31):

$m = 0$: pure rotation $(M_s = R)$
$m = t, \theta \longrightarrow 0$: pure translation $(M_s = T)$
$m \perp l^*$: general rotation $(M_s = M)$
$m \not\perp l^*$: general screw motion.

A motor M transforms covariantly any entity $\underline{u} \in \mathbb{R}_{4,1}$ according to

$$\underline{u}' = M\underline{u}\widetilde{M} \tag{32}$$

with $\underline{u}' \in \mathbb{R}_{4,1}$. An equivalent equation is valid for the dual entity $\underline{U} \in \mathbb{R}_{4,1}$. Because motors concatenate multiplicatively, a multiple-motor transformation of $\underline{u}$ resolves recursively. Let be $M = M_2 M_1$, then

$$\underline{u}'' = M\underline{u}\widetilde{M} = M_2 M_1 \underline{u} \widetilde{M}_1 \widetilde{M}_2 = M_2 \underline{u}' \widetilde{M}_2. \tag{33}$$

It is a feature of any GA that also composed entities, which are built by the outer product of other ones, transform covariantly by a linear transformation. This is called outermorphism [12] and it means the preservation of the outer product under linear transformations. Let $\underline{z} \in \langle \mathbb{R}_{4,1} \rangle_2$ be a circle, which is composed by two spheres $\underline{s}_1, \underline{s}_2 \in \langle \mathbb{R}_{4,1} \rangle_1$ according to $\underline{z} = \underline{s}_1 \wedge \underline{s}_2$. Then the transformed circle computes as

$$\underline{z}' = M\underline{z}\widetilde{M} = M(\underline{s}_1 \wedge \underline{s}_2)\widetilde{M} = \langle M(\underline{s}_1 \underline{s}_2)\widetilde{M} \rangle_2 \tag{34}$$

$$= \langle M\underline{s}_1 \widetilde{M} M \underline{s}_2 \widetilde{M} \rangle_2 = M\underline{s}_1 \widetilde{M} \wedge M\underline{s}_2 \widetilde{M} \tag{35}$$

$$= \underline{s}_1' \wedge \underline{s}_2'. \tag{36}$$

Following Section 1, this is an important feature of the chosen algebraic embedding that will be demonstrated in Section 3.

3   Shape Models from Coupled Twists

In this section we will approach step by step the kinematic design of algebraic and transcendental curves and surfaces by coupling a certain set of twists as generators of a multiple-parameter Lie group action.

3.1   Constrained Motion in a Kinematic Chain

In the preceding section we argued that each entity $\underline{u}_i$ contributing to the rigid model of another entity $\underline{u}$ is performing the same transformation, represented by the motor M. Now we assume an ordered set of non-rigidly coupled rigid components of an object. This is for example a model of bar-shaped mechanisms [18] if the components are coupled by either revolute or prismatic joints. Such model is called a kinematic chain [17]. In a kinematic chain the task is to formulate the net movement of the end-effector at the n-th joint by movements of the j-th joints, $j = 1, ..., n - 1$, if the 0-th joint is fixed coupled with a world coordinate system. These movements are discribed by the motors M_j. Let T_j

be the transformation of an attached joint j with respect to the base coordinate system, then for $j = 1, ..., n$ the point $\underline{x}_{j,i_j}, i_j = 1, ..., m_j$, transforms according to

$$T_j(\underline{x}_{j,i_j}, M_j) = M_1...M_j\underline{x}_{j,i_j}\widetilde{M}_j...\widetilde{M}_1 \tag{37}$$

and

$$T_0(\underline{x}_{0,i_0}) = \underline{x}_{0,i_0}. \tag{38}$$

The motors M_j are representing the flexible geometry of the kinematic chain very efficiently. This results in an object model $\mathcal{O}$ defined by a kinematic chain with n segments and described by any geometric entity $\underline{u}_{j,i_j} \in \mathbb{R}_{4,1}$ attached to the j-th segment,

$$\mathcal{O} = \left\{ T_0(\underline{u}_{0,i_0}), T_1(\underline{u}_{1,i_1}, M_1), ..., T_n(\underline{u}_{n,i_n}, M_n) | n, i_0, ..., i_n \in \mathbb{N} \right\}. \tag{39}$$

If $\underline{u}_{j,i_j}$ is performing a motion caused by the motor M, then

$$\underline{u}'_{j,i_j} = M\left(T_j(\underline{u}_{j,i_j}, M_j)\right)\widetilde{M} \tag{40}$$

$$= M(M_1...M_j\underline{u}_{j,i_j}\widetilde{M}_j...\widetilde{M}_1)\widetilde{M}. \tag{41}$$

Obviously, this equation describes a constrained motion of the considered entities.

3.2   The Operational Model of Shape

We will now introduce another type of constrained motion, which can be realized by physical systems only in special cases but should be understood as a generalization of a kinematic chain. This is our proposed model of operational or kinematic shape [24]. An operational shape means that a shape results from the net effect, that is the orbit, of a point under the action of a set of coupled operators. So the operators at the end are the representations of the shape. A kinematic shape means the shape for which these operators are the motors as representations of $SE(3)$ in $\mathbb{R}_{4,1}$. The principle is simple. It goes back to the interpretation of any $g \in SE(3)$ as a Lie group action [18], see equation (1). But only in $\mathbb{R}_{4,1}$ we can take advantage of its representation as rotation around the axis $\underline{l}$, see equations (21), (25) and (26).

In Section 2.1 we introduced the sphere and the circle from IPNS and OPNS, respectively. We call these definitions the canonical ones. On the other hand, a circle has an operational definition which is given by the following.
Let $\underline{x}_\phi$ be a point which is a mapping of another point $\underline{x}_0$ by $g \in SE(3)$ in $\mathbb{R}_{4,1}$. This may be written as

$$\underline{x}_\phi = M_\phi\underline{x}_0\widetilde{M}_\phi \tag{42}$$

with M_ϕ being the motor which rotates $\underline{x}_0$ by an angle ϕ,

$$M_\phi = \exp\left(-\frac{\phi}{2}\Psi\right). \tag{43}$$

Here again is Ψ the twist as a generator of the rotation around the axis $\underline{l}$, see equation (21). Note that $\Psi = \alpha \underline{l}, \alpha \in \mathbb{R}$. If ϕ covers densely the whole span $[0, ..., 2\pi]$, then the generated set of points $\{\underline{x}_\phi\}$ is also dense. The infinite set $\{\underline{x}_\phi\}$ is the orbit of a rotation caused by the infinite set $\{M_\phi\}$, which has the shape of a circle in $\mathbb{R}^3$. The set $\{\underline{x}_\phi\}$ represents the well-known subset concept in a vector space of geometric objects in analytic geometry. In fact, that circle is on the horosphere N_e^3 because it is composed only by points. We will write for the circle $\underline{z}_{\{1\}}$ instead of $\{\underline{x}_\phi\}$ to indicate the different nature of that circle in comparison to either $\underline{z}$ or $\underline{Z}$ of Section 2.1. The index $\{1\}$ means that the circle is generated by one twist from a continuous argument ϕ. So the circle, embedded in $\mathbb{R}_{4,1}$, is defined by

$$\underline{z}_{\{1\}} = \left\{ \underline{x}_\phi \mid \text{for all } \phi \in [0, ..., 2\pi] \right\}. \tag{44}$$

Its radius is given by the distance of the chosen point $\underline{x}_0$ to the axis $\underline{l}$ and its orientation and position in space depends on the parameterization of $\underline{l}$. That $\underline{z}_{\{1\}}$ is defined by an infinite set of arguments is no real problem in the case of computational geometry or applications where only discretized shape is of interest. More interesting is the fact that in the canonical definitions of Section 2.1 the geometric entities are all derived from either spheres or points. In the case of the operational definition of shape, the circle is the basic geometric entity instead, respectively rotation is the basic operation.

A sphere results from the coupling of two motors, M_{ϕ_1} and M_{ϕ_2}, whose twist axes meet at the center of the sphere and which are perpendicularly arranged. The following twists are possible generators, but any other orientation is even good,

$$\Psi_1 = e_{12} + e(c \cdot e_{12}) \tag{45}$$
$$\Psi_2 = e_{31} + e(c \cdot e_{31}) \tag{46}$$

with the sphere center c, and e_{12}, e_{31} are two orthogonal planes.

The resulting constrained motion of a point $\underline{x}_{0,0}$ performs a rotation on a sphere given by $\phi_1 \in [0, ..., 2\pi]$ and $\phi_2 \in [0, ..., \pi]$,

$$\underline{x}_{\phi_1, \phi_2} = M_{\phi_2} M_{\phi_1} \underline{x}_{0,0} \widetilde{M}_{\phi_1} \widetilde{M}_{\phi_2}. \tag{47}$$

The complete orbit of a sphere is given by

$$\underline{s}_{\{2\}} = \left\{ \underline{x}_{\phi_1, \phi_2} \mid \text{for all } \phi_1 \in [0, ..., 2\pi], \phi_2 \in [0, ..., \pi] \right\}. \tag{48}$$

Let us come back to the point of generalization of the well-known kinematic chains. These models of linked bar mechanisms have to be physically feasible. Instead, our model of coupled twists is not limited by that constraint. Therefore, the sphere expresses a virtual coupling of twists. This includes both location and orientation in space, and the possibility of fixating several twists at the same location, for any dimension of the space $\mathbb{R}^n$. There are several extensions of the introduced kinematic model which are only possible in CGA.

First, while the group $SE(3)$ can only act on points, its representation in $\mathbb{R}_{4,1}$ may act in the same way on any entity $\underline{u} \in \mathbb{R}_{4,1}$ derived from either points or spheres. This results in high complex free-form shapes caused from the motion of relatively simple generating entities and low order sets of coupled twists.

Second, only by coupling a certain set of twists, high complex free-form shapes may be generated from a complex enough constrained motion of a point.

Let $\underline{u}_{\{n\}}$ be the shape generated by n motors $M_{\phi_1}, ..., M_{\phi_n}$. We call it the n-twist model,

$$\underline{u}_{\{n\}} = \left\{ \underline{x}_{\phi_1,...,\phi_n} \mid \text{for all } \phi_1, ..., \phi_n \in [0, ..., 2\pi] \right\} \tag{49}$$

with

$$\underline{x}_{\phi_1,...,\phi_n} = M_{\phi_n}...M_{\phi_1}\underline{x}_{0,...,0}\widetilde{M}_{\phi_1}...\widetilde{M}_{\phi_n}. \tag{50}$$

Then the last equation may also be written

$$\underline{x}_{\phi_1,...,\phi_n} = M_{\phi_n}...M_{\phi_2}\underline{x}_{\{\phi_1,0,...,0\}}\widetilde{M}_{\phi_2}...\widetilde{M}_{\phi_n}. \tag{51}$$

By continuing that reformulation, we get finally

$$\underline{x}_{\phi_1,...,\phi_n} = M_{\phi_n}\underline{x}_{\{\phi_1,...,\phi_{n-1},0\}}\widetilde{M}_{\phi_n}. \tag{52}$$

This corresponds also to

$$\underline{x}_{\phi_1,...,\phi_n} = M\underline{x}_{0,...,0}\widetilde{M} \tag{53}$$

with $M = M_{\phi_n}...M_{\phi_1}$. The set $\{\underline{x}_{\phi_1,0,...0}\}$ represents a circle and the set $\{\underline{x}_{\phi_1,...,\phi_{n-1},0}\}$ represents an entity which, coupled with a circle, will result in $\{\underline{x}_{\phi_1,...,\phi_n}\}$. While equation (50) is representing a multiple-parameter Lie group form of $SE(3)$, where the nested motors are carrying the complexity of $\underline{u}_{\{n\}}$, the complexity is stepwise shifted to $\underline{u}_{\{n-1\}}$ in equation (52). Furthermore, while both equations (52) and (53) are linear in the motors, equation (53) looks so simple because the parameters of the resulting motor M are now a function in the space spanned by the parameters of the generating twists $\Psi_1, ..., \Psi_n$ and the arguments of the motors $\phi_1, ..., \phi_n$. Instead of using the single-parameter form (53), we prefer equation (50).

3.3   Free-Form Objects

There are a lot of more degrees of freedom to design free-form objects embedded in $\mathbb{R}_{4,1}$ by the motion of a point caused by coupled twists. While a single rotation-like motor generates a circle, a single translation-like motor generates a line as a root of non-curved objects. Of course, several of both variants can be mixed.

Other degrees of freedom of the design result from the following extensions:

- Introducing an individual angular frequency λ_i to the motor M_{ϕ_i} also influences the synchronization of the rotation angles ϕ_i.

- Rotation within limited angular segments $\phi_i \in [\alpha_{i_1}, ..., \alpha_{i_2}]$ with $0 \le \alpha_{i_1} < \alpha_{i_2} \le 2\pi$ is possible.

Let us consider the simple example of a 2-twist model of shape,

$$\underline{u}_{\{2\}} = \left\{ \underline{x}_{\phi_1, \phi_2} \mid \text{for all } \phi_1, \phi_2 \in [0, ..., 2\pi] \right\} \tag{54}$$

with

$$\underline{x}_{\phi_1, \phi_2} = M_{\lambda_2 \phi_2} M_{\lambda_1 \phi_1} \underline{x}_0 \widetilde{M}_{\lambda_1 \phi_1} \widetilde{M}_{\lambda_2 \phi_2}, \tag{55}$$

$\lambda_1, \lambda_2 \in \mathbb{R}$ and $\phi_1 = \phi_2 = \phi \in [0, ..., 2\pi]$.

That model can generate not only a sphere, but an ellipse ($\lambda_1 = -2, \lambda_2 = 1$), several well-known algebraic curves (in space), see [24], such as cardioid, nephroid or deltoid, transcendental curves like a spiral, or surfaces. For the list of examples see Table 1.

Table 1. Simple geometric entities generated from up to three twists

Entity	Generation	Class
point	twist axis intersected with a point	0twist curve
circle	twist axis non-collinear with a point	1twist curve
line	twist axis is at infinity	1twist curve
conic	2 parallel non-collinear twists	2twist curve $\lambda_1 = 1, \lambda_2 = -2$
line segment	2 twists, building a degenerate conic	2twist curve $\lambda_1 = 1, \lambda_2 = -2$
cardioid	2 parallel non-collinear twists	2twist curve $\lambda_1 = 1, \lambda_2 = 1$
nephroid	2 parallel non-collinear twists	2twist curve $\lambda_1 = 1, \lambda_2 = 2$
rose	2 parallel non-collinear twists, j loops	2twist curve $\lambda_1 = 1, \lambda_2 = -j$
spiral	1 finite and 1 infinite twist	2twist curve $\lambda_1 = 1, \lambda_2 = 1$
sphere	2 perpendicular twists	2twist surface $\lambda_1 = 1, \lambda_2 = 1$
plane	2 parallel twists at infinity	2twist surface
cylinder	2 twists, one at infinity	2twist surface
cone	2 twists, one at infinity	2twist surface
quadric	a conic rotated with a third twist	3twist surface

Interestingly, the order of nonlinearity of algebraic curves grows faster than the number of the generating motors.

By replacing the initial point $\underline{x}_0$ by any other geometric entity, $\underline{u}_0$, built from either points or spheres by applying the outer product, the concepts remain the same. This makes the kinematic object model in conformal space a recursive one.

The infinite set of arguments ϕ_i of the motor M_{ϕ_i} to generate the entity $\underline{u}_{\{n\}}$ will in practice reduce to a finite one, which results in a discrete entity $\underline{u}_{[n]}$. The index $[n]$ indicates that n twists are used with a finite set of arguments $\{\phi_{i,j_i} \mid j_i \in \{0, ..., m_i\}\}$.

The previous formulations of free-form shape did assume a rigid model. As in the case of the kinematic chain, the model can be made flexible. This happens by encapsulating the entity $\underline{\boldsymbol{u}}_{[n]}$ into a set of motors $\left\{ \boldsymbol{M}_j^d | j = J, ..., 1 \right\}$, which results in a deformation of the object.

$$\underline{\boldsymbol{u}}_{[n]}^d = \boldsymbol{M}_J^d ... \boldsymbol{M}_1^d \underline{\boldsymbol{u}}_{[n]} \widetilde{\boldsymbol{M}}_1^d ... \widetilde{\boldsymbol{M}}_J^d \tag{56}$$

Finally, the entity $\underline{\boldsymbol{u}}_{[n]}^d$ may perform a motion under the action of a motor $\boldsymbol{M}$, which itself may be composed by a set of motors $\{ \boldsymbol{M}_i | i = I, ..., 1 \}$ according to equation (4),

$$\underline{\boldsymbol{u}}_{[n]}^{d'} = \boldsymbol{M} u_{[n]}^d \widetilde{\boldsymbol{M}}. \tag{57}$$

But a twist is not only an operator, it may also play in CGA the role of an operand,

$$\Psi' = \boldsymbol{M} \Psi \boldsymbol{M}. \tag{58}$$

This causes a dynamic shape model as an alternative to (56). If the angular argument of $\boldsymbol{M}$ is specified by, e.g. $\theta = 2\pi t$, $t_* \leq t \leq t^*$, then the twist axis $\underline{l}$ may move along the arc of a circle. An interesting application is the so-called ball-and-socket joint required to accurately model shoulder and hip joints of articulated persons [10].

So far, the entity $\boldsymbol{u}_{\{n\}}$ was embedded in the Euclidean space. Lifting up the entity to the conformal space, $\underline{\boldsymbol{u}}_{\{n\}} \in \mathbb{R}_{4,1}$, is simply done by

$$\underline{\boldsymbol{u}}_{\{n\}} = \boldsymbol{e} \wedge \left(\boldsymbol{u}_{\{n\}} + \boldsymbol{e}_- \right) = \boldsymbol{e} \wedge \boldsymbol{U}_{\{n\}} \tag{59}$$

with $\boldsymbol{U}_{\{n\}}$ being the shape in the projective space $\mathbb{R}_{3,1}$.

4   Twist Models and Fourier Representations

The message of the last subsection is the following. A finite set of coupled twist (or nested motors) performs a constrained motion of any set of geometric entities, whose orbit uniquely represents either a curve, a surface or a volume of arbitrary complexity. This needs a parameterized model of the generators of the shape. In some applications the reverse problem may be of interest. That is to find a parameterized twist model for a given shape. That task can be solved: Any curve, surface or volume of arbitrary complexity can be mapped to a finite set of coupled twists, but in a non-unique manner. That means, that there are different models which generate the same shape.

We will show here that there is a direct and intuitive relation between the twist model of shape and the Fourier representations. The Fourier series decomposition and the Fourier transforms in their different representations are well-known techniques of signal analysis and image processing [20]. The interesting fact that this equivalence of representations results in a fusion of concepts from geometry, kinematics, and signal theory is of great importance in engineering.

Furthermore, because the presented modelling of shape is embedded in a conformal space, there is also a single access for embedding the Fourier representations
in either conformal or projective geometry. This is quite different from the recent
publications [29], [30]. It will hopefully enable image processing which is conformally embedded and, in the case of image sequences, the pose in space can
be coupled with image analysis in a better way than in [31]. Our first attempt
to formulate projective Fourier coefficients [24] showed some serious problems
which have to be overcome in future work. We will not go into details here.

4.1   The Case of a Closed Planar Curve

Let us consider a closed curve $c \in \mathbb{R}^2$ in a parametric representation with $t \in \mathbb{R}$.
Then its Fourier series representation is given by

$$c(t) = \sum_{\nu=-\infty}^{\infty} \gamma_\nu \exp\left(\frac{j2\pi\nu t}{T}\right) \tag{60}$$

with the Fourier coefficients γ_ν, $\nu \in \mathbb{Z}$ as frequency and j, $j^2 = -1$, as the
imaginary unit and T as the curve length.

This model of a curve has been used for a long time in image processing for
shape analysis by Fourier descriptors (these are the Fourier coefficients) [33],
[9]. Furthermore, affine invariant Fourier descriptors can be used [1] to couple a
space curve to its affine image.

We will translate this spectral representation into the model of an infinite
number of coupled twists by following the method presented in [25]. Because
equation (60) is valid in an Euclidean space, the twist model has to be reformu-
lated accordingly. This will be shown for the case of a 2-twist curve $\underline{c}_{\{2\}}$. Then
equation (55) can be written in $\mathbb{R}_3$ for $\phi_1 = \phi_2 = \phi$ as

$$x_\phi = R_{\lambda_2\phi}\left(\left(R_{\lambda_1\phi}(x_0 - t_1)\widetilde{R}_{\lambda_1\phi} + t_1\right) - t_2\right)\widetilde{R}_{\lambda_2\phi} + t_2 \tag{61}$$

$$= p_0 + V_{1,\phi}p_1\widetilde{V}_{1,\phi} + V_{2,\phi}p_2\widetilde{V}_{2,\phi}. \tag{62}$$

Here the translation vectors have been absorbed by the vectors p_i and the V_i
are built by certain products of the rotors $R_{\lambda_i\phi}$. We call the p_i the phase vectors.
Next, for the aim of interpreting that equation as a Fourier series expansion, we
rewrite the Fourier basis functions as rotors of an angular frequency $i \in \mathbb{Z}$, in
the plane $l \in \mathbb{R}_2$, $l^2 = -1$,

$$R_{\lambda_i\phi} = \exp\left(-\frac{\lambda_i\phi}{2}l\right) = \exp\left(-\frac{\pi i\phi}{T}l\right). \tag{63}$$

All rotors of a planar curve lie in the same plane as the phase vectors p_i. After
some algebra, see [25], we get for the transformed point

$$x_\phi = \sum_{i=0}^{2} p_i \exp\left(\frac{2\pi i\phi}{T}l\right) \tag{64}$$

and for the curve as subspace of $\mathbb{R}^3$ the infinite set of points

$$c_{\{2\}} = \{x_\phi| \text{ for all } \phi \in [0, ..., 2\pi] \text{ and for all } i \in \{0, 1, 2\}\} . \tag{65}$$

A general (planar) curve is given by

$$c_{\{\infty\}} = \{x_\phi| \text{ for all } \phi \in [0, ..., 2\pi] \text{ and for all } i \in \mathbb{Z}\} , \tag{66}$$

respectively as Fourier series expansion, written in the language of kinematics

$$c_{\{\infty\}} = \left\{ \lim_{n \to \infty} \sum_{i=-n}^{n} p_i \exp\left(\frac{2\pi i\phi}{T}l\right) \right\} \tag{67}$$

$$= \left\{ \lim_{n \to \infty} \sum_{i=-n}^{n} R_{\lambda_i\phi} p_i \widetilde{R}_{\lambda_i\phi} \right\} . \tag{68}$$

A discretized curve is called a contour. In that case equation (67) has to consider a finite model of n twists and the Fourier series expansion becomes the inverse discrete Fourier transform. Hence, a planar contour is given by the finite sequence $c_{[n]}$ with the contour points $c_k, -n \leq k \leq n$, in parametric representation

$$c_k = \sum_{i=-n}^{n} p_i \exp\left(\frac{2\pi ik}{2n+1}l\right) , \tag{69}$$

and the phase vectors are computed as a discrete Fourier transform of the contour

$$p_i = \frac{1}{2n+1} \sum_{k=-n}^{n} c_k \exp\left(-\frac{2\pi ik}{2n+1}l\right) . \tag{70}$$

These equations imply that the angular argument ϕ_k is replaced by k. If the contour can be interpreted as a satisfactory sampled curve [20], the curve $c_{\{\infty\}}$ can be reconstructed from $c_{[n]}$.

4.2   Extensions of the Concepts

The extension of the modelling of a planar curve, embedded in $\mathbb{R}^3$, to a 3D curve is easily done. This happens by taking its projections to either e_{12}, e_{23}, or e_{31} as periodic planar curves. Hence, we get the superposition of these three components. Let $c_{[n]}^j$ be these components in the case of a 3D contour with the rotation axes l_j^* perpendicular to the rotation planes l_j. Then

$$c_{[n]} = \sum_{j=1}^{3} c_{[n]}^j \tag{71}$$

with the contour points of the projections c_k^j, $j = 1, 2, 3$ and $-n \leq k \leq n$,

$$c_k^j = \sum_{i=-n}^{n} p_i^j \exp\left(\frac{2\pi ik}{2n+1}l_j\right) . \tag{72}$$

Another useful extension is with respect to surface representations, see [25]. If this surface is a 2D function orthogonal to a plane spanned by the bivectors e_{ij}, then the twist model corresponds to the 2D inverse FT. In the case of an arbitrary orientation of the rotation planes l_j instead, or in the case of the surface of a 3D object, the procedure is comparable to that of equation (72). The surface is represented as a two-parametric surface $s(t_1, t_2)$ generated as a superposition of the three projections $s^j(t_1, t_2)$.

In the case of a discrete surface in a two-parametric representation we have the finite surface representation $s_{[n_1, n_2]}$,

$$s_{[n_1, n_2]} = \sum_{j=1}^{3} s^j_{[n_1, n_2]} \tag{73}$$

with the surface points of the projections $s^j_{k_1, k_2}$, $j = 1, 2, 3$ and $-n_1 \leq k_1 \leq n_1$, $-n_2 \leq k_2 \leq n_2$,

$$s^j_{k_1, k_2} = \sum_{i_1=-n_1}^{n_1} \sum_{i_2=-n_2}^{n_2} p^j_{i_1, i_2} \exp\left(\frac{2\pi i_1 k_1}{2n_1 + 1} l_j\right) \exp\left(\frac{2\pi i_2 k_2}{2n_2 + 1} l_j\right) \tag{74}$$

and the phase vectors

$$p^j_{i_1, i_2} = \frac{1}{2n_1 + 1} \frac{1}{2n_2 + 1} p^{j'}_{i_1, i_2} \tag{75}$$

$$p^{j'}_{i_1, i_2} = \sum_{k_1=-n_1}^{n_1} \sum_{k_2=-n_2}^{n_2} s^j_{k_1, k_2} \exp\left(-\frac{2\pi i_1 k_1}{2n_1 + 1} l_j\right) \exp\left(-\frac{2\pi i_2 k_2}{2n_2 + 1} l_j\right). \tag{76}$$

Finally, we will formulate an alternative model of a curve $\underline{c} \in \mathbb{R}_{4,1}$ [24]. While equation (67) expresses the additive superposition of rotated phase vectors in Euclidean space, the following model expresses a multiplicative coupling of the twists directly in conformal space.

$$\underline{c}_{\{\infty\}} = \left\{ \lim_{n \longrightarrow \infty} \left(\prod_{i=n}^{-n} T_{\lambda_i \phi} \right) \underline{O} \left(\prod_{i=-n}^{n} \tilde{T}_{\lambda_i \phi} \right) \right\}. \tag{77}$$

This equation results from the assumption that the point $x_0 = (0, 0, 0) \in \mathbb{R}^3$, expressed as the affine point $\underline{O} \in \mathbb{R}_{4,1}$, $\underline{O} = e \wedge e_0$, is translated $2n + 1$ times by the translators

$$T_{\lambda_i \phi} = \frac{1 + e t_{\lambda_i \phi}}{2} \tag{78}$$

with the Euclidean vector

$$t_{\lambda_i \phi} = R_{\lambda_i \phi} p_i \tilde{R}_{\lambda_i \phi}. \tag{79}$$

It turns out that equation (77) has some numeric advantages in application [24].

The discussed equivalence of the twist model and the Fourier representation has several advantages in practical use of the model. The most important may be the applicability to low-frequency approximations of the shape. For instance in pose estimation [24] the estimations of the motion parameters of non-convex objects can be regularized efficiently in that way. Instead of estimating motors, the parameters of the twists are estimated because of numeric reasons.

It turned out that the paper [14] already proposed an elliptic approximation to a contour. The authors call their Fourier transform (FT) the elliptic FT. The generating model of shape is that of coupled ellipses. Later on, after Bracewell [2] rediscovered that this type of FT has been already proposed in [11] as real-valued FT, it is well-known as Hartley transform. Taking the Hartley transform instead of the complex-valued FT has the advantage of reducing the computational complexity by a factor of two in both contour and surface based free-form objects. In the case of a surface model equation (73) involves three complex-valued 2D Fourier transforms. Instead of that one single quaternion-valued 3D Fourier transform can be applied. This leads to a slight reduction of the computational complexity. But finally, neither of the presented versions of the Fourier transform must be applied in real time. Instead, precomputing and table look up is the fastest version.

In some applications it is not necessary to have at hand the global shape of an object as an inverse discrete Fourier transform. Instead, a local spectral representation of the shape would be sufficient. The Gabor transform [8] could be a candidate. But a better choice is the monogenic signal [6]. This is computed by scale adaptive filters [7] for getting the spectral representations of oriented lines in the plane. That approach is comparable to the model of coupled twists. But in contrast to our former assumption, the orientation of the orientation plane $\underline{l}$ is not fixed but adapted to the orientation of the shape tangents in the plane. An extension to lines in $\mathbb{R}^3$ and to its coupling to the twist model is work in progress.

5   Summary and Conclusions

We presented an operational or kinematic model of shape in $\mathbb{R}^3$. This model is based on the Lie group $SE(3)$, embedded in the conformal geometric algebra $\mathbb{R}_{4,1}$ of the Euclidean space. While the modelling of shape in $\mathbb{R}^3$ caused by actions of $SE(3)$ is limited, a lot of advantages result from the chosen algebraic embedding in real applications. One of these is the possibility of conformal (and projective) shape models. We did not discuss any applications in detail. Instead, we refer the reader to the website http://www.ks.informatik.uni-kiel.de with respect to the problem of pose estimation. In that work we could show that the pose estimation based on the presented shape model can cope with incomplete and noisy data. In addition to that robustness, pose estimation is numerically stable and fast.

Because the chosen twist model is equivalent to the Fourier representation (in some aspects it overcomes that), the proposed shape representation unifies geometry, kinematics, and signal theory. It can be expected that this will have a great impact on both theory and practice in computer vision, computer graphics and modelling of mechanisms.

References

1. K. Arbter. Affine-invariant fourier descriptors. In J.C. Simon, editor, *From Pixels to Features*, pages 153–164. Elsevier Science Publishers, 1989.
2. R.N. Bracewell. *The Fourier Transform and its Applications*. McGraw Hill Book Company, New York, 1984.
3. L. Dorst and D. Fontijne. An algebraic foundation for object-oriented euclidean geometry. In E.M.S. Hitzer and R. Nagaoka, editors, *Proc. Innovative Teaching in Mathematics with Geometric Algebra, Kyoto*, pages 138–153. Research Institute for Mathematics, Kyoto University, May, 2004.
4. R.T. Farouki. Curves from motion, motion from curves. In P.-J. Laurent, P. Sablonniere, and L.L. Schumaker, editors, *Curve and Surface Design*. Vanderbilt University Press, Nashville, TN, 2000.
5. O. Faugeras. Stratification of three-dimensional vision: projective, affine and metric representations. *Journ. of the Optical Soc. of America*, 12:465–484, 1995.
6. M. Felsberg and G. Sommer. The monogenic signal. *IEEE Trans. Signal Processing*, 49(12):3136–3144, 2001.
7. M. Felsberg and G. Sommer. The monogenic scale-space: A unified approach to phase-based image processing in scale-space. *Journal of Mathematical Imaging and Vision*, 21:5–26, 2004.
8. G. Granlund and H. Knutsson. *Signal Processing for Computer Vision*. Kluwer Academic Publ., Dordrecht, 1995.
9. G.H. Granlund. Fourier processing for hand print character recognition. *IEEE Trans. Computers*, 21:195–201, 1972.
10. F.S. Grassia. Practical parameterization of rotations using the exponential map. *Journal of Graphics Tools*, 3(3):29–48, 1998.
11. R.V.L. Hartley. A more symmetrical fourier analysis applied to transmission problems. *Proc. IRE*, 30:144–150, 1942.
12. D. Hestenes. The design of linar algebra and geometry. *Acta Appl. Math.*, 23:65–93, 1991.
13. D. Hestenes and G. Sobczyk. *Clifford Algebra to Geometric Calculus*. D. Reidel Publ. Comp., Dordrecht, 1984.
14. F.P. Kuhl and C.R. Giardina. Elliptic fourier features of a closed contour. *Computer Graphics and Image Processing*, 18:236–258, 1982.
15. A. Lasenby and J. Lasenby. Surface evolution and representation using geometric algebra. In R. Cipolla and R. Martin, editors, *The Mathematics of Surfaces IX*, pages 144–168. Springer-Verlag, London, 2000.
16. H. Li, D. Hestenes, and A. Rockwood. Generalized homogeneous coordinates for computational geometry. In G. Sommer, editor, *Geometric Computing with Clifford Algebras*, pages 27–59. Springer-Verlag, Heidelberg, 2001.
17. J.M. McCarthy. *Introduction to Theoretical Kinematics*. The MIT Press, Cambridge, MA, 1990.
18. R.M. Murray, Z. Li, and S.S. Sastry. *A Mathematical Introduction to Robotic Manipulation*. CRC Press, Boca Raton, 1994.
19. T. Needham. *Visual Complex Analysis*. Clarendon Press, Oxford, 1997.
20. A.V. Oppenheim and R.W. Schafer. *Discrete-Time Signal Processing*. Prentice Hall, Inc., Upper Saddle River, 1989.
21. C. Perwass and D. Hildenbrand. Aspects of geometric algebra in euclidean, projective and conformal space. Technical Report 0310, Technical Report, Christian-Albrechts-Universität zu Kiel, Institut für Informatik und Praktische Mathematik, 2003.

22. C. Perwass and G. Sommer. Numerical evaluation of versors with clifford algebra. In L. Dorst, C. Doran, and J. Lasenby, editors, *Applications of Geometric Algebra in Computer Science and Engineering*, pages 341–350. Birkhäuser, Boston, 2002.

23. J. Rooney. A comparison of representations of general screw displacements. *Environment and Planning B*, 5:45–88, 1978.

24. B. Rosenhahn. Pose estimation revisited. Technical Report 0308, PhD thesis, Christian-Albrechts-Universität zu Kiel, Institut für Informatik und Praktische Mathematik, 2003.

25. B. Rosenhahn, C. Perwass, and G. Sommer. Free-form pose estimation by using twist representations. *Algorithmica*, 38:91–113, 2004.

26. C.G. Small. *The Statistical Theory of Shape*. Springer-Verlag, New York, 1996.

27. G. Sommer. Algebraic aspects of designing behavior based systems. In G. Sommer and J.J. Koenderink, editors, *Algebraic Frames for the Perception-Action Cycle*, volume 1315 of *Lecture Notes in Computer Science*, pages 1–28. Proc. Int. Workshop AFPAC'97, Kiel, Springer–Verlag, Heidelberg, 1997.

28. G. Sommer, editor. *Geometric Computing with Clifford Algebras*. Springer-Verlag, Heidelberg, 2001.

29. J. Turski. Projective fourier analysis for patterns. *Pattern Recognition*, 33:2033–2043, 2000.

30. J. Turski. Geometric fourier analysis of the conformal camera for active vision. *SIAM Review*, 46(2):230–255, 2004.

31. T.P. Wallace and O.R. Mitchell. Analysis of three-dimensional movements using fourier descriptors. *IEEE Trans. Pattern Analysis and Machine Intell.*, 2(6):583–588, 1980.

32. M. Yaglom. *Felix Klein and Sophus Lie*. Birkhäuser, Boston, 1988.

33. C.T. Zahn and R.Z. Roskies. Fourier descriptors for plane closed curves. *IEEE Trans. Computers*, 21:269–281, 1972.

Recent Applications of Conformal Geometric Algebra

Anthony Lasenby

Cavendish Laboratory, University of Cambridge,
Cambridge CB3 0HE, UK
a.n.lasenby@mrao.cam.ac.uk
http://www.mrao.cam.ac.uk/~anthony

Abstract. We discuss a new covariant approach to geometry, called conformal geometric algebra, concentrating particularly on applications to projective geometry and new hybrid geometries. In addition, a new method of working, which can achieve similar results, but using only one extra dimension instead of two, is also discussed.

1   Introduction

Our aim is to show how useful the covariant approach is in geometry. Like Klein, we will consider a particular geometry as a class of transformations which leave a particular object invariant. However, by starting with null vectors and using a conformal representation (as in the conformal representation of Euclidean geometry introduced by Hestenes [1]), we will see that covariance (retention of same physical content after transformation) is also of fundamental importance. Keeping covariance, and using not just null vectors but the whole representation space, we will be able to step outside conformally flat geometries to include projective geometry as well. This occurs in a fashion different from that which Klein envisaged and retains, at least in this approach, a fundamental role for null vectors. In the final sections, we discuss a new approach to the area of conformal geometric algebra, which needs only one extra dimension in which to carry out computations, as against the two needed in the Hestenes approach. This has potential savings in computational time, and in physics applications, seems a more natural framework in which to work. The examples we use are mainly drawn from geometry and physics. However, as already demonstrated in [2,3] and other papers, the conformal geometric algebra approach has many applications in computer graphics and robotics. Thus the methods developed here may similarly be expected to be useful in application to this area, and we illustrate with the specific example of interpolation.

In order to allow space here to adequately describe new material, we refer the reader to [3] (this volume), for a summary of the conformal geometric algebra approach to Euclidean geometry. We also assume that the reader is familiar with

H. Li, P. J. Olver and G. Sommer (Eds.): IWMM-GIAE 2004, LNCS 3519, pp. 298–328, 2005.

the extension of this to non-Euclidean spaces. This was first carried out in [4] and [5, 6, 7, 8]. The full synthesis, whereby exactly the same techniques and notation can be used seamlessly in all of spherical, hyperbolic and Euclidean geometry, has been described in [2, 9, 10, 11].

Despite not giving a beginning summary of the mathematics, it is important, to avoid any confusion, to stress what notation we will be using. We follow here the notation for the two additional vectors that are adjoined to the base space, which was used by Hestenes & Sobczyk in [12], Chapter 8. These two vectors are e, with $e^2 = 1$, and $\bar{e}$, with $\bar{e}^2 = -1$. The null vectors which are formed from them are $n = e + \bar{e}$ and $\bar{n} = e - \bar{e}$, and in the conformal approach to Euclidean geometry play the roles of the point at infinity, and the origin, respectively. This differs from the notation used subsequently by Hestenes in [1].

We should note, as a global comment, that no pretence of historical accuracy or completeness is made as regards the descriptions of geometry and geometrical methods given here. Almost certainly, all the geometrical material contained here, even when described as 'new', has been discovered and discussed many times before, in a variety of contexts. The aim of this contribution is therefore to try to show the utility, and unification of approach, that conformal geometric algebra can bring.

2   Comparison of Kleinian and Covariant Views of Geometry

The Kleinian view of geometry is illustrated schematically in Fig. 1. Here the primary geometry, from which others are derived, is projective geometry, as exemplified for example by the geometry of the real projective plane in n dimensions, $\mathbb{RP}^n$ (see e.g. [13]). Projective geometry is primary, in that the largest

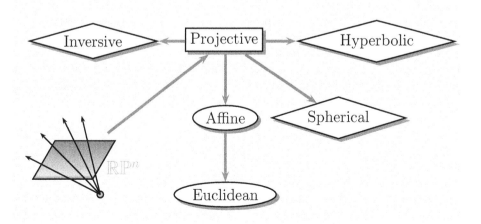

Fig. 1. The Kleinian view of geometry

group of transformations is allowed for it. Other geometries are then derived from it via restrictions which are imposed on the allowed transformations. For example spherical geometry is the subgroup of projective geometry which preserves the Euclidean angle between two lines through the origin, while affine and then Euclidean geometry represent increasing degrees of restriction in the behaviour under transformation of 'ideal points' at infinity (see [13, 14, 15]). The view of geometry which we begin with here, however, puts Euclidean, spherical and hyperbolic in a different relationship. As stated above, let us denote by e and $\bar{e}$, the additional vectors which we adjoin to the three basis vectors, e_1, e_2 and e_3 of ordinary Euclidean space, in order to construct the conformal geometric algebra, and let $n = e + \bar{e}$. Then, in the null vector representation

$$
\begin{aligned}
&\text{Euclidean geometry} \leftrightarrow \text{set of transformations which preserve } n \\
&\text{Hyperbolic geometry} \leftrightarrow \text{set of transformations which preserve } e \\
&\text{Spherical geometry} \leftrightarrow \text{set of transformations which preserve } \bar{e} \\
&\text{Inversive geometry} \leftrightarrow \text{set of transformations of the form } S\mathcal{O}S,
\end{aligned}
\tag{1}
$$

where in the last line S represents a sphere in one of the above geometries, and $\mathcal{O}$, the object being reflected in the sphere is one of a point, line, circle, plane or sphere (see [2]). Euclidean, hyperbolic and spherical geometry thus appear as much more similar to each other in this approach — effectively on the same level as each other, rather than being quite different specialisations of projective geometry. Also inversive geometry is reduced to the study of generalised reflections in each space.

What is the significance of *covariance* in this approach? The key idea is that we can freely write down expressions composed of elementary objects in the geometry, such as lines, planes, spheres and circles (including non-Euclidean varieties of each), and that these remain geometrically meaningful after a transformation of the appropriate kind for the geometry. An example can illustrate this. Let P be a point and L a line. Consider the new point $Q = LPL$, the (possibly non-Euclidean) reflection of P in the line L. Suppose we carry out a transformation in the geometry, given by the rotor R (e.g. this might be a translation, rotation, or dilation) to get a new point $P' = RP\tilde{R}$, and line $L' = RL\tilde{R}$. The question is, does the new point we obtain by carrying out the reflection process with the transformed P' and L', correspond to the transformation of the old Q? That is, is the following true:

$$
RQ\tilde{R} = L'P'L'?
\tag{2}
$$

The r.h.s. is $(RL\tilde{R})(RP\tilde{R})(RL\tilde{R})$, which since the algebra is associative and $R\tilde{R} = \tilde{R}R = 1$ for a rotor, means that we obtain $RQ\tilde{R}$ as required. This may seem trivial, but is the essence of the advantages of conformal geometric algebra for geometry. We can compose 'objects' into expressions in the same way as words are composed into sentences, and the expressions will be geometrically meaningful if they are covariant.

3   Extending the Framework

We have seen that with this conformal framework, where null vectors in our 5d algebra represent points, we can nicely represent Euclidean, spherical, inverse and hyperbolic geometry. But, is there any way of extending it to encompass Affine and Projective geometries? This would be very desirable in many fields. For example, in computer vision, projective geometry is used a great deal, partly for the obvious reason that it is related to the images projected down from the 3-dimensional world to the camera screen or retina, but also used because a projective approach allows the linearization of transformations that would otherwise be non-linear. In particular, by adding a third coordinate to the usual two in a camera plane, translations in 2-space become homogeneous linear transformations in 3-space. Also, incidence relations between points, lines and planes can be expressed in an efficient manner. However, using a purely linear algebra projective approach, one cannot extend these incidence relations to include conic sections, and moreover *metric* relations between objects, for example relations between conic sections and sums and differences of lengths, are lost, since a metric is not accessible, except by going to a Euclidean specialization.

If it were possible to extend the conformal approach to include projective and affine geometry, however, it might be possible to remedy these deficiencies, and to include incidence relations between lines and conic sections, and metric relations between points. We now show, that this is indeed possible, and that the method of achieving it has a natural structure which emphasizes the covariance properties of objects. The resulting theory does not quite achieve all we might ideally want, since although lines can be intersected with conic sections in a useful way, conic sections cannot be intersected with each other in the same fashion which is possible for the intersection of circles and spheres. An extension of this kind does not seem to be possible. However, that we have still done something significant is clear when one applies the principles we set up in the section, with an underlying Euclidean geometry, to the case where the underlying geometry is hyperbolic or spherical. Amazingly, all the same constructions and types of results still hold, thus generating what appear to be completely new geometries, with possible applications in many areas.

To start with then, we consider how to encode projective geometry in the null-vector description of Euclidean space already summarised.

Clearly we cannot embed projective geometry in our conformal setup as it stands. The essence of conformal mappings is that they preserve angles, and the essence of projective and affine transformations is that they do not preserve angles! Thus we have to step outside the framework we have set up so far. This framework has at its core the assumption that points in a Euclidean space $\mathbb{R}^{(p,0)}$ are represented by null vectors in the space $\mathbb{R}^{(p+1,1)}$, two dimensions higher, A projective transformation cannot be represented by a function taking null vectors to null vectors in this space, since then it would be a conformal transformation in the Euclidean space. The only possible conclusion, is that projective transformations must involve non-null vectors. But then how can we relate the results back to points in Euclidean space?

The key to the resolution of this is the following result. Let Y be any vector in $\mathbb{R}^{(p+1,1)}$ such that $Y \cdot n \neq 0$. Then there exists a unique decomposition of Y of the form

$$Y = \alpha X + \beta n \qquad (3)$$

where α and β are scalars, X is null, and $X \cdot n = -1$. The proof of this is constructive. Assume it can be done, then

$$Y^2 = 2\alpha\beta X \cdot n = -2\alpha\beta$$
$$Y \cdot n = \alpha X \cdot n = -\alpha \qquad (4)$$

This entails

$$\alpha = -Y \cdot n, \qquad \beta = \frac{Y^2}{2Y \cdot n}$$

$$X = -\left(\frac{Y}{Y \cdot n} - \frac{Y^2}{2(Y \cdot n)^2} n\right) = -\frac{YnY}{2(Y \cdot n)^2} \qquad (5)$$

The values just given for α, β and X, satisfy $X \cdot n = -1$ and $\alpha X + \beta n = Y$, and so this proves the decomposition can be carried out and that the answer is unique. Up to scale, we see that X is the reflection of the point at infinity n, in the (generally non-null) vector Y.

What we have found then, is a unique mapping from any non-null vector Y satisfying $Y \cdot n \neq 0$, to a null vector X. Thus any non-null vector of this kind can be taken as representing a point in the finite Euclidean plane. Projective transformations can therefore work by mapping null vectors representing Euclidean points to non-null vectors — thereby avoiding the problem that 'null' $\mapsto$ 'null' would be a conformal transformation — and then these non-null vectors can be taken as representing Euclidean points, thus overall achieving a mapping from points to points. Note if the vector Y is such that $Y \cdot n = 0$, so that we cannot employ the above construction, then we will need some supplementary rule for what to do, and this is discussed further below.

The mappings we use will be linear mappings h taking $\mathbb{R}^{(p+1,1)} \mapsto \mathbb{R}^{(p+1,1)}$. Our main assertion is as follows:

Projective transformations correspond to linear functions which preserve the point at infinity up to scale, i.e. for which $h(n) \propto n$.

Using functions of this kind, points in Euclidean space are mapped to points in Euclidean space via

$$h(X) = Y = \alpha X' + \beta n \to X' \qquad (6)$$

which defines a mapping from null X to null X'. If the value of α is independent of X, the mapping is an *Affine* transformation.

The mapping we have described,

$$h(X) = \alpha X' + \beta n \qquad (7)$$

is implicit as regards obtaining X', but using the result above we can get a properly normalised point explicitly via

$$X' = -\frac{1}{[h(X) \cdot n]^2} h(X) n h(X) \qquad (8)$$

i.e. (up to scale) we just need to reflect the point at infinity in $h(X)$. This is just an easy way of carrying out the construction; the key question is the covariance of the $h(X) = \alpha X' + \beta n$ mapping itself. As already described, covariant means: do we get the same result if we transform first then carry out our mapping as if we map first then transform? One finds:

$$Rh(X)\tilde{R} = \alpha RX'\tilde{R} + \beta Rn\tilde{R}$$
$$= \alpha RX'\tilde{R} + \beta n \tag{9}$$

So this works due to the invariance of the point at infinity under the allowed rotors. We note that we could allow dilations as well, which rescale n. These lead to the same output vector $RX'\tilde{R}$, since β would then just be rescaled.

3.1   Boundary Points

Before giving examples of how all this works, we first consider the question raised above: what null vector should we associate with a non-null Y if $Y \cdot n = 0$, and what then happens to our projective mapping?

Now in the case of hyperbolic geometry, we know that the boundary of the space (the edge of the Poincaré disc in 2d) is given by the set of points satisfying the covariant constraint $X \cdot e = 0$. This is covariant since the allowed rotors keep e invariant in the hyperbolic case. Thus the boundary points are mapped into each other under translations and rotations. In [10] it is shown that for this hyperbolic case, as well as for de Sitter and anti-de Sitter universes, the boundary points can be represented by null vectors. For example, for the Poincaré disc, the boundary points are of the form

$$X = \lambda(\cos\phi e_1 + \sin\phi e_2 + \bar{e}) \tag{10}$$

where λ is a scalar, and we have a whole set of them — one for each 'direction' in the underlying space.

In the Euclidean case, however, we have not yet achieved an equivalent to this, having only the single vector n available to represent points at infinity. The resolution to this is now clear — the vectors representing the boundary points must be the set of vectors satisfying the covariant constraint $Y \cdot n = 0$, but we must admit as solutions non-null as well as null vectors. Thus the boundary points are the set of vectors

$$Y = a + \beta n \tag{11}$$

where a is a Euclidean vector. If we impose the further (covariant) constraint $Y^2 = 1$, then this reduces the set of vectors to just a direction in Euclidean space, plus an arbitrary multiple of n. This seems a very sensible definition for a set of 'points at infinity'.

We can summarize this setup as follows.

$$\begin{array}{lll} \text{Interior points satisfy:} & X \cdot n = -1, & X^2 = 0 \\ \text{Boundary points satisfy:} & Y \cdot n = 0, & Y^2 = 1 \end{array} \tag{12}$$

There is quite a neat symmetry in this. Note that the Euclidean transformations of rotation, translation and even dilation, when carried out using the rotor approach, all preserve the condition $Y^2 = 1$. (This is because only rotation affects the Euclidean a part of the vector.) In the case of non-Euclidean geometry, e.g. for the Poincaré disc, the boundary points, being null, are only preserved by the allowed transformations up to scale.

It is now obvious what to do if the mapping $h(X)$ produces a point Y such that $Y \cdot n = 0$. We just accept this non-null point Y as the output of the mapping, instead going through a further stage of finding an associated null vector X', for which the construction is impossible in this case. The non-null point Y contains more information than simply assigning a null vector output of n, since as well as seeing that we are at infinity, we can also tell from what direction infinity is approached.

3.2   Specialization to Affine Geometry

Above we made the brief comment that restricting α to a constant in equation (6), was the same as restricting the projective transformations to affine transformations. Armed with the new concept of 'boundary points' we can now understand this further. We give an alternative definition of affine transformations as follows:

> Affine transformations correspond to the subset of projective transformations which map boundary points to boundary points, up to scale. Thus h is affine if $Y \cdot n = 0 \implies h(Y) \cdot n = 0$.

In this form we can recognize the classical approach to restricting projective transformations to affine form. E.g. in 2d, one would ask that the so called 'line at infinity' be preserved , while in 3d it would be the 'plane at infinity' [15, 14]. These objects correspond to our 'boundary points'.

To see the link with our previous version of an affine transformation, we can argue as follows. Let Z be a vector in $\mathbb{R}^{(p+1,1)}$ which is general except that it satisfies $Z \cdot n = 0$. A fully general vector in $\mathbb{R}^{(p+1,1)}$ can be derived from this by adding an arbitrary multiple of $\bar{n}$, and we set $Y = Z + \theta \bar{n}$, where θ is a scalar. We then have the following results:

$$h(Y) = h(Z) + \theta h(\bar{n})$$
$$h(Y) \cdot n = \theta h(\bar{n}) \cdot n$$
$$Y \cdot n = 2\theta \tag{13}$$
$$h(Y) \cdot n = \left(\frac{1}{2} h(\bar{n}) \cdot n \right) Y \cdot n$$

We note the pre-factor in front of $Y \cdot n$ in the last equation is independent of Y for a given h. Now in equation (6), we can see that $\alpha = (h(X) \cdot n)/(X' \cdot n)$. Thus providing X and X' are both normalised, which we are assuming, we can deduce that for an affine transformation α is constant, as stated.

3.3 Specialization to Euclidean Geometry

An aspect of the conventional approach to projective geometry which often appears rather mysterious, particularly in applications to computer vision, is the use of ideal points, i.e. points lying in the line or plane at infinity, with complex coordinates. These are used in particular for encoding the specialization of affine geometry to Euclidean geometry, where the latter is defined to include the set of similitudes, as well as rotations and translations.

In this conventional approach, the Euclidean geometry of the plane is recovered from projective geometry as the set of affine transformations which leave the points $(1, i, 0)$ and $(1, -i, 0)$ invariant. These are called the ideal points $\mathcal{I}$ and $\mathcal{J}$ (where we use calligraphic letters to avoid confusion with our pseudoscalar I).

Similarly the Euclidean geometry of 3d space, is recovered by asking that an affine transformation leaves invariant a certain quadratic form involving complex ideal points, known as the *absolute conic* [15, 14]. This form is

$$\sum_{i=1}^{4} x_i^2 = 0 \tag{14}$$

where $x_4 = 0$ in order to place the set of points in the ideal plane at infinity.

The form of geometric algebra we use restricts itself to an underlying real space only, and does not make use of an uninterpreted scalar imaginary i. By taking this approach, one can often find a geometrical meaning for the various occurrences of i in the conventional approaches — a meaning that is often obscured without geometric algebra. It is thus interesting to see if the geometric algebra approach can shed light on the points $\mathcal{I}$ and $\mathcal{J}$, and give some increased understanding of the absolute conic.

Since $h(n)$ is proportional to n and maps boundary points to boundary points (since it is assumed affine), and also since the length of $a + \beta n$ does not depend on β when a is Euclidean, we can specialize our discussion of the ideal points $\mathcal{I}$ and $\mathcal{J}$ to a simple linear function f say, taking $\mathbb{R}^2$ to $\mathbb{R}^2$, which is the restriction of h to these spaces.

In terms of the orthogonal basis $\{e_1, e_2\}$, this has the matrix equivalent

$$f_{ij} = f(e_i) \cdot e_j \tag{15}$$

We now write down the conventional matrix relation for the ideal points and reconstitute from this what it means in a GA approach.

Saying that $\mathcal{I}$ is invariant under f means conventionally that

$$\begin{pmatrix} f_{11} & f_{12} \\ f_{21} & f_{22} \end{pmatrix} \begin{pmatrix} 1 \\ i \end{pmatrix} = z \begin{pmatrix} 1 \\ i \end{pmatrix} \tag{16}$$

where z is some complex number. Now complex numbers are the spinors of the geometric algebra of $\mathbb{R}^2$, and we accordingly represent $z = x + iy$ by $Z = x + I_2 y$ where $I_2 = e_1 e_2$ is the pseudoscalar in 2d. Our translation of the invariance condition is thus that

$$f(e_1) \cdot e_1 + I_2 f(e_1) \cdot e_2 = Z$$
$$f(e_2) \cdot e_1 + I_2 f(e_2) \cdot e_2 = Z I_2 \qquad (17)$$

Now $I_2 f(e_1) \cdot e_2 = \langle I_2 f(e_1) e_2 \rangle_2 = \langle -f(e_1) I_2 e_2 \rangle_2 = -f(e_1) \wedge e_1$. We deduce

$$e_1 f(e_1) = Z, \quad e_1 f(e_2) = Z I_2 \qquad (18)$$

Premultiplying by e_1 we obtain

$$f(e_1) = \tilde{Z} e_1 \quad \text{and} \quad f(e_2) = \tilde{Z} e_2 \qquad (19)$$

from which we can deduce that

$$f(a) = \tilde{Z} a = a Z \quad \forall \text{ Euclidean vectors } a \text{ in } \mathbb{R}^2 \qquad (20)$$

A general spinor Z or $\tilde{Z}$ applied to the left or right of a dilates a by $|Z|$ and rotates it through $\tan^{-1}(\langle Z \rangle_2 / \langle Z \rangle_0)$, so we have indeed found that demanding that the ideal point $\mathcal{I}$ is preserved is equivalent to specializing to Euclidean transformations. We note that requiring that $\mathcal{J}$ is preserved as well as $\mathcal{I}$, is redundant (which it is also in the conventional approach, in fact) since the invariance condition

$$f(e_1) \cdot e_1 - I_2 f(e_1) \cdot e_2 = Z'$$
$$f(e_2) \cdot e_1 - I_2 f(e_2) \cdot e_2 = -Z' I_2 \qquad (21)$$

reconstitutes to give

$$f(a) = a \tilde{Z}' = Z' a \quad \forall \text{ Euclidean vectors } a \text{ in } \mathbb{R}^2 \qquad (22)$$

just identifying Z' as $\tilde{Z}$, and giving no new information.

So having seen that (16) is directly equivalent to (20) can we say whether the GA formulation offers any improvement over the conventional one? We think the answer is yes, in that (20) more directly reveals the geometrical origins of what is going on, and also clearly refers homogeneously to all vectors in the space, rather than singling out particular points as being preserved. It also avoids the need for introducing complex coordinates of course.

The Absolute Conic. We now consider a similar process in 3d. We will again let f be the specialization of h to the Euclidean portion of the space, and we will let z_i and z_i' ($i = 1, 2, 3$) be general complex numbers. In the conventional approach, 'leaving the absolute conic invariant' amounts to the following:

The matrix $||f_{ij}||$ is such that the output vector (z_1', z_2', z_3') in the relation

$$\begin{pmatrix} z_1' \\ z_2' \\ z_3' \end{pmatrix} = \begin{pmatrix} f_{11} & f_{12} & f_{13} \\ f_{21} & f_{22} & f_{23} \\ f_{31} & f_{32} & f_{33} \end{pmatrix} \begin{pmatrix} z_1 \\ z_2 \\ z_3 \end{pmatrix} \qquad (23)$$

satisfies $(z_1')^2 + (z_2')^2 + (z_3')^2 = 0$ if $z_1^2 + z_2^2 + z_3^2 = 0$.

Framed like this, the statement looks rather odd, since we have clearly introduced a non-Hermitian metric on the space of complex vectors. In the GA approach, we can make more sense of things since we have available entities which are naturally null. For example, consider the vector and bivector combination $e_1 + I_3 e_2$, where I_3 is the pseudoscalar for $\mathbb{R}^3$. Then $(e_1 + I_3 e_2)^2 = 0$. Such combinations arise naturally in electromagnetism for example, where e_1 would be interpreted as a 'relative vector' and $I_3 e_2$ as a relative bivector', with both actually being bivectors in spacetime. (For example, in a spacetime basis $\{\gamma_0, \gamma_i\}$ we would identify $e_i = \gamma_i \gamma_0$ and $I_3 = e_1 e_2 e_3 = \gamma_0 \gamma_1 \gamma_2 \gamma_3$.) Taking the sum of them is natural in forming the Faraday electromagnetic bivector F, where the two parts would then correspond to the electric and magnetic fields, and the null condition is that $F^2 = 0$.

Following this line, our reformulation of the notion of leaving the absolute conic invariant is as follows.

Let a and b be Euclidean vectors. Then we require the linear function f to satisfy the relation

$$(f(a) + I_3 f(b))^2 = 0 \iff (a + I_3 b)^2 = 0 \tag{24}$$

In this version it is easy to see the link with specializing to Euclidean transformations, since

$$\begin{aligned}(f(a) + I_3 f(b))^2 &= f^2(a) - f^2(b) + 2 I_3 f(a) \cdot f(b) \\ (a + I_3 b)^2 &= a^2 - b^2 + 2 I_3 a \cdot b\end{aligned} \tag{25}$$

$(a + I_3 b)^2$ is thus only zero for vectors a and b which are the same length and orthogonal, and f is required to preserve both these features. This pins down f to a combination of rotation and dilatation.

To see how (23) is equivalent to (24) is also easy. If we set $z_i = x_i + i y_i$, $(i = 1, 2, 3)$, we simply need to make the identification $a = x_i e_i$, $b = y_i e_i$, and we see the same result is achieved in both cases.

To judge here whether the GA formulation is superior to the conventional one, is more difficult than in the 2d case, since the availability of a pseudoscalar which squares to -1 and commutes with all vectors in 3d, means that the GA and conventional formulations are closely parallel. However, the GA formulation does reveal that the complex points made use of in the notion of the absolute conic, which many have found rather mysterious, are actually vector plus bivector combinations, and that the idea of these being null takes on a familiar look when we see the parallel with electromagnetism, where the analogue of the second equation of (25) is

$$F^2 = 0, \quad \text{with } F = E + I_3 B \quad \Longrightarrow \quad E^2 - B^2 = 0 \quad \text{and} \quad E \cdot B = 0 \tag{26}$$

which is the statement that the Faraday bivector is null if the electric and magnetic fields have the same magnitude and are orthogonal (as in an electromagnetic plane wave for example).

4   Lines and Conics

Having considered the specialization to Affine and Euclidean geometry in some detail, we now give a short, schematic account of some remaining major features of projective geometry. A more comprehensive account will be given in a future publication, and some applications and further details are given in [3].

We need to see what happens to lines under an h transformation. This works well, since the requirement $h(n) = n$ means that if we are transforming the line $L = A \wedge B \wedge n$, we obtain

$$h(L) = h(A) \wedge h(B) \wedge n$$
$$= \alpha_A \alpha_B A' \wedge B' \wedge n, \tag{27}$$

with only α appearing since the wedge kills off the β parts.

Thus we get a line through the transformed points, while the factor $\alpha_A \alpha_B$ contains information that can be expressed via cross-ratios.

For conics, if C is a multivector representing a circle or sphere in Euclidean space, then the equation $Y \wedge h(C) = 0$, where $Y = \alpha X' + \beta n$, and $h^{-1}(Y)$ is null, successfully produces a solution space corresponding to a general conic, which we can think of as the image of C under h. However, one cannot use this approach to intersect two different conics, unless it happens that they correspond to the same h function acting on different initial circles or spheres.

h itself for an object transforms covariantly as

$$h(X) \mapsto R h(\tilde{R} X R) \tilde{R}. \tag{28}$$

This means that we can use rotors to achieve transformation of any conic, or quadric in 3d, thus achieving one of our main objectives. For example rotation and translation of a quadric in 3d via use of rotors, is shown in Fig. 2(a) and (b).The rotors can encompass all of rotation, translation and dilation, and thus we can position and scale any conic or quadric using rotor techniques.

The other success of this method is that one can use the machinery which enables intersection of any line or plane with a circle or sphere, to look at intersections of lines and planes with conics and quadrics. There are many applications of this in computer graphics.

The way we can do this is to use reflection of $h^{-1}(X')$ in n to bring back points X' to where they originated before application of h:

$$X = -\frac{1}{[h^{-1}(X') \cdot n]^2} h^{-1}(X') n h^{-1}(X') \tag{29}$$

A line transforms back under this to another line, whilst a conic transforms back to the underlying circle or sphere. We can then intersect the new line with the resulting circle or sphere, using the standard intersection techniques in conformal geometric algebra, and then forward transform again to find the solution to the original problem. An example of this is shown schematically Fig. 3, which whilst not looking very exciting, was worked out using this method, and thus constitutes proof that we can indeed intersect lines with arbitrary conics.

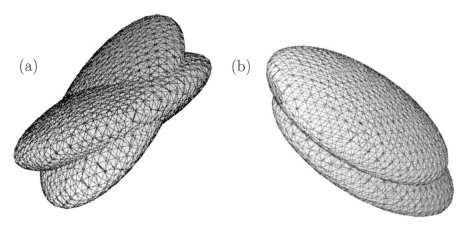

Fig. 2. Illustration of the result of using rotors to (a) rotate and (b) translate a quadric surface, defined by an h function, in $\mathbb{R}^3$

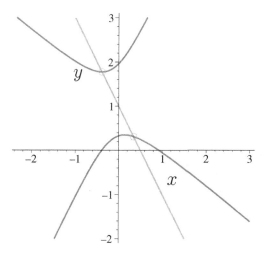

Fig. 3. Intersection of a line and a hyperbola using conformal method described in text

5   Creating New Geometries

We now reach our second overall topic in this brief survey. This is the use of conformal geometry to create new, hybrid geometries, which combine the features of both projective and non-Euclidean geometries. In particular, using this method can we carry out projective transformations *within* hyperbolic or spherical spaces. This is quite different from, for example, the representation of hyperbolic geometry in terms of projective geometry, as in the Klein approach.

The way we do this is illustrated in Fig. 4. We apply linear functions which now preserve e, or $\bar{e}$, rather than n, and wind back to null vectors using

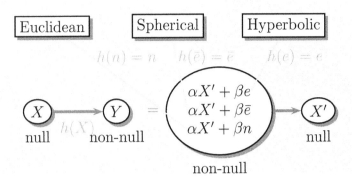

Fig. 4. Schematic diagram showing how the new hybrid geometries are created

$$h(X) = \alpha X' + \beta e \tag{30}$$

for the hyperbolic case, and

$$h(X) = \alpha X' + \beta \bar{e} \tag{31}$$

for the spherical case. Considering for example the hyperbolic case, we see that (30) implicitly defines a null vector X' for each X, and that this happens in a fully covariant manner, since now e is the object preserved under the relevant rotor operations.

So what might we expect to be able to do using this structure? Most importantly, we will want to see if it is possible to obtain a non-Euclidean version of the family of conics (quadrics in 3d), which are familiar in projective geometry as applied in Euclidean space. For the latter there are two separate ways of defining them. For example, an ellipse may be defined as the locus of points which have a constant summed distance to two fixed points (the focii) or alternatively via the projective transformation of a circle. It is a theorem for conics that these two approaches lead to the same object. Ideally then, for non-Euclidean conics, we will want both of the following being true and leading to the same object:

1. The sum or difference of non-Euclidean lengths being constant defines the locus;
2. The shape arises via a projective transformation (linear mapping in one dimension higher) of e.g. a circle.

So can we form for example a non-Euclidean ellipse? This would have constant sum of non-Euclidean distance to two given points (the non-Euclidean foci). Such an object can of course be defined, and we will end up with a certain definite locus. But will it also correspond to a projective transformation applied to a circle? Here is an example. Let us take as our h-function:

$$h(a) = a + \frac{1}{2}\delta\, a\cdot e_1 \bar{e} \qquad (32)$$

where δ is a scalar. This looks very simple, but the actual expressions implied by it are quite complicated. Temporarily writing $\boldsymbol{x} = (x, y)$ for the real space coordinates, and $\boldsymbol{x}' = (x', y')$ for those they are transformed to via (30), then, taking the appropriate branches of the square root functions, one finds:

$$\boldsymbol{x}' = 2\lambda^2\boldsymbol{x}/\left(x^2 + y^2 + \lambda^2 + \delta\,\lambda x + \right.$$
$$\left. \sqrt{x^4 + 2\,x^2y^2 - 2\,\lambda^2x^2 + 2\,x^3\delta\,\lambda + y^4 - 2\,\lambda^2y^2 + 2\,y^2\delta\,\lambda x + \lambda^4 + 2\,\lambda^3\delta\,x + \delta^2\lambda^2x^2}\right)$$
$$(33)$$

Thus there is a radial dilation, but with a factor which is a complicated function of position. Choose $\delta = 2.25$, and map the circle $r = 1/4$ using the above h function in the basic construction

$$h(X) = \alpha X' + \beta e \qquad (34)$$

This gives the result shown in Fig.5.

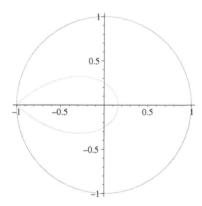

Fig. 5. Mapping of the circle $r = 1/4$ using the h-function of equation (32) with $\delta = 2.25$. As discussed in the text, this is actually a non-Euclidean parabola

This particular case is actually a limiting case of an non-Euclidean ellipse — one of the foci has migrated to infinity! Thus it should be regarded as a non-Euclidean parabola. We will not substantiate this here, but instead, to get to a more easily understandable version of an ellipse, let us try a slightly different form of transformation which still preserves e, namely

$$h(a) = a + \frac{1}{2}\delta\, a\cdot e_1\, e_1 \qquad (35)$$

This form leads to centered objects. As an example, choose $\delta = 2.24$, and map the circle $r = 1/4$ again. This leads to the locus shown in Fig.6. The explicit equations for (x', y') in this case are

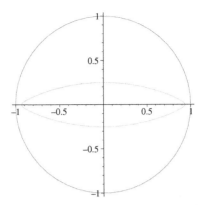

Fig. 6. Mapping of the circle $r = 1/4$ using the h-function of equation (35) with $\delta = 2.24$. This is a non-Euclidean ellipse

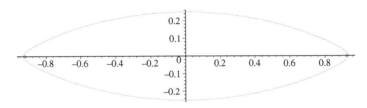

Fig. 7. The same non-Euclidean ellipse as in the previous figure, but with the locations of the non-Euclidean foci marked

$$
\begin{aligned}
x' &= \lambda^2 x (2 + \delta) / \left(x^2 + y^2 + \lambda^2 \right. \\
&\quad + \sqrt{x^4 + 2x^2 y^2 - 2x^2 \lambda^2 + y^4 - 2\lambda^2 y^2 + \lambda^4 - 4x^2 \lambda^2 \delta - \delta^2 \lambda^2 x^2} \Big) \\
y' &= 2\lambda^2 y / \left(x^2 + y^2 + \lambda^2 \right. \\
&\quad + \sqrt{x^4 + 2x^2 y^2 - 2x^2 \lambda^2 + y^4 - 2\lambda^2 y^2 + \lambda^4 - 4x^2 \lambda^2 \delta - \delta^2 \lambda^2 x^2} \Big)
\end{aligned}
\tag{36}
$$

Now can we tie these structures into the concept of constant sum of non-Euclidean distances from two foci positions? If so where are they? One finds that indeed foci can be found such that sums of non-Euclidean distances agree with the h function method. However, this is quite complicated to prove, so we shall not go through this here, but instead just illustrate in the context of the example just given. The foci are in fact at $x = \pm 0.925$, whilst the intersections of the 'ellipse' with the x axis are at $x = \pm 0.934$. The relationship of the foci to the ends of the axes is illustrated in Fig.7, and of course in terms of our normal view of the relationship between an ellipse and its foci looks rather odd. However, the total sum of non-Euclidean distances to the two foci is genuinely constant (at 3.372).

Fig. 8. A non-Euclidean ellipse/hyperbola displayed on the sphere. The curve is the image of the circle r=1/4, and uses $\delta = 10$ in the h-function given in equation (37)

The same procedures can also be carried out for spherical, as against hyperbolic geometry. In this case we need to preserve $\bar{e}$ rather than e. Also, here it is possible to visualise the results rather more easily, by directly lifting the loci up onto a sphere for display purposes. Shown in Fig. 8 is the typical type of locus on the sphere that can result. A very interesting feature of such curves on the sphere is that they are simultaneously ellipses and hyperbolae! That is, the curve shown (in multiple views) in Fig. 8 satisfies simultaneously that the sum of distances (in a great circle sense on the surface of the sphere) to two given points is constant, and also that the difference of distances to two points is also constant. One can see that indeed, from some viewpoints the curve looks like an ellipse, whilst for others it looks like a hyperbola. Defining these curves this way, can of course be carried out without any recourse to geometric algebra. What is new here, is that these curves are actually constructed via a 'projectivity' acting on a circle in a non-Euclidean spherical space. The particular projectivity used corresponds to an h function

$$h(a) = a + \frac{1}{2}\delta\, a{\cdot}e_1 e \tag{37}$$

and the curve plotted is the image of a circle of radius 1/4, with $\delta = 10$.

Overall, as regards projective geometry, we see that the GA approach provides a flexible framework enabling a wide range of geometries to be treated in a unified fashion. Everything said here extends seamlessly to 3d and higher dimensions, although a lot of work remains to be done to explore the structures which become available this way.

6   The '1d Up' Approach to Conformal Geometry

In the approach to conformal geometry advocated by David Hestenes, two extra dimensions are necessary. However, do we really need these? From investigations that are still ongoing, it appears that for physics problems, just one extra dimension may be sufficient, and indeed more natural, while for pure geometry aspects, occasionally having in mind a further dimension is useful for abstract manipulation, but again one may not need it for actual calculation.

In terms of the physics applications, what makes this possible is using de Sitter or anti-de Sitter space as the base space, rather than Lorentzian spacetime. The rotor structure in these spaces allows us to do rotations and translations as for the 'Euclidean' (here Lorentzian) case, so that as regards the physics we can do everything we want, aside perhaps from dilations. (These, however, can be taken care of by an h-function, if necessary.) This can be achieved using only one extra dimension, namely $\bar{e}$ in the de-Sitter case, and e in the anti-de Sitter case, as we show below.

However, in the current conformal geometric algebra approach to Euclidean space, then conceptually one is always using both extra dimensions, since $\bar{n}$ is required as the origin, and n is integral in the translation formulae.

It may be objected that we do not wish to work with e.g. de Sitter space as a base space, since then there is intrinsic curvature in everything, and while this may be alright, maybe even desirable for physics purposes, it does not make any sense for engineering applications.

However, it turns out that in applications we can take a Euclidean limit at the end of the calculations, such that the curvature is removed, but we have nevertheless derived what we wanted whilst still only going up 1 dimension, so actually this is not a problem.

To be more specific on what we are going to do here, we shall start with a physical context, and embed spacetime in a 5d space, in which conceptually we are keeping e constant, therefore it is de Sitter space. However, we never actually need to bring e in to do the physics. Our basis vectors are thus

$$e_0(\equiv \gamma_0), \quad e_1(\equiv \gamma_1), \quad e_2(\equiv \gamma_2), \quad e_3(\equiv \gamma_3), \quad \text{and} \quad \bar{e} \tag{38}$$

with squares $+1, -1, -1, -1, -1$ respectively.

Let $Y = y^\mu e_\mu$ be a general vector in this space. For our purposes here, it is sufficient to restrict ourselves to the part of Y space in which $Y^2 < 0$. Our fundamental representation can be expressed via the null vector X used in our previous 6d embedding space for the de Sitter case [10]. This was scaled so that $X \cdot e = -1$. Our new representation works via

$$\hat{Y} = X + e \tag{39}$$

which then via X corresponds to a point in 4d. $\hat{Y}$ is a scaled version of Y satisfying

$$\hat{Y}^2 = -1 \tag{40}$$

We note that $\hat{Y} \cdot e = X \cdot e + e \cdot e = -1 + 1 = 0$, so that indeed $\hat{Y}$ does not contain e. Moreover,

$$\hat{Y}^2 = (X + e)^2 = 2X \cdot e + e^2 = -1 \tag{41}$$

consistent with the definition of $\hat{Y}$.

Of course we could also go directly from a general Y point to a point in 4d de Sitter space, without using X as an intermediary. The formulae for this, which we give assuming $y^4 > 0$, are quite simple:

$$x^\mu = \frac{\lambda y^\mu}{l + y^4}, \qquad \mu = 0, 1, 2, 3 \tag{42}$$

where $l = \sqrt{-Y^2}$. We can see that the x^μ so defined are homogeneous of degree 0 in the y^μ, and therefore depend only on the direction of Y, $\hat{Y}$, in agreement with the definition via X.

6.1 Geometry

As a start, we need to reassure ourselves that the conformal geometry we carried out in 6 dimensions in e.g. [10], is all still possible and works here in 5d.

We have seen so far that points in spacetime are no longer represented by null vectors, but by *unit* vectors, the $\hat{Y}$ discussed above. However, null vectors still turn out to be very important — these are the vectors which represent the boundary points. In physical terms, they are *null momenta*, and we shall label boundary points as P to emphasise this. To convince ourselves that the boundary points are null, as in the treatment in [10], even though the interior points have unit square (actually negative unit square, but the same principle), we can take a limit as an interior point approaches the boundary.

Using the notation for coordinates as in [10], consider as an example the $\hat{Y}$ corresponding to the point

$$x = (t, x, y, z) = (-\lambda(1 - \delta), 0, 0, 0) \tag{43}$$

This is

$$\hat{Y} = \frac{1}{\delta(2 - \delta)}((2 - 2\delta + \delta^2)\bar{e} - 2(1 - \delta)e_0) \tag{44}$$

It is easy to verify that this indeed satisfies $\hat{Y}^2 = -1$. Taking the limit as $\delta \to 0$ we obtain

$$\hat{Y} \to \frac{1}{\delta}(\bar{e} - e_0) \tag{45}$$

as the first term of an expansion of $\hat{Y}$ in δ. This tells us that indeed $\hat{Y}$ is tending to a null vector as x tends to the lower boundary. We decide on a choice of scale for this particular null vector by writing

$$P = P_0 = \bar{e} - e_0 \tag{46}$$

for this point. We will define the corresponding point on the top boundary by $P_0' = \bar{e} + e_0$ (see Fig. 9).

One of the main features of the Hestenes approach to conformal geometry, is that it encompasses *distance geometry* [16], in that inner products of null vectors correspond to Euclidean distances between points. In [10] this was extended to the non-Euclidean case. He we can achieve exactly the same by noting that for

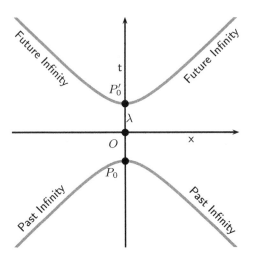

Fig. 9. The boundaries of de Sitter space (blue lines) in our conformal setup. The positions of the 4d points corresponding to the null momenta $P_0 = \bar{e}-e_0$ and $P_0' = \bar{e}+e_0$ are also shown

two points $\hat{Y}_1$ and $\hat{Y}_2$, with corresponding null vectors X_1 and X_2, we have from equation (39) that

$$
\begin{aligned}
\hat{Y}_1{\cdot}\hat{Y}_2 &= (X_1 + e){\cdot}(X_2 + e) \\
&= X_1{\cdot}X_2 - 1
\end{aligned}
\tag{47}
$$

Thus the new version of equation (3.6) from [10] is

$$
d(x_1, x_2) = \begin{cases}
\lambda \sinh^{-1}\sqrt{-(\hat{Y}_1{\cdot}\hat{Y}_2 + 1)/2} & \text{if } \hat{Y}_1{\cdot}\hat{Y}_2 < -1, \\
0 & \text{if } \hat{Y}_1{\cdot}\hat{Y}_2 = -1, \\
\lambda \sin^{-1}\sqrt{(\hat{Y}_1{\cdot}\hat{Y}_2 + 1)/2} & \text{if } \hat{Y}_1{\cdot}\hat{Y}_2 > -1.
\end{cases}
\tag{48}
$$

Note that $\hat{Y}_1{\cdot}\hat{Y}_2 = -1$ is the condition for x_1 and x_2 to lie on each others light cones.

We now need to decide what operations can be carried out on points in our new representation. Equation (39) shows that any operation in the old null vector approach which preserves e will transfer covariantly to the new approach. This means that the three basic operations discussed in [10] will all work here, namely:

$$
\begin{array}{c}
\text{Rotations, carried out with rotors with bivector generators} \quad e_\mu e_\nu \\
\text{Translations, carried out with rotors with bivector generators} \quad \bar{e}\, e_\nu \\
\text{Inversions, carried out via} \quad \text{'object'} \mapsto e\,\text{'object'}\, e
\end{array}
\tag{49}
$$

(Here μ and ν belong to the set $\{0, 1, 2, 3\}$.) We see that in terms of their effect on a point, inversions correspond to $\hat{Y} \mapsto -\hat{Y}$. The geometrical significance of this is important physically, but would take us too far afield to discuss further here.

In terms of building up a geometry, the next step is to find out what lines (geodesics) correspond to. In the conformal geometry approach using two extra dimensions, a d-line in de Sitter space through the points with null vector representatives A and B is $A \wedge B \wedge e$ (see [10]). For us, lines become bivectors, and if A and B are now the *unit* vector representatives, then the (un-normalised) line through them is represented by

$$L = A \wedge B \tag{50}$$

In particular, if the point Y lies on the line, then

$$Y \wedge L = Y \wedge A \wedge B = 0 \tag{51}$$

This corresponds to how we would represent a line in projective geometry. However, even though we only use the same dimensionality as would be used in projective geometry, the availability of the operations in equation (49) means we can go far beyond the projective geometry in terms of how we can manipulate lines and points. E.g. the rotor for translation through a timelike 4d vector a is (see [10])

$$R_T = \frac{1}{\sqrt{\lambda^2 - a^2}} (\lambda + \bar{e}a) \tag{52}$$

Applying this to L, we get a translated line, in exactly the same way as we would get if we applied it to the null vector version, except here we only have to work in 5d geometry rather than 6d. This is a computational advantage to having lines as bivectors rather than trivectors. Two more profound advantages are as follows.

A problem of the null vector representation is that if two null vectors are added together, then we do not in general get another null vector. Thus we cannot form weighted sums of points, in the same way as one could in projective geometry. However, here we can. If we take

$$Y_3 = \alpha \hat{Y}_1 + \beta \hat{Y}_2 \tag{53}$$

then the resulting Y_3 is generally normalisable, so that we can define a new point this way, exactly as in the projective case. The set of points formed this way (i.e. as α and β are varied over some range) corresponds to the line $\hat{Y}_1 \wedge \hat{Y}_2$, since of course

$$(\alpha \hat{Y}_1 + \beta \hat{Y}_2) \wedge (\hat{Y}_1 \wedge \hat{Y}_2) = 0 \tag{54}$$

for all α and β. Exactly the same comments apply to *d-planes*, formed via

$$\Phi = \hat{Y}_1 \wedge \hat{Y}_2 \wedge \hat{Y}_3 \tag{55}$$

and linear combinations of the form

$$Y_4 = \alpha \hat{Y}_1 + \beta \hat{Y}_2 + \gamma \hat{Y}_3 \tag{56}$$

etc.

The second advantage of the new approach is as follows. Consider a d-line L formed from two normalised points A and B as in equation (50). Let us contract L with a point on the line, say A. We obtain

$$L \cdot A = (A \wedge B) \cdot A = (B \cdot A)A - A^2 B = (B \cdot A)A + B \qquad (57)$$

We claim this is the *tangent vector* to the line at A. The point is that if we add a multiple of this vector to A, we will get a point along the line joining A to B. This is a very direct definition of tangent vector. We shall see below that if we normalise things appropriately, then with the path along the geodesic L written as $\hat{Y}(s)$, with s a scalar parameter, then

$$\dot{\hat{Y}} = \frac{d\hat{Y}(s)}{ds} = L \cdot \hat{Y}(s) \qquad (58)$$

gives a unit tangent vector at each point of the path, which we can take as the *velocity* of traversing the path. Thus we can see that the line itself acts as a *bivector generator of motion along the line*. Clearly one will also be able to form a rotor version of this equation as well. A novelty here is that *the bivector generator acts on the position to give the velocity*. This is in contrast to similar equations in 4d, e.g. the Lorentz force law, where the bivector generator (the electromagnetic Faraday) acts on the velocity to give the acceleration.

Within the null vector approach to conformal geometry, if we attempt to take a tangent to a line we obtain a bivector which must be wedged with e again before it becomes meaningful. This then recovers the line itself again [2]. This is sensible, at one level, but precludes anything to do with the bivector generator approach, which we have found to be very useful in physical applications.

We next return to the boundary points. These give a very good way of encoding geodesics. Let P be a point on the lower boundary, and P' a point on the upper boundary, as in Fig. 10.

If we form a linear combination of P and P', with positive coefficients, say

$$Y = \alpha P + \beta P' \qquad (59)$$

then this defines an interior point Y on the geodesic (d-line) joining them. Furthermore, the bivector representing this line is $P \wedge P'$. Since P and P' are null, then we see that the tangent vector at Y is given up to scale by

$$T_Y = \beta P' - \alpha P \qquad (60)$$

We will discuss elsewhere how to relate the direction of this tangent vector in Y space, with the direction of the corresponding tangent vector (e.g. as shown on the plot) in x space, but note here that it works in a simple and straightforward manner.

6.2   Non-euclidean Circles and Hyperbolae

Obviously a key feature of the Hestenes approach to conformal geometry is the way it deals with circles and spheres. This is discussed in the context of non-Euclidean geometry in, for example, [2, 9]. This approach is obviously very useful,

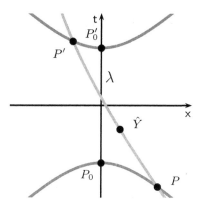

Fig. 10. Plot illustrating the formation of an interior point by linear combination of two boundary points. For the example shown, $Y = 3P + P'$. The part of the geodesic between the boundary points is given by considering the set of points $\alpha P + \beta P'$ for all positive α and β. The direction in Y space of the tangent vector at $\hat{Y}$, is given in general by $\beta P' - \alpha P$ (see text), and so is in the direction $P' - 3P$ at the point shown

and may be preferable to what we are going to discuss here in some applications. However, what we will show here, is that the functionality of this approach can be effectively reproduced using one less dimension for the actual computations. Taking the specific case of de Sitter spacetime, we show that while the availability of the basis vector e is very useful for conceptual and analytic purposes, it is actually better not to have any e component in the vectors representing the points. This is because it is then possible to separate out cleanly the parts of an equation containing e and those that do not, and then results can be obtained from each part separately. These tend to give expressions which achieve what we want, but with many fewer computations than in the 'two dimensions up' case.

To take a specific case, we consider finding the centre and radius of a non-Euclidean circle in the (x, y, z) hyperplane within a de Sitter spacetime. (This transfers through directly to the approach to 3d Euclidean space which we discuss below.)

We take the circle to be defined by three unit vector points A, B and C. We therefore know from the null vector approach that the circle is represented by

$$\Sigma = (A - e)\wedge(B - e)\wedge(C - e) \tag{61}$$

and that points $\hat{Y}$ on the circle must satisfy

$$(\hat{Y} - e)\wedge(A - e)\wedge(B - e)\wedge(C - e) = 0 \tag{62}$$

Expanding out (61) we obtain

$$\Sigma = V - eS \tag{63}$$

where

$$V = A\wedge B\wedge C \quad \text{and} \quad S = A\wedge B + B\wedge C + C\wedge A \tag{64}$$

Now none of $\hat{Y}$, V or S contains e. Thus (62) rearranges to

$$\hat{Y} \wedge V - eV + e\hat{Y} \wedge S = 0 \qquad (65)$$

and thus we get the two conditions

$$\hat{Y} \wedge V = 0 \qquad (66)$$
$$\hat{Y} \wedge S = V \qquad (67)$$

In fact (67) implies (66), so we need just (67), plus the fact that $\hat{Y}$ is a unit vector, for the circle to be specified.

Now, to get the centre of the circle passing through A, B and C, we know that in the null vector approach the quantity $\Sigma e \Sigma$ will be important. This is a covariant construction yielding a vector, but of course not a null vector, and finding the explicit null vector representing the centre from this is slightly messy (see [2, 9]). Here we know straightaway that the point we want will be the 'non-e' part of $\Sigma e \Sigma$. Since S and V commute, we get

$$\Sigma e \Sigma = e(S^2 - V^2) - 2VS \qquad (68)$$

from which we can deduce that the unit vector representing the non-Euclidean centre of the circle is

$$D = -\frac{SV}{|S||V|} \qquad (69)$$

(The choice of sign here corresponds to keeping $y^4 > 0$.) Furthermore, it is easy to show that the non-Euclidean 'radius' (in the sense of minus the dot product of a point on the circle with the centre), is given by

$$\rho = -\hat{Y} \cdot D = \frac{|V|}{|S|} \qquad (70)$$

($\hat{Y}$ being any point on the circle).

These operations (particularly the one for finding the centre) seem likely to be much easier to compute than the equivalent ones in the 'two dimensions up' approach. Moreover, they are susceptible of a further conceptual simplification. The fact that V and S commute is due to V acting as a pseudoscalar within the (non-Euclidean) plane spanned by the triangle. Viewed this way, we can see that *the centre of the circumcircle is the 'dual' of the sum of the sides*. This is a nice geometrical result.

More generally, for any point $\hat{Y}$ in the plane of the triangle, we can form its dual in V, which will be a d-line, and for any d-line we can get a corresponding dual point. To see this, let us take e.g. the dual of the 'side' $A \wedge B$. We will denote this by C', i.e.

$$C' = -A \wedge B \, E_V \qquad (71)$$

where $E_V = V/|V|^2$ and we have adopted a convenient normalisation for the dual point. We see that $\hat{Y} \cdot C' = 0$ for any $\hat{Y}$ on the d-line $A \wedge B$. Thus all points on this line are equidistant from C', which provides another definition of this dual point. Defining further

$$A' = -B \wedge C \, E_V \quad \text{and} \quad B' = -C \wedge A \, E_V \tag{72}$$

we can see that D satisfies

$$D = \rho(A' + B' + C') \tag{73}$$

i.e., in order to find the circumcircle centre, we take a linear combination, with equal weights, of the points dual to the sides.

A yet further useful way to understand this, is via the notion of *reciprocal frame*. Defining

$$\begin{aligned} A_1 &= A, & A_2 &= B, & A_3 &= C \\ A^1 &= A', & A^2 &= B', & A^3 &= C' \end{aligned} \tag{74}$$

then we have

$$A^i \cdot A_j = -\delta^i_j \quad \text{for} \quad i, j = 1, 2, 3 \tag{75}$$

which (modulo a sign) is the usual defining relation for a reciprocal frame. This relation immediately makes clear how (73) is consistent with (70). For example, dotting (73) with A, we immediately get $A \cdot D = -\rho$, as expected.

This then suggests perhaps the simplest way of finding the centre and radius of a non-Euclidean circle defined by three points. Let the three points be called A_1, A_2 and A_3, and define a reciprocal frame to these via (75). Then the centre and radius of the (non-Euclidean) circle through the three points are given by

$$D' = A^1 + A^2 + A^3 \tag{76}$$

and

$$\rho^2 = -\frac{1}{D'^2} \tag{77}$$

respectively.

The calculations to get the reciprocal frame are of course the same as in (71) and (72), but this route seems very clean conceptually, and gets both the centre and radius in one go. Moreover, we can immediately generalise to a sphere and its centre.

Let A_i, $i = 1 \ldots 4$, be four points through which we wish to find a non-Euclidean sphere. Form the reciprocal frame A^i and then let

$$D' = \sum_{i=1}^{4} A^i \tag{78}$$

Then with $\rho^2 = -1/D'^2$, we have that ρ is the radius of the sphere, and $\rho D'$ is the (normalised) point representing its centre. At this point we have a method which is entirely independent of the 'null vector' approach and of the use of e. This statement follows since the result can be proved immediately in its own terms — in particular, by virtue of the definition of the reciprocal frame, the D' defined by (78) is clearly equidistant from each A_i, with 'distance' given by ρ.

7   Electromagnetism, Point Particle Models and Interpolation

The remark after equation (58) suggests an analogy with the Lorentz force law of electromagnetism in GA form: this is $dv/ds = F \cdot v$ where v is the particle 4-velocity and F is the Faraday bivector $\boldsymbol{E} + I\boldsymbol{B}$ (the I here being the pseudoscalar for spacetime). The details will be the subject of a further paper, but it is possible to set up electromagnetism in an interesting way in the 5d space described above. Using this approach, standard problems take on an intriguingly geometrical character, and then one can move down to 4d spacetime in order to find their physical consequences. An important feature is that solutions can be 'moved around' with rotors in 5d, which then correspond to translated and rotated solutions in 4d. This can be expected to be useful e.g. in numerical calculations of the scattered fields from conductors which can be decomposed into surface facets. Fields can then be moved round between facets, themselves described with conformal geometric algebra, by using appropriate rotors in 5d.

We content ourselves with one illustration here, which shows clearly the geometrical character which electromagnetism takes on in 5d. After this we show how the type of point particle model which this view leads to, can give us an insight into something more directly relevant for computer graphics, namely interpolation.

It is possible to set up a point particle Lagrangian which leads to an equation for the position of a particle that is same as for d-line motion above, namely $dY/ds = L \cdot Y$ where L is the line the particle is currently moving along. This would be constant d-line in the absence of any forces. The effect of an electromagnetic field is that now the line L, instead of being constant, satisfies

$$\frac{dL}{ds} = F \times L \qquad (79)$$

where $\times$ is the GA commutator product $(A \times B = (1/2)(AB - BA))$ and F is the 5d Faraday bivector. As an example of the latter, the field at position Y due to a point charge moving along a geodesic (d-line) L' is (ignoring some constant factors)

$$F = \frac{Y \cdot (Y \wedge L')}{|Y \wedge L'|^3} \qquad (80)$$

This is clearly singular, as one would expect, if Y lies on the line L' itself. In the last two equations we see a very interesting feature of the physics that emerges in 5d, which is that it is not points which are important, but the whole world line which a particle travels over.

One can also set up a rotor version of the above in which the rotor encodes the rotation of a fixed set of fiducial axes, in a similar way to the GA formulation of rigid body mechanics. The fixed axes are $\bar{e}$ (the origin) and e_0 (the reference timelike velocity of the body). These get rotated to $\hat{Y}$ (the particle position) and $\hat{T}$ (the instantaneous velocity), via

$$\hat{Y} = \psi \bar{e} \tilde{\psi}, \quad \hat{T} = \psi e_0 \tilde{\psi} \qquad (81)$$

where ψ is the rotor. It turns out that ψ itself then satisfies the rotor equation

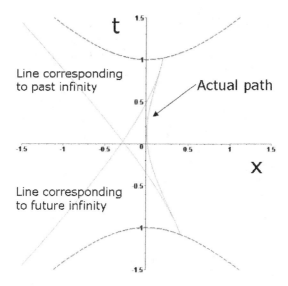

Fig. 11. Motion of a charged particle in the 5d version of a constant electric field. The particle emanates from past infinity, initially along a d-line (free-particle geodesic), but is then accelerated by the field onto a different d-line before reaching future infinity

$$\frac{d\psi}{ds} = \frac{1}{2}L\psi - \frac{1}{2}F(\hat{Y})\cdot\hat{T}\,\hat{T}\psi \tag{82}$$

where $L = \hat{Y}\hat{T}$. This is very novel. We now have a axis pointing out of the particle, $\hat{Y}$, which represents where it is, rather than its velocity. An actual example of such a computation is shown in Fig. 11. This is for the 5d version of a constant electric field in the x direction. At the lower edge of the diagram is the past infinity surface from which the particle emanates. The geodesic motion that would be implied by its initial motion is shown as the 'line corresponding to past infinity'. Similarly its asymptotic final motion is shown as the 'line corresponding to future infinity'. Inbetween, one can see the actual path of the particle, which suffers acceleration in the x direction, due to the action of the electric field. A very intriguing feature of this diagram, is that the asymptotic motion of the particle towards past and future follows free-particle geodesic lines, rather than becoming ever more accelerated, which is what happens in the standard approach for motion under a constant field in 4d.

The point particle model we are using prompts the question of whether one could use bivector interpolation to interpolate motion between any two ψs. This will then be performing interpolation of position and rotation simultaneously, but just using one dimension up. In addition to reduced computational complexity, we suggest that this removes an ambiguity which would otherwise be present in the 2d up approach. In particular, we claim that the '1d up' approach has exactly the right number of degrees of freedom to simultaneously represent the position and rigid body attitude of a body in n dimensions. We see this as follows. Starting with e.g. a 3d body we have that the rigid body attitude accounts for 3 d.o.f., while rigid body position accounts for another 3, implying overall 6. This is the same as the number of components of a 4d bivector and thus of a rotor in 4d. Generally, to describe rotation in n dimensions, we need $1/2n(n-1)$ components to describe rotation, whilst position needs another n. Therefore the total is $1/2n(n+1)$, which is the same as the number of bivector (and therefore independent rotor) components in $n+1$ dimensions. In the 2d up approach, rotors have more components than we need, and thus there is an inherent ambiguity in linking rotors with the actual attitude and position of a rigid body. Of course, it may be objected that we have only achieved the possibility of working with one dimension extra by going to curved space. However, one can explicitly demonstrate for this case of interpolation, that one can bring everything back unambiguously to Euclidean geometry, by letting $\lambda \to \infty$ at the end of the calculations.

A further problem, which we believe this approach solves, is that so far no proper Lagrangian formulation or motivation has been found for interpolation in the '2d up' approach. A Lagrangian approach would be very desirable, since it would then give a rationale for preferring a particular method of interpolation over others. For modelling rigid body motion in the 1d up approach, we propose the rotor Lagrangian

$$\mathcal{L} = -\frac{1}{2}(-\tilde{\psi}\dot{\psi} + \dot{\tilde{\psi}}\psi) \cdot (-\tilde{\psi}\dot{\psi} + \dot{\tilde{\psi}}\psi) + \theta(\psi\tilde{\psi} - 1) \tag{83}$$

Here θ is a Lagrange multiplier enforcing the rotor character of ψ. This Lagrangian is the specialisation of that for rigid body dynamics given on page 426 of [9], to the case where the inertia tensor is spherically symmetric. Its first term is effectively the 'rotational energy' of the spinning point particle we are using to represent the axes in our higher dimensional space. The differentiation is w.r.t a path parameter s. One finds the solution to the equations of motion is $\psi = \exp(Bs)$, where B is a constant bivector. This therefore indeed gives bivector interpolation as the way in which one set of axes transforms into another. Two examples of this type of motion are shown in Fig. 12. In these examples, we are dealing with rigid body motion in 3d, and using a rotor in 4d. In dealing with 3d space and using the '1d up' approach, we have a choice of signature of the extra axis. Assuming we take the 3 space axes as having positive square, then if we add an extra axis with negative square, we will be working with hyperbolic 3d space, whilst if we add an axis with positive square, we will have spherical space. The example given above for finding the centre of a (non-Euclidean) sphere through 4 points, was in a spherical space, which is the natural restriction of our version of de Sitter space to spatial directions only. For some other applications, hyperbolic space might be more appropriate. The question would then be, for engineering applications, whether the Euclidean limit differs, depending on whether one works in spherical or hyperbolic space in 1d up. Both the examples shown in Fig. 12 are for spherical space, as is evident in the picture at the left, which shows a complete closed curve as the orbit under $\psi = \exp(Bs)$ interpolation. At the right, we see a different case, which is in the Euclidean limit in which $\lambda \to \infty$. Here the position interpolation reduces to a constant pitch helix,

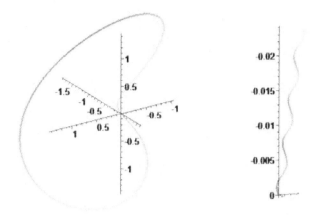

Fig. 12. Examples of interpolation based upon the 'spinning point particle' Lagrangian (83). The position of the rigid body, based upon the direction of the $\hat{Y}$ axes, is shown. Both are calculated in 4d 'spherical space', and at the left a complete closed path is shown. At the right, the Euclidean limit is shown, resulting in a helical path near the origin

and one can show (details will be given elsewhere), that exactly the same helix is obtained whether one reduces to the Euclidean limit from spherical space or hyperbolic space, which is reassuring as to the meaningfulness of the result.

8   Non-euclidean Projective Geometry Revisited, and the Role of Null Vectors

Having established the utility of the 1d up approach, it is worth returning to what we did in Section 5 and asking whether the role of null vectors in 2d up is crucial in this. The answer is no! We can achieve all the results given in that section by only moving 1d up. This is to some extent obvious from the fact that we insisted that $h(n)$, $h(e)$ and $h(\bar{e})$, left n, e or $\bar{e}$ invariant, respectively, in the Euclidean, hyperbolic or spherical cases. We can thus 'throw away' e or $\bar{e}$ as appropriate, and work only in the 1d up space left. As a specific example, let us say that we want to carry out projective geometry in the hyperbolic 2d plane (the Poincaré disc). Then our basis vectors in the 1d up conformal representation are e_1, e_2 and $\bar{e}$, with $e_1^2 = 1$, $e_2^2 = 1$ and $\bar{e}^2 = -1$. We let $h(a)$ be a general, constant linear transformation in this 3d space, and define for a general point Y, that its image under the 'projectivity' h is

$$Y' = h(Y) \tag{84}$$

There is no need to try to go back to a null vector here. Assuming Y^2 is non-zero and has the right sign, a general point Y immediately represents a point in the disc, and if it is null it represents a point on the boundary. It is then easy to

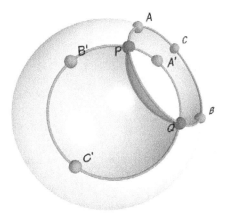

Fig. 13. Diagram illustrating the relation between intersection of hyperplanes in the Poincaré ball, and the intersection of circles, defined via null vectors, in the boundary surface of the ball. See text for details

show that an h which leaves the boundary invariant corresponds to 'rigid' (rotor) motions in the interior, whilst if it does not, it corresponds precisely to the non-Euclidean projective transformations we studied above. As an example, the h given in equation (32) can be used directly in (84) and again yields the non-Euclidean figure shown in Fig. 5. We can see that it will be a projectivity, since e.g. the boundary point $e_1 + \bar{e}$ is mapped to the (non-null) vector representing an interior point $e_1 + (1 + (1/2)\delta)\bar{e}$.

Where then, can one see a natural role for the 2d up approach, and the null vectors it uses? The most natural way this seems to appear is shown in the example of Fig. 13. Here we have the 'Poincaré ball' in 3d. The boundary of this consists of a set of null vectors (the surface of the sphere). This boundary is a 2d space, corresponding with the set of null vectors (with scale ignored) in the 4d space spanned by e_1, e_2, e_3, $\bar{e}$. Thus in this boundary space we are automatically working with null vectors in the 2d up approach! The example shown of this, is for the intersection of two d-planes within the unit ball. Each d-plane is specified by 3 boundary points, A, B, C and A', B', C' respectively. When we intersect the d-planes, we get a d-line in the interior, which intersects the boundary at the points labelled P and Q. On the boundary itself, the interpretation is that we have intersected the two *circles* $A \wedge B \wedge C$ and $A' \wedge B' \wedge C'$ and the result is the bivector formed from the points of intersection P and Q. This construction throws some light on the role of wedges of two null points, which occurs very frequently in the 2d up approach. We see that we can reinterpret them as the end points of a d-line in one dimension higher, and in fact this bivector is precisely the representation of the d-line in the 1d up approach.

Acknowledgements

I would like to thank Joan Lasenby for much help with earlier versions of this paper, and Joan Lasenby, Chris Doran and Richard Wareham for collaboration on several of the topics discussed here. In addition I would like to thank the organizer of the meeting in China where this talk was given, Dr. Hongbo Li, very much for his excellent organization and hospitality in China, and for his patience in dealing with a very tardy author.

References

1. D. Hestenes. Old wine in new bottles: a new algebraic framework for computational geometry. In E. Bayro-Corrochano and G. Sobczyk, editors, *Geometric Algebra with Applications in Science and Engineering*, page 3. Birkauser, Boston, 2001.
2. A.N. Lasenby, J. Lasenby, and R.J. Wareham. A Covariant Approach to Geometry Using Geometric Algebra. Technical Report CUED/F-INFENG/TR-483, Cambridge University Engineering Department, 2004. See
   http://www -sigproc.eng.cam.ac.uk/ga.
3. R. Wareham, J. Cameron, J. Lasenby, and P. Kaimakis. Applications of conformal geometric algebra in computer vision and graphics. In *International Workshop on Geometric Invariance and Applications in Engineering*, Xi'an, China, May 2004.
4. H. Li. Hyperbolic conformal geometry with Clifford algebra. *Int. J. Theor. Phys.*, 40(1):81, 2001.
5. David Hestenes, Hongbo Li, and Alyn Rockwood. New Algebraic Tools for Classical Geometry. In G. Sommer, editor, *Geometric Computing with Clifford Algebra*, 3–26. Springer, Heidelberg, 2001.
6. Hongbo Li, David Hestenes, and Alyn Rockwood. Generalized Homogeneous Coordinates for Computational Geometry. In G. Sommer, editor, *Geometric Computing with Clifford Algebra*, 27–60. Springer, Heidelberg, 2001.
7. Hongbo Li, David Hestenes, and Alyn Rockwood. Spherical Conformal Geometry with Geometric Algebra. In G. Sommer, editor, *Geometric Computing with Clifford Algebra*, 61–76. Springer, Heidelberg, 2001.
8. Hongbo Li, David Hestenes, and Alyn Rockwood. A Universal Model for Conformal Geometries of Euclidean, Spherical and Double-Hyperbolic Spaces. In G. Sommer, editor, *Geometric Computing with Clifford Algebra*, 77–104. Springer, Heidelberg, 2001.
9. C.J.L Doran and A.N. Lasenby. *Geometric Algebra for Physicists*. Cambridge University Press, 2003.
10. A.N. Lasenby. Conformal geometry and the universe. *Phil. Trans. R. Soc. Lond. A*, to appear. See http://www.mrao.cam.ac.uk/~clifford/publications/abstracts/anl_ima2002.html.
11. A.N. Lasenby. Modeling the cosmos: The shape of the universe , Keynote address. SIGGRAPH, 2003, San Diego. See http://www.mrao.cam.ac.uk/~anthony/recent_ga.php.
12. D. Hestenes and G. Sobczyk. *Clifford Algebra to Geometric Calculus*. Reidel, Dordrecht, 1984.
13. D.A. Brannan, M.F. Espleen, and J.J. Gray. *Geometry*. Cambridge University Press, 1999.

14. J.G. Semple and G.T. Kneebone. *Algebraic Projective Geometry*. Oxford University Press, 1952.
15. O. Faugeras. Stratification of 3-D vision: projective, affine, and metric representations. *J.Opt.Soc.America A*, 12:465, 1995.
16. A.W.M. Dress and T.F. Havel. Distance geometry and geometric algebra. *Found. Phys.*, 23(10):1357–1374, 1993.

Applications of Conformal Geometric Algebra in Computer Vision and Graphics

Rich Wareham, Jonathan Cameron, and Joan Lasenby

Cambridge University Engineering Department,
Trumpington St., Cambridge, CB2 1PZ, United Kingdom

Abstract. This paper introduces the mathematical framework of conformal geometric algebra (CGA) as a language for computer graphics and computer vision. Specifically it discusses a new method for pose and position interpolation based on CGA which firstly allows for existing interpolation methods to be cleanly extended to pose and position interpolation, but also allows for this to be extended to higher-dimension spaces and all conformal transforms (including dilations). In addition, we discuss a method of dealing with conics in CGA and the intersection and reflections of rays with such conic surfaces. Possible applications for these algorithms are also discussed.

1 Introduction

Since its inception in the mid-1970s, Computer Graphics (CG) has almost universally used linear algebra as its mathematical framework. This may be due to two factors; most early practitioners of computer graphics were mathematicians familiar with it and linear algebra provided a compact, efficient way of representing points, transformations, lines, etc.

As computing power becomes cheaper, the opportunity arises to investigate new frameworks for CG which, although not providing the time/space efficiency of linear algebra, may provide a conceptually simpler system or one of greater analytical power.

We intend to introduce a system that makes use of Clifford Algebras and in particular the geometric interpretation introduced by [14], termed Geometric Algebra. We then describe the use of this system for both pose and position interpolation and for the reflection and intersection of rays with conics.

1.1 Brief Overview of Geometric Algebra

We shall assume a basic familiarity with Clifford Algebras and merely describe the mechanism whereby they may be used to perform geometric operations. There already exist many introductions to Geometric Algebra that may be consulted [13, 9, 17].

We first write down a basis which can generate all elements of the Clifford Algebra with vector elements in $\mathbb{R}^2$ through linear combination

$$\{1, e_1, e_2, e_1 e_2\} \tag{1}$$

H. Li, P. J. Olver and G. Sommer (Eds.): IWMM-GIAE 2004, LNCS 3519, pp. 329–349, 2005.

where e_1, e_2 are the usual orthonormal Euclidean basis vectors and hence $e_1 e_2 = e_1 \cdot e_2 + e_1 \wedge e_2 = e_1 \wedge e_2$. For convenience we shall denote $e_i e_j$ as e_{ij} and, generally, $e_i e_j \ldots e_k$ as $e_{ij\ldots k}$. A general linear sum of these components is termed a *multivector* whereas a sum of only grade-n components is called a *n-vector*. This paper will use the convention of writing multivectors, bivectors, trivectors and higher-grade elements in upper-case and vectors and scalars in lower-case.

Firstly, note that

$$(e_1 e_2)^2 = e_1 e_2 e_1 e_2 = -e_1 e_2 e_2 e_1 = -1 \tag{2}$$

and thus we have an element of the algebra which squares to -1. As shown by [12, 9] this can be identified with the unit imaginary $i = \sqrt{-1}$ in $\mathbb{C}$ and, as such, the highest-grade basis-element in a geometric algebra is often denoted I and referred to as the *pseudoscalar*.

Note that in spaces with dimension n the maximum grade object possible has grade n and the n-vector $e_1 \wedge \ldots \wedge e_n = I$ is therefore the pseudoscalar. The result of the product xI is termed the *dual* of x, denoted as x^*.

In [12] it was also shown that many of the identities and theorems dealing with complex numbers have a direct analogue in Clifford Algebra. Therefore a geometrical interpretation, which was named *Geometric Algebra*, was suggested.

It can be shown [17, 14] that there is a general method for rotating a vector x involving the formation of a multivector of the form $R = e^{-B\phi/2}$. This represents an anticlockwise rotation ϕ in a plane specified by the bivector B. The transformation is given by

$$x \mapsto RxR^{-1}. \tag{3}$$

We refer to these bivectors which have a rotational effect as *rotors*. The computation of the inverse of a multivector is rather computationally intensive in general (and indeed can require a full 2^n-dimension matrix inversion for a space of dimension n). To combat this we define the *reversion* of a n-vector $X = e_i e_j \ldots e_k$ as $\tilde{X} = e_k \ldots e_j e_i$, i.e. the literal reversion of the components. Since the reversion of $e_i e_j$ is $e_j e_i = -e_i e_j$, and by looking at the expression for R, it is clear that $\tilde{R} \equiv R^{-1}$ for rotors. Computing $\tilde{R}$ is easier since it involves only a sign change for some orthogonal elements of a multivector.

It is worth comparing this method of rotation to quaternions. The three bivectors $B_1 = e_{23}, B_2 = e_{31}$ and $B_3 = e_{12}$ act identically to the three imaginary components of quaternions, $\mathbf{i}, -\mathbf{j}$ and $\mathbf{k}$ respectively. The sign difference between B_2 and $\mathbf{j}$ is due to the fact that the quaternions are not derived from the usual right-handed orthogonal co-ordinate system.

A particular rotation is represented via the quaternion q given by

$$q = q_0 + q_1 \mathbf{i} + q_2 \mathbf{j} + q_3 \mathbf{k}$$

where $q_0^2 + q_1^2 + q_2^2 + q_3^2 = 1$. Interpolation between rotations is then performed by interpolating each of the q_i over the surface of a four-dimensional hypersphere. If we are interpolating between unit quaternions q_0 and q_1 the SLERP interpolation is

$$q = \begin{cases} q_0(q_0^{-1}q_1)^\lambda & \text{if } q_0 \cdot q_1 \geq 0 \\ q_0(q_0^{-1}(-q_1))^\lambda & \text{otherwise} \end{cases} \tag{4}$$

where λ varies in the range $(0, 1)$ [19, 23].

Recalling that, in complex numbers, the locus of $\exp(i\frac{\phi}{2})$ is the unit circle, it is somewhat simple to show that, for some bivector B, where $B^2 = -1$, the locus of the action of $\exp(-B\frac{\phi}{2})$ upon a point with respect to varying ϕ is also a circle in the plane of B. Therefore if we consider some rotations $R_1, R_2 = \exp(kB)R_1$, where k is a scalar and B is some normalised bivector, it is easy to see after some thought that the quaternionic interpolation is exactly given by

$$R_\lambda = (R_2\tilde{R}_1)^\lambda R_1 = \exp(\lambda k B)R_1$$

where λ, the interpolation parameter, varies in the range $(0, 1)$. A further moment's thought will reveal that this method, unlike quaternionic interpolation, is not confined to three dimensions but instead readily generalises to higher-dimensions.

Reflections. Reflections are particularly easy to represent in Geometric Algebra. Reflecting a vector a in a plane with unit normal $\hat{n}$ we obtain a reflected vector a' given by

$$a' = -\hat{n}a\hat{n}$$

which may easily be verified by expanding out to

$$a' = a - 2(\hat{n} \cdot a)\hat{n}$$

and noting that this does indeed give the reflection of a. This pattern of 'sandwiching' an object between two others is often found in GA-based algorithms and can be thought of as representing the reflection of the central object in the 'sandwiching' object.

1.2   Conformal Geometric Algebra (CGA)

In the Conformal Model [14, 13, 18, 17, 20] we extend the space by adding two additional basis vectors. We first define the *signature*, (p, q) of a space $\mathcal{A}(p, q)$ with basis vectors, $\{e_i\}$, such that $e_i^2 = +1$ for $i = 1, ..., p$ and $e_j^2 = -1$ for $j = p + 1, ..., p + q$. For example $\mathbb{R}^3$ would be denoted as $\mathcal{A}(3, 0)$. We extend $\mathcal{A}(3, 0)$ so that it becomes mixed signature and is defined by the basis

$$\{e_1, e_2, e_3, e, \bar{e}\}$$

where e and $\bar{e}$ are defined so that

$$e^2 = 1, \quad \bar{e}^2 = -1, \quad e \cdot \bar{e} = 0$$

$$e \cdot e_i = \bar{e} \cdot e_i = 0 \quad \forall\, i \in \{1, 2, 3\}.$$

This space is denoted as $\mathcal{A}(4,1)$. In general a space $\mathcal{A}(p,q)$ is extended to $\mathcal{A}(p+1, q+1)$. We may now define the vectors n and $\bar{n}$:

$$n = e + \bar{e}, \quad \bar{n} = e - \bar{e}$$

It is simple to show by direct substitution that both n and $\bar{n}$ are *null vectors* (i.e. $n^2 = \bar{n}^2 = 0$). It can be shown [14, 17] that rigid body transforms may be represented conveniently via rotors in this algebra.

We shall also make use of a transform based upon that proposed by Hestenes in [14]. The original transform is dimensionally inconsistent however so we introduce a fundamental unit length scale, λ, into the equation to make it dimensionally consistent

$$F(x) = \frac{1}{2\lambda^2} \left[x^2 n + 2\lambda x - \lambda^2 \bar{n} \right] \equiv X \tag{5}$$

where λ is usually set to be unity. At this point, notice that we can identify the origin with $\bar{n}$ since $F(0) \propto \bar{n}$ and that as $x \to \infty$, $F(x)$ becomes n. We may also find the inverse transform

$$F^{-1}(X) = \lambda(X \wedge n) \cdot e \tag{6}$$

Rotations. As one might expect from their relation to complex numbers, there exists an element of the algebra which performs rotation in the plane. These pure-rotation rotors all have the form e^{-B} where B has only components of the form e_{ij}, $i, j \in \{1, 2, 3\}$.

A useful property of this mapping is that pure-rotation rotors retain their properties as can be shown by considering the effect of a rotor $R = \exp(-\frac{\phi}{2}e_{ij})$, $i, j \in \{1, 2, 3\}$ upon $F(x)$. Setting $\lambda = 1$ for the moment,

$$RF(x)\tilde{R} = \frac{1}{2}R(x^2 n + 2x - \bar{n})\tilde{R} \tag{7}$$

$$= \frac{1}{2}\left(x^2 Rn\tilde{R} + 2Rx\tilde{R} - R\bar{n}\tilde{R} \right) = F(Rx\tilde{R}) \tag{8}$$

since rotors leave n and $\bar{n}$ invariant and $(Rx\tilde{R})^2 = x^2$.

Translations. The translation rotor T_a is defined [17, 18] as

$$T_a = \exp\left[\frac{na}{2} \right] = 1 + \frac{na}{2}$$

and will transform a null-vector representation of the vector x to the null-vector representation of $x_a = x + a$ in the following manner:

$$F(x_a) = T_a F(x)\tilde{T}_a$$

Dilations. It can also be shown [17, 18] that the rotor $D_\alpha = \exp(\alpha e\bar{e}/2)$ has the effect of dilating x by a factor of $e^{-\alpha}$, i.e. $D_\alpha F(x)\tilde{D}_\alpha \propto F(e^{-\alpha}x)$

Table 1. Representations of various geometric objects

Line $- L = X_1 \wedge X_2 \wedge n$	Circle $- C = X_1 \wedge X_2 \wedge X_3$
Plane $- \Phi = X_1 \wedge X_2 \wedge X_3 \wedge n$	Sphere $- \Sigma = X_1 \wedge X_2 \wedge X_3 \wedge X_4$

Inversions. Finally, inversions $(x \mapsto x^2/x)$ may be represented [17, 18] as $F(x) \mapsto eF(x)e$. Although this will not be discussed here, this becomes particularly important when considering non-Euclidean geometry.

This mapping has provided a similar advantage to that of homogeneous coordinates, namely that rigid body transforms become multiplicative and any such transform may be represented by a rotor. In addition, transforms followed by, or preceded by, a dilation or inversion may also be represented multiplicatively.

Representation of Geometric Objects. We have shown how rigid body transformations may be performed on a vector x by operating on its null-vector representation $F(x)$. We may also ask what form the multivector M takes if the solutions of $F(x) \wedge M = 0$ lie on a circle, sphere, line or plane. It has been shown in [17, 18] that the form of M depends only on the null-vector representation of points which lie on the object. The forms of M for various objects are summarised in Table 1. Note that if we identify the vector n with the point at infinity, a line is just a special case of a circle which passes through infinity and a plane is equally a special case of a sphere. It becomes convenient, therefore, to group planes and spheres by the collective term *generalised spheres* and, similarly, to define a *generalised circle* as either a circle or a line.

Suppose we have a generalised sphere which passes through the points $x_1, \cdots,$ x_4. Hence, for all points x on the object, $F(x) \wedge M = 0$ where

$$M = F(x_1) \wedge F(x_2) \wedge F(x_3) \wedge F(x_4). \tag{9}$$

Now consider the effect of a rotor R upon M

$$RM\tilde{R} = RF(x_1)\tilde{R} \wedge RF(x_2)\tilde{R} \wedge RF(x_3)\tilde{R} \wedge RF(x_4)\tilde{R} \tag{10}$$

as $R(a \wedge b)\tilde{R} = R(a)\tilde{R} \wedge R(b)\tilde{R}$. Furthermore we can say

$$RM\tilde{R} = F(Rx_1\tilde{R}) \wedge F(Rx_2\tilde{R}) \wedge F(Rx_3\tilde{R}) \wedge F(Rx_4\tilde{R}) \tag{11}$$

which represents a generalised sphere passing through the transformed points $Y_i = RX_i\tilde{R}$. Clearly, if R represents a conformal transformation, the generalised sphere represented by M is similarly transformed.

Intersections may be performed efficiently using the *meet* operator. The general meet operation can be found by noting that for r-grade and s-grade blades M_r and M_s, a point, X, on the intersection must satisfy $X \wedge M_r = X \wedge M_s = 0$ which can then be shown to be equivalent to

$$X \wedge \left[\langle M_r M_s \rangle_{2l-r-s}\right]^* = 0$$

where $[\,\cdot\,]^*$ denotes multiplication by the pseudoscalar, l is the dimension of the space (in the case of $A(4,1)$, $l = 5$) and $\langle X \rangle_i$ denotes the extraction of the i-grade component from X. If we define the meet operator

$$M_r \vee M_s = \left[\langle M_r M_s \rangle_{2l-r-s} \right]^*$$

then we can interpret the meet of two objects as their intersection in many cases.

The key feature of this approach is that we have placed few constraints on the form of M_r and M_s and thus we can intersect objects in a fairly general manner instead of using object-specific algorithms.

2   Pose and Position Interpolation

In this section we shall use the term *displacement rotor* to refer to a rotor which performs some rigid-body transform. Referring to the displacement rotors presented above, we see that all of them have a common form; they are all ex-ponentiated bivectors. Rotations are generated by bivectors with no component parallel to n and translations by a bivector with no components perpendicular to n. We may therefore postulate that all displacement rotors (we shall deal with rotors including a dilation later) can be expressed as

$$R = \exp(B)$$

where B is the sum of two bivectors, one formed from the outer product of two vectors which have no components parallel to e or $\bar{e}$. The other is formed from the outer product with n of vectors with no components parallel to e or $\bar{e}$. The effect of this is to separate the basis bivectors of B into bivectors with components of the form $e_i \wedge e_j$, $e_i \wedge e$ and $e_i \wedge \bar{e}$.

We shall proceed assuming that all displacement rotors can be written as the exponentiation of a bivector of the form $B = ab + cn$ where a, b and c are *spatial* vectors, i.e. if $n \in \mathcal{A}(m+1,1)$ then $\{a,b,c\} \in \mathbb{R}^m$. It is clear that the set of all B is some linear subspace of all the bivectors.

We now suppose that we may interpolate rotors by defining some function $\ell(R)$ which acts upon rotors to give the generating bivector element. We then perform direct interpolation of this generator. We postulate that direct interpola-tion of such bivectors, as in the reformulation of quaternionic interpolation above, will give some smooth interpolation between the displacements. It is therefore a defining property of $\ell(R)$ that

$$R \equiv \exp(\ell(R)) \tag{12}$$

and so $\ell(R)$ may be considered to act as a logarithm-like function in this context. It is worth noting that $\ell(R)$ does not possess all the properties usually associated with logarithms, notably that, since $\exp(A)\exp(B)$ is not generally equal to $\exp(B)\exp(A)$ in non-commuting algebras, $\ell(\exp(A)\exp(B))$ cannot be equal to $A + B$ except in special cases.

To avoid the the risk of assigning more properties to $\ell(R)$ than we have shown, we shall resist the temptation to denote the function $\log(R)$. The most obvious property of $\log(\cdot)$ that $\ell(\cdot)$ doesn't possess is $\log(AB) = \log(A) + \log(B)$. This is clear since the geometric product is not commutative in general whereas addition is.

2.1   Form of exp(B) in Euclidean Space

The form of $\exp(B)$ may be derived after a little work. We start by assuming that B is of the form $B = \phi P + tn$ where $t \in \mathbb{R}^n$, ϕ is some scalar and P is a 2-blade where $P^2 = -1$ and it is formed from spatial vectors. We then define $t_\parallel$ to be the component of t lying in the plane of P and $t_\perp = t - t_\parallel$. From this assumption it can be shown [26, 25] that the form of $\exp(B)$ is

$$\exp(B) = [\cos(\phi) + \sin(\phi)P][1 + t_\perp n] + \mathrm{sinc}(\phi)t_\parallel n$$

It is also shown in [25] that this can only represent an Euclidean translation and rotation rotor. It is worth exploring the geometric interpretation of this expression. me vector satisfying $a \cdot n = a \cdot P = 0$.

In [25] we discuss how to obtain a geometrical description of the action of the rotor in terms of the bivector B. We shall state this here without proof. We state that the action of the rotor

$$R = \exp\left(\frac{\psi}{2}P + \frac{tn}{2}\right)$$

is to translate along a vector $t_\perp$ which is the component of t which does not lie in the plane of P, rotate by ψ in the plane of P and finally translate along $t'_\parallel$ which is given by

$$t'_\parallel = -\mathrm{sinc}\left(\frac{\psi}{2}\right)t_\parallel\left(\cos\left(\frac{\psi}{2}\right) - \sin\left(\frac{\psi}{2}\right)P\right)$$

which is the component of t lying in the plane of P, rotated by $\psi/2$ in that plane.

2.2   Method for Evaluating $\ell(R)$

We have found a form for $\exp(B)$ given that B is in a particular form. Now we seek a method to take an arbitrary displacement rotor, $R = \exp(B)$ and reconstruct the original B. Should there exist a B for all possible R, we will show that our initial assumption that all displacement rotors can be formed from a single exponentiated bivector of special form is valid. We shall term this initial bivector the *generator* bivector (to draw a parallel with Lie algebras).

We can obtain the following identities for $B = (\psi/2)P + tn/2$ by simply considering the grade of each component of the exponential:

$$\langle R \rangle_0 = \cos\left(\frac{\psi}{2}\right)$$

$$\langle R \rangle_2 = \sin\left(\frac{\psi}{2}\right) P + \cos\left(\frac{\psi}{2}\right) t_\perp n + \operatorname{sinc}\left(\frac{\psi}{2}\right) t_\parallel n$$

$$\langle R \rangle_4 = \sin\left(\frac{\psi}{2}\right) P t_\perp n$$

It is somewhat straightforward to reconstruct $\psi, t_\perp$ and $t_\parallel$ from these components by partitioning a rotor as above. Once we have a method which gives the generator B for any displacement rotor R we have validated our assumption.

Theorem 1. *The inverse-exponential function $\ell(R)$ is given by*

$$\ell(R) = ab + c_\perp n + c_\parallel n$$

where

$$\|ab\| = \sqrt{|(ab)^2|} = \cos^{-1}(\langle R \rangle_0)$$

$$ab = \frac{(\langle R \rangle_2 \, n) \cdot e}{\operatorname{sinc}(\|ab\|)}$$

$$c_\perp n = -\frac{ab \, \langle R \rangle_4}{\|ab\|^2 \operatorname{sinc}(\|ab\|)}$$

$$c_\parallel n = -\frac{ab \, \langle ab \, \langle R \rangle_2 \rangle_2}{\|ab\|^2 \operatorname{sinc}(\|ab\|)}$$

Proof. It is clear from the above that the form of $\|ab\|$ is correct. We thus proceed to show the remaining equations to be true

$$\langle R \rangle_2 = \cos(\|ab\|) \, c_\perp n + \operatorname{sinc}(\|ab\|) \left[ab + c_\parallel n \right]$$
$$\langle R \rangle_2 \, n = \operatorname{sinc}(\|ab\|) \, abn$$
$$(\langle R \rangle_2 \, n) \cdot e = \operatorname{sinc}(\|ab\|) \, ab$$

and hence the relation for ab is correct.

$$\langle R \rangle_4 = \operatorname{sinc}(\|ab\|) \, abc_\perp n$$
$$ab \, \langle R \rangle_4 = -\|ab\|^2 \operatorname{sinc}(\|ab\|) \, c_\perp n$$

and hence the relation for $c_\perp n$ is correct.

$$\langle R \rangle_2 = \cos(\|ab\|) \, c_\perp n + \operatorname{sinc}(\|ab\|) \left[ab + c_\parallel n \right]$$
$$ab \, \langle R \rangle_2 = \cos(\|ab\|) \, abc_\perp n + \operatorname{sinc}(\|ab\|) \left[abc_\parallel n - \|ab\|^2 \right]$$
$$\langle ab \, \langle R \rangle_2 \rangle_2 = \operatorname{sinc}(\|ab\|) \, abc_\parallel n$$

and hence the relation for $c_\parallel n$ is correct.

3   Interpolation via Logarithms

We have shown that any displacement of Euclidean geometry may be mapped smoothly onto a linear subspace of the bivectors. This immediately suggests applications to smooth interpolation of displacements. Consider a set of poses we wish to interpolate, $\{P_1, P_2, ..., P_n\}$ and a set of rotors which transform some origin pose to these target poses, $\{R_1, R_2, ..., R_n\}$. We may map these rotors onto the set of bivectors $\{\ell(R_1), \ell(R_2), ..., \ell(R_n)\}$ which are simply points in some linear subspace of the bivectors. We may now choose any interpolation of these bivectors which lies in this space and for any bivector on the interpolant, B'_λ, we can compute a pose, $\exp(B'_\lambda)$. We believe this method is more elegant and conceptually simpler than many other approaches based on Lie-algebras [11, 21, 22, 5].

Another interpolation scheme is to have the poses defined by a set of chained rotors so that $\{P_1, P_2, ..., P_n\}$ is represented by

$$\{R_1, \Delta R_1 R_1, \Delta R_2 R_2, ..., \Delta R_n R_n\}$$

where $R_i = \Delta R_{i-1} R_{i-1}$ as in figure 1. Using this scheme the interpolation between pose R_i and R_{i+1} involves forming the rotor $R_{i,\lambda} = \exp(B_{i,\lambda}) R_{i-1}$ where $B_{i,\lambda} = \lambda \ell(\Delta R_{i-1})$ and λ varies between 0 and 1 giving $R_{i,0} = R_{i-1}$ and $R_{i,1} = R_i$.

We now investigate two interpolation schemes which interpolate through target poses, ensuring that each pose is passed through. This kind of interpolation is often required for key-frame animation techniques. The first form of interpolation is piece-wise linear interpolation of the relative rotors (the latter case above). The second is direct quadratic interpolation of the bivectors representing the final poses (the former case).

3.1   Piece-Wise Linear Interpolation

Direct piece-wise linear interpolation of the set of relative bivectors is one of the simplest interpolation schemes we can consider. Consider the example shown in figure 1. Here there are three rotors to be interpolated. We firstly find a rotor, ΔR_n which takes us from rotor R_n to the next in the interpolation sequence, R_{n+1}.

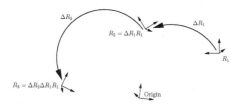

Fig. 1. Rotors used to piece-wise linearly interpolate between key-rotors

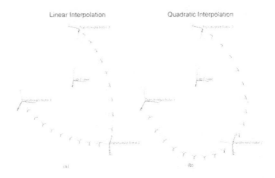

Fig. 2. Examples of (a) piece-wise linear and (b) quadratic interpolation for 3 representative poses

$$R_{n+1} = (\Delta R_n)R_n$$
$$\Delta R_n = R_{n+1}\tilde{R}_n.$$

We then find the bivector, ΔB_n which generates $\Delta R_n = \exp(\Delta B_n)$. Finally we form a rotor interpolating between R_n and R_{n+1}:

$$R_{n,\lambda} = \exp(\lambda \Delta B_n)R_n$$

where λ is in the range $[0,1]$ and $R_{n,0} = R_n$ and $R_{n,1} = R_{n+1}$. Clearly this interpolation scheme changes abruptly at interpolation points, something which is reflected in the resulting interpolation as shown in figure 2. It is interesting to note, after a moment's thought, that for pure-rotation rotors this reduces *exactly* to the quaternionic SLERP interpolation in equation 4.

3.2 Quadratic Interpolation

Another simple form for interpolation is the quadratic interpolation where a quadratic is fitted through three interpolation points, $\{B_1, B_2, B_3\}$ with an interpolation parameter varying in the range $(-1, +1)$:

$$B'_\lambda = \left(\frac{B_3 + B_1}{2} - B_2\right)\lambda^2 + \frac{B_3 - B_1}{2}\lambda + B_2$$

giving

$$B'_{-1} = B_1, \quad B'_0 = B_2 \text{ and } B'_{+1} = B_3$$

This interpolation varies smoothly through B_2 and is reflected in the final interpolation, as shown in figure 2. Extensions to the quadratic interpolation for more than three interpolation points, such as smoothed quadratic interpolation [7] or even a traditional cubic spline or Bézier interpolation are readily available.

It is worth noting that each of the methods described above may be performed using either direct interpolation of the bivector $\ell(R)$ corresponding to a rotor R or by interpolating the relative rotors which take one rotor to another. It is not yet clear which will give the best results and indeed it is probably application dependent.

3.3   Form of the Interpolation

We now derive a clearer picture of the precise form of a simple linear interpolation between two rotors in order to relate the interpolation to existing methods used in mechanics and robotics. We will consider the method used above whereby the rotor being interpolated takes one pose to another.

Path of the Linear Interpolation.   Since we have shown that $\exp(B)$ is indeed a rotor, it follows that any Euclidean pure-translation rotor will commute with it. Thus we only need consider the interpolant path when interpolating from the origin to some other point since any other interpolation can be obtained by simply translating the origin to the start point. This location independence of the interpolation is a desirable property in itself but also provides a powerful analysis mechanism.

We have identified in section 2.1 the action of the $\exp(B)$ rotor in terms of $\psi, P, t_\parallel$ and $t_\perp$. We now investigate the resulting interpolant path when interpolating from the origin. We shall consider the interpolant $R_\lambda = \exp(\lambda B)$ where λ is the interpolation co-ordinate and varies from 0 to 1. For any values of $\psi, P, t_\parallel$ and $t_\perp$,

$$\lambda B = \frac{\lambda \psi}{2}P + \frac{\lambda(t_\perp + t_\parallel)n}{2}$$

from our expansion of $\exp(B)$ we see that the action of $\exp(\lambda B)$ is a translation along $\lambda t_\perp$, a rotation by $\lambda \psi$ in the plane of P and finally a translation along

$$t'_\parallel = -\mathrm{sinc}\left(\frac{\lambda \psi}{2}\right)\lambda t_\parallel \left(\cos\left(\frac{\lambda \psi}{2}\right) - \sin\left(\frac{\lambda \psi}{2}\right)P\right).$$

We firstly resolve a three dimensional orthonormal basis, $\{a, b, t_\perp\}$, relative to P where a and b are orthonormal vectors in the plane of P and hence $P = ab$. We may now express $t_\parallel$ as $t_\parallel = t^a a + t^b b$ where $t^{\{a,b\}}$ are suitably valued scalars.

The initial action of $\exp(B)$ upon a frame centred at the origin is therefore to translate it to $\lambda t_\perp$ followed by a rotation in the plane of P. Due to our choice of starting point, this has no effect on the frame's location (but will have an effect on the pose, see the next section).

Finally there is a translation along $t'_\parallel$ which, using $c = \cos\left(\frac{\lambda \psi}{2}\right)$ and $s = \sin\left(\frac{\lambda \psi}{2}\right)$, can be expressed in terms of a and b as

$$t'_\parallel = -\frac{2s}{\lambda \psi}\lambda(t^a a + t^b b)(c - sab)$$

$$= -\frac{2s}{\psi}\left[c(t^a a + t^b b) + s(t^b a - t^a b)\right]$$

$$\equiv -\frac{2s}{\psi}\left[a(t^a c + t^b s) + b(t^b c - t^a s)\right].$$

The position, r_λ, of the frame at λ along the interpolation is therefore

$$r_\lambda = -\frac{2s}{\psi}(a(t^a c + t^b s) + b(t^b c - t^a s)) + \lambda t_\perp$$

which can easily be transformed via the harmonic addition theorem to

$$r_\lambda = -\frac{2s}{\psi}\alpha \left[a\cos\left(\frac{\lambda\psi}{2} + \beta_1\right) + b\cos\left(\frac{\lambda\psi}{2} + \beta_2\right)\right] + \lambda t_\perp$$

where $\alpha^2 = (t^a)^2 + (t^b)^2$, $\tan\beta_1 = -\frac{t^b}{t^a}$ and $\tan\beta_2 = -\frac{-t^a}{t^b}$. It is easy, via geometric construction or otherwise, to verify that this implies that $\beta_2 = \beta_1 + \frac{\pi}{2}$. Hence $\cos(\theta + \beta_2) = -\sin(\theta + \beta_1)$. We can now express the frame's position as

$$r_\lambda = -\frac{2\alpha}{\psi}\left[a\sin\left(\frac{\lambda\psi}{2}\right)\cos\left(\frac{\lambda\psi}{2} + \beta_1\right) - b\sin\left(\frac{\lambda\psi}{2}\right)\sin\left(\frac{\lambda\psi}{2} + \beta_1\right)\right] + \lambda t_\perp$$

which can be re-arranged to give

$$r_\lambda = -\frac{\alpha}{\psi}\left[a\left(\sin\left(\lambda\psi + \beta_1\right) - \sin\beta_1\right) + b\left(\cos\left(\lambda\psi + \beta_1\right) - \cos\beta_1\right)\right] + \lambda t_\perp$$

$$= -\frac{\alpha}{\psi}\left[a\sin\left(\lambda\psi + \beta_1\right) + b\cos\left(\lambda\psi + \beta_1\right)\right] + \frac{\alpha}{\psi}\left[a\sin\beta_1 + b\cos\beta_1\right] + \lambda t_\perp$$

noting that in the case $\psi \to 0$, the expression becomes $r_\lambda = \lambda t_\perp$ as one would expect. Since a and b are defined to be orthonormal, the path is clearly some cylindrical helix with the axis of rotation passing through $\alpha/\psi\left[a\sin\beta_1 + b\cos\beta_1\right]$.

It is worth noting a related result in screw theory, Chasles' theorem [2], which states that a general displacement may be represented using a screw motion (cylindrical helix) such as we have derived. Screw theory is widely used in mechanics and robotics and the fact that the naïve linear interpolation generated by this method is indeed a screw motion suggests that applications of this interpolation method may be wide-ranging, especially since this method allows many other forms of interpolation, such as Bézier curves or three-point quadratic to be performed with equal ease. Also the pure rotation interpolation given by this method reduces exactly to the quaternionic or Lie group interpolation result allowing the method to easily extend existing ones based upon these interpolations.

Pose of the Linear Interpolation. The pose of the transformed frame is unaffected by pure translation and hence the initial translation by $\lambda t_\perp$ has no effect. The rotation by $\lambda\psi$ in the plane, however, now becomes important. The subsequent translation along $t'_\parallel$ also has no effect on the pose. We find, therefore, that the pose change λ along the interpolant is just the rotation rotor $R_{\lambda\psi,P}$.

3.4   Interpolation of Dilations

In certain circumstances it is desirable to add in the ability to interpolate dilations. It can be shown [6] that this can be done by extending the form of the bivector, B, which we exponentiate, as follows

$$B = \phi P + tn + \omega N$$

where $N = e\bar{e}$. This bivector form is now sufficiently general [6] to be able to represent dilations as well. In this case obtaining the exponentiation and logarithm function is somewhat involved [6]. We obtain finally that

$$\exp(\phi P + tn + \omega N)$$
$$= (\cos(\phi) + \sin(\phi)P)(\cosh(\omega) + \sinh(\omega)N + \sinh c(\omega)t_\perp n)$$
$$+ (\omega^2 + \phi^2)^{-1}[-\omega\sin(\phi)\cosh(\omega) + \phi\cos(\phi)\sinh(\omega)]P$$
$$+ (\omega^2 + \phi^2)^{-1}[\omega\cos(\phi)\sinh(\omega) + \phi\sin(\phi)\cosh(\omega)]t_\parallel n$$

where $\sinh c(\omega) = \omega^{-1}\sinh(\omega)$. Note that this expression reduces to the original form for $\exp(B)$ when $\omega = 0$, as one would expect.

It is relatively easy to use the above expansion to derive a logarithm-like inverse function.

If we let $R = \exp(B)$ then we may recreate B from R using the method presented below. Here we use $\langle R|e_i\rangle$ to represent the component of R parallel to e_i, i.e. $\langle R|N\rangle = \langle R|e_{45}\rangle$ in 3-dimensions. We also use $\langle R\rangle_i$ to represent the i-th grade-part of R and $S(X)$ to represent the 'spatial' portion of X (i.e. those components not parallel to e and $\bar{e}$).

$$\omega = \tanh^{-1}\left(\frac{\langle R|N\rangle}{\langle R|1\rangle}\right) \qquad W = \langle R\rangle_2 - \cos(\phi)\sinh(\omega)N$$
$$\qquad\qquad\qquad\qquad\qquad - \sin(\phi)\cosh(\omega)P$$
$$\phi = \cos^{-1}\left(\frac{\langle R|N\rangle}{\sinh(\omega)}\right) \qquad\qquad - \cos(\phi)\sinh c(\omega)t_\perp n$$

$$P = \frac{S(\langle R\rangle_2)}{\sin(\phi)\cosh(\omega)} \qquad X = -\omega\sin(\phi)\cosh(\omega) + \phi\cos(\phi)\sinh(\omega)$$

$$t_\perp = -\frac{\langle R\rangle_4 - \sin(\phi)\sinh(\omega)PN}{\sin(\phi)\sinh c(\omega)}\left(\frac{P\bar{n}}{2}\right) \quad Z = \omega\cos(\phi)\sinh(\omega) + \phi\sin(\phi)\cosh(\omega)$$

$$t = t_\parallel + t_\perp \qquad t_\parallel = \frac{(-XP+Z)}{\sin^2(\phi)\cosh^2(\omega) + \cos^2(\phi)\sinh^2(\omega)}W$$

Path of Interpolation. We now consider the path resulting from the simple linear interpolation of both pose and dilation using the results derived above. In [6] a full derivation is given and here we present only the result. As before, the derivation proceeds by considering the position, r_λ, of the frame formed by applying the interpolation rotor $R = \exp(\lambda(\phi P + tn + \omega N))$ to $\bar{n}$ (the origin), with the interpolation parameter λ varying from 0 to 1. After a little work we obtain the path of the interpolation as

$$r_\lambda = -\frac{1}{\omega}|t_\perp|\hat{t}_\perp - \frac{\omega}{\omega^2 + \phi^2}|t_\parallel|\hat{t}_\parallel + \frac{\phi}{\omega^2 + \phi^2}|t_\parallel|\hat{t}_\perp + e^{-2\lambda\omega}.$$

$$\left(\frac{|t_\perp|}{\omega}\hat{t}_\perp + \frac{\cos\left(2\lambda\phi + \tan^{-1}\left(\frac{\phi}{\omega}\right)\right)}{\sqrt{\omega^2 + \phi^2}}|t_\parallel|\hat{t}_\parallel - \frac{\sin\left(2\lambda\phi + \tan^{-1}\left(\frac{\phi}{\omega}\right)\right)}{\sqrt{\omega^2 + \phi^2}}|t_\parallel|\hat{t}_\perp\right)$$

where $\hat{t}_\parallel$ is the normalised component of t parallel to P, $\hat{t}_\perp$ is the normalised component perpendicular to P and $\hat{t}_\perp$ is a unit vector perpendicular to both $\hat{t}_\parallel$ and $\hat{t}_\perp$. This is the equation of a conical helix.

It can be shown [6] that when setting the dilation component, ω, to zero this path is equivalent to that derived for pose interpolation.

3.5   Applications

The method outlined is applicable to any problem which requires the smooth interpolation of pose. We have chosen to illustrate an application in mesh deformation. Smooth mesh deformation is often required in medical imaging applications [8] or video coding [15]. Here we use it to deform a 3d mesh in a manner which has a visual effect similar to 'grabbing' a corner of the mesh and 'twisting' it into place. We do this by specifying a set of 'key-rotors' which certain parts of

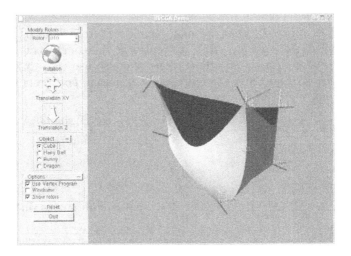

Fig. 3. A cube distorted via the linear interpolation of rotors specifying its corner vertices

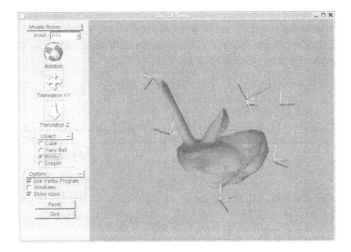

Fig. 4. The Stanford Bunny[24] distorted by the same rotors

the mesh must rotate and translate to coincide with. Our implementation takes advantage of the hardware acceleration offered by the Graphics Processing Units (GPUs) available on today's consumer-level graphics hardware. A full discussion of the method will appear elsewhere.

We believe this method leads to images which are intuitively related to the rotors specifying the deformation. Furthermore, the method need not only be applied to meshes with simple geometry and can readily be applied to meshes with complex geometry without any tears or other artifacts appearing in the mesh. Figures 3 and 4 illustrate the method in the case of a cube and a mesh with approximately 36,000 vertices.

4   Conics

So far we have been limited in which objects we can manipulate and intersect in CGA. Specifically we can only deal with planes, circles, spheres, lines and points. In this section we consider a method of extending our approach to conics and intersections thereof. We shall proceed by considering the 3-dimensional Euclidean case but this method may be generalised to higher dimensions if required; here we follow the procedure outlined in [16]. We shall begin by considering the set of points on the unit sphere at the origin:

$$\{c_\theta e_1 + s_\theta c_\phi e_2 + s_\theta c_\phi e_3 : \theta \in (0, 2\pi], \phi \in (0, \pi]\}$$

where $c_\theta = \cos(\theta)$, $s_\theta = \sin(\theta)$ and similarly for s_ϕ and c_ϕ. We apply the transform $F(\cdot)$ to obtain the set, S, of representations for these points:

$$S = \left\{ \frac{n}{2} + c_\theta e_1 + s_\theta c_\phi e_2 + s_\theta c_\phi e_3 - \frac{\bar{n}}{2} : \theta \in (0, 2\pi], \phi \in (0, \pi] \right\}$$

and consider the effect of the transform from [16] below

$$C_\alpha(X) = \beta \left[X + \frac{\alpha}{2}(X \cdot e_1) \wedge \bar{n} \right]$$

which we shall term the *conic transform*, upon the elements of S where β is chosen so that our normalisation constraint $C_\alpha(X) \cdot n = -1$ still holds:

$$\{C_\alpha(X) : X \in S\}$$
$$\equiv \left\{ \frac{1}{1 - \alpha \cos \theta} \left(\frac{n}{2} + c_\theta e_1 + s_\theta c_\phi e_2 + s_\theta c_\phi e_3 \right) - \frac{\bar{n}}{2} : \theta \in (0, 2\pi], \phi \in (0, \pi] \right\}$$

The elements of this set are no-longer null-vectors (they have extra components parallel to n) but one may still apply the inverse mapping $F^{-1}(\cdot)$ and extract a spatial vector (i.e. one with no components parallel to e or $\bar{e}$). It is found [16, 6] that the spatial vectors extracted thus lie on the surface of a quadric with a rotational cross-section corresponding to a conic. The form of this conic cross-section is determined by the parameter α which can be interpreted as the eccentricity:

- $\alpha = 0$: The set of points lie on the unit sphere.
- $0 < \alpha < 1$: The set of points lie on an ellipsoid formed by rotating an ellipse in the e_{12} plane about e_1. The origin lies at one of the foci and α is the eccentricity.
- $\alpha = 1$: The set of points lie on a paraboloid specified by the points

$$\{x : 2x \cdot e_1 = [(x \cdot e_2)^2 + (x \cdot e_3)^2] - 1\}.$$

- $\alpha > 1$: The set of points lie on a two-sheeted hyperboloid formed by rotating an hyperbola in the e_{12} plane about e_1. The origin is once again one of the foci and the eccentricity is α.

We may therefore form a set of representations for points lying on these conics by simply applying appropriate translation, dilation and rotation rotors after the conic transform.

4.1   Properties of the Conic Transform

It can further be shown [6] that the conic transform preserves the outer product but not the inner (and hence geometric) products. That is to say $C_\alpha(X \wedge Y) = C_\alpha(X) \wedge C_\alpha(Y)$. It also preserves the 'special' vectors n and $\bar{n}$. From this we can easily show that $C_\alpha(\cdot)$ does not change the nature of 'flat' objects (i.e. planes and lines transform to other planes and lines) since

$$C_\alpha(X_1 \wedge X_2 \wedge \cdots \wedge X_i \wedge n) = C_\alpha(X_1) \wedge C_\alpha(X_2) \wedge \cdots \wedge C_\alpha(X_i) \wedge C_\alpha(n)$$
$$= C_\alpha(X_1) \wedge C_\alpha(X_2) \wedge \cdots \wedge C_\alpha(X_i) \wedge n.$$

It is also easy to verify by direct substitution that the conic transform preserves direction of points from the origin, i.e. that

$$F^{-1}(C_\alpha(F(a_i e_i))) = \frac{a_i}{1 - \alpha a_1} e_i$$

where we have adopted the usual summation convention.

4.2   Intersections

We may now consider how to apply this transform to intersecting conics with lines. Suppose that we have found a rotor R and value of α such that the set of points represented by

$$\{RC_\alpha(X)\tilde{R} : X \in S\}$$

lie on the conic we wish to intersect. Suppose further that we wish to intersect it with the line represented by the trivector L (as in figure 5a). We shall denote the points of intersection A and B. Now consider the effect of applying $C_\alpha^{-1}(\tilde{R}XR)$ to each of our objects and points (where X is the appropriate object). The points on the conic are transformed to the unit sphere. The line, L, is transformed to $C_\alpha^{-1}(\tilde{R}LR)$ (figure 5) which is still a line since the conic transform maps lines to

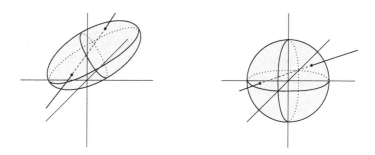

Fig. 5. Intersection of line L with (a) conic and (b) transformation back to intersection with unit sphere

lines. The points of intersection between this new line and our unit sphere may be found via the usual meet formulation:

$$A' \wedge B' = \Sigma \vee C_\alpha^{-1}(\tilde{R}LR)$$

where Σ is the multivector representing the unit sphere which may be formed as

$$\Sigma = F(e_1) \wedge F(e_2) \wedge F(e_3) \wedge F(-e_1)$$

or similar.

Finally we consider the transformed points $A = RC_\alpha(A')\tilde{R}$ and $B = RC_\alpha(B') \cdot \tilde{R}$. They must lie upon our original line L since A' and B' clearly lie on $C_\alpha^{-1}(\tilde{R}LR)$ and

$$RC_\alpha(C_\alpha^{-1}(\tilde{R}LR))\tilde{R} = L.$$

They must also lie upon our conic since A' and B' lie on the unit sphere and the transformation is exactly that which generated our original conic. If they lie upon our conic and upon L they must therefore be the points of intersection. We have therefore formulated a method for intersecting lines and conics.

4.3   Reflections

We now consider the reflection of rays from conics. In order to reflect a ray we wish to find the tangent plane for a conic at the intersection point of the incoming ray with the conic. Again we make use of the fact that $C_\alpha(\cdot)$ maps planes to planes and apply $C_\alpha^{-1}(\tilde{R}XR)$ to all objects moving the points on the conic to the unit sphere. We then find the intersection point and tangent plane for the unit sphere. It can be shown [6] that when transformed back via $RC_\alpha(\cdot)\tilde{R}$ the tangent plane of the sphere maps to the tangent plane for the conic. Reflection can then be performed by simply reflecting the incoming ray L in this plane, Φ:

$$L' = \Phi L \Phi.$$

5 Line Images in a Para-Catadioptric Camera

As an illustration of the power of the techniques described above, let us consider the simple but useful question of what the image of a straight line in a parabolic mirror based Single View Point catadioptric (SVP Para-catadioptric) camera is. This problem has been considered in a number of papers e.g. [10, 27, 3, 4] and is useful for calibration and scene reconstruction.

The setup in question is illustrated in figure 6. With a parabolic mirror the SVP constraint requires that all the rays from the mirror imaged by the camera are parallel to the mirror axis and hence the camera is orthographic. Let us consider the source of light forming an image at the point $xe_2 + ye_3$. As we are imaging with an orthographic camera, for the setup shown, this ray is in the e_1 direction, which is axis of both the camera and the parabolic mirror. The focus of the parabolic mirror is situated at the origin.

$$S_1 = [e_{23} + (xe_3 - ye_2)n]^*$$

This result is easily established if we consider the form of the dual of a line [17].

$$L^* = \hat{m}I_3 + [(a \wedge \hat{m})I_3]n$$

with $\hat{m} = e_1$ (the ray's direction), $a = xe_2 + ye_3$ (a point on the ray) and $I_3 = e_{123}$. This form is analogous to writing the line in terms of Plücker coordinates where 3 of the coordinates give the line's direction and the other 3 give its moment about the origin.

We are interested in the reflection of this ray in the mirror. This reflected ray, S_2 is found as described in Section 4.3.

$$S_2 = \frac{1 - x^2 - y^2}{2}e_{145} - xe_{245} - ye_{345} \tag{13}$$

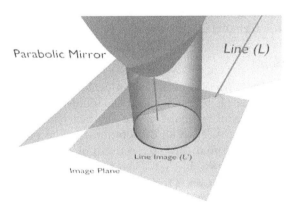

Fig. 6. Para-catadioptric Line Image Formation

For this ray to be the image ray of a point on a line L they must intersect and hence

$$L \vee S_2 = 0 \tag{14}$$

As stated above, a line, L, with unit direction vector $\hat{m} = m_1 e_1 + m_2 e_2 + m_3 e_3$ (such that $\hat{m}^2 = 1$) passing through the point $a = a_1 e_1 + a_2 e_2 + a_3 e_3$, can be expressed as

$$L = [\hat{m} I_3 + ((a \wedge \hat{m}) I_3) n]^* \tag{15}$$

Substituting equations 15 and 13 into 14 and rearranging, the condition obtained is

$$\left(x + \frac{a_3 m_1 - a_1 m_3}{a_2 m_3 - a_3 m_2} \right)^2 + \left(y + \frac{a_1 m_2 - a_2 m_1}{a_2 m_3 - a_3 m_2} \right)^2$$
$$- \left(1 + \left(\frac{a_3 m_1 - a_1 m_3}{a_2 m_3 - a3_m 2} \right)^2 + \left(\frac{a_1 m_2 - a_2 m_1}{a_2 m_3 - a_3 m_2} \right)^2 \right) = 0 \tag{16}$$

As others have observed [10], this is a circle (L' in figure 6). A similar derivation can be used to establish the image of a sphere in such a camera the only significant change being that the resulting condition is the product of two separate circle equations, indicating an ambiguity which can trivially be resolved.

6   Conclusions

In this paper we have developed and discussed a natural, extensible method for interpolating pose and position in both 3-dimensions and higher. The method allows for any traditional path interpolation method to be extended to encompass both pose and position whilst retaining its desirable properties. We further extend this method to deal with dilations. A method of handling conics within the CGA framework together with algorithms for ray-conic intersections and reflections was also discussed.

These algorithms could form a useful basis for many applications in computer graphics and computer vision. For example the use of jointed ellipsoidal models in marker-less motion capture requires computing the intersection of a ray with an ellipse which may be calculated using the method above. In the case of single viewpoint catadioptric cameras [1, 10] the reflection of rays from conics is often used and the method above may be used to form analytic solutions to a number of problems [6].

References

1. S. Baker and S. Nayar. A theory of single-viewpoint catadioptric image formation. *International Journal of Computer vision*, pages 1–22, 1999.
2. Sir R. Ball. *A Treatise on the Theory of Screws*. Cambridge University Press, 1900.

3. J. Barreto and H. Araujo. Paracatadioptric camera calibration using lines. In *Proceedings of ICCV*, pages 1359–1365, 2003.
4. E. Bayro-Corrochano and C. López-Franco. Omnidirectional vision: Unified model using conformal geometry. In *Proceedings of ECCV*, pages 536–548, 2004.
5. S. Buss and J. Fillmore. Spherical averages and applications to spherical splines and interpolation. *ACM Transactions on Graphics*, pages 95–126, 2001.
6. J. Cameron. Applications of Geometric Algebra, August 2004. PhD First Year Report, Cambridge University Engineering Department.
7. Z. Cendes and S. Wong. C1 quadratic interpolation over arbitrary point sets. *IEEE Computer Graphics and Applications*, pages 8–16, Nov 1987.
8. S. Cotin, H. Delingette, and N. Ayache. Real-time elastic deformations of soft tissues for surgery simulation. *IEEE Transactions on Visualization and Computer Graphics*, 5(1):62–73, 1999.
9. C. Doran and A. Lasenby. *Geometric Algebra for Physicists*. CUP, 2003.
10. C. Geyer and K. Daniilidis. Catadioptric projective geometry. *International Journal of Computer Vision*, pages 223–243, 2001.
11. V. Govindu. Lie-algebraic averaging for globally consistent motion estimation. In *Proceedings of CVPR*, pages 684–691, 2004.
12. D. Hestenes. *New Foundations for Classical Mechanics*. Reidel, Second Edition, 1999.
13. D. Hestenes. Old wine in new bottles: A new algebraic framework for computational geometry. In E. Bayro-Corrochano and G. Sobczyk, editors, *Geometric Algebra with Applications in Science and Engineering*, chapter 1. Birkhäuser, 2001.
14. D. Hestenes and G. Sobczyk. *Clifford Algebra to Geometric Calculus: A unified language for mathematics and physics*. Reidel, 1984.
15. S. Kshirsagar, S. Garchery, and N. Magnenat-Thalmann. Feature point based mesh deformation applied to mpeg-4 facial animation. In *Proceedings of the IFIP TC5/WG5.10 DEFORM'2000 Workshop and AVATARS'2000 Workshop on Deformable Avatars*, pages 24–34. Kluwer, 2001.
16. A. Lasenby. Recent applications of conformal geometric algebra. In *International Workshop on Geometric Invariance and Applications in Engineering*, Xi'an, China, May 2004.
17. J. Lasenby, A. Lasenby, and R. Wareham. A Covariant Approach to Geometry using Geometric Algebra. Technical Report CUED/F-INFENG/TR-483, Cambridge University Engineering Department, 2004.
18. H. Li, D. Hestenes, and A. Rockwood. Generalized homogeneous co-ordinates for computational geometry. In G. Sommer, editor, *Geometric Computing with Clifford Algebra*, pages 25–58. Springer, 2001.
19. M. Lillholm, E.B. Dam, and M. Koch. Quaternions, interpolation and animation. Technical Report DIKU-TR-98/5, University of Copenhagen, July 1998.
20. S. Mann and L. Dorst. Geometric algebra: A computational framework for geometrical applications (part 2). *IEEE Comput. Graph. Appl.*, 22(4):58–67, 2002.
21. M. Moakher. Means and averaging in the group of rotations. *SIAM Journal of Applied Matrix Analysis*, pages 1–16, 2002.
22. F. Park and B. Ravani. Smooth invariant interpolation of rotations. *ACM Transactions on Graphics*, pages 277–295, 1997.
23. K. Shoemake. Animating rotation with quaternion curves. *Computer Graphics*, 19(3):245–251, 1985.
24. G. Turk and M. Levoy. Zippered polygon meshes from range images. In *SIGGRAPH 1996 Proceedings*, pages 331–318, 1994.

25. R. Wareham and J. Lasenby. Rigid body pose and position interpolation using geometric algebra. *Submitted to ACM Transactions on Graphics*, September 2004.
26. R. Wareham, J. Lasenby, and A. Lasenby. Computer Graphics using Conformal Geometric Algebra. *Philosophical Transactions A of the Royal Society, special issue*. To appear soon.
27. X. Ying and Z. Hu. Spherical objects based motion estimation for catadioptric cameras. In *Proceedings of ICPR*, pages 231–234, 2004.

Conic Sections and Meet Intersections in Geometric Algebra

Eckhard M.S. Hitzer

Department of Physical Engineering, University of Fukui, Japan
hitzer@mech.fukui-u.ac.jp

Abstract. This paper first gives a brief overview over some interesting descriptions of conic sections, showing formulations in the three geometric algebras of Euclidean spaces, projective spaces, and the conformal model of Euclidean space. Second the conformal model descriptions of a subset of conic sections are listed in parametrizations specific for the use in the main part of the paper. In the third main part the meets of lines and circles, and of spheres and planes are calculated for all cases of real and virtual intersections. In the discussion special attention is on the hyperbolic carriers of the virtual intersections.

1   Introduction

Previous Work

D. Hestenes used geometric algebra to give in his textbook New Foundations for Classical Mechanics [1] a range of descriptions of conic sections. The basic five ways of construction there are:

- the semi-latus rectum formula
- with polar angles (ellipse)
- two coplanar circles (ellipse)
- two non-coplanar circles (ellipse)
- second order curves depending on three vectors

Animated and interactive online illustrations for all this can be found in [2]. Reference [2] also treats plane conic sections defined via Pascal's Theorem by five general points in a plane in

- the geometric algebra of the 2+1 dimensional projective plane
- the conformal geometric algebra of the 2+2 dimensional conformal model of the Euclidean plane.

This was inspired by Grassmann's treatment of plane conic sections in terms of five general points in a plane [3]. In both cases the meet operation is used in an essential way. The resulting formulas are quadratic in each of the five conformal points.

H. Li, P. J. Olver and G. Sommer (Eds.): IWMM-GIAE 2004, LNCS 3519, pp. 350–362, 2005.

By now it is also widely known that the conformal geometric algebra model of Euclidean space allows for direct linear product representations [7, 12] of the following subset of conics: Points, pairs of points, straight lines, circles, planes and spheres. It is possible to find direct linear product representations with 5 constitutive points for general plane conics by introducing the geometric algebra of a six dimensional Euclidean vector space [4].

Beyond this the meet [8] operation allows to e.g. generate a circle from the intersection of two spheres or a sphere and a plane. The meet operation is well defined no matter whether two spheres truly intersect each other (when the distance of the centers is less then the sum of the radii but greater than their difference), but also when they don't (when the distance of the centers is greater than the sum of the radii or less than their difference).

The meet of two non-intersecting circles in a plane can be interpreted as a virtual point pair with a distance that squares to a negative real number [5]. (If the circles intersect, the square is positive.)

This leads to the following set of questions:

- How does this virtual point pair depend on the locations of the centers?
- What virtual curve is generated if we continuously increase the center to center distances?
- What is the dependence on the radii of the circles?
- Does the meet of a straight line (a cirle with infinite radius) with a circle also lead to virtual point pairs and a virtual locus curve (depending on the distance of straight line and circle)?
- How is the three dimensional situation of the meet of two spheres or a plane and a sphere related to the two dimensional setting?

All these questions will be dealt with in this paper.

2   Background

2.1   Clifford's Geometric Algebra

Clifford's *geometric algebra* $Cl(n - q, q) = \mathbb{R}_{n-q,q}$ of a real n-dimensional vector space $\mathbb{R}^{n-q,q}$ can be defined with four *geometric product* axioms [9] for a canonical vector basis, which satisfies

1. $e_k^2 = +1$ ($1 \le k \le n - q$), $e_k^2 = -1$ ($n - q < k \le n$).
2. The square of a vector $\mathbf{x} = x^k e_k$ ($1 \le k \le n$) is given by the reduced quadratic form

$$\mathbf{x}^2 = (x^k e_k)^2 = \sum_k (x^k)^2 e_k^2,$$

which supposes $e_k e_l + e_l e_k = 0$, $k \ne l$, $1 \le k, l \le n$.
3. Associativity: $(e_k e_l)e_m = e_k(e_l e_m), 1 \le k, l, m \le n$.
4. $\alpha e_k = e_k \alpha$ for all scalars $\alpha \in \mathbb{R}$.

A geometric algebra is an example of a graded algebra with a basis of real scalars, vectors, bivectors, ... , n-vectors (pseudoscalars), i.e. the grades range from $k = 0$ to $k = n$. The grade k elements form a $\binom{n}{k}$ dimensional k-vector space. Each k-vector is in one-to-one correspondence with a k dimensional subspace of $\mathbb{R}^{n-q,q}$. A general multivector A of $\mathbb{R}_{n-q,q}$ is a sum of its grade k parts

$$A = \sum_{k=0}^{n} \langle A \rangle_k.$$

The grade zero index is often dropped for brevity: $\langle A \rangle = \langle A \rangle_0$. Negative grade parts $k < 0$ or elements with grades $k > n$ do not exist, they are zero. By way of grade selection a number of practically useful products of multivectors A, B, C is derived from the geometric product:

1. The scalar product
$$A * B = \langle AB \rangle. \tag{1}$$

2. The outer product
$$A \wedge B = \sum_{k,l=0}^{n} \langle \langle A \rangle_k \langle B \rangle_l \rangle_{k+l}. \tag{2}$$

3. The left contraction
$$A \lrcorner B = \sum_{k,l=0}^{n} \langle \langle A \rangle_k \langle B \rangle_l \rangle_{l-k} \tag{3}$$

which can also be defined by

$$(C \wedge A) * B = C * (A \lrcorner B) \tag{4}$$

for all $C \in \mathbb{R}_{n-q,q}$.
4. The right contraction

$$A \llcorner B = \sum_{k,l=0}^{n} \langle \langle A \rangle_k \langle B \rangle_l \rangle_{k-l} = (\tilde{B} \lrcorner \tilde{A})^{\sim} \tag{5}$$

or defined by
$$A * (B \wedge C) = (A \llcorner B) * C \tag{6}$$

for all $C \in \mathbb{R}_{n-q,q}$. The tilde sign $\tilde{A}$ indicates the reverse order of all elementary vector products in every grade component $\langle A \rangle_k$.
5. Hestenes and Sobczyk's [8] inner product generalization

$$\langle A \rangle_k \cdot \langle B \rangle_l = \langle \langle A \rangle_k \langle B \rangle_l \rangle_{|k-l|} \quad (k \neq 0, l \neq 0), \tag{7}$$
$$\langle A \rangle_k \cdot \langle B \rangle_l = 0 \quad (k = 0 \text{ or } l = 0). \tag{8}$$

The scalar and the outer product (already introduced by H. Grassmann) are well accepted. There is some debate about the use of the left and right contractions on one hand or Hestenes and Sobczyk's "minimal" definition on the other hand as the preferred generalizations of the inner product of vectors [6]. For many practical purposes (7) and (8) are completely sufficient. But the exception for grade zero factors (8) needs always to be taken into consideration when deriving formulas involving the inner product. Hestenes and Sobczyk's book [8] shows this in a number of places. The special consideration for grade zero factors (8) also becomes necessary in software implementations. Beyond this, (4) and (5) show how the salar and the outer product already fully imply left and right contractions. It is therefore infact possible to begin with a Grassmann algebra, introduce a scalar product for vectors, induce the (left or right) contraction and thereby define the geometric product, which generates the Clifford geometric algebra. It is also possible to give direct definitions of the (left or right) contraction [10].

2.2   Conformal Model of Euclidean Space

Euclidean vectors are given in an orthonormal basis $\{e_1, e_2, e_3\}$ of $\mathbb{R}^{3,0} = \mathbb{R}^3$ as

$$\mathbf{p} = p_1 \mathbf{e}_1 + p_2 \mathbf{e}_2 + p_3 \mathbf{e}_3, \ \mathbf{p}^2 = p^2 \ . \tag{9}$$

One-to-one corresponding conformal points in the 3+2 dimensions of $\mathbb{R}^{4,1}$ are given as

$$P = \mathbf{p} + \frac{1}{2}p^2 \mathbf{n} + \bar{\mathbf{n}}, \ P^2 = \mathbf{n}^2 = \bar{\mathbf{n}}^2 = 0, \ P * \bar{\mathbf{n}} = \mathbf{n} * \bar{\mathbf{n}} = -1 \tag{10}$$

with the special conformal points of $\mathbf{n}$ infinity, $\bar{\mathbf{n}}$ origin. This is an extension of the Euclidean space similar to the projective model of Euclidean space. But in the conformal model extra dimensions are introduced both for origin and infinity. $P^2 = 0$ shows that the conformal model first restricts $\mathbb{R}^{4,1}$ to a four dimensional null cone (similar to a light cone in special relativity) and second the normalization condition $P * \bar{\mathbf{n}} = -1$ further intersects this cone with a hyperplane.

We define the Minkowski plane pseudoscalar (bivector) as

$$N = \mathbf{n} \wedge \bar{\mathbf{n}}, \ N^2 = 1 \ . \tag{11}$$

By joining conformal points with the outer product (2) we can generate the subset of conics mentioned above: pairs of points, straight lines, circles, planes and spheres [7, 11, 12, 13, 14]. Detailed formulas to be used in the rest of the paper are given in the following subsections [15].

2.3   Point Pairs

$$P_1 \wedge P_2 = \mathbf{p}_1 \wedge \mathbf{p}_2 + \frac{1}{2}(p_2^2 \mathbf{p}_1 - p_1^2 \mathbf{p}_2)\mathbf{n} - (\mathbf{p}_2 - \mathbf{p}_1)\bar{\mathbf{n}} + \frac{1}{2}(p_1^2 - p_2^2)N \tag{12}$$

$$= \ldots = 2r\{\hat{\mathbf{p}} \wedge \mathbf{c} + \frac{1}{2}[(c^2 + r^2)\hat{\mathbf{p}} - 2\mathbf{c} * \hat{\mathbf{p}} \ \mathbf{c}]\mathbf{n} + \hat{\mathbf{p}}\bar{\mathbf{n}} + \mathbf{c} * \hat{\mathbf{p}}N\} \ , \tag{13}$$

Fig. 1. Pair of intersection points P_1, P_2 with distance $2r$, midpoint C and unit direction vector $\hat{\mathbf{p}}$ of the connecting line segment

with distance $2r$, unit direction of the line segment $\hat{\mathbf{p}}$, and midpoint $\mathbf{c}$ (comp. Fig. 1):

$$2r = |\,\mathbf{p}_1 - \mathbf{p}_2\,|, \quad \hat{\mathbf{p}} = \frac{\mathbf{p}_1 - \mathbf{p}_2}{2r}, \quad \mathbf{c} = \frac{\mathbf{p}_1 + \mathbf{p}_2}{2} \ . \tag{14}$$

For $\hat{\mathbf{p}} \wedge \mathbf{c} = 0$ ($\hat{\mathbf{p}} \,\|\, \mathbf{c}$) we get

$$P_1 \wedge P_2 = 2r\{C - \frac{1}{2}r^2\mathbf{n}\}\hat{\mathbf{p}}N \tag{15}$$

for $\hat{\mathbf{p}} * \mathbf{c} = 0$ ($\hat{\mathbf{p}} \perp \mathbf{c}$) we get

$$P_1 \wedge P_2 = -2r\{C + \frac{1}{2}r^2\mathbf{n}\}\hat{\mathbf{p}} \tag{16}$$

with conformal midpoint

$$C = \mathbf{c} + \frac{1}{2}c^2\mathbf{n} + \bar{\mathbf{n}} \ . \tag{17}$$

2.4   Straight Lines

Using the same definitions the straight line through P_1 and P_2 is given by

$$P_1 \wedge P_2 \wedge \mathbf{n} = 2r\hat{\mathbf{p}} \wedge C \wedge \mathbf{n} = 2r\{\hat{\mathbf{p}} \wedge \mathbf{c}\,\mathbf{n} - \hat{\mathbf{p}}N\} \tag{18}$$

2.5   Circles

$$P_1 \wedge P_2 \wedge P_3 = \alpha\{\mathbf{c} \wedge I_c + \frac{1}{2}[(c^2 + r^2)I_c - 2\mathbf{c}(\mathbf{c} \,\lrcorner\, I_c)]\mathbf{n} + I_c\bar{\mathbf{n}} - (\mathbf{c} \,\lrcorner\, I_c)N\}$$

$$= \alpha(C + \frac{1}{2}r^2\mathbf{n}) \wedge \{I_c + \mathbf{n}(C \,\lrcorner\, I_c)\} \tag{19}$$

describes a circle through the three points P_1, P_2 and P_3 with center $\mathbf{c}$, radius r, and circle plane bivector

$$I_c = \frac{(\mathbf{p}_1 - \mathbf{p}_2) \wedge (\mathbf{p}_2 - \mathbf{p}_3)}{\alpha} \ , \tag{20}$$

where the scalar $\alpha > 0$ is chosen such that

$$I_c^2 = -1 \ . \tag{21}$$

For the inner product we use the left contraction $\rfloor$ in (19) as discussed in section 2.1. For $\mathbf{c} \wedge I_c = 0$ (origin $\bar{\mathbf{n}}$ in circle plane) and the conformal center (17) we get

$$P_1 \wedge P_2 \wedge P_3 = -\alpha\{C - \frac{1}{2}r^2\mathbf{n}\}I_cN. \tag{22}$$

2.6   Planes

Using the same α, C and I_c as for the circle

$$P_1 \wedge P_2 \wedge P_3 \wedge \mathbf{n} = \alpha\, C \wedge I_c \wedge \mathbf{n} = \alpha\{\mathbf{c} \wedge I_c\mathbf{n} - I_cN\} \tag{23}$$

defines a plane through P_1, P_2, P_3 and infinity. For $\mathbf{c} \wedge I_c = 0$ (origin $\bar{\mathbf{n}}$ in plane) we get

$$P_1 \wedge P_2 \wedge P_3 \wedge \mathbf{n} = -\alpha I_cN. \tag{24}$$

2.7   Spheres

$$P_1 \wedge P_2 \wedge P_3 \wedge P_4 = \beta(C - \frac{1}{2}r^2\mathbf{n})IN \tag{25}$$

defines a sphere through P_1, P_2, P_3 and P_4 with radius r, conformal center C, unit volume trivector $I = \mathbf{e}_1\mathbf{e}_2\mathbf{e}_3$, and scalar

$$\beta = (\mathbf{p}_1 - \mathbf{p}_2) \wedge (\mathbf{p}_2 - \mathbf{p}_3) \wedge (\mathbf{p}_3 - \mathbf{p}_4)\, I^{-1}\ . \tag{26}$$

3   Full Meet of Two Circles in One Plane

The meet of two circles (comp. Fig. 2)

$$V_1 = (C_1 - \frac{1}{2}r_1^2\mathbf{n})I_cN, \quad V_2 = (C_2 - \frac{1}{2}r_2^2\mathbf{n})I_cN\ , \tag{27}$$

with conformal centers C_1, C_2, radii r_1, r_2, in one plane I_c (containing the origin $\bar{\mathbf{n}}$), and join four-vector $J = I_cN$ is[1]

$$M = (V_1 \rfloor J^{-1}) \rfloor V_2 \tag{28}$$

$$= \frac{1}{2}[(\mathbf{c}_1^2 - r_1^2) - (\mathbf{c}_2^2 - r_2^2)]I_c + \frac{1}{2}[(\mathbf{c}_2^2 - r_2^2)\mathbf{c}_1 - (\mathbf{c}_1^2 - r_1^2)\mathbf{c}_2]I_c\mathbf{n} + (\mathbf{c}_2 - \mathbf{c}_1)I_c\bar{\mathbf{n}} + \mathbf{c}_1 \wedge \mathbf{c}_2 I_cN \tag{29}$$

$$= \ldots = \frac{d}{2r}P_1 \wedge P_2 \tag{30}$$

with

---

[1] The dots ($\ldots$) in (13), (30), (41) and (47) indicate nontrivial intermediate algebraic calculations whose details are omitted here because of lack of space.

$$d = \mid \mathbf{c}_2 - \mathbf{c}_1 \mid , \tag{31}$$

$$r^2 = \frac{M^2}{(M \wedge \mathbf{n})^2} = d^2 \{ \frac{r_1^2 r_2^2}{d^4} - \frac{1}{4}(1 - \frac{r_1^2}{d^2} - \frac{r_2^2}{d^2})^2 \} \tag{32}$$

and [like in (14)]

$$\mathbf{p}_1 = \mathbf{c} + r\hat{\mathbf{p}}, \quad \mathbf{p}_2 = \mathbf{c} - r\hat{\mathbf{p}} , \tag{33}$$

$$\mathbf{c} = \mathbf{c}_1 + \frac{1}{2}(1 + \frac{r_1^2 - r_2^2}{d^2})(\mathbf{c}_2 - \mathbf{c}_1), \quad \hat{\mathbf{p}} = \frac{\mathbf{c}_2 - \mathbf{c}_1}{d} I_c . \tag{34}$$

We further get *independent* of r^2

$$M \wedge \mathbf{n} = d\hat{\mathbf{p}} \wedge C \wedge \mathbf{n} , \tag{35}$$

which is in general a straight line through P_1, P_2, and in particular for $r^2 = 0$ ($M^2 = 0$, $\mathbf{p}_1 = \mathbf{p}_2 = \mathbf{c}$) the tangent line at the intersection point.[2] Note that r^2 may become *negative* (depending on r_1 and r_2, for details compare Fig. 4). The vector from the first circle center $\mathbf{c}_1$ to the middle $\mathbf{c}$ of the point pair is

$$\mathbf{c} - \mathbf{c}_1 = \frac{1}{2}(1 + \frac{r_1^2 - r_2^2}{d^2})(\mathbf{c}_2 - \mathbf{c}_1) , \tag{36}$$

with (oriented) length

$$d_1 = \frac{1}{2}(d + \frac{r_1^2 - r_2^2}{d}) \tag{37}$$

This length d_1, half the intersection point pair distance r and the circle radius r_1 are related by

$$r^2 + d_1^2 = r_1^2 . \tag{38}$$

We therefore observe (comp. Fig. 2, 3 and 4) that (38)

- describes for $r^2 > 0$ all points of real intersection (on the circle V_1) of the two circles.
- For $r^2 < 0$ (38) becomes the locus equation of the virtual points of intersection, i.e. two hyperbola branches that extend symmetrically on both sides of the circle V_1 (assuming e.g. that we move V_2 relative to V_1).
- The sequence of circle meets of Fig. 3 clearly illustrates, that as e.g. circle V_2 moves from the right side closer to circle V_1, also the virtual intersection points approach along the hyperbola branch ($r^2 < 0$) on the same side of V_1, until the point of outer tangence ($d = r_1 + r_2$, $r = 0$). Then we have real intersection points ($r^2 > 0$) until inner tangence occurs ($d = r_2 - r_1$, $r = 0$). Reducing $d < r_2 - r_1$ even further leads to virtual intersection points, wandering outwards on the same side hyperbola branch as before until d becomes infinitely small. Moving C_2 over to the other side of C_1 repeats the phenomenon just described on the other branch of the hyperbola (symmetry to the vertical symmetry axis of the hyperbola through C_1).

---

[2] We actually have $\lim_{r \to 0} M = d\{\hat{\mathbf{p}} + C * \hat{\mathbf{p}} \, \mathbf{n}\} \wedge C$, which can be interpreted [5] as the tangent vector of the two tangent circles, located at the point C of tangency.

- The transverse symmetry axis line of the two hyperbola branches is given by $C_1 \wedge C_2 \wedge \mathbf{n}$, i.e. the straight line through the two circle centers.
- The assymptotics are at angles $\pm\frac{\pi}{4}$ to the symmetry axis.
- The radius r_1 is the semitransverse axis segment.

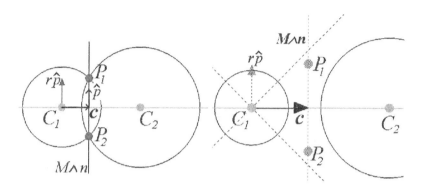

Fig. 2. Two intersecting circles with centers C_1, C_2, intersecting in points P_1, P_2 at distance $2r$, with midpoint $\mathbf{c}$ and unit direction vector $\hat{\mathbf{p}}$ of the connecting line segment. Left side: Real intersection ($r_2 < d < r_1 + r_2$), right side: Virtual intersection ($d > r_1 + r_2$)

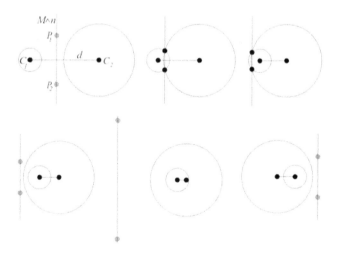

Fig. 3. Real and virtual intersection points (33) and vertical carrier line $M \wedge \mathbf{n}$ of (35) for two circles with radii $r_1 < r_2$ and centers C_1, C_2 at central distances d (31). Top left: $d > r_1 + r_2$, $r^2 < 0$. Top center: $r_2 < d < r_2 + r_1$, $r^2 > 0$. Top right: $r_2 - r_1 < d < r_2$, $r^2 > 0$. Bottom left: $d + r_1 < r_2$, $r^2 < 0$. Bottom center: smaller d. Bottom right: similar to bottom left, but C_1 on other side of C_2

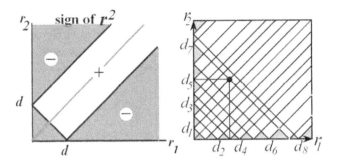

Fig. 4. Left side: Positive $(+)$ and negative $(-)$ signs of r^2 depending on the radii of the two circles and the circle center distance d. $r = 0$ (0) on the border lines (shape of an open U tilted in the $r_1 = r_2$ direction), which separate $(-)$ and $(+)$ regions. Right side: Black dot for case $r_1 < r_2$. Eight different values of d are indicated, showing the tilted U-shaped (0) border lines between the $(-)$ and $(+)$ regions. $d_1 < r_2 - r_1$ $(-)$, $d_2 = r_2 - r_1$ (0), $d_3 = r_1$ $(+)$, $r_2 - r_1 < d_4 < r_2$ $(+)$, $d_5 = r_2$ $(+)$, $r_2 < d_6 < r_1 + r_2$ $(+)$, $d_7 = r_1 + r_2$ (0) and $d_8 > r_1 + r_2$ $(-)$

4   Full Meet of Circle and Straight Line in One Plane

Now we turn our attention to the meet of a circle with center C_1, radius r_1 in plane I_c, $I_c^2 = -1$ (including the origin $\bar{\mathbf{n}}$)

$$V_1 = (C_1 - \frac{1}{2}r_1^2\mathbf{n})I_cN \ , \tag{39}$$

and a straight line through C_2, with direction $\hat{\mathbf{p}}$ and in the same plane I_c,

$$V_2 = \hat{\mathbf{p}} \wedge C_2 \wedge \mathbf{n} = \hat{\mathbf{p}} \wedge \mathbf{c}_2\mathbf{n} - \hat{\mathbf{p}}N \ . \tag{40}$$

(For convenience C_2 be selected such that $d =\mid \mathbf{c}_2 - \mathbf{c}_1 \mid$ is the distance of the circle center C_1 from the line V_2. See Fig. 5.)

The meet of V_1 and V_2 is

$$M = (V_1 \,\lrcorner\, J^{-1}) \,\lrcorner\, V_2 = \ldots = \frac{-1}{2r}P_1 \wedge P_2 \tag{41}$$

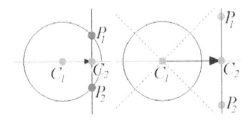

Fig. 5. Real and virtual intersections of circle and straight line

with the join $J = I_c N$,
$$r^2 = M^2 = r_1^2 - d^2 \ , \tag{42}$$
and
$$\mathbf{p}_1 = \mathbf{c}_2 + r\hat{\mathbf{p}}, \quad \mathbf{p}_2 = \mathbf{c}_2 - r\hat{\mathbf{p}} \ . \tag{43}$$
r and $\hat{\mathbf{p}}$ have the same meaning as in (14). Note that $r^2 < 0$ for $d > r_1$.

We observe that

- Equations (38) and (42) are remarkably similar.
- The point pair $P_1 \wedge P_2$ is **now always** on the straight line V_2, and has center C_2!
- For $r^2 > 0$ ($d < r_1$)
$$r^2 + d^2 = r_1^2 \tag{44}$$
  describes the real intersections of circle and straight line.
- For $r^2 < 0$ ($d > r_1$)
$$r^2 + d^2 = r_1^2 \tag{45}$$
  describes the virtual intersections of circle and straight line.
- The general formula $M \wedge \mathbf{n} = -V_2$ holds for all values of r, even if $r^2 = M^2 = 0$ ($\mathbf{p}_1 = \mathbf{p}_2 = \mathbf{c}_2$). In this special case V_2 is tangent to the circle.[3]
- In all other repects, the virtual intersection locus hyperbola has the same properties (symmetry, transverse symmetry axis, assymptotics and semi-transverse axis segment) as that of the meet of two circles in one plane.

5   Full Meet of Two Spheres

Let us assume two spheres (see Fig. 6)
$$V_1 = (C_1 - \tfrac{1}{2}r_1^2 \mathbf{n})IN, \quad V_2 = (C_2 - \tfrac{1}{2}r_2^2 \mathbf{n})IN \ . \tag{46}$$

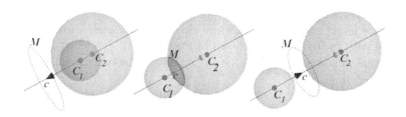

Fig. 6. Real and virtual intersections of two spheres ($r_1 < r_2$). Left: $d < r_2 - r_1 < r_2$, center: $r_2 < d < r_1 + r_2$, right: $d > r_1 + r_2$

with centers C_1, C_2, radii r_1, r_2, and (3+2)-dimensional pseudoscalar join $J = IN$. The meet of these two spheres is

---

[3] For the case of tangency we have now $\lim_{r \to 0} M = -\{\hat{\mathbf{p}} + C_2 * \hat{\mathbf{p}} \ \mathbf{n}\} \wedge C_2$, which can be interpreted [5] as a vector in the line V_2 attached to C_2, tangent to the circle.

$$M = (V_1 \lrcorner J^{-1}) \lrcorner V_2$$

$$= \frac{1}{2}[(\mathbf{c}_1^2 - r_1^2) - (\mathbf{c}_2^2 - r_2^2)]I - \frac{1}{2}[(\mathbf{c}_2^2 - r_2^2)\mathbf{c}_1 - (\mathbf{c}_1^2 - r_1^2)\mathbf{c}_2]In + (\mathbf{c}_1 - \mathbf{c}_2)I\bar{\mathbf{n}} + \mathbf{c}_1 \wedge \mathbf{c}_2 IN$$

$$= \ldots = d(C + \frac{1}{2}r^2\mathbf{n}) \wedge \{I_c + \mathbf{n}(C \lrcorner I_c)\} \,, \tag{47}$$

where r and d are defined as for the case of intersecting two circles $[r^2 = -M^2/(M \wedge \mathbf{n})^2$, note the sign!] We further introduced in (47) the plane bivector

$$I_c = \frac{(\mathbf{c}_1 - \mathbf{c}_2)}{d}I \tag{48}$$

and the vector (see Fig. 6)

$$\mathbf{c} = \mathbf{c}_1 + \frac{1}{2}(1 + \frac{r_1^2 - r_2^2}{d^2})(\mathbf{c}_2 - \mathbf{c}_1) \,. \tag{49}$$

Comparing (19) and (47) we see that M is a conformal circle multivector with radius r, oriented parallel to I_c in the plane $M \wedge \mathbf{n} = dC \wedge I_c \wedge \mathbf{n}$, and with center C.

Regarding the formula

$$r^2 + d_1^2 = r_1^2 \tag{50}$$

with $d_1 = |\mathbf{c} - \mathbf{c}_1|$ it remains to observe that

- equation (50) describes for $r^2 > 0$ $(d_1 < r_1)$ the real radius r circles of intersection of two spheres.
- These intersection circles are centered at $C = \mathbf{c} + \frac{1}{2}c^2\mathbf{n} + \bar{\mathbf{n}}$ in the plane $M \wedge \mathbf{n}$ perpendicular to the center connecting straight line $C_1 \wedge C_2 \wedge \mathbf{n}$, i.e. parallel to the bivector of (48).
- For $r^2 = 0$, $M \wedge \mathbf{n}$ gives still the (conformal) tangent plane trivector of the two spheres.[4]
- We have for $r^2 < 0$ $(d_1 > r_1)$ virtual circles of intersection forming a hyperboloid with **two sheets**, as shown in Fig. 7.
- The transverse symmetry axis (straight) line of the two sheet hyperboloid is $C_1 \wedge C_2 \wedge \mathbf{n}$.
- The discussion of the meet of two circles in section 3 related to Fig. 3 and Fig. 4 applies also to the case of the meet of two spheres. r^2 is now the squared radius of real and virtual meets (circles instead of point pairs).
- The asymptotic double cone has angle $\pi/4$ relative to the transverse symmetry axis.
- The sphere radius (e.g. r_1) is again the semitransverse axis segment of the two sheet hyperboloid (assuming e.g. that we move V_2 relative to V_1).

---

[4] For the case of tangency ($r^2 = 0$) we have now $M = d\, C \wedge \{I_c + \mathbf{n}(C \lrcorner I_c)\}$, which can be interpreted [5] as tangent direction bivector I_c of the two tangent spheres, located at the point C of tangency.

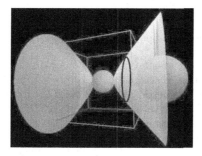

Fig. 7. Two sheet hyperboloid of virtual intersections of two spheres

6 Full Meet of Sphere and Plane

Let V_1 be a conformal sphere four-vector [as in (46)] and V_2 a conformal plane four-vector [(23) with normalization $\alpha = 1$]. The analogy to the case of circle and straight line is now obvious. For the virtual ($r^2 < 0$) intersections we get the same two sheet hyperboloid as for the case of sphere and sphere, but now both real and virtual intersection circles are always on the plane V_2, as shown in Fig. 8. The general formula $M \wedge \mathbf{n} = V_2$ holds for all values of r, even if $r^2 = M^2 = 0$. In this special case V_2 is tangent to the sphere.

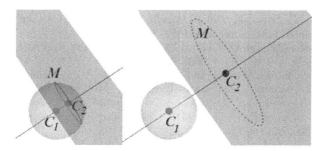

Fig. 8. Real and virtual intersections of sphere and plane

Acknowledgement

Soli Deo Gloria. I thank my wife and my children. I further thank especially Prof. Hongbo Li and his colleagues for organizing the GIAE workshop. The GAViewer [16] was used to create Figs. 1,2,3,5,6 and 8 and to probe many of the formulas. In Fig. 4 the interactive geometry software Cinderella [17] was used. Fig. 7 was created by C. Perwass with CLUCalc [18]. I thank the Signal Processing Group of the CUED for its hospitality in the period of finishing this paper and L. Dorst for a number of important comments.

References

1. D. Hestenes, New Foundations for Classical Mechanics (2nd ed.), Kluwer, Dordrecht, 1999.
2. E.M.S. Hitzer, Learning about Conic Sections with Geometric Algebra and Cinderella, in E.M.S. Hitzer, R. Nagaoka, H. Ishi (eds.), Proc. of Innovative Teaching of Mathematics with Geometric Algebra Nov. 2003, Research Institute for Mathematical Sciences (RIMS), Kyoto, Japan, RIMS **1378**, pp. 89–104 (2004). Interactive online presentation: http://sinai.mech.fukui-u.ac.jp/ITM2003/presentations/Hitzer/page1.html
3. H. Grassmann, A new branch of math., tr. by L. Kannenberg, Open Court, 1995. H. Grassmann, Extension Theory, tr. by L. Kannenberg, AMS, Hist. of Math., 2000.
4. C. Perwass, Analysis of Local Image Structure using Intersections of Conics, Technical Report, University of Kiel, (2004) http://www.perwass.de/published/perwass_tr0403_v1.pdf
5. L. Dorst, Interactively Exploring the Conformal Model, Lecture at *Innovative Teaching of Mathematics with Geometric Algebra 2003*, Nov. 20-22, Kyoto University, Japan. L. Dorst, D. Fontijne, An algebraic foundation for object-oriented Euclidean geometry, in E.M.S. Hitzer, R. Nagaoka, H. Ishi (eds.), Proc. of Innovative Teaching of Mathematics with Geometric Algebra Nov. 2003, Research Institute for Mathematical Sciences (RIMS), Kyoto, Japan, RIMS **1378**, pp. 138–153 (2004).
6. L. Dorst, The Inner Products of Geometric Algebra, in L. Dorst et. al. (eds.), Applications of Geometric Algebra in Computer Science and Engineering, Birkhaeuser, Basel, 2002.
7. C. Doran, A. Lasenby, J. Lasenby, Conformal Geometry, Euclidean Space and Geometric Algebra, in J. Winkler (ed.), Uncertainty in Geometric Computations, Kluwer, 2002.
8. D. Hestenes, G. Sobczyk, Clifford Algebra to Geometric Calculus, Kluwer, Dordrecht, reprinted with corrections 1992.
9. G. Casanova, L'Algebre De Clifford Et Ses Applications, Special Issue of Adv. in App. Cliff. Alg. Vol 12 (S1), (2002).
10. P. Lounesto, Clifford Algebras and Spinors, 2nd ed., CUP, Cambridge, 2001.
11. G. Sobczyk, Clifford Geometric Algebras in Multilinear Algebra and Non-Euclidean Geometries, Lecture at *Computational Noncommutative Algebra and Applications*, July 6-19, 2003, http://www.prometheus-inc.com/asi/algebra2003/abstracts/sobczyk.pdf
12. D. Hestenes, H. Li, A. Rockwood, New Algebraic Tools for Classical Geometry, in G. Sommer (ed.), Geometric Computing with Clifford Algebras, Springer, Berlin, 2001.
13. E.M.S. Hitzer, KamiWaAi - Interactive 3D Sketching with Java based on Cl(4,1) Conformal Model of Euclidean Space, Advances in Applied Clifford Algebras 13(1), pp. 11-45 (2003).
14. E.M.S. Hitzer, G. Utama, The Geometric Algebra Java Package – Novel Structure Implementation of 5D GeometricAlgebra $\mathbb{R}_{4,1}$ for Object Oriented Euclidean Geometry, Space-Time Physics and Object Oriented Computer Algebra, to be published in Mem. Fac. Eng. Univ. Fukui, 53(1) (2005).
15. E.M.S. Hitzer, Euclidean Geometric Objects in the Clifford Geometric Algebra of {Origin, 3-Space, Infinity}, to be published in Bulletin of the Belgian Mathematical Society - Simon Stevin.
16. GAViewer homepage http://www.science.uva.nl/ga/viewer/index.html
17. Cinderella website, http://www.cinderella.de/
18. C. Perwass, CLUCalc website, http://www.clucalc.info/

nD Object Representation and Detection from Single 2D Line Drawing*

Hongbo Li[1], Quan Wang[1], Lina Zhao[2], Ying Chen[1], and Lei Huang[1]

[1] Mathematics Mechanization Key Laboratory,
Academy of Mathematics and Systems Science,
Chinese Academy of Sciences,
Beijing 100080, P. R. China
[2] Department of Mathematics and Computer Science,
Science of School, Bejing University of Chemical Technology

Abstract. In this paper, we propose the *wireframe representation* of nD object, which is a single 2D line drawing. A *wireframe model* of an nD object is composed of a set of edges connecting a set of vertices, a subset of closed r-chains called boundary r-chains which surround the $(r+1)$D pieces of the object, and a set of filling patterns for the boundary r-chains for $0 < r < n$. The wireframe representation is the perspective projection of the wireframe model of the object from its surrounding space to the image plane. Combining the projective geometric constraints of the wireframe representation with the idea of local construction and deletion, we propose an algorithm for high dimensional object detection from single 2D line drawing, under the most general assumption that neither the dimension of the object nor the dimension of the surrounding space is known, two neighboring faces can be coplanar, and whether or not the object is a manifold is unknown. Our algorithm outperforms any other algorithm in 2D face identification in that it generally does not generate redundant cycles that are not assigned as faces, and can handle 3D solids of over 10,000 faces.

1   Introduction

Representing and perceiving nD object has been a very fascinating problem in both science and art [14], [11]. If the representation media is a monitor, a typical technique is called grand tour [2], in which the nD object is projected onto a moving 2D plane, and by visually perceiving coherence in the contiguous 2D images, one can get an idea of the actual structure of the nD object. If the representation media is a piece of paper, then for $n = 3$ the simplest representation is a line drawing which is the projection of the wireframe of the object, like drafting in geometric design and mathematical diagram. To perceive the nD object one needs to rebuild the nD structure from its 2D projection.

* Supported partially by NKBRSF 2004CB318001 and NSFC 10471143.

H. Li, P. J. Olver and G. Sommer (Eds.): IWMM-GIAE 2004, LNCS 3519, pp. 363–382, 2005.

How can the n dimensions be recovered from a representation in which almost all dimensions are lost? To start with, let us analyze how a solid in 3D space is perceived. No one can direct his eyesight to pierce through the solid. The only perceived object is the boundary of the solid, which is a 2D *closed manifold*. It is the closedness that allows us to fill the boundary with solid content to achieve one more dimension. When we watch a line drawing of the wireframe of a solid, which is essentially one dimensional, we extract each cycle of edges, either fill it by a plane or by some other surface to improve its dimension by one. Then we detect if any closed manifold is formed by the planes and surfaces, and if so, gain one more dimension by filling the closed manifold with solid content. The closedness of a manifold and a pattern to fill it are the two essential things in our 3D perception from low dimensional data.

From the topological point of view, a closed manifold is a *closed chain* in homology, and the boundary of a non-closed manifold is a *boundary chain*. While all boundary chains are closed, the converse is not true. For any dimension r, the quotient of the closed r-chains over the boundary r-chains is called the r-*th homology* of the object. Determining the boundary chains from the closed ones is equivalent to determining the homology.

The *wireframe model* of an nD object consists of (1) a set of edges connecting a finite set of points of the object, (2) a subset of closed r-chains, called *boundary r-chains*, which are the boundaries of the $(r+1)$D pieces of the object for $0 < r < n$, (3) a set of filling patterns, one for each boundary r-chain. Below are some examples.

Example 1. nD simplices.

Fig. 1. 4D simplex

The wireframe representation of an nD simplex is the complete graph of $n+1$ vertices, i.e., any two vertices are connected by an edge. Every triangular cycle in the graph is the boundary of a 2D simplex, and every complete subgraph of r vertices is the boundary of an $(r + 1)$D simplex. Any boundary r-chain is filled with an $(r + 1)$D flat patch bounded by the chain. The nD reconstruction is unique topologically. As a consequence, the reconstruction of any triangulated topological space from its wireframe is unique.

Example 2. nD cubes.

The wireframe of a 1D cube is an edge. The wireframe of a 2D cube is a square obtained by connecting the corresponding vertices of two copies of a 1D cube. Recursively, the wireframe of an nD cube is an edge-vertex graph obtained by connecting the corresponding vertices of two copies of an $(n-1)$D cube. In the graph, every square cycle is the boundary of a 2D cube, and every closed r-chain composed of $2r+2$ different rD cubes is the boundary of an $(r+1)$D cube. Any boundary r-chain is filled with an $(r+1)$D flat patch bounded by the chain. The reconstruction is unique topologically.

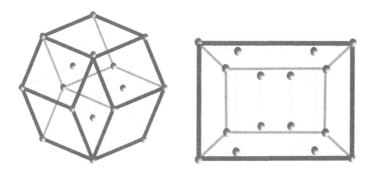

Fig. 2. 4D cube.   Left: bird view. Right: window view

Example 3. nD spheres.

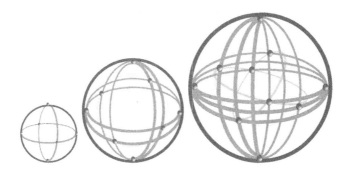

Fig. 3. nD spheres. Here $n = 2, 3, 4$

An nD sphere is the boundary of an $(n+1)$D ball, so its wireframe can be induced from that of the $(n+1)$D ball. The classical triangulation of the unit ball provides the following wireframe model: let $\mathbf{e}_1, \ldots, \mathbf{e}_{n+1}$ be a basis of $\mathcal{R}^{n+1}$, then for any $1 \leq i < j \leq n+1$, the unit circle is drawn on the coordinate plane of $\mathbf{e}_i, \mathbf{e}_j$. All together there are $n(n+1)/2$ circles connecting $2n+2$ points $\pm\mathbf{e}_k$

for $1 \leq k \leq n+1$, and there are $2n+2$ rectilinear edges connecting the center
of the sphere with $\pm e_k$. Every triangular cycle is the boundary of a 2D cell,
and every closed r-chain composed of $r+2$ different rD cells is the boundary of
an $(r+1)$D cell. Any boundary r-chain is filled with an $(r+1)$D flat patch if
it contains any rectilinear edge, or an $(r+1)$D spherical patch otherwise. The
$(n+1)$D ball is composed of 2^{n+1} cells of dimension $(n+1)$. The reconstruction
from the wireframe is unique topologically.

Example 4. nD spheres, another wireframe model.

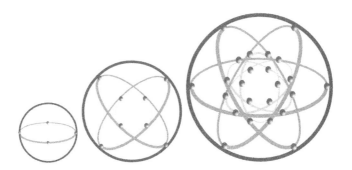

Fig. 4. nD spheres. Here $n = 2, 3, 4$

The following wireframe model of nD sphere is not induced from $(n+1)$D
ball. It consists of $(n-1)2^{n-2}$ pairs of antipodal points on the sphere and 2^{n-1}
great circles connecting them. The great circles are divided into n layers. The
layers, numbered from 1 to n, are from the outermost of the wireframe to the
innermost. The i-th layer for $1 \leq i \leq n$, contains C_{n-1}^{i-1} circles; each circle in the
layer contains $n-i$ circles of layer $i+1$ and is contained in $i-1$ circles of layer
$i-1$. So each circle contacts $n-1$ other circles, and all its contacting circles are
in the neighboring layers.

The above model can be interpreted as follows: the image of any 2D sphere
under a parallel projection onto the plane can be represented by a circle circum-
scribing an ellipse at two antipodal points. The circle is the contour of the sphere,
and the ellipse is any great circle on the sphere. So we can make the regulation
that for a pair of great circles on the nD sphere to represent a 2D sphere, it is
necessary and sufficient that their images under a fixed parallel projection are
a closed 1-chain composed of two ellipses in contact at two antipodal points, in
which one ellipse is contained in another. In the wireframe, every closed 1-chain
of this type is the boundary of a 2D sphere, and every closed r-chain composed
of 2^r different rD spheres is the boundary of an $(r+1)$D sphere. Any boundary
r-chain is filled with an $(r+1)$D sphere. The reconstruction from the wireframe
is unique topologically.

For over two decades, the wireframe representation of 3D object has been an
active research topic in computer-aided design [1] and [4], computer graphics [3],

[10], [15] and computer vision [13], [9], [6]. A 2D line drawing is the projection of a wireframe object where all the edges and vertices are visible, which makes it possible to reconstruct the complete 3D model. The reconstruction consists of two steps: face identification and 3D geometry reconstruction. Face identification is to find those closed cycles which are boundaries of faces of a prescribed filling pattern. 3D geometry reconstruction is to determine the spatial positions and orientations of the faces.

In this paper, we study the wireframe representation of nD object from the projective geometric point of view. In particular, we study the face identification of a high dimensional wireframe model from its single 2D line drawing, under the most general assumption that the dimension of the object is unknown, the dimension of the surrounding space is unknown, two neighboring faces of the object can be in the same plane, and whether or not the object is a manifold is unknown.

When $n > 3$, we have not found any publication on face identification. When $n = 3$, most algorithms in the literature consist of two steps: searching for cycles in the line drawing, and searching for faces from the cycles. In searching for cycles, a graph-based method is proposed by [12], and further improved by a depth-first searching algorithm of [7]. If the object is a 2D manifold, i.e., the boundary of a 3D solid, the method proposed by [8] is very efficient in reducing the number of redundant cycles which cannot be assigned as faces. All these methods are *global* in that the searching is within the whole wireframe. A consequence is that the number of cycles potential for being faces is usually much larger than the number of correct faces. In searching for faces from the cycles, [12] used an optimization-based method, [7] used a maximal weight clique finding algorithm, and [8] used an algorithm of backtracking the state-space tree. Since the searching for faces from the cycles is a NP-complete problem, the fastest algorithm can only handle wireframe models of less than 30 faces.

In this paper we propose a very efficient algorithm for face identification. When the dimension of the surrounding space is three, the algorithm outperforms all other algorithms on face identification in both speed and range of application. For all the examples in [12], [7], [8], our algorithm can generate all the solutions for ambiguous wireframe models, and generally do not produce any redundant cycles that are not assigned as faces. In other words, the step of searching for faces from the cycles is generally skipped. Moreover, our algorithm can handle complicated 3D objects of over 10,000 faces.

The idea is to *locally construct* the $(r + 1)$D faces and *delete* useless rD faces. Starting from a vertex called the origin, we *localize* the wireframe by considering only the subgraph of the origin and its neighboring vertices. Within the local wireframe, we look for rD cycles corresponding to $(r + 1)$D faces based on topological and geometric consideration. The identified faces can reduce the searching scope by blocking off some branches. For an rD face, the number of $(r + 1)$D faces containing it is sufficient to prevent any more new $(r + 1)$D faces from passing through it, then it can be *deleted*. After the construction and deletion, we set the origin to be the current local wireframe, and repeat the previous localization, construction and deletion procedure. The older origins are gradually deleted, so that the local wireframe remains a medium size.

In Section 2 we investigate some properties of the wireframe representation. In Section 3 we propose some techniques for high dimensional face identification. In Section 4 we propose an object detection algorithm together with some experimental results.

2   Wireframes and Line Drawings

We assume that the wireframe representation of an nD object is obtained by a perspective projection from an mD projective space to a projective plane, where $m \geq n$, such that all the edges and vertices are revealed, and if three vertices are not collinear in the mD space, nor should their images. Let us see if such projection exists.

2.1 nD Perspective Projection

The perspective projection from $\mathcal{P}^n$ to $\mathcal{P}^2$ can be invariantly represented in Grassmann algebra or Clifford algebra [5]. In Grassmann algebra, the exterior product of r vectors is called a *decomposable r-extensor*, or *r-blade*. The product is denoted by juxtaposition of elements. Let X_{n-2} be a nonzero $(n-2)$-blade in the Grassmann algebra over the base vector space $\mathcal{R}^{n+1}$, i.e., $X_{n-2} = \mathbf{x}_1\mathbf{x}_2\cdots\mathbf{x}_{n-2}$ where the $\mathbf{x}$'s are linearly independent vectors. The perspective projection with center X_{n-2} is

$$\mathbf{x} \mapsto X_{n-2}\mathbf{x}, \text{ for } \mathbf{x} \in \mathcal{R}^{n+1}. \tag{1}$$

It can also be taken as the composition of $n-2$ perspective projections with centers $\mathbf{x}_1, \mathbf{x}_2, \ldots, \mathbf{x}_{n-2}$ respectively.

In coordinate representation, let $\mathbf{e}_1, \ldots, \mathbf{e}_{n+1}$ be a basis of $\mathcal{R}^{n+1}$. We can always find three vectors $\mathbf{e}_{i_1}, \mathbf{e}_{i_2}, \mathbf{e}_{i_3}$ from the basis such that they form a new basis with vectors $\mathbf{x}_1, \mathbf{x}_2, \ldots, \mathbf{x}_{n-2}$, i.e., the $(n+1) \times (n+1)$ determinant $f = \det(X_{n-2}\mathbf{e}_{i_1}\mathbf{e}_{i_2}\mathbf{e}_{i_3}) \neq 0$. Rearrange the indices so that the three vectors in f have indices $1, 2, 3$. Then

$$\mathbf{e}'_1 = X_{n-2}\mathbf{e}_1/f, \ \mathbf{e}'_2 = X_{n-2}\mathbf{e}_2/f, \ \mathbf{e}'_3 = X_{n-2}\mathbf{e}_3/f \tag{2}$$

is a basis of the projective image plane. Any $X_{n-2}\mathbf{e}_i$ for $i > 3$ can be written as a linear combination of $\mathbf{e}'_1, \mathbf{e}'_2, \mathbf{e}'_3$:

$$\begin{aligned} X_{n-2}\mathbf{e}_i = \det(X_{n-2}\mathbf{e}_i\mathbf{e}_2\mathbf{e}_3)\mathbf{e}'_1 + \det(X_{n-2}\mathbf{e}_i\mathbf{e}_3\mathbf{e}_1)\mathbf{e}'_2 \\ + \det(X_{n-2}\mathbf{e}_i\mathbf{e}_1\mathbf{e}_2)\mathbf{e}'_3. \end{aligned} \tag{3}$$

The coordinate representation of the perspective projection (1) is that for any $\mathbf{x} = x_1\mathbf{e}_1 + x_2\mathbf{e}_2 + \cdots x_{n+1}\mathbf{e}_{n+1}$,

$$X_{n-2}\mathbf{x} = (x_1 \det(X_{n-2}\mathbf{e}_1\mathbf{e}_2\mathbf{e}_3) + \sum_{i=4}^{n+1} x_i \det(X_{n-2}\mathbf{e}_i\mathbf{e}_2\mathbf{e}_3))\mathbf{e}'_1$$

$$+(x_2 \det(X_{n-2}\mathbf{e}_1\mathbf{e}_2\mathbf{e}_3) + \sum_{i=4}^{n+1} x_i \det(X_{n-2}\mathbf{e}_i\mathbf{e}_3\mathbf{e}_1))\mathbf{e}'_2$$

$$+(x_3 \det(X_{n-2}\mathbf{e}_1\mathbf{e}_2\mathbf{e}_3) + \sum_{i=4}^{n+1} x_i \det(X_{n-2}\mathbf{e}_i\mathbf{e}_1\mathbf{e}_2))\mathbf{e}'_3.$$

Given finitely many vertices $\mathbf{x}_1, \ldots, \mathbf{x}_m$ which do not lie in the same projective line, we want to find a perspective center X_{n-2} such that (1) every vertex has an image, (2) no two vertices are projected onto the same image point, (3) no three non-collinear vertices are projected onto the same image line. It is easily seen that the above constraints are equivalent to

$$\det(X_{n-2}\mathbf{x}_i\mathbf{x}_j\mathbf{x}_k) \neq 0, \text{ for any } \mathbf{x}_i, \mathbf{x}_j, \mathbf{x}_k \text{ not on the same line.} \qquad (4)$$

Almost all $(n-2)$-blades satisfy (4).

2.2   Constraints on Wireframe Models

We consider the wireframe models whose fillings are affine flat patches unanimously, called *rigid wireframe models*. The filling has a strong constraint, called the rD *face adjacency constraint* [12]: if two rD faces share two $(r-1)$D faces of different $(r-1)$D planes, they must belong to the same rD plane.

By a perspective projection, a 2D face is projected onto a patch in the image plane whose boundary can only go across itself at a vertex. The 2D *non-self-intersection constraint* [6] says that (1) if a cycle has two edges intersecting not at a vertex, then it can not be a 2D face, (2) if two edges intersect not at a vertex then they cannot be coplanar.

In a wireframe model, it often occurs that three vertices are collinear. Such a *permanent collinearity constraint* must be provided additionally in its wireframe representation. Similarly, any rD *permanent coplanarity constraint* must be provided additionally. In this paper we only consider the case $r = 1$. The edges in a permanent collinearity constraint form a *line*. The connected components of a line are the *real parts*, while the virtual connections between real parts are the *virtual parts* [8].

If the enclosed regions of two 2D faces intersect, the faces must be coplanar. Let there be two 1D cycles sharing exactly two vertices and the vertices are not connected either by an edge or by a line. If in the image plane, the line segment between the two vertices intersects both the enclosed regions of the cycles, then the two cycles must be coplanar to be assigned as faces. This is the 2D *non-interior-intersection constraint* [8]. It has no rD generalization, because the interior and the exterior of an rD face for $r > 2$ cannot be distinguished in a line drawing.

If an rD face F_r is not in an rD cycle but their intersection is exactly the constituent $(r-1)$D faces of F_r, then F_r is a *chord* of the cycle [12]. A face with chord can always be decomposed along its chord into two coplanar faces sharing one common side. On the other hand, if a cycle with chord is not assigned as a face, then it can be decomposed into two shorter ones, and by assigning them to be faces they do not need to be coplanar. To gain more degree of freedom, an rD cycle with chord is not assigned as a face. This is the rD *non-chord constraint*.

2.3   Face Identification

A 0D face is a vertex, a 0D *cycle* is a pair of vertices of an edge, a 1D face is an edge. For $r > 0$, an rD *cycle* is a set of rD faces such that (1) if two faces intersect, the intersection belong to their iD faces for $0 \le i < r$, (2) any $(r-1)$D face of one rD face is shared by exactly one other rD face in the set, (3) not all rD faces belong to the same rD plane.

All edges are 1D faces. For $r > 1$, the rD *face identification* is to select a subset called rD *faces*, from the $(r-1)$D cycles satisfying the constraints in the previous subsection, and fill them with a prescribed rD pattern. In previous work on face identification, it is generally assumed that any two neighboring faces are not coplanar. We feel that this is too strong a constraint to include many interesting models, so we discard it.

Even when $n = 3$, face identification is a difficult problem. The number of cycles of a graph is generally exponential in the number of vertices [8]. Given that all the cycles satisfying the constraints are found, the number of possible combinations of potential faces is exponential in the number of the cycles. When n is unknown, face identification becomes increasingly more difficult. As a result, the solution is generally not unique.

To design an algorithm for this double-exponential complexity problem, the goal should be first on the detection of the highest dimension n, second on the finding of as many different nD pieces as possible, third on the derivation of as many different solutions in dimension n as possible. To achieve this, the **design guideline** is *to produce as many rD faces on different rD planes as possible.*

2.4   Some Terminology on Wireframe Models

Neighbor: Two rD faces are *neighbors* if their intersection is $(r-1)$-dimensional.

Simple cycle: An rD cycle is said to be *simple* if its rD faces satisfy the rD non-chord constraint and 2D non-self-intersection constraint.

Rigid cycle and face: An rD cycle is said to be *rigid* if when its rD faces are in general position in the surrounding space, the minimal affine space containing the cycle is $(r+1)$-dimensional.

Any 2D rigid cycle is triangular. Any rD rigid simple cycle is the boundary of an $(r+1)$D flat patch. A face assigned to a rigid cycle is said to be *rigid.*

Elastic cycle and face: An rD cycle is said to be *elastic* if when its rD faces are in general position, the minimal affine space containing the cycle is $(r+2)$-dimensional.

Any 2D elastic cycle is square. Any rD rigid simple cycle is generally the boundary of two $(r+1)$D flat patches in an $(r+2)$D affine space. A face assigned to a elastic cycle is said to be *elastic.*

Plastic cycle and face: An rD cycle is *plastic* if it is neither rigid nor elastic. Its assigned face is said to be *plastic.*

Grid: A 1D grid is a line. For $r > 1$, an rD *grid* consists of at least two rD faces. It is obtained from a set C originally containing only one rD face B_r by repeatedly putting into C all the iD faces $(0 \leq i < r)$ in the same $(r-1)$D plane with an $(r-1)$D face of C.

Polyface: If two or more rD faces or grids belong to the same rD plane but are not in the same rD grid, they can merge together to form an rD *polyface*, which is a patch of the plane. Whenever a new vertex is joined, the polyface is said to be a *new* one.

Contradictory grid and polyface: A grid is said to be *contradictory* if it violates the 2D non-self-intersection constraint. A new polyface is said to be *contradictory* if it violates either the non-self-intersection constraint or the non-interior-intersection constraint, or it includes two existing faces not belonging to the same rD plane.

Degree: The *degree* of an rD face F_r with respect to a set of $(r+1)$D faces is the number of $(r+1)$D faces containing F_r.

Stretch: An rD *stretch* is either an rD face containing no $(r-1)$D face of degree two, or a maximal number of connected rD faces in which any rD face shares at least one $(r-1)$D face of degree two with another rD face. The *boundary* of an rD stretch is the $(r-1)$D faces included in only one rD face of the stretch. In [8], a 1D stretch is called an unambiguous path.

Saturation: An rD face F_r is said to be *saturated* in a wireframe model if it has an $(r-1)$D face B_{r-1} such that every rD face of the wireframe model passing through B_{r-1} is in an $(r+1)$D face containing F_r. An rD stretch not lying in an rD plane is said to be *saturated* if it is contained in an $(r+1)$D face.

Fork: Let F_r be an rD face, grid or polyface of a wireframe model. Let B_{r-2} be an $(r-2)$D face of F_r. If not every $(r-1)$D face of the wireframe model passing through B_{r-2} is in an rD face, grid or polyface having $(r-1)$D intersection with F_r, then B_{r-2} is called an $(r-2)$D *fork* of F_r.

When F_r is an rD grid and C_{r-1} is an $(r-1)$D face of F_r, if not every rD face of the wireframe model passing through C_{r-1} is in an $(r+1)$D face, grid or polyface having rD intersection with F_r, then C_{r-1} is called an $(r-1)$D *fork* of F_r.

3   Face Identification Techniques

3.1   Deletion

When $n = 3$, the idea of deleting edges in cycle searching can be found in [8]. There the object is a 3D solid in which any edge occurs in exactly two faces, so if two faces containing the edge have been found, the edge can be deleted.

In high dimensional cycle searching for a general object, *deletion* is very important because it reduces the search scope. An rD stretch can be deleted if the deletion does not influence the assignment result of $(r+1)$D faces, grids and polyfaces.

Theorem 1. [Deletion Theorem] Let F_r be a saturated rD stretch. In searching for $(r+1)$D faces, F_r can be deleted if one of the following conditions is satisfied:

(1) F_r is in an rD plane, but neither in any rD grid or polyface nor in any $(r+1)$D grid or polyface. Through every rD face neighboring F_r there is an $(r+1)$D face which contains F_r and in which there is at most one rD face not neighboring F_r.

(2) $r = 1$. After other previous deletions, in the remaining part of any 1D grid or 2D face, grid or polyface containing F_1, the deletion of F_1 does not disconnect any two path-connected fork vertices.

Proof. (1) Let the number of $(r-1)$D faces in F_r be k. Let $B_1, \ldots B_p$ be all the rD faces neighboring F_r. Since F_r is not in any rD or $(r+1)$D grid or polyface, the B's and F_r are pairwise not in the same rD plane. Let $C_1, \ldots, C_q$ be all the $(r+1)$D faces containing F_r. Then each C_i contains exactly k of the faces B's, and the C's are pairwise not in the same $(r+1)$D plane. If any new rD cycle passing through F_r is a face, it has to form a new polyface with one of the C's, say C_i. Then the new cycle has to pass through all the neighboring rD faces of F_r in C_i. As a result, the cycle must have the chord which is the unique rD face of C_i not neighboring F_r. By the non-chord constraint, the cycle cannot be assigned as a face.

(2) If F_1 is part of a line, since all fork vertices of the line are at the same side of F_1, any new chordless cycle containing part of the line does not passes through F_1. Assume that F_1 is not on any line. If any new cycle passing through F_1 is a face, it has to form a new polyface with one of the faces, grids and polyfaces containing F_1. If F_1 is deleted, since in the remaining part of any of the above faces, grids and polyfaces, any two fork vertices are still connected, the deletion does not influence the forming of the new polyfaces.

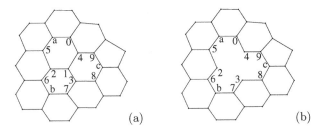

Fig. 5. Deletion and localization

Example 5. In Figure 5(a), three cycles $125a04$, $138c94$ and $126b73$ are already 2D faces. Then $12, 13, 14$ are all saturated stretches, and vertex 1 is not a fork of any face. Edge 12 is contained in two faces, and its deletion does not disconnect any two forks in either face. So edge 12 can be deleted. Similarly, edges $13, 14$ can be deleted. Vertex 1 can also be deleted because it is no longer connected with any other vertex. The result is Figure 5(b).

3.2   Localization

Let W be a wireframe model. A *local wireframe model* of W is a subset of the vertices of W together with all the rD faces formed by the vertices. A *localization filter*, or *localization*, of W is a sequence of local wireframe models $S_1 \subset S_2 \subset \ldots \subset S_k = W$ in which each successor introduces more vertices than its predecessor.

Localization is very important in face identification. With the introduction of new vertices, all the edges among them and the existing vertices are introduced, so are all the higher dimensional connections. By introducing new vertices according to the closeness of their relations with the existing ones, the complexity of cycle searching can be reduced. Below are some typical localizations.

The first is the *neighbor localization*: (1) Start from a vertex and put it in an empty set C. (2) Put all the neighboring vertices of C into C. (3) Repeat step 2 until all vertices are in C. In neighbor localization, the older elements in C may be deleted earlier because their contributions to cycle generation are exhausted earlier.

The second is the *family localization*: (1) Start from a set C containing only one vertex, put all the neighboring vertices of C into C. (2) Put the vertices which are common neighbors of at least two vertices of C into C, and if no such vertices then use step 2 of the neighbor localization. (3) Repeat step 2 until all vertices are in C. This localization makes easier the generation of new cycles.

The third is the *couple localization*: Steps 1 and 3 are the same with those in the family localization. Step 2 is to put the following vertices into C: each vertex is a common neighbor of at least two vertices of C, and the vertex forms a triangular or square simple cycle with some vertices in C. If there is no such vertex then use step 2 of the family localization.

The fourth is the *rigid couple* (or *family*, or *neighbor*) *localization*: Steps 1 and 3 are the same with those in the couple (or family, or neighbor) localization. Step 2 is to put all the vertices collinear with at least two vertices of C into C, and if there is no such vertex then use step 2 of the couple (or family, or neighbor) localization.

Example 6. In Figure 5(a) there is neither square nor triangular cycle, nor collinearity constraint. In the neighbor localization, we have

$$\{1\} \subset \{1,2,3,4\} \subset \{1,2,3,4,5,6,7,8,9,0\}$$
$$\subset \{1,2,3,4,5,6,7,8,9,0,a,b,c\}.$$

By this step the edges $12, 13, 14$ can be deleted, so is the isolated vertex 1. The next step will be another "wave propagation" starting from the cycle of vertices $26b738c940a5$.

3.3   Cycle Searching and Blocking

rD cycles are classified into rigid, elastic and plastic cycles according to their rigidity. Rigid cycles are always faces, so they should be found first. rD rigid cycles can be found by the following algorithm. For every pair of rD faces sharing one $(r-1)$D face:

(1) Let C be a set containing only the pair.

(2) Repeatedly put into C all the rD faces, grids and polyfaces sharing with C at least two $(r-1)$D faces of different $(r-1)$D planes.

(3) Search in C for an rD cycle containing the pair.

Elastic cycles are most likely to be faces because they have the next strongest rigidity. In the above rigid cycle searching algorithm, if the pair of rD faces is replaced by a triplet of rD faces F_1, F_2, F_3 such that F_1, F_2 and F_2, F_3 are neighbors but F_1, F_3 are not, the algorithm can also be used to find rD elastic cycles.

To find a general rD cycle passing through a fixed rD face, the depth-first searching strategy is necessary. Finding all cycles, or less ambitiously, finding all simple cycles, is a NP-complete problem. For a wireframe of 3D solid, [8] proposed to find only those cycles satisfying properties of real faces by the depth-first-searching-with-properties (DFSP) strategy, i.e, to use constraints to reduce branches in the searching.

2D non-self-intersection constraint can block some branches permanently. The blocks are called *intersection blocks*. 2D non-interior-intersection constraint can block some branches from including the interior of an existing 2D face. The blocks are called *face blocks*. rD non-chord constraint can permanently block some branches from generating cycles with chords. The blocks are called *chord blocks*. To further reduce the number of redundant cycles, the above blocks are far from being enough.

In [8], the models are 3D solids in which no two neighboring faces are coplanar. If a cycle is identified as a face then no other path through two edges of it can generate a face. This blocking is based on 2D face-adjacency constraint, and is extremely efficient in reducing the number of branches: among the 11 examples they tested, 3 have no redundant cycles at all, and 7 each have less than four redundant cycles.

The idea can be readily generalized to rD case by the rD face adjacency constraint. If an rD branch intersects an rD face, grid or polyface F and the intersection has rD affine closure, then the branch is blocked by F. The block is called a *face, grid* or *polyface block*. There are two typical cases: (1) the intersection is two $(r-1)$D faces of different $(r-1)$D planes, (2) the intersection contains an $(r-1)$D face and another face not in the same $(r-1)$D plane.

If the blocks of a branch are all face (grid, polyface) blocks of the same (grid, polyface), then the branch and the (grid, polyface) form an new rD polyface. If the polyface is not contradictory then it is accepted. This cycle searching strategy can be named *depth-first searching with blocks* (**DFSB**).

Example 7. Vertices $1, 4, 9$ and $5, 8, 0$ are respectively collinear. Assume that 1364 and 4687 are already identified as faces. Find all faces passing through edge 90.

First extend 90 to a stretch 4908. Then $41, 47, 85, 87$ are blocked by face 4687 according to the 2D non-interior-intersection constraint. At vertex 4, edge 46 is

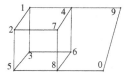

Fig. 6. Face blocks

blocked by faces 1364 and 4687; at vertex 8, edge 86 is blocked by face 4687. So every path through the stretch is blocked by face 4687; the stretch has to be coplanar with the face. This coplanarity is forbidden by the two different face blocks over edge 46. The stretch cannot be in any face.

3.4 Combination

Let $S_1 \subset S_2$ be two local wireframe models, and assume that all $(r+1)$D faces in S_1 have been identified. In S_2, there are two kinds of new rD cycles: a cycle containing only one vertex in $S_2 - S_1$ is called a *singleton* of the vertex; a cycle containing at least two vertices in $S_2 - S_1$ is called a *twin*. In face identification, elastic cycles have priority over plastic cycles, and singletons have priority over twins. Thus, there are four **levels of priority** in new cycles which are not rigid.

Assume that all cycles satisfy the constraints in Subsection 2.2. Within the same level, if multiple cycles are generated, it must be determined if they are degrading or not. A set of rD cycles is said to be *degrading*, if when assigning all of them to be new faces, one cycle in the set forms a new polyface with an existing face or another cycle in the set. For a set of degrading cycles, a face assignment with minimal number of new polyfaces is optimal.

The technique *combination* is to find an optimal face assignment for a level of degrading cycles. Assume that all new elastic cycles are generated at the same time, only after their face identification are the plastic cycles generated. Below we consider only elastic cycles. The plastic ones are similar.

Step 1. Singletons. Let V be the vertices in $S_2 - S_1$ each of which has degrading rD singleton set. Let there be l elements in V, and let the number of singletons of vertex i be k_i. By choosing one singleton for each vertex, there are $k_1 k_2 \cdots k_l$ different combinations. Delete the combinations which are a degrading set. Among the combinations left, find those leaving the maximal number of new elastic cycles from forming new polyfaces. Such combinations are optimal.

If there is only one optimal combination, then the singletons in the combination are assigned as faces. If there are multiple optimal combinations, they must be saved as nodes in the *state-space tree* [8]. Starting from each node, the cycle searching continues, with all singletons in the node assigned as faces.

Step 2. Singletons and twins. The singletons and twins not in the optimal combination are selected one by one according to the principle that its acceptance as a face does not produce any contradictory polyface. Among the selected set T, find a subset whose assignment as faces produces from T the maximal number of $(r + 1)$D faces on different $(r + 1)$D planes. If there are multiple such subsets then the state-space tree extends to accept them as new nodes. The cycles not assigned as faces are redundant.

Example 8. A diagonal is drawn in a cube, which makes the face identification very difficult to reach dimension three [12]. The family localization produces new vertex series

$$1 \longrightarrow 2, 3, 4, 5 \longrightarrow 6, 7, 8. \tag{5}$$

Only in the last local model do the cycles appear. The elastic simple singletons are

$$(6 : 1264, 1364); \quad (7 : 1275, 1273, 1375); \quad (8 : 1284, 1485). \tag{6}$$

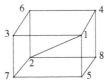

Fig. 7. Combination

The elastic simple twins are $2637, 2648, 2758$. The singleton set of each new vertex is degrading. There are 12 combinations, among which only the combination $(1364, 1375, 1485)$ prevents the maximal number of elastic cycles from forming polyfaces. This number is three, which is exactly what we need to build the 3D model. No state-space tree is generated.

3.5   Repair and Recovery

The virtual parts of lines usually lead to the generation of plastic cycles. The *repair and recovery* technique is to fill each virtual part with a virtual edge, make face identification in the repaired wireframe, and then recover the original wireframe by merging among the faces containing the virtual edges. The technique is very helpful in simplifying 1D cycle searching and avoiding the generation of "topologically correct but geometrically impossible" solutions [12].

The time to recover an original line is when the repaired line is to be completely deleted. First find all faces containing any virtual edge of the line, called *virtual faces*. For every virtual edge, its virtual faces on the same side of the line are merged in pairs, and new cycles are formed by deleting the virtual edge. There may be multiple combinations, so the state-space tree may be extended.

After the merging there are two possibilities: (1) If there are two faces unmerged, one on each side of the virtual edge, then they can merge to form a new cycle. (2) If there is only one face unmerged, then delete it. The newly formed cycles are accepted as new faces if they are simple and do not produce contradictory polyfaces.

Example 9. In Figure 8, there is one line 1563 with virtual part 56. Without repairing, cycles 123685 and 143675 may be identified as faces, against the 2D non-interior-intersection constraint. After repairing, the eight triangular cycles are easily found, which block the generation of any other cycle. In recovery, there is a unique pattern to recover two faces from the four virtual faces of edge 56.

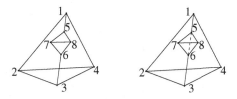

Fig. 8. Repair and recovery

4   Object Detection Algorithm and Practice

Input: (1) a 2D line drawing composed of vertices and edges. A vertex is represented by 2D coordinates, an edge by two vertices.
(2) A set of lines. A line is represented by a sequence of edges.
(3) A vertex as the origin of localization. The default is any vertex of maximal degree.

Output: Objects of dimension > 1: faces, grids and polyfaces.

Initialization: (1) Find all pairs of edges and lines intersecting not at a vertex.
(2) Repair lines.

Step 1. Localization:(1) Start from the origin, use rigid family localization to generate a set of new vertices S. (2) Set $r = 1$. Set the root node of the state-space tree to be empty.
Step 2. Cycle searching:
(1) Find all new rD rigid simple cycles.
(2) Find all new rD elastic simple cycles and make $(r + 1)$D face identification. Use the combination technique if necessary. If there are multiple optimal combinations then put current information of all the dimensions into the current node of the state-space tree, generate descendant nodes to save the optimal combinations.
(3) Find all new rD plastic simple cycles and make $(r + 1)$D face identification.
(4) In the end of face identification for every level of new cycles, use the Deletion Theorem to delete rD stretches. If a line with virtual edges is completely deleted, then recover the real faces of the original wireframe. Delete any rD stretch having degree-one $(r - 1)$D face repeatedly.
(5) If S does not contain all the vertices of the line drawing, then set the origin to be S, and go back to Step 1.
Step 3. Exploring the state-space tree: If the state-space tree is not empty, find a new node by depth-first searching. Start from the node to make another interpretation of the line drawing.

Remark. When Step 2 finishes, a complete interpretation of the line drawing is obtained, which may not reach the highest dimension. Further exploration of the state-space tree gives different interpretations. An alternative is to change the origin, run the algorithm again to generate a new interpretation. In our experiments, the state-space tree is generally always empty, and changing the origin is generally the only way to get a different interpretation; if a default origin is used, then an interpretation reaching the highest dimension is always obtained without further exploring the state-space tree.

Example 10. In Figure 9(a) there are two lines 1584 and 2673. As a 2D manifold, it has 16 faces. By the algorithm in [8], 21 redundant cycles are produced and further selection is needed to figure out the real ones. By our algorithm, 1 grid (composed of 3 faces) and 18 faces outside the grid are found, besides 4 rigid faces of dimension three. No redundant cycle is generated.

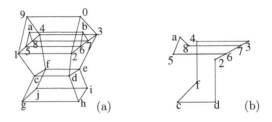

Fig. 9. Manifold reconstruction

The results are different because we do not assume that the surrounding space is 3D and the object is a single solid. The 2D manifold interpretation can be easily derived from our 3D result once their assumption is provided. Without the assumption, the triangular cycles $149, 58a, 67b, 230$ and the square cycle $cdef$ are all faces, while the cycles $15a849, 26b730$ are not.

Below we show the procedure of 2D face identification in Figure 9(a). The rigid family localization produces new vertex series:

$$1 \longrightarrow 2, 5, 8, 4, 9, c \longrightarrow 6, 7, 3, a, d, 0, f \longrightarrow b, e \longrightarrow g, h, i, j.$$

New cycles in the second local wireframe model: rigid cycle 149.

New cycles in the third local wireframe model:

(1) Rigid cycles $58a, 230$.
(2) Vertex 6 has one elastic singleton 1256. Grid 12345678 is formed.
(3) Vertex d has one elastic singleton $12cd$.
(4) Vertex 0 has one elastic singleton 1290.
(5) Vertex f has one elastic singleton $14cf$.
(6) There is one elastic twin 9034.
(7) The deletable stretches are $12, 1c, 02, 03, 09, 19, 49, 1584$. The fork vertices on the grid is $2, 3$. No plastic cycle is found. The local wireframe after the deletion is shown in Figure 9(b).

New cycles in the fourth local wireframe model:

(1) Rigid cycle $67b$.
(2) Vertex b has two elastic singletons: $56ba, ab78$.
(3) Vertex c has three elastic singletons: $cdfe, 43fe, 23de$.
(4) The deletable stretches are $2673, ab, 5a, 8a, 6b, 7b, 34, 56, 78$.
No plastic cycle is found. The local wireframe after the deletion is a square $cdef$.

New cycles in the last local wireframe model:

(1) No elastic singleton.
(2) Five elastic twins: $cdhg, cfjg, feij, deih, ijgh$.
(3) All edges are deleted.

We have implemented the algorithm in VC++ 6.0, have tested all the examples in [7], [8], [12], [1], in addition to higher dimensional examples made by ourselves. They amount to 32 examples of dimension within three and 12 examples of dimension over three.

The most striking example is k copies of Figure 9(a) connected by cubes, see Figure 10. When $k = 100$, there are 2496 faces of dimension two, among which 300 are in 100 grids and 2296 are not. They are found within 55 seconds by an IBM PC of Intel 2.60GHz CPU and 248MB RAM. When $k = 401$, there are 10021 faces, among which 1203 are in 401 grids and 9219 are not. They take 76 minutes to be found.

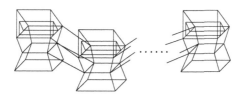

Fig. 10. k copies of Figure 9(a) connected by $k - 1$ cubes

The table in the end of this paper lists some examples we tested and their high dimensional interpretations. In the table, V, E, L denote the number of vertices, edges and lines respectively; F_i, G_i, P_i denote the iD faces, grids and polyfaces, where only the faces not belonging to any grid or polyface are counted.

Figure 11 (1): This is an example for which the algorithm fails to find all 3D interpretations. because of the two collinearity constraints, the cube on the left cannot coexist with the column of pentagon top and bottom on the right. Since a pentagon cycle is plastic and a square cycle is elastic, the current priority regulation disallows the column to be found first. Indeed, in the column interpretation, at least three square cycles are redundant, while in the cube interpretation, only one pentagon cycle is redundant.

Figure 11 (13): A left vertex of a cube is dragged to the right. It is no longer 3D geometrically.

Figure No.	V, E, L	F_2, G_2, P_2	F_3, G_3, P_3	F_4	F_5
6	10, 15, 2	6, 0, 0	1, 0, 0		
7	8, 13, 0	6, 0, 0	1, 0, 0		
8	8, 12, 1	6, 0, 0	1, 0, 0		
9	20, 36, 2	18, 1, 0	4, 0, 0		
11(1)	15, 25, 2	12, 0, 0	1, 0, 0		
11(2)	16, 25, 1	12, 0, 0	2, 0, 0		
11(3)	20, 30, 5	10, 0, 1	2, 0, 0		
11(4)	16, 24, 1	10, 0, 0	1, 0, 0		
11(5)	22, 33, 10	7, 0, 2	0, 1, 0		
11(6)	20, 30, 2	10, 0, 0	1, 0, 0		
11(7)	23, 37, 3	11, 3, 0	0, 0, 1		
11(8)	18, 29, 0	14, 0, 0	2, 0, 0		
11(9)	24, 28, 1	2, 0, 0			
11(10)	17, 26, 2	11, 0, 0	1, 0, 0		
11(11)	48, 72, 0	26, 0, 0	1, 0, 0		
11(12)	48, 72, 0	26, 0, 0	1, 0, 0		
11(13)	8, 12, 0	4, 0, 0			
11(14)	16, 28, 2	10, 1, 2	1, 1, 0		
11(15)	26, 39, 2	12, 1, 0	1, 0, 0		
11(16)	12, 20, 0	7, 0, 2	0, 0, 1		
11(17)	10, 21, 0	18, 0, 0	7, 0, 0	1	
11(18)	14, 21, 2	7, 1, 0	1, 0, 0		
11(19)	16, 24, 1	$\begin{cases} 10, 0, 0 \\ 10, 0, 0 \end{cases}$	$\begin{matrix} 1, 0, 0 \\ 1, 0, 0 \end{matrix}$		
11(20)	9, 15, 2	$\begin{cases} 4, 1, 0 \\ 3, 0, 1 \\ 2, 0, 1 \end{cases}$			
11(21)	8, 16, 0	14, 0, 0	6, 0, 0	1	
11(22–24)	8, 24, 0	38, 0, 0	40, 0, 0	25	3
11(25)	16, 32, 0	24, 0 ,0	8, 0, 0	1	
11(26)	24, 38, 5	9, 3, 0	1, 0, 0		
11(27)	9, 18, 0	15, 0, 0	6, 0, 0	1	
11(28)	40, 50, 0	54, 0, 0	14, 0, 0	1	
11(29)	42, 68, 2	32, 0, 2	7, 0, 0		
11(30)	15, 50, 25	0, 10, 0	0, 1, 0		

Figure 11 (19): There are two different interpretations having the same number of 2D and 3D objects. This is a classical example of ambiguous wireframe [12].

Figure 11 (20): Diagonal 28 intersects side 14 in a cube. This is no longer 3D geometrically. There are various 2D interpretations, depending on where to start the localization.

Figure 11 (21): The diagonals in two faces of a cube cross but do not intersect. This is impossible in 3D, but explicable in 4D.

Figure 11 (22–24): They are wireframes of the same 5D object by different perspective projections. The interpretations are identical.

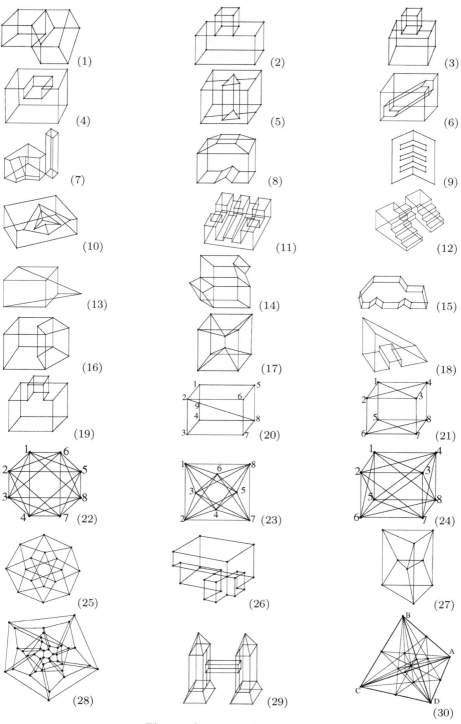

Fig. 11. Some tested examples

Figure 11 (28): Two copies of a dodecahedron are connected by connecting their corresponding vertices. The resulting wireframe has a 4D interpretation.
Figure 11 (30): This is the barycentric subdivision of tetrahedron $ABCD$.

5   Conclusion

This may be the first time that the wireframe representation of nD object is studied algorithmically. Various properties of the representation are studied from the projective geometric point of view. For wireframe models with rigid filling pattern, a very efficient algorithm for face identification is designed and implemented, under the most general assumption.

For high dimensional interpretation of a single 2D curvilinear line drawing with non-rigid filling pattern, with the aid of virtual vertices and virtual rectilinear edges in [8], the techniques developed in this paper can be readily applied.

References

1. Agarwel, S., and Waggenspack, W. :Decomposition method for extracting face topologies from wireframe models. Computer Aided Design **24**, 3, 123–140, 1992.
2. Asimov, D. : The grand tour: A tool for viewing multidimensional data. SIAMJ. Scientific and Statistical Computing **6**, 1, 128–143, 1985.
3. Courter, S., and Brewer, J. : Automated conversion of curvilinear wireframe models to surface boundary models: A topological approach. Comput. Graph. **20**, 4, 171–178, 1986.
4. Ganter, M., and Uicker, J. : From wireframe to solid geometric: Auromated conversion of data representations. Computer in Mechanical Eng. **2**, 2, 40–45, 1983.
5. Hestenes, D., and Sobczyk, G. : Clifford algebra to geometric calculus. D. Reidel, Dordrecht, 1984.
6. Leclerc, Y., and Fischler, M. : An optimization based approach to the interpretation of single line drawings as 3d wireframes. Int'l J. Computer Vision **8**, 2, 113–136, 1992.
7. Liu, J., and Lee, Y. : A graph-based method for face identification from a single 2d line drawing. IEEE Trans. on PAMI **23**, 10, 1106–1119, 2001.
8. Liu, J., Lee, Y., and Cham, W. : Identifying faces in a 2d line drawing representing a manifold object. IEEE Trans. on PAMI **24**, 12, 1579–1593, 2002.
9. Marill, T. : Emulation the human interpretation of line drawings as 3d objects. Int'l J. Computer Vision **6**, 2, 147–161, 1991.
10. Hanrahan, P. : Creating volume models from edge-vertex Graphs. Computer Graphics. **16**, 3, 77–84, 1982.
11. Miyazaki, K. : An adventure in multidimensional space: The art and geometry of polygons, polyhedra, and polytopes. Wileyinterscience Publ, 1983.
12. Shpitalni, M., and Lipson, H. :Identification of faces in a 2d line drawing projection of a wireframe object. IEEE Trans. on PAMI **18**, 10, 1000–1012, 1996.
13. Sugihara, K. : Machine interpretation of line drawings. MIT Press, 1986.
14. Thurston, W. : Three-dimensional geometry and topology. Vol 1. Princeton Univ. Press, Princeton, 1997.
15. Turner, A., Chapman, D., and Penn, A. : Sketching space. Computers and Graphics **24**, 869–879, 2000.

Polyhedral Scene Analysis Combining Parametric Propagation with Calotte Analysis[*]

Hongbo Li[1], Lina Zhao[2], and Ying Chen[1]

[1] Mathematics Mechanization Key Laboratory,
Academy of Mathematics and Systems Science,
Chinese Academy of Sciences, Beijing 100080, P.R. China
[2] Department of Mathematics and Computer Science,
Science of School, Bejing University of Chemical Technology

Abstract. Polyhedral scene analysis studies whether a 2D line drawing of a 3D polyhedron is realizable in the space, and if so, parameterizing the space of all possible realizations. For generic 2D data, symbolic computation with Grassmann-Cayley algebra is necessary. In this paper we propose a general method, called parametric calotte propagation, to solve the realization and parameterization problems in polyhedral scene analysis at the same time. Starting with the fundamental equations of Sugihara in the form of bivector equations, we can parameterize all the bivectors by introducing new parameters. The realization conditions are implied in the scalar equations satisfied by the new parameters, and can be derived by further analysis of the propagation result. The propagation procedure generally does not bifurcate, and the result often contains equations in factored form, thus makes further algebraic manipulation easier. In application, the method can be used to find linear construction sequences for non-spherical polyhedra.

Keywords: polyhedral scene analysis, imaging algebra, Grassmann-Cayley algebra, constraint satisfaction, projective reconstruction.

1   Introduction

Polyhedral scene analysis is a classical problem in artificial intelligence, robotics and computer vision. The main topic is to determine whether a 2D line drawing of a 3D polyhedron is realizable in the space, and if so, give a parameterization of the space of all possible realizations.

In the computer vision society, the realization problem was first studied by Mackworth (1973), Huffman (1977) using labeling schemes and reciprocal diagrams. Their approach can provide necessary but not sufficient conditions. Sugihara (1982, 1984a, 1984b, 1986) established a necessary and sufficient condition

---

[*] Supported partially by NKBRSF 2004CB318001 and NSFC 10471143.

H. Li, P. J. Olver and G. Sommer (Eds.): IWMM-GIAE 2004, LNCS 3519, pp. 383–402, 2005.
© Springer-Verlag Berlin Heidelberg 2005

on the general realizability, which is the existence of solutions in a linear programming problem. Sugihara (1999) further proposed the concept "resolvable sequences" for polyhedra and proved that they always exist for spherical polyhedra, i.e., polyhedra which are homeomorphic to a sphere. This result is used by Ros and Thomas (2002) to overcome the super-strictness in scene analysis.

In the combinatorial geometry society, the realization problem was studied in a different approach. Whiteley (1987, 1989, 1991, 1992), Crapo and Ryan (1986), Crapo (1991), Crapo and Whiteley (1993) studied the realization problem with structure geometry (Crapo and Whiteley, 1982; White and Whiteley, 1983), invariant theory (Doubilet et al., 1974; White, 1975) and synthetic geometry (Bokowski and Sturmfels, 1989; Sturmfels, 1993; Richter-Gebert, 1996). Various necessary and sufficient conditions are established in terms of either the existence of compatible cross-sections, or strict self-stresses in frameworks, or Grassmann-Cayley algebraic or bracket algebraic equalities.

According to Crapo (1991), the polyhedral scene analysis is an example in which a system of linear equations whose rank controls the critical qualitative features. The combinatorial pattern determines a sparse matrix pattern that has both a generic rank, for general independent values of the nonzero entries, and a geometric rank, for special values for the coordinates of the points and planes in the scene. While considerable work has been carried out on the generic rank and the parameterization problem therein, only a few work deals with the geometric rank, and ever fewer on the parameterization.

In our opinion, the major difference between Sugihara's linear programming approach and the invariant approach of Crapo, Whiteley, et al., is that the former is numerical while the latter is symbolic. For symbolic computation, Grassmann-Cayley algebra and bracket algebra are a good help. Although the problem itself is linear, the realization conditions for a polyhedron whose 2D coordinates are generic, can be very involved, bifurcated, and difficult to obtain. To make this clear let us raise an example.

Example 1. In the space there is a torus composed of 9 vertices and 9 faces. Each vertex is in four faces and each face has four vertices. Assume that no three vertices in a face are collinear, reconstruct the 3D torus from its 2D image.

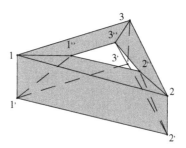

Fig. 1. Example 1

This example is taken from Sugihara (1986) as an example in which the realization conditions are difficult to find. Sugihara (1999) further showed that the torus does not have any resolvable sequence. His fundamental equations reflect the incidence relations of all pairs of incident vertices and faces. In this example, there are 36 fundamental equations in 36 unknowns. To reduce the size of the system, Crapo (1991) established a set of syzygy equations, which can be taken as the result of eliminating all the faces from the fundamental equations. For this example, there are 9 syzygy equations in 9 unknowns. Although the latter system is linear and sparse, it is still too difficult to solve symbolically. Following a different approach, we obtain the rank classification of the coefficient matrix M of the syzygy equations and the associated realization conditions as follows. It can be seen that the conditions are highly bifurcated.

Let $N = \{[\mathbf{123}], [\mathbf{1'2'3'}], [\mathbf{1''2''3''}], [\mathbf{11'1''}], [\mathbf{22'2''}], [\mathbf{33'3''}]\}$, and let

$$\begin{aligned}
E_1 &= \{\mathbf{12} \wedge \mathbf{1'2'} \wedge \mathbf{1''2''}, \mathbf{13} \wedge \mathbf{1'3'} \wedge \mathbf{1''3''}, \mathbf{23} \wedge \mathbf{2'3'} \wedge \mathbf{2''3''}\}, \\
E_2 &= \{\mathbf{11'} \wedge \mathbf{22'} \wedge \mathbf{33'}, \mathbf{11''} \wedge \mathbf{22''} \wedge \mathbf{33''}, \mathbf{1'1''} \wedge \mathbf{2'2''} \wedge \mathbf{3'3''}\}.
\end{aligned} \tag{1.1}$$

For any element $x \in E_1 \cup E_2$, let $\tilde{x}$ be the unique bracket in N whose three elements do not occur in x.

1. $4 \leq \operatorname{rank}(M) \leq 6$. The generic rank is 6.
2. $\operatorname{rank}(M) = 4$ if and only if $E_1 = E_2 = \{0\}$.
   When $\operatorname{rank}(M) = 4$, the line drawing can be lifted to 9 distinct spatial planes, i.e., no two faces are coplanar.
3. $\operatorname{rank}(M) = 5$ if and only if either
   (1) $E_1 = \{0\}$, E_2 has only one element x equal to zero, and $\tilde{x} = 0$, or
   (2) $E_2 = \{0\}$, E_1 has only one element x equal to zero, and $\tilde{x} = 0$.
   When $\operatorname{rank}(M) = 5$, the line drawing can be lifted to 3 or 5 distinct planes.
4. $\operatorname{rank}(M) = 6$ if and only if there are at most 3 elements in $E_1 \cup E_2$ equal to zero.
   When $\operatorname{rank}(M) = 6$, the line drawing can only be lifted to a spatial plane. The torus is said to be *trivial*, or *unrealizable*.

In the literature, there are three strategies that can be used to find the realization conditions symbolically. The first is Crapo (1991)'s syzygy equations. When the realization conditions contain only one equality, it can be obtained from the syzygy equations directly. Two important cases are covered by this strategy: (1) the truncated pyramid, (3) n-calotte. However, as in Example 1, this strategy may be difficult to employ for polyhedra whose realization conditions are bifurcated and contain multiple equalities.

The second strategy is the compatible cross-sections of Whiteley (1991). It is applicable only to spherical or disk polyhedra (Ros, 2000). The third strategy is the star-delta reductions of Ros and Thomas (1998). It is applicable only to spherical polyhedra. By either strategy, the realization conditions can be represented by Grassmann-Cayley algebraic equalities, but are not invariant due to the introduction of extraneous objects.

It remains an open problem to find a general and efficient strategy for the classification of the geometric ranks and the parameterization of the corresponding solution spaces. Our first attack to this problem is to use some elimination techniques for polynomial systems to manipulate the fundamental equations. We write the equations in terms of vectors, bivectors and their brackets for easy manipulation by vectorial equation solving (Li, 2000). By a suitable choice of an order for elimination, we can arrive at an algebraic classification. However, bifurcation in the procedure of elimination will inevitably occur. For complicated problems, elimination methods often fail to be efficient.

Our second attack is to use the *global parametric propagation* technique based on vectorial equation solving. In some sense, parametric propagation is the reverse of elimination. While in elimination the leading unknown is to be eliminated first, in parametric propagation the least leading unknown is labeled as the "origin", and new parameters are introduced to solve the unknown next to the origin. The origin is gradually enlarged by encompassing more and more solved unknowns, and the solving procedure propagates from the ever-refreshing origin to its surroundings. The system is ultimately transformed into such a system that the unknowns become the parameters introduced in every round of propagation. Compared with elimination, global parametric propagation has the following advantages: (1) the order of propagation is easy to determine, (2) the propagation procedure generally does not bifurcate.

The result of global parametric propagation is generally much easier to deal with than the original one. For further analysis of the result, Cayley factorization among bracket polynomial equations is required. However, even Cayley factorization for a single bracket polynomial equation remains an open problem. Crapo (1991) suggested that for bracket polynomials produced in geometric computation, the geometric background can help finding rational Cayley factorization. Tay (1994) successfully employed this idea in the rational Cayley factorization of Calotte equations.

Our third attack is stimulated by this idea of algebraic factorization with geometric means. The technique *local calotte analysis* is proposed to replace as many syzygy equations as possible by an equivalent system of calotte equations in factored form. It appears that a significant number of factored equations can be produced in this way.

Our method for polyhedral scene analysis is a combination of global parametric propagation with local calotte analysis. We have implemented it with Maple 8, and have tested it by 20 examples. Surprising results include the discovery of *linear construction sequences* for some non-spherical polyhedra whose resolvable sequences do not exist. The linear construction sequences have all the properties of resolvable sequences needed in application.

This paper is organized as follows. In Section 2 Grassmann-Cayley algebra is briefly introduced. In Section 3 a novel algebraic formulation of polyhedral scene analysis is introduced. In Section 4 the global parametric propagation by vectorial equation solving is introduced. Section 5 is on local calotte analysis. Section 6 contains some further examples.

2   Grassmann-Cayley Algebra

Grassmann-Cayley algebra is the most general structure in which projective properties are expressed in a coordinate-free way. The following is a coordinate-free definition of this algebra.

Let $\mathcal{V}$ be an n-dimensional vector space over a field $\mathcal{F}$ whose characteristic is not 2. Then $\mathcal{V}$ generates the *Grassmann algebra* $(\Lambda(\mathcal{V}), \vee)$ which is composed of the *Grassmann space* $\Lambda(\mathcal{V})$ and the *outer product* "$\vee$". The outer product of k vectors is called a *k-extensor*, and k is called the *grade*. The linear subspace spanned by all k-flats is called the *k-vector space*. Its elements are called *k-vectors*. The whole space $\Lambda(\mathcal{V})$ is the direct sum of the k-vector spaces with k ranging from 0 to n. An element in $\Lambda(\mathcal{V})$ is called a *multivector*. The k-graded part of a multivector x is denoted by $\langle x \rangle_k$. In application, the outer product is usually denoted by the juxtaposition of elements.

Let I_n be a fixed nonzero n-vector in $\Lambda(\mathcal{V})$. The *bracket* of a multivector x with respect to I_n is defined by

$$[x] = \langle x \rangle_n / I_n. \tag{2.1}$$

The bilinear form $(x, y) \mapsto [xy]$ for $x, y \in \Lambda(\mathcal{V})$ is nonsingular, and induces a linear invertible mapping i from $\Lambda(\mathcal{V})$ to its dual space $\Lambda(\mathcal{V}^*)$, where $\mathcal{V}^*$ is the dual space of $\mathcal{V}$. The pullback of the outer product from $\Lambda(\mathcal{V}^*)$ to $\Lambda(\mathcal{V})$ through i is called the *wedge product*, or *meet product*, in $\Lambda(\mathcal{V})$, and is denoted by "$\wedge$". The Grassmann space $\Lambda(\mathcal{V})$ equipped with the two products is called the *Grassmann-Cayley algebra* over $\mathcal{V}$.

The following are representations of some geometric objects and incidence relations in 2D projective geometry by Grassmann-Cayley algebra.

(1) A point is represented by a nonzero vector, which is unique up to scale. In this paper, a point is always denoted by a bold-faced integer or character.

(2) A line passing through points $\mathbf{1}, \mathbf{2}$ is represented by $\mathbf{12}$. Three points $\mathbf{1}, \mathbf{2}, \mathbf{3}$ are collinear if and only if their outer product is zero, or equivalently, $[\mathbf{123}] = 0$.

(3) The intersection of two lines $\mathbf{12}, \mathbf{1'2'}$ is represented by $\mathbf{12} \wedge \mathbf{1'2'}$.

(4) Three lines $\mathbf{12}, \mathbf{1'2'}, \mathbf{1''2''}$ are collinear if and only if their wedge product is zero, i.e., $\mathbf{12} \wedge \mathbf{1'2'} \wedge \mathbf{1''2''} = 0$.

Bracket algebra refers to the ring of brackets. *Cayley factorization* is the transformation from a bracket polynomial to an expression in Grassmann-Cayley algebra called *Cayley expression*. It is a crucial step from algebraic characterization to geometric description. White (1991) developed a general Cayley factorization algorithm for multilinear bracket polynomials. Li and Wu (2003) developed a set of Cayley factorization formulas for non-multilinear bracket polynomials based on *Cayley expansion*, the inverse of Cayley factorization.

3   A New Formulation of Polyhedral Scene Analysis

A polyhedral line drawing is either a perspective or a parallel projection of a polyhedron. A polyhedron can be described by its vertices $\{V_1, \ldots, V_m\}$, faces $\{F_1, \ldots, F_n\}$ and the incidence structure composed of all the incidence pairs (V_i, F_j). Let $\mathbf{n}$ be either a point or a direction in the space. Then by either the perspective projection centered at the point $\mathbf{n}$, or the parallel projection along the direction $\mathbf{n}$, the vertices of a polyhedron are projected into an image plane $\mathcal{I}$, with 2D coordinates (x_i, y_i) for the image $\mathbf{i}$ of vertex V_i. Both $\mathbf{n}$ and the image plane are already given in the space.

Conventional Assumption:
(1) Any face of the polyhedron has at least three non-collinear vertices;
(2) $\mathbf{n}$ is not incident to any plane supporting a non-triangular face.

The 3D *projective reconstruction*, or *scene analysis* of the polyhedron, is to recover under the Conventional Assumption the 3D positions of the vertices and faces from the 2D coordinates so that the incidence structure is preserved.

In 3D projective geometry, the vector space is 4D in which the image plane is a 3D vector subspace $\mathcal{I}$, and $\mathbf{n}$ is a 1D subspace outside $\mathcal{I}$. Let $\mathbf{e}_1, \mathbf{e}_2, \mathbf{e}_4$ be a basis of $\mathcal{I}$. When $\mathbf{n} = (n_1, n_2, n_3)$ is a space point, let $\mathbf{e}_3 = (n_1, n_2, n_3, 1)$; when $\mathbf{n}$ is a spatial direction, let $\mathbf{e}_3 = (n_1, n_2, n_3, 0)$. Then $\mathbf{e}_1, \mathbf{e}_2, \mathbf{e}_3, \mathbf{e}_4$ form a basis of the 4D vector space. The homogeneous coordinates of image $\mathbf{i}$ with respect to the basis $\mathbf{e}_1, \mathbf{e}_2, \mathbf{e}_4$ are $(x_i, y_i, 1)$, so the homogeneous coordinates of point V_i with respect to the basis $\mathbf{e}_1, \mathbf{e}_2, \mathbf{e}_3, \mathbf{e}_4$ are $(x_i, y_i, h_i, 1)$, where h_i is the unknown "height". The homogeneous coordinates of face F_j is $(a_j, b_j, -1, c_j)$. Now point V_i is in face F_j if and only if

$$a_j x_i + b_j y_i + c_j = h_i. \tag{3.2}$$

When (i, j) ranges over all incidence pairs, we obtain a set of linear homogeneous equations in the unknowns a, b, c, h's, called *Sugihara's fundamental equations*.
Let

$$\mathbf{B}_j = a_j \mathbf{e}_2 \mathbf{e}_4 - b_j \mathbf{e}_1 \mathbf{e}_4 + c_j \mathbf{e}_1 \mathbf{e}_2. \tag{3.3}$$

Then $\mathbf{B}_j$ is a bivector in the Grassmann space generated by $\mathcal{I}$, called the *inhomogeneous coordinates* of face F_j. Now (3.2) can be written as an equality in the Grassmann-Cayley algebra generated by $\mathcal{I}$:

$$h_i = [\mathbf{i}\mathbf{B}_j]. \tag{3.4}$$

For any 4-tuple of vertices V_1, V_2, V_3, V_4 in face F_j, their 2D coordinates satisfy the GP syzygy:

$$[1\mathbf{B}_j][234] - [2\mathbf{B}_j][134] + [3\mathbf{B}_j][124] - [4\mathbf{B}_j][123] = 0. \tag{3.5}$$

When j ranges over all faces with more than three vertices, and the 4-tuples of vertices range over all vertices in the same face, we obtain a set of linear equations of the h's by substituting (3.4) into (3.5):

$$h_1[\mathbf{234}] - h_2[\mathbf{134}] + h_3[\mathbf{124}] - h_4[\mathbf{123}] = 0. \tag{3.6}$$

They are called *Crapo's syzygy equations*.

Theorem 1. Under the Conventional Assumption, the fundamental equations and the syzygy equations have the same solutions for the h's.

In practice, the following stronger assumption covers all the interesting cases in scene analysis. Henceforth, it will be adopted throughout this paper.

Practical Assumption:
(1) No three neighboring vertices in a face are collinear;
(2) $\mathbf{n}$ is not incident to any plane supporting a non-triangular face.

Although the syzygy equations have much less unknowns than the fundamental equations, they are often difficult to solve. In this paper, we split the fundamental equations into two systems. The first system is that the height of a vertex V_i computed from any incident face $F_1, \ldots, F_k$ is the same:

$$[\mathbf{iB}_1] = [\mathbf{iB}_2] = \cdots = [\mathbf{iB}_k], \text{ for any } 1 \le j \le k. \tag{3.7}$$

It is called the $\mathbf{B}$-*system*, or *bivector system*. The second system is that one face F_1 is sufficient to characterize the height of vertex V_i:

$$h_i = [\mathbf{iB}_1]. \tag{3.8}$$

It is called the h-*system*, or *height system*. The h-system is obviously trivial. The $\mathbf{B}$-*system* is much easier to solve than both the fundamental equations and the syzygy equations.

4   Vectorial Equation Solving and Parametric Propagation

In this paper, by a *vectorial equation* we mean a Grassmann-Cayley algebraic or bracket algebraic equality in which the leading variable is either a vector or a bivector (2-vector). To solve a set of vectorial equations with the same leading variable is to find all solutions and their existence conditions.

The bivector system is a set of linear bracket equations with bivector unknowns in 3D vector space. The idea *global parametric propagation* is to construct the solutions parametrically. New parameters are introduced every time the number of equations is not sufficient in the construction. The parameters may be either free or constrained, and the constraints may not be obtained until all the unknowns are constructed. The goal is to transform the system into one in which the unknowns are the new parameters, and which is usually much easier to solve. The propagation is realized by solving the system parametrically following a dynamically changing order of the bivector unknowns.

Below is a list of equations with single unknown bivectors and their solutions. In the list, the $\mathbf{V}$'s are vectors, the $\mathbf{B}$'s are bivectors, the μ's are scalars, and the ω's are new parameters.

Type B.1.

$$\begin{cases} [\mathbf{V}_1\mathbf{B}] = \mu_1 \\ [\mathbf{V}_2\mathbf{B}] = \mu_2 \\ [\mathbf{V}_3\mathbf{B}] = \mu_3 \\ \mathbf{V}_1\mathbf{V}_2 \neq 0 \end{cases} \tag{4.1}$$

Solution: either

$$\begin{cases} [\mathbf{V}_1\mathbf{V}_2\mathbf{V}_3] \neq 0, \\ [\mathbf{V}_1\mathbf{V}_2\mathbf{V}_3]\,\mathbf{B} = \mu_1\mathbf{V}_2\mathbf{V}_3 - \mu_2\mathbf{V}_1\mathbf{V}_3 + \mu_3\mathbf{V}_1\mathbf{V}_2; \end{cases}$$

or

$$\begin{cases} [\mathbf{V}_1\mathbf{V}_2\mathbf{V}_3] = 0, \\ \mu_1\mathbf{V}_2\mathbf{V}_3 - \mu_2\mathbf{V}_1\mathbf{V}_3 + \mu_3\mathbf{V}_1\mathbf{V}_2 = 0, \\ [\mathbf{V}_1\mathbf{B}] = \mu_1, \\ [\mathbf{V}_2\mathbf{B}] = \mu_2. \end{cases}$$

In particular, if $\mu_i = [\mathbf{V}_i\mathbf{B}']$ for $i = 1, 2, 3$, then either (1) $[\mathbf{V}_1\mathbf{V}_2\mathbf{V}_3] \neq 0$ and $\mathbf{B} = \mathbf{B}'$, or (2) $[\mathbf{V}_1\mathbf{V}_2\mathbf{V}_3] = 0$ and $\mathbf{B} = \mathbf{B}' + \omega\mathbf{V}_1\mathbf{V}_2$.

Type B.2. For $m \geq 4$,

$$\begin{cases} [\mathbf{V}_1\mathbf{B}] = \mu_1 \\ [\mathbf{V}_2\mathbf{B}] = \mu_2 \\ \cdots \quad \cdots\cdots \\ [\mathbf{V}_k\mathbf{B}] = \mu_m \\ \mathbf{V}_1\mathbf{V}_2 \neq 0 \end{cases} \tag{4.2}$$

Solution: either

$$\begin{cases} [\mathbf{V}_1\mathbf{V}_2\mathbf{V}_k] \neq 0 \text{ for some } 3 \leq k \leq m, \\ [\mathbf{V}_1\mathbf{V}_2\mathbf{V}_k]\,\mathbf{B} = \mu_1\mathbf{V}_2\mathbf{V}_k - \mu_2\mathbf{V}_1\mathbf{V}_k + \mu_k\mathbf{V}_1\mathbf{V}_2, \\ \mu_1[\mathbf{V}_2\mathbf{V}_k\mathbf{V}_l] - \mu_2[\mathbf{V}_1\mathbf{V}_k\mathbf{V}_l] + \mu_k[\mathbf{V}_1\mathbf{V}_2\mathbf{V}_l] - \mu_l[\mathbf{V}_1\mathbf{V}_2\mathbf{V}_k] = 0, \; \forall l \neq 1, 2, k; \end{cases}$$

or

$$\begin{cases} [\mathbf{V}_1\mathbf{V}_2\mathbf{V}_k] = 0, \text{ for any } 3 \leq k \leq m, \\ \mu_1\mathbf{V}_2\mathbf{V}_k - \mu_2\mathbf{V}_1\mathbf{V}_k + \mu_k\mathbf{V}_1\mathbf{V}_2 = 0, \text{ for any } 3 \leq k \leq m, \\ [\mathbf{V}_1\mathbf{B}] = \mu_1, \\ [\mathbf{V}_2\mathbf{B}] = \mu_2. \end{cases}$$

Type B.3.

$$\begin{cases} [\mathbf{V}_1\mathbf{B}] = [\mathbf{V}_1\mathbf{B}'] \\ [\mathbf{V}_2\mathbf{B}] = [\mathbf{V}_2\mathbf{B}'] \\ \mathbf{V}_1\mathbf{V}_2 \neq 0 \end{cases} \tag{4.3}$$

Solution: $\mathbf{B} = \mathbf{B}' + \omega\mathbf{V}_1\mathbf{V}_2$.

As an example, we solve the bivector system in Example 1. Let i, j, k be an even permutation of $1, 2, 3$. The face with vertices $V_i, V_j, V_{i'}, V_{j'}$ is denoted by $F_{k''}$, with coordinates $\mathbf{B}_{k''}$. Similarly, the face with vertices $V_i, V_j, V_{i''}, V_{j''}$ is denoted by $F_{k'}$, and the face with vertices $V_{i'}, V_{j'}, V_{i''}, V_{j''}$ is denoted by F_k.

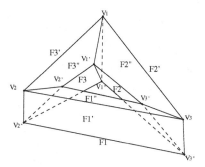

Fig. 2. Example 1 continued

The bivector system is

$$
\begin{cases}
[\mathbf{1B}_{2'}] = [\mathbf{1B}_{2''}] = [\mathbf{1B}_{3'}] = [\mathbf{1B}_{3''}] \\
[\mathbf{2B}_{1'}] = [\mathbf{2B}_{1''}] = [\mathbf{2B}_{3'}] = [\mathbf{2B}_{3''}] \\
[\mathbf{3B}_{2'}] = [\mathbf{3B}_{2''}] = [\mathbf{3B}_{1'}] = [\mathbf{3B}_{1''}] \\
[\mathbf{1'B}_{2}] = [\mathbf{1'B}_{2''}] = [\mathbf{1'B}_{3}] = [\mathbf{1'B}_{3''}] \\
[\mathbf{2'B}_{1}] = [\mathbf{2'B}_{1''}] = [\mathbf{2'B}_{3}] = [\mathbf{2'B}_{3''}] \\
[\mathbf{3'B}_{1}] = [\mathbf{3'B}_{1''}] = [\mathbf{3'B}_{2}] = [\mathbf{3'B}_{2''}] \\
[\mathbf{1''B}_{2}] = [\mathbf{1''B}_{2'}] = [\mathbf{1''B}_{3}] = [\mathbf{1''B}_{3'}] \\
[\mathbf{2''B}_{1}] = [\mathbf{2''B}_{1'}] = [\mathbf{2''B}_{3}] = [\mathbf{2''B}_{3'}] \\
[\mathbf{3''B}_{1}] = [\mathbf{3''B}_{1'}] = [\mathbf{3''B}_{2}] = [\mathbf{3''B}_{2'}]
\end{cases}
\tag{4.4}
$$

Choose $\mathbf{B}_{3''}$ as the origin by setting $\mathbf{B}_{3''} = 0$. The following are different rounds of propagation.

Round 1: Propagation towards $\mathbf{B}_3$.

$$[\mathbf{1'B}_3] = 0, \ [\mathbf{2'B}_3] = 0 \xrightarrow{B.3} \mathbf{B}_3 = \omega_{1'2'}\mathbf{1'2'}.$$

Round 2: Propagation towards $\mathbf{B}_{3'}$.

$$[\mathbf{1B}_{3'}] = 0, \ [\mathbf{2B}_{3'}] = 0, \ [\mathbf{1''B}_{3'}] = \omega_{1'2'}[\mathbf{1'2'1''}], \ [\mathbf{2''B}_{3'}] = \omega_{1'2'}[\mathbf{1'2'2''}]$$

$$\xrightarrow{B.2} \begin{cases} [\mathbf{121''}]\mathbf{B}_{3'} = \omega_{1'2'}[\mathbf{1'2'1''}]\mathbf{12}, \\ \omega_{1'2'}\mathbf{12} \wedge \mathbf{1'2'} \wedge \mathbf{1''2''} = 0. \end{cases}$$

Round 3: Propagation towards $\mathbf{B}_{1''}$.

$$[\mathbf{2B}_{1''}] = 0, \ [\mathbf{2'B}_{1''}] = 0 \xrightarrow{B.3} \mathbf{B}_{1''} = \omega_{22'}\mathbf{22'}.$$

Round 4: Propagation towards $\mathbf{B}_{2''}$.

$$[\mathbf{1B}_{2''}] = 0, \ [\mathbf{1'B}_{2''}] = 0, \ [\mathbf{3B}_{2''}] = -\omega_{22'}[\mathbf{232'}], \ [\mathbf{3'B}_{2''}] = \omega_{22'}[\mathbf{22'3'}]$$

$$\xrightarrow{B.2} \begin{cases} [\mathbf{131'}]\mathbf{B}_{2''} = \omega_{22'}[\mathbf{232'}]\mathbf{11'}, \\ \omega_{22'}\mathbf{11'} \wedge \mathbf{22'} \wedge \mathbf{33'} = 0. \end{cases}$$

Round 5: Propagation towards $\mathbf{B}_1$.

$$[\mathbf{3'B}_1] = \omega_{22'}[\mathbf{22'3'}], \ [\mathbf{2''B}_1] = \omega_{1'2'}[\mathbf{1'2'2''}], \ [\mathbf{2'B}_1] = 0$$

$$\xrightarrow{B.1} [\mathbf{2'3'2''}]\mathbf{B}_1 = \omega_{1'2'}[\mathbf{1'2'2''}]\mathbf{2'3'} - \omega_{22'}[\mathbf{22'3'}]\mathbf{2'2''}.$$

Round 6: Propagation towards $\mathbf{B}_2$.

$$[\mathbf{3'B}_2] = \omega_{22'}[\mathbf{22'3'}], \ [\mathbf{1''B}_2] = \omega_{1'2'}[\mathbf{1'2'1''}], [\mathbf{1'B}_2] = 0,$$

$$[\mathbf{2'3'2''}][\mathbf{3''B}_2] = \omega_{1'2'}[\mathbf{1'2'2''}][\mathbf{2'3'3''}] - \omega_{22'}[\mathbf{22'3'}][\mathbf{2'2''3''}]$$

$$\xrightarrow{B.2} \begin{cases} [\mathbf{1'3'1''}]\mathbf{B}_2 = \omega_{1'2'}[\mathbf{1'2'1''}]\mathbf{1'3'} - \omega_{22'}[\mathbf{22'3'}]\mathbf{1'1''}, \\ (\mathbf{1'1''} \wedge \mathbf{2'2''} \wedge \mathbf{3'3''})(\omega_{1'2'}[\mathbf{1'2'3'}] - \omega_{22'}[\mathbf{22'3'}]) = 0. \end{cases}$$

Round 7: Propagation towards $\mathbf{B}_{1'}$.

$$[\mathbf{3B}_{1'}] = -\omega_{22'}[\mathbf{232'}], \ [\mathbf{2''B}_{1'}] = \omega_{1'2'}[\mathbf{1'2'2''}], \ [\mathbf{2B}_{1'}] = 0,$$

$$[\mathbf{2'3'2''}][\mathbf{3''B}_{1'}] = \omega_{1'2'}[\mathbf{1'2'2''}][\mathbf{2'3'3''}] - \omega_{22'}[\mathbf{22'3'}][\mathbf{2'2''3''}]$$

$$\xrightarrow{B.2} \begin{cases} [\mathbf{232''}]\mathbf{B}_{1'} = \omega_{22'}[\mathbf{232'}]\mathbf{22''} + \omega_{1'2'}[\mathbf{1'2'2''}]\mathbf{23}, \\ (\mathbf{23} \wedge \mathbf{2'3'} \wedge \mathbf{2''3''})(\omega_{1'2'}[\mathbf{1'2'2''}] - \omega_{22'}[\mathbf{22'2''}]) = 0. \end{cases}$$

Round 8: Propagation towards $\mathbf{B}_{2'}$.

$$[\mathbf{3B}_{2'}] = -\omega_{22'}[\mathbf{232'}], \ [\mathbf{1''B}_{2'}] = \omega_{1'2'}[\mathbf{1'2'1''}], \ [\mathbf{1B}_{2'}] = 0,$$

$$[\mathbf{2'3'2''}][\mathbf{3''B}_{2'}] = \omega_{1'2'}[\mathbf{1'2'2''}][\mathbf{2'3'3''}] - \omega_{22'}[\mathbf{22'3'}][\mathbf{2'2''3''}]$$

$$\xrightarrow{B.2} \begin{cases} [\mathbf{131''}]\mathbf{B}_{2'} = \omega_{1'2'}[\mathbf{1'2'1''}]\mathbf{13} + \omega_{22'}[\mathbf{232'}]\mathbf{11''}, \\ \omega_{22'}([\mathbf{11''3''}][\mathbf{232'}][\mathbf{2'3'2''}] + [\mathbf{131''}][\mathbf{22'3'}][\mathbf{2'2''3''}]) \\ = \omega_{1'2'}([\mathbf{131''}][\mathbf{1'2'2''}][\mathbf{2'3'3''}] - [\mathbf{133''}][\mathbf{1'2'1''}][\mathbf{2'3'2''}]). \end{cases}$$

Solution: With new parameters $\omega_{1'2'}, \omega_{22'}$ and by setting $\mathbf{B}_{3''} = 0$, (4.4) is changed into

$$\begin{cases} \mathbf{B}_{3''} = 0, \\ \mathbf{B}_3 = \omega_{1'2'}\mathbf{1'2'}, \\ [\mathbf{121''}]\mathbf{B}_{3'} = \omega_{1'2'}[\mathbf{1'2'1''}]\mathbf{12}, \\ \mathbf{B}_{1''} = \omega_{22'}\mathbf{22'}, \\ [\mathbf{131'}]\mathbf{B}_{2''} = \omega_{22'}[\mathbf{232'}]\mathbf{11'}, \\ [\mathbf{2'3'2''}]\mathbf{B}_1 = \omega_{1'2'}[\mathbf{1'2'2''}]\mathbf{2'3'} - \omega_{22'}[\mathbf{22'3'}]\mathbf{2'2''}, \\ [\mathbf{1'3'1''}]\mathbf{B}_2 = \omega_{1'2'}[\mathbf{1'2'1''}]\mathbf{1'3'} - \omega_{22'}[\mathbf{22'3'}]\mathbf{1'1''}, \\ [\mathbf{232''}]\mathbf{B}_{1'} = \omega_{1'2'}[\mathbf{1'2'2''}]\mathbf{23} + \omega_{22'}[\mathbf{232'}]\mathbf{22''}, \\ [\mathbf{131''}]\mathbf{B}_{2'} = \omega_{1'2'}[\mathbf{1'2'1''}]\mathbf{13} + \omega_{22'}[\mathbf{232'}]\mathbf{11''}, \end{cases} \tag{4.5}$$

together with

$$\begin{cases} \omega_{22'}\mathbf{11'} \wedge \mathbf{22'} \wedge \mathbf{33'} = 0, \\ \omega_{1'2'}\mathbf{12} \wedge \mathbf{1'2'} \wedge \mathbf{1''2''} = 0, \\ (\omega_{1'2'}[\mathbf{1'2'3'}] - \omega_{22'}[\mathbf{22'3'}])\mathbf{1'1''} \wedge \mathbf{2'2''} \wedge \mathbf{3'3''} = 0, \\ (\omega_{1'2'}[\mathbf{1'2'2''}] - \omega_{22'}[\mathbf{22'2''}]) \ \mathbf{23} \wedge \mathbf{2'3'} \wedge \mathbf{2''3''} = 0, \\ \omega_{22'}([\mathbf{11''3''}][\mathbf{232'}][\mathbf{2'3'2''}] + [\mathbf{131''}][\mathbf{22'3'}][\mathbf{2'2''3''}]) \\ \quad = \omega_{1'2'}([\mathbf{131''}][\mathbf{1'2'2''}][\mathbf{2'3'3''}] - [\mathbf{133''}][\mathbf{1'2'1''}][\mathbf{2'3'2''}]). \end{cases} \tag{4.6}$$

It is clear that the order of bivectors in the propagation is very important. In this example, the propagation sequence is $\mathbf{B}_{3''} = 0 \xrightarrow{\omega_{1'2'}} \mathbf{B}_3 \xrightarrow{B.2} \mathbf{B}_{3'} \xrightarrow{\omega_{22'}} \mathbf{B}_{1''} \xrightarrow{B.2} \mathbf{B}_{2''} \xrightarrow{B.1} \mathbf{B}_1 \xrightarrow{B.2} \mathbf{B}_2, \mathbf{B}_{1'}, \mathbf{B}_{2'}$.

In our implementation, the origin can be set to be any face, and the order of bivectors is determined by the *construction difficulty* of the faces they represent. First, the *construction level* of a vertex is the minimum of all the construction levels of the faces incident to it. Second, the *construction level* of a face equals one plus the maximum of all the construction levels of the vertices used in constructing the face. The origin of construction is of level zero. Third, the *construction difficulty* of a face is the sum of the construction levels of the vertices used in constructing the face. In the above example:

Round 1. The origin $F_{3''}$ has four neighboring faces each sharing one edge with it. They all have zero construction difficulty. The propagation is towards F_3.

Round 2. Only face $F_{3'}$ has more than two constructed vertices. It is then constructed.

Round 3–4. The six remaining faces each have two constructed vertices. Only faces $F_{1''}, F_{2''}$ each have zero construction difficulty.

Round 5–8. Faces $F_1, F_2, F_{1'}, F_{2'}$ have the same construction difficulty.

5   Local Calotte Analysis

Consider the last equation in (4.6). Is it factorable? The answer is no if only this single equation is considered, and yes if the other four equations are taken into account. In fact, the last equation can be replaced by any of the following factored equations:

$$\begin{cases} (\omega_{1'2'}[\mathbf{131'}][\mathbf{1'2'1''}] - \omega_{22'}[\mathbf{11'1''}][\mathbf{232'}])\mathbf{13} \wedge \mathbf{1'3'} \wedge \mathbf{1''3''} = 0, \\ (\omega_{1'2'}[\mathbf{11'3'}][\mathbf{1'2'1''}] - \omega_{22'}[\mathbf{11'1''}][\mathbf{22'3'}])\mathbf{13} \wedge \mathbf{1'3'} \wedge \mathbf{1''3''} = 0. \end{cases} \quad (5.1)$$

On one hand, (5.1) is much better for further algebraic manipulation. On the other hand, it is too difficult to obtain algebraically from the equations in (4.6). In this section, we propose a geometric method to realize the factorization. The method is based on the geometric concept "calotte" originally proposed by Crapo (1991). The following is a generalization of this concept.

Definition 1. An (n, m)-*calotte* in the space, where $3 \leq m \leq n$, is composed of two sequences of vertices $A_n = P_1 \ldots P_n$ and $B_m = Q_1 \ldots Q_m$ such that (1) the intersection of the two sets of vertices is empty, (2) the P's are pairwise distinct, (3) P_iP_j is an edge for $i < j$ if and only if $|i - j| = 1 \bmod n$, (3) for any edge P_iP_{i+1}, there is a unique pair of vertices Q_{l_i} and Q_{l_i+1} such that if $Q_{l_i} \neq Q_{l_i+1}$ then they are coplanar with P_iP_{i+1}. The face containing the four vertices is called a *circumface* of the calotte.

A calotte is said to be *regular* if the P's are the vertices of a face. An (n, n)-calotte is said to be *symmetric*, called an n-*calotte*.

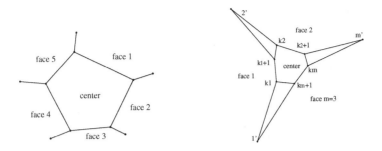

Fig. 3. A 5-calotte and a (6,3)-calotte

Theorem 2. For calotte $(P_1 \ldots P_n, Q_1 \ldots Q_m)$, let the homogeneous coordinates of vertices P_i, Q_j be $(\mathbf{i}, h_i), (\mathbf{j'}, h_{j'})$ respectively. For $1 \leq i \leq m$, let $\mathbf{B}_i$ be the inhomogeneous coordinates of circumface $P_{k_i} P_{k_i+1} Q_{(i+1)} Q_i$, and let

$$
\begin{aligned}
\mathbf{C}_i &= \mathbf{k_i}(\mathbf{k_i} + 1), \\
g_i &= -h_{k_i}[\mathbf{i'}(\mathbf{i}+1)'(\mathbf{k_i}+1)] + h_{k_i+1}[\mathbf{i'}(\mathbf{i}+1)'\mathbf{k_i}],
\end{aligned}
\tag{5.2}
$$

Assume that the P's and $m-1$ circumfaces have been constructed. Then the last circumface exists if and only if the following *calotte equation* in $h_{1'}$ has solution:

$$
c\, h_{1'} = \sum_{i=1}^{m} g_i \prod_{j=1}^{i-1}[\mathbf{j'C}_j] \prod_{l=i+1}^{m}[(\mathbf{l}+1)'\mathbf{C}_l],
\tag{5.3}
$$

where

$$
\begin{aligned}
c = &-[\mathbf{1'C}_1][\mathbf{2'C}_2]\cdots[(\mathbf{m}-1)'\mathbf{C}_{m-1}][\mathbf{m'C}_m] \\
&+[\mathbf{2'C}_1][\mathbf{3'C}_2]\cdots[\mathbf{m'C}_{m-1}][\mathbf{1'C}_m]
\end{aligned}
\tag{5.4}
$$

is the *Crapo binomial* of the calotte.

If the calotte is regular, let $\mathbf{B}_0$ be the inhomogeneous coordinates of the face of the P's, then the calotte equation can be written as $(\mathbf{B}_i - \mathbf{B}_0)c = 0$, or equivalently, $(\mathbf{B}_{i+1} - \mathbf{B}_i)c = 0$ for some $1 \leq i \leq m$.

For 3-calotte, the calotte equation is $(\mathbf{B}_2 - \mathbf{B}_1)\mathbf{11'} \wedge \mathbf{22'} \wedge \mathbf{33'} = 0$.

In vectorial equation-solving, the equation set of type **B.2** (4.2) is most often encountered. The solution is composed of the explicit expression of the unknown bivector, together with $m-3$ syzygy equations which are usually not factorizable. The above theorem can be used to replace the syzygy equations with calotte equations, and even in factored form in some cases. The task is then to search for the suitable calottes locally. This procedure is called *local calotte analysis*. We show this with Example 1.

During Rounds 2, 4, 6, 7 of the parametric propagation, the calotte equations of calottes $(V_1 V_{1'} V_{1''}, V_2 V_{2'} V_{2''})$, $(V_1 V_2 V_3, V_{1'} V_{2'} V_{3'})$, $(V_{1'} V_{2'} V_{3'}, V_{1''} V_{2''} V_{3''})$, $(V_2 V_{2'} V_{2''}, V_3 V_{3'} V_{3''})$ occur. In Round 8, both calottes $(V_1 V_{1'} V_{1''}, V_3 V_{3'} V_{3''})$ and $(V_1 V_2 V_3, V_{1''} V_{2''} V_{3''})$ occur. The equation of either calotte can be used to replace the last equation in (4.6). For calotte $(V_1 V_{1'} V_{1''}, V_3 V_{3'} V_{3''})$, the calotte equation is

$$(\omega_{1'2'}[\mathbf{131'}][\mathbf{1'2'1''}]\mathbf{1'3'} - \omega_{22'}[\mathbf{131'}][\mathbf{22'3'}]\mathbf{1'1''} - \omega_{22'}[\mathbf{1'3'1''}][\mathbf{232'}]\mathbf{11'})$$
$$\mathbf{13} \wedge \mathbf{1'3'} \wedge \mathbf{1''3''} = 0. \quad (5.5)$$

It has two equivalent scalar forms (5.1).

6   Further Analysis of the Propagation Result

The technique combining global parametric propagation with local calotte analysis is called *parametric calotte propagation*. Its result is a set of explicit expressions of the bivectors and a system of linear homogeneous equations satisfied by the new parameters. After eliminating the new parameters, we obtain a variety of systems of bracket polynomials formed by the 2D coordinates of vertices. *Further analysis* of the propagation result includes (1) solving the system of new parameters, (2) solving the systems of bracket polynomials, (3) explaining the results geometrically. In general, the last step needs Cayley factorization.

As an example, let us analyze the result of Example 1. (5.5) has a scalar factor and a bivector factor, denoted by A, which is a multiple of $\mathbf{1'3'}$. A can be replaced by $[A\mathbf{V}]$ for any vector $\mathbf{V}$ as long as $[A\mathbf{V}] \neq 0$. Here $\mathbf{V}$ can be any of $\mathbf{1}, \mathbf{3}, \mathbf{1''}, \mathbf{3''}$. Vectors $\mathbf{1}, \mathbf{1''}$ each change A into a bracket binomial, which are given in (5.1); vectors $\mathbf{3}, \mathbf{3''}$ each change A into a 3-termed bracket polynomial. If necessary, A can be replaced by either of its two shortest scalar forms.

The parametric system $\mathcal{P}$ composed of (4.6) with the last equation replaced by (5.5) is a set of linear equations in factored form. Below we discuss all the branches of the solutions of $\omega_{1'2'}, \omega_{22'}$.

Case 1. If $\omega_{1'2'} = \omega_{22'} = 0$, then $\mathcal{P}$ is trivially satisfied. Since all the $\mathbf{B}$'s are zero, the corresponding 3D realization is trivial in that all faces are coplanar.

Case 2. If $\omega_{1'2'} = 0 \neq \omega_{22'}$, then $\mathbf{B}_3 = \mathbf{B}_{3'} = \mathbf{B}_{3''}$, i.e., faces $F_3, F_{3'}, F_{3''}$ are coplanar. System P becomes

$$\begin{cases} \mathbf{11'} \wedge \mathbf{22'} \wedge \mathbf{33'} = 0, \\ \mathbf{1'1''} \wedge \mathbf{2'2''} \wedge \mathbf{3'3''} = 0, \\ [\mathbf{22'2''}]\mathbf{23} \wedge \mathbf{2'3'} \wedge \mathbf{2''3''} = 0, \\ ([\mathbf{131'}][\mathbf{22'3'}]\mathbf{1'1''} + [\mathbf{1'3'1''}][\mathbf{232'}]\mathbf{11'})\mathbf{13} \wedge \mathbf{1'3'} \wedge \mathbf{1''3''} = 0. \end{cases} \quad (6.6)$$

The bivector factor A can be replaced by $[\mathbf{11'1''}]$, if either $\mathbf{1}$ or $\mathbf{1''}$ is used to change it to a scalar factor.

Case 2.1. If both $\mathbf{13} \wedge \mathbf{1'3'} \wedge \mathbf{1''3''} = 0$ and $\mathbf{23} \wedge \mathbf{2'3'} \wedge \mathbf{2''3''} = 0$, then by the following lemma, the six tuples

$$(\mathbf{12}, \mathbf{1'2'}, \mathbf{1''2''}), \ (\mathbf{13}, \mathbf{1'3'}, \mathbf{1''3''}), \ (\mathbf{23}, \mathbf{2'3'}, \mathbf{2''3''}),$$
$$(\mathbf{11'}, \mathbf{22'}, \mathbf{33'}), \ (\mathbf{11''}, \mathbf{22''}, \mathbf{33''}), \ (\mathbf{1'1''}, \mathbf{2'2''}, \mathbf{3'3''}) \quad (6.7)$$

are all concurrent lines. The corresponding configuration is called the *triple Desargues configuration*. By Desargues Theorem, the six intersections $\mathbf{A}, \mathbf{B}, \mathbf{C}, \mathbf{A'},$

$\mathbf{B}', \mathbf{C}'$ of the six tuples lie on two lines, with each line passing through three intersections. According to (4.5), The line drawing can be lifted to a spatial torus supported by 7 distinct planes.

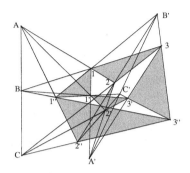

Fig. 4. Triple Desargues configuration

Lemma 1. If

$$\begin{cases} \mathbf{12} \wedge \mathbf{1'2'} \wedge \mathbf{1''2''} = 0 \\ \mathbf{23} \wedge \mathbf{2'3'} \wedge \mathbf{2''3''} = 0 \\ \mathbf{11'} \wedge \mathbf{22'} \wedge \mathbf{33'} = 0 \\ \mathbf{1'1''} \wedge \mathbf{2'2''} \wedge \mathbf{3'3''} = 0 \end{cases} \tag{6.8}$$

then

$$\mathbf{13} \wedge \mathbf{1'3'} \wedge \mathbf{1''3''} = 0, \ \ \mathbf{11''} \wedge \mathbf{22''} \wedge \mathbf{33''} = 0. \tag{6.9}$$

If

$$\begin{cases} \mathbf{12} \wedge \mathbf{1'2'} \wedge \mathbf{1''2''} = 0 \\ \mathbf{13} \wedge \mathbf{1'3'} \wedge \mathbf{1''3''} = 0 \\ \mathbf{23} \wedge \mathbf{2'3'} \wedge \mathbf{2''3''} = 0 \\ \mathbf{11''} \wedge \mathbf{22''} \wedge \mathbf{33''} = 0 \end{cases} \tag{6.10}$$

then under the condition $[\mathbf{1'2'3'}] \neq 0$,

$$\mathbf{11'} \wedge \mathbf{22'} \wedge \mathbf{33'} = 0, \ \ \mathbf{1'1''} \wedge \mathbf{2'2''} \wedge \mathbf{3'3''} = 0. \tag{6.11}$$

Case 2.2. If $[\mathbf{11'1''}] = [\mathbf{22'2''}] = 0$, then $\mathbf{3}, \mathbf{3'}, \mathbf{3''}$ must be collinear. Lines $\mathbf{11'1''}, \mathbf{22'2''}, \mathbf{33'3''}$ concur. By (4.5), the lift lies in 3 planes, because

$$\mathbf{B}_1 = \mathbf{B}_{1'} = \mathbf{B}_{1''} = \mathbf{22'}, \ \ \mathbf{B}_2 = \mathbf{B}_{2'} = \mathbf{B}_{2''} = \frac{[\mathbf{232'}]}{[\mathbf{131'}]} \mathbf{11'}. \tag{6.12}$$

Case 2.3. If $[\mathbf{11'1''}] = 0 \neq [\mathbf{22'2''}]$, then $\mathbf{3}, \mathbf{3'}, \mathbf{3''}$ are not collinear. For two nondegenerate triangles $\mathbf{22'2''}, \mathbf{33'3''}$, since lines $\mathbf{23}, \mathbf{2'3'}, \mathbf{2''3''}$ concur, by Desargues Theorem, the three distinct intersections $\mathbf{22'} \cap \mathbf{33'}, \mathbf{22''} \cap \mathbf{33''}, \mathbf{2'2''} \cap \mathbf{3'3''}$

are not collinear, i.e., $\mathbf{11}'' \wedge \mathbf{22}'' \wedge \mathbf{33}'' = 0$. By (4.5), $\mathbf{B}_2 = \mathbf{B}_{2'} = \mathbf{B}_{2''}$. The lift lies in 5 distinct planes.

Case 2.4. If $[\mathbf{11'1''}] \neq 0 = [\mathbf{22'2''}]$, the case is similar to Case 2.3.

Case 3. If $\omega_{22'} = 0 \neq \omega_{1'2'}$, then $\mathbf{B}_{1''} = \mathbf{B}_{2''} = \mathbf{B}_{3''}$. The system P becomes

$$
\begin{cases}
\mathbf{12} \wedge \mathbf{1'2'} \wedge \mathbf{1''2''} = 0, \\
[\mathbf{1'2'3'}](\mathbf{1'1''} \wedge \mathbf{2'2''} \wedge \mathbf{3'3''}) = 0, \\
\mathbf{23} \wedge \mathbf{2'3'} \wedge \mathbf{2''3''} = 0, \\
\mathbf{13} \wedge \mathbf{1'3'} \wedge \mathbf{1''3''} = 0.
\end{cases}
\tag{6.13}
$$

Case 3.2. If $[\mathbf{1'2'3'}] = 0$, then $\mathbf{B}_1 = \mathbf{B}_2 = \mathbf{B}_3$. Furthermore, $\mathbf{B}_{1'} = \mathbf{B}_{2'} = \mathbf{B}_{3'} = 0$ if and only if $[\mathbf{123}] = 0$.

Case 3.1. If $\mathbf{1'1''} \wedge \mathbf{2'2''} \wedge \mathbf{3'3''} = 0$, then when $[\mathbf{123}] \neq 0$, by Lemma 1, the corresponding configuration is the triple Desargues configuration; when $[\mathbf{123}] = 0$, $\mathbf{B}_{1'} = \mathbf{B}_{2'} = \mathbf{B}_{3'} = 0$. Furthermore, $\mathbf{B}_1 = \mathbf{B}_2 = \mathbf{B}_3$ if and only if $[\mathbf{1'2'3'}] = 0$.

Case 4. Suppose that $\omega_{1'2'}, \omega_{22'}$ are both nonzero. We prove that at most one of the three equalities holds:

$$\omega_{1'2'}[\mathbf{131'}][\mathbf{1'2'1''}]\mathbf{1'3'} - \omega_{22'}([\mathbf{131'}][\mathbf{22'3'}]\mathbf{1'1''} + [\mathbf{1'3'1''}][\mathbf{232'}]\mathbf{11'}) = 0 \quad (6.14)$$

$$\omega_{1'2'}[\mathbf{1'2'2''}] - \omega_{22'}[\mathbf{22'2''}] = 0 \quad (6.15)$$

$$\omega_{1'2'}[\mathbf{1'2'3'}] - \omega_{22'}[\mathbf{22'3'}] = 0 \quad (6.16)$$

If (6.15) and (6.16) hold, then

$$[\mathbf{1'2'3'}][\mathbf{22'2''}] - [\mathbf{1'2'2''}][\mathbf{22'3'}] = [\mathbf{21'2'}][\mathbf{2'3'2''}] = 0,$$

which violates the Practical Assumption. If (6.14) and (6.16) hold, then by substituting (6.16) into the second of the following two shortest scalar forms of (6.14),

$$\omega_{1'2'}[\mathbf{131'}][\mathbf{1'2'1''}] - \omega_{22'}[\mathbf{11'1''}][\mathbf{232'}] = 0,$$
$$\omega_{1'2'}[\mathbf{11'3'}][\mathbf{1'2'1''}] - \omega_{22'}[\mathbf{11'1''}][\mathbf{22'3'}] = 0,$$

we get

$$[\mathbf{11'3'}][\mathbf{1'2'1''}] - [\mathbf{11'1''}][\mathbf{1'2'3'}] = [\mathbf{11'2'}][\mathbf{1'3'1''}] = 0,$$

which violates the Practical Assumption. Similarly, (6.14) and (6.15) cannot hold at the same time.

By Lemma 1, if system P is not the triple Desargues configuration, then it must be

$$
\begin{cases}
\mathbf{11'} \wedge \mathbf{22'} \wedge \mathbf{33'} = 0, \\
\mathbf{12} \wedge \mathbf{1'2'} \wedge \mathbf{1''2''} = 0, \\
\mathbf{23} \wedge \mathbf{2'3'} \wedge \mathbf{2''3''} = 0, \\
\mathbf{13} \wedge \mathbf{1'3'} \wedge \mathbf{1''3''} = 0, \\
[\mathbf{1''2''3''}] = 0, \\
\omega_{1'2'}[\mathbf{1'2'3'}] = \omega_{22'}[\mathbf{22'3'}].
\end{cases}
\tag{6.17}
$$

By (4.5), $\mathbf{B}_1 = \mathbf{B}_2 = \mathbf{B}_3$. The case is similar to Cases 2.2 and 2.3.

From the above analysis, we obtain the rank classification of the syzygy equations of Example 1 presented in Section 1.

7   Linear Construction Sequence

In Sugihara (1999), a *resolvable sequence* of a polyhedron is defined to be a permutation of all its vertices and faces such that (1) when a vertex occurs, it is incident to at most three previous faces; (2) when a face occurs, it is incident to at most three previous vertices; (3) when two faces share three or more vertices, they appear earlier than the third of the common vertices; (4) when two vertices are incident to three or more faces, they appear earlier than the third of the common faces. Sugihara proved that all spherical polyhedra have resolvable sequences. Ros and Thomas (2002) used such sequences to parameterize the 2D coordinates of the vertices in a correct line drawing, which in turn enables them to choose a subset of numerical data that is non redundant and that is still enough to specify the polyhedron unambiguously.

In our viewpoint, a resolvable sequence is nothing but a *linear construction sequence* in which the constructions are among the given incidence relations. In a general linear construction sequence, the constructions are not restricted to the given incidence relations, but also include some derived incidence relations. All the properties of resolvable sequences needed in application are occupied by linear construction sequences.

Theorem 3. In Example 1, under the assumption that (1) no three neighboring vertices in a face are collinear, (2) no two faces sharing a common edge are coplanar, the torus has multiple linear construction sequences. Let $\mathbf{2}, \mathbf{3}, \mathbf{1'}, \mathbf{2'}, \mathbf{3'}, \mathbf{1''}, \mathbf{2''}$ be free points in the plane, let $\mathbf{B}_{3''}$ be a free bivector in the plane and let $\omega_{1'2'}, \omega_{22'}$ be free parameters, then the following is a sequence of linear constructions:

$$
\begin{cases}
\mathbf{1} & = \mathbf{2}(\mathbf{1'2'} \wedge \mathbf{1''2''}) \wedge \mathbf{1'}(\mathbf{22'} \wedge \mathbf{33'}), \\
\mathbf{3''} & = (\mathbf{23} \wedge \mathbf{2'3'})\mathbf{2''} \wedge (\mathbf{1'1''} \wedge \mathbf{2'2''})\mathbf{3'}, \\
\mathbf{B}_3 & = \mathbf{B}_{3''} + \omega_{1'2'}\mathbf{1'2'}, \\
\mathbf{B}_{3'} & = \mathbf{B}_{3''} + \omega_{1'2'}([\mathbf{1'2'1''}]/[\mathbf{121''}])\mathbf{12}, \\
\mathbf{B}_{1''} & = \mathbf{B}_{3''} + \omega_{22'}\mathbf{22'}, \\
\mathbf{B}_{2''} & = \mathbf{B}_{3''} + \omega_{22'}([\mathbf{232'}]/[\mathbf{131'}])\mathbf{11'}, \\
\mathbf{B}_1 & = \mathbf{B}_{3''} + \omega_{1'2'}([\mathbf{1'2'2''}]/[\mathbf{2'3'2''}])\mathbf{2'3'} - \omega_{22'}([\mathbf{22'3'}]/[\mathbf{2'3'2''}])\mathbf{2'2''}, \\
\mathbf{B}_2 & = \mathbf{B}_{3''} + \omega_{1'2'}([\mathbf{1'2'1''}]/[\mathbf{1'3'1''}])\mathbf{1'3'} - \omega_{22'}([\mathbf{22'3'}]/[\mathbf{1'3'1''}])\mathbf{1'1''}, \\
\mathbf{B}_{1'} & = \mathbf{B}_{3''} + \omega_{1'2'}([\mathbf{1'2'2''}]/[\mathbf{232''}])\mathbf{23} + \omega_{22'}([\mathbf{232'}]/[\mathbf{232''}])\mathbf{22''}, \\
\mathbf{B}_{2'} & = \mathbf{B}_{3''} + \omega_{1'2'}([\mathbf{1'2'1''}]/[\mathbf{131''}])\mathbf{13} + \omega_{22'}([\mathbf{232'}]/[\mathbf{131''}])\mathbf{11''}, \\
h_1 & = [\mathbf{1B}_{3''}], \quad h_2 = [\mathbf{2B}_{3''}], \quad h_3 = [\mathbf{3B}_{1''}], \\
h_{1'} & = [\mathbf{1'B}_{3''}], \quad h_{2'} = [\mathbf{2'B}_{3''}], \quad h_{3'} = [\mathbf{3'B}_{1''}], \\
h_{1''} & = [\mathbf{1''B}_3], \quad h_{2''} = [\mathbf{2''B}_3], \quad h_{3''} = [\mathbf{3''B}_1].
\end{cases}
$$

8   Further Examples

The parametric calotte propagation algorithm is implemented with Maple 8, and has been tested by 20 examples. The examples include both trihedral and non-trihedral, both spherical and non-spherical, polyhedral scenes. In the experiment, no bifurcation occurs during the propagation. However, bifurcation occurs immediately after the elimination is invoked to solve the new parameters. This suggests that by fewer parameters, no matter if they are independent or not, only part of the solution space may be parameterized.

Example 2. Two tori in Example 1 are merged together by identifying two of their outward faces. As in Figure 5, such a polyhedron has 14 vertices and 16 faces. Each face has four vertices. 10 vertices are each on four faces and 4 vertices are each on six faces. Vertices V_1, V_3, V_8, V_{10} are no longer coplanar.

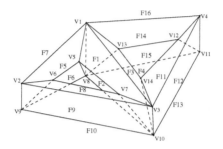

Fig. 5. Example 2

Propagation:

$$\mathbf{B}_1 = 0 \xrightarrow{\omega_{57}} \mathbf{B}_2 \xrightarrow{\omega_{15}} \mathbf{B}_5 \xrightarrow{B.2} \mathbf{B}_8 \xrightarrow{B.1} \mathbf{B}_6 \xrightarrow{B.2} \mathbf{B}_7, \mathbf{B}_9, \mathbf{B}_{10} \xrightarrow{\omega_{13}} \mathbf{B}_3 \xrightarrow{B.2} \mathbf{B}_4$$
$$\xrightarrow{\omega_{1,13}} \mathbf{B}_{14} \xrightarrow{B.2} \mathbf{B}_{11} \xrightarrow{B.1} \mathbf{B}_{12} \xrightarrow{B.2} \mathbf{B}_{13}, \mathbf{B}_{15}, \mathbf{B}_{16}.$$

Four parameters $\omega_{13}, \omega_{15}, \omega_{1,13}, \omega_{57}$ are introduced. They satisfy 9 equations, among which only 8 are algebraically independent.

$$\begin{cases}
\omega_{15}\mathbf{V}_1\mathbf{V}_5 \wedge \mathbf{V}_2\mathbf{V}_6 \wedge \mathbf{V}_3\mathbf{V}_7 = 0, \\
\omega_{1,13}\mathbf{V}_1\mathbf{V}_{13} \wedge \mathbf{V}_3\mathbf{V}_{14} \wedge \mathbf{V}_4\mathbf{V}_{12} = 0, \\
(\omega_{15}[\mathbf{V}_1\mathbf{V}_5\mathbf{V}_8] - \omega_{57}[\mathbf{V}_5\mathbf{V}_7\mathbf{V}_8])\mathbf{V}_1\mathbf{V}_2 \wedge \mathbf{V}_5\mathbf{V}_6 \wedge \mathbf{V}_8\mathbf{V}_9 = 0, \\
(\omega_{15}[\mathbf{V}_1\mathbf{V}_5\mathbf{V}_6] + \omega_{57}[\mathbf{V}_5\mathbf{V}_6\mathbf{V}_7])\mathbf{V}_5\mathbf{V}_8 \wedge \mathbf{V}_6\mathbf{V}_9 \wedge \mathbf{V}_7\mathbf{V}_{10} = 0, \\
\omega_{57}\mathbf{V}_5\mathbf{V}_7 \wedge \mathbf{V}_8\mathbf{V}_{10} \wedge \mathbf{V}_{13}\mathbf{V}_{14} - \omega_{13}\mathbf{V}_1\mathbf{V}_3 \wedge \mathbf{V}_8\mathbf{V}_{10} \wedge \mathbf{V}_{13}\mathbf{V}_{14} = 0, \\
(\omega_{57}[\mathbf{V}_5\mathbf{V}_7\mathbf{V}_8] - \omega_{13}[\mathbf{V}_1\mathbf{V}_3\mathbf{V}_8] + \omega_{113}[\mathbf{V}_1\mathbf{V}_8\mathbf{V}_{13}]) \\
\qquad\qquad\qquad \mathbf{V}_1\mathbf{V}_4 \wedge \mathbf{V}_8\mathbf{V}_{11} \wedge \mathbf{V}_{12}\mathbf{V}_{13} = 0, \\
(\omega_{15}[\mathbf{V}_1\mathbf{V}_5\mathbf{V}_6][\mathbf{V}_3\mathbf{V}_7\mathbf{V}_{10}] + \omega_{57}[\mathbf{V}_3\mathbf{V}_6\mathbf{V}_7][\mathbf{V}_5\mathbf{V}_7\mathbf{V}_{10}]) \\
\qquad\qquad\qquad \mathbf{V}_2\mathbf{V}_3 \wedge \mathbf{V}_6\mathbf{V}_7 \wedge \mathbf{V}_9\mathbf{V}_{10} = 0, \\
(\omega_{57}[\mathbf{V}_5\mathbf{V}_7\mathbf{V}_{10}][\mathbf{V}_{12}\mathbf{V}_{13}\mathbf{V}_{14}] + \omega_{1,13}[\mathbf{V}_1\mathbf{V}_{12}\mathbf{V}_{13}][\mathbf{V}_{10}\mathbf{V}_{13}\mathbf{V}_{14}] \\
\quad +\omega_{13}[\mathbf{V}_1\mathbf{V}_3\mathbf{V}_{10}][\mathbf{V}_{12}\mathbf{V}_{13}\mathbf{V}_{14}])\mathbf{V}_8\mathbf{V}_{13} \wedge \mathbf{V}_{10}\mathbf{V}_{14} \wedge \mathbf{V}_{11}\mathbf{V}_{12} = 0, \\
(\omega_{57}\mathbf{V}_1\mathbf{V}_4 \wedge \mathbf{V}_5\mathbf{V}_7 \wedge \mathbf{V}_8\mathbf{V}_{10} - \omega_{1,13}[\mathbf{V}_1\mathbf{V}_4\mathbf{V}_{13}][\mathbf{V}_1\mathbf{V}_8\mathbf{V}_{10}] \\
\quad -\omega_{13}[\mathbf{V}_1\mathbf{V}_3\mathbf{V}_4][\mathbf{V}_1\mathbf{V}_8\mathbf{V}_{10}])\mathbf{V}_3\mathbf{V}_4 \wedge \mathbf{V}_{10}\mathbf{V}_{11} \wedge \mathbf{V}_{12}\mathbf{V}_{14} = 0.
\end{cases}$$

In the list, there are 8 factored calotte equation. The other equation can also be replaced by a calotte equation. The 9 calottes are

$$
\begin{array}{lll}
(V_1V_2V_3,\ V_5V_6V_7) & (V_1V_5V_8,\ V_2V_6V_9) & (V_2V_6V_9,\ V_3V_7V_{10}) \\
(V_5V_6V_7,\ V_8V_9V_{10}) & (V_{12}V_{13}V_{14},\ V_1V_3V_4) & (V_3V_{10}V_{14},\ V_4V_{11}V_{12}) \\
(V_4V_{12}V_{13},\ V_8V_{10}V_{11}) & (V_4V_{11}V_{12},\ V_1V_8V_{13}) & (V_1V_5V_8V_{13},\ V_3V_7V_{10}V_{14})
\end{array}
$$

This example also has multiple linear construction sequences.

Example 3. Two trihedral spherical polyhedra S_1 and S_2 are connected through two tubes T_1, T_2 at two pairs of five-vertexed faces. The two pairs of faces are then removed so that the corresponding vertices are no longer coplanar. Thus, S_1 has 60 vertices and 30 faces, among which 20 faces are six-vertexed and 10 faces are five-vertexed. S_1 has 20 vertices and 10 faces, and each face has five vertices. Both tubes are composed of three columns, and each column is composed of five square faces. All together there are 100 vertices and 70 faces.

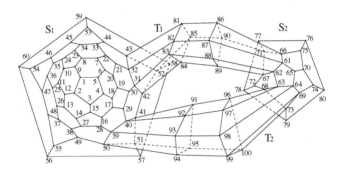

Fig. 6. Example 3

We decompose the polyhedron into four smaller parts S_1, S_2, T_1 and T_2, run parametric calotte propagation in each part, then assemble the polyhedron by joining the parts at the junction loops.

Step 1. Start from a face of S_1, run parametric propagation in S_1. One parameter is introduced, which satisfies 24 equations. The equations can be replaced by calotte ones, among which 22 are in factored form.

Step 2. Start from the loop $S_1 \cap T_1$, run parametric propagation in T_1. Three parameters are introduced, which satisfy 3 non-factored calotte equations.

Step 3. Start from the loop $S_1 \cap T_2$, run parametric propagation in T_2. Three parameters are introduced, which satisfy 3 non-factored calotte equations.

Step 4. Start from the two loops $S_2 \cap T_1$ and $S_2 \cap T_2$, run parametric propagation in S_2. No parameter is introduced. There are 10 parametric equations, 2 of which are factored calotte equations. After the four parts are put together, the 8 equations can be replaced by 8 calotte equations in non-factored form.

References

1. J. Bokowski and B. Sturmfels. *Computational Synthetic Geometry,* **LNM 1355**, Springer, Berlin, Heidelberg, 1989.
2. H. Crapo and W. Whiteley. Stresses on Frameworks and Motions of Panel Structures: a Projective Geometric Introduction. *Structural Topology* 6: 42-82, 1982.
3. H. Crapo and J. Ryan. Spatial Realizations of Linear Scenes. *Structural Topology* **13**: 33-68, 1986.
4. H. Crapo. Invariant-Theoretic Methods in Scene Analysis and Structural Mechanics. *J. of Symbolic Computation* 11: 523-548, 1991.
5. H. Crapo and W. Whiteley. Plane Self Stresses and Projected Polyhedra I: The Basic Pattern. *Structural Topology* **20**: 55-78, 1993.
6. D. Huffman. Realizable Configurations of Lines in Pictures of Polyhedra, *Machine Intelligence* 8: 493-509, 1977.
7. H. Li. Vectorial Equation-Solving for Mechanical Geometry Theorem Proving. *J. Automated Reasoning* 25: 83-121, 2000.
8. H. Li and Y. Wu. Automated Short Proof Generation for Projective Geometric Theorems with Cayley and Bracket Algebras, I. Incidence Geometry. To appear in *J. of Symbolic Computation.*
9. A. Mackworth. Interpreting Pictures of Polyhedral Scenes, *Artificial Intelligence* 4: 121-137, 1973.
10. J. Richter-Gebert. *Realization Spaces of Polytopes.* **LNM 1643**, Springer, Berlin, Heidelberg, 1996.
11. L. Ros and F. Thomas. Analysing Spatial Realizability of Line Drawings through Edge-Concurrence Tests. In: *Proc. IEEE Int'l Conf. Robotics and Automation,* vol IV, pp. 3559-3566, May 1998.
12. L. Ros. A Kinematic-Geometric Approach to Spatial Interpretation of Line Drawings, Ph. D. Thesis, Polytechnic Univ. of Catalonia, May 2000.
13. L. Ros, F. Thomas. Overcoming Superstrictness in Line Drawing Interpretation. *IEEE Trans. PAMI* **24(4)**: 456-466, 2002.
14. B. Sturmfels and W. Whiteley. On the Synthetic Factorization of Projectively Invariant Polynomials. *J. of Symbolic Computation* 11: 439-453, 1991.
15. K. Sugihara. Mathematical Structures of Line Drawings of Polyhedrons – Towards Man-Machine Communication by Means of Line Drawings. *IEEE Trans. PAMI* **4**: 458-469, 1982.
16. K. Sugihara (1984a). An Algebraic and Combinatorial Approach to the Analysis of Line Drawings of Polyhedra. *Discrete Appl. Math.* **9**: 77-104, 1984.
17. K. Sugihara (1984b). An Algebraic Approach to Shape-from-Image Problems, *Artificial Intelligence* **23**: 59-95, 1984.
18. K. Sugihara. *Machine Interpretation of Line Drawings,* MIT Press, Cambridge, Mass, 1986.
19. K. Sugihara. Resolvable Representation of Polyhedra. *Discrete Computational Geometry* **21(2)**: 243-255, 1999.
20. T. Tay. On the Cayley Factorization of Calotte Conditions. *Discrete Computational Geometry* 11: 97-109, 1994.
21. N. White and W. Whiteley. The Algebraic Geometry of Stresses in Frameworks, *SIAM J. Alg. Discrete Math.* **4**: 481-511, 1983.
22. N. White. Multilinear Cayley Factorization. *J. of Symbolic Computation* 11: 421-438, 1991.
23. W. Whiteley. From a Line Drawing to a Polyhedron. *J. Math. Psych.* **31**: 441-448, 1987.

24. W. Whiteley. A Matroid on Hypergraphs with Applications in Scene Analysis and Geometry. *Discrete Computational Geometry* **4**: 75-95, 1989.

25. W. Whiteley. Weavings, Sections and Projections of Spherical Polyhedra. *Discrete Appl. Math.* **32**: 275-294, 1991.

26. W. Whiteley. Matroids and Rigid Structures. In *Matroid Applications*, N. White *ed.*, Encyclopedia of Mathematics and Its Applications **40**, pp. 1-53, Cambridge University Press, Cambridge, 1992.

A Unified and Complete Framework of Invariance for Six Points

Yihong Wu and Zhanyi Hu

National Laboratory of Pattern Recognition, Institute of Automation,
Chinese Academy of Sciences, P.O. Box 2728, Beijing 100080, P.R. China
{yhwu, huzy}@nlpr.ia.ac.cn

Abstract. Projective geometric invariants play an important role in computer vision. To set up the invariant relationships between spatial points and their images from a single view, at least six pairs of spatial and image points are required. In this paper, we establish a unified and complete framework of the invariant relationships for six points. The framework covers the general case already developed in the literature and two novel cases. The two novel cases describe that six spatial points and the camera optical center lie on a quadric cone or a twisted cubic, called quadric cone case or twisted cubic case. For the general case and quadric cone case, camera parameters can be determined uniquely. For the twisted cubic case, camera parameters cannot be determined completely; this configuration of camera optical center and spatial points is called a critical configuration. The established unified framework may help to effectively identify the type of geometric information appearing in certain vision tasks, in particular critical geometric information. An obvious advantage using this framework to obtain geometric information is that no any explicit estimation on camera projective matrix and optical center is needed.

1 Introduction

Projective geometric invariants play an important role in computer vision [13, 14, 19]. Bracket algebra is a basic algebraic tool for studying projective geometric invariance [22, 23]. Due to the powerful computations of geometric algebra, bracket algebra, as a subalgebra of geometric algebra on scalars, has found more and more applications in computer vision [2, 3, 4, 5, 6, 7, 11, 20, 24, 25]. Applications of geometric algebra and its subalgebras [10, 22, 12, 21] are developing in many fields of engineering. This paper presents a computation of bracket algebra on invariance of six points in 3D reconstruction, an important problem of computer vision.

There have been many studies on the invariant relationship, also called the invariance, of six spatial points and their images [2, 4, 5, 9, 15, 16, 17, 18]. In 1994, Quan [15, 16] gave the invariance of six spatial points and their images for the general non-degenerate configuration under the canonical coordinates, from which the invariance under any coordinate system is obtained by a coordinate

H. Li, P. J. Olver and G. Sommer (Eds.): IWMM-GIAE 2004, LNCS 3519, pp. 403–417, 2005.

transformation. In 1995, Carlsson [4,5] derived the invariance for the general non-degenerate configuration and the invariance for a specific non-degenerate configuration of the four-point-coplanar case by bracket algebra. The resulting invariants are coordinate-free. In 2000, Roh and Kweon [17] provided the invariance of six spatial points and their images under the same specific non-degenerate configuration as the one in [4,5] under the canonical coordinates, then the invariance under any coordinate system is obtained by a coordinate transformation. In 2002, Bayro-Corrochano and Banarer [2] re-derived the invariance of six spatial points and their images for the general non-degenerate configuration, and reported some applications of the invariance. As shown in their work, the invariance can be applied to 3D reconstruction, object recognition, robot vision and further tasks.

Motivated by their work and the importance of invariance, in this paper we set up invariant relationships for two specific configurations, i.e. when six spatial points and camera optical center lie on a quadric cone or a twisted cubic. We call them quadric cone case and twisted cubic case respectively. It is interesting that the number of equations describing the invariant relationship for the twisted cubic case is two rather than one as for the general case developed in the previous literature. In addition, the invariant relationship for the general case reported in the previous work, and the ones here for the quadric cone case and twisted cubic case constitute a complete framework, unified by brackets, in which the invariant relationships are coordinate-free, free of camera optical center, and free of camera projective matrix.

Estimation of camera parameters is a key problem for 3D reconstruction. One of the popular methods for this problem is to recover the camera parameters from at least six pairs of corresponding spatial and image points [1]. By using this method, for the general case and the quadric cone case, camera parameters can be determined uniquely; for the twisted cubic case, camera parameters cannot be recovered completely. This configuration of camera and spatial points in the twisted cubic case is critical for camera parameter estimation [8]. The established complete framework of invariant relationships between six spatial points and their images could help researchers to clearly and efficiently identify the type of geometric information, in particular, the critical geometric information, when they calibrate a camera from six points. An obvious advantage using this framework to detect the relative position information of camera and scene is that no any reconstruction on camera projective matrix or optical center is needed. A detailed algorithm applying the novel invariant relationship of the twisted cubic case to recognize critical configuration of 3D reconstruction, and experiments on simulated and real data, are reported in our recent technical report [25].

The organization of the paper is as follows. Some preliminaries are listed in Section 2. Section 3 gives the invariant relationships between six spatial points and their images when camera optical center and the six spatial points lie on either a quadric cone or a twisted cubic. The complete and unified framework of invariant relationships for six points is elaborated in Section 4. Some experimental results are shown in Section 5. Section 6 gives some conclusions.

2   Preliminaries

In this paper, a bold capital letter denotes either a homogeneous 4-vector or a matrix; a bold small letter denotes a homogeneous 3-vector; a bracket "[] " denotes the determinant of vectors in it. Different columns in the brackets are separated with commas, and the commas are omitted if no ambiguity could arise. In addition, throughout this paper, we assume that no three image points are collinear (then neither three spatial points can be collinear nor coplanar with the optical center), and no two image points are coincident (i.e. different spatial points have different image points respectively, or the optical center cannot be collinear with any two spatial points).

Bracket algebra is in fact the algebra on determinants, which is the basic algebra tool for computing projective geometric invariants [22, 23]. Exchanging two vectors in a bracket will change the sign of the bracket. For example: $[\mathbf{a}_1\mathbf{a}_2\mathbf{a}_3] = -[\mathbf{a}_2\mathbf{a}_1\mathbf{a}_3]$. $[\mathbf{a}_1\mathbf{a}_2\mathbf{a}_3] = 0$ means that $\mathbf{a}_1, \mathbf{a}_2, \mathbf{a}_3$ are collinear. $[\mathbf{A}_1\mathbf{A}_2\mathbf{A}_3\mathbf{A}_4] = 0$ means that $\mathbf{A}_i, i = 1..4$ are coplanar, especially if three of them are collinear.

There exist relations among determinants. One kind of these relations are the following Grassmann-Plücker relations:

$$[\mathbf{a}_1\mathbf{a}_2\mathbf{a}_5][\mathbf{a}_3\mathbf{a}_4\mathbf{a}_5] - [\mathbf{a}_1\mathbf{a}_3\mathbf{a}_5][\mathbf{a}_2\mathbf{a}_4\mathbf{a}_5] + [\mathbf{a}_1\mathbf{a}_4\mathbf{a}_5][\mathbf{a}_2\mathbf{a}_3\mathbf{a}_5] = 0, \tag{1}$$

$$\begin{aligned} &[\mathbf{A}_1\mathbf{A}_2\mathbf{A}_5\mathbf{A}_6][\mathbf{A}_3\mathbf{A}_4\mathbf{A}_5\mathbf{A}_6] - [\mathbf{A}_1\mathbf{A}_3\mathbf{A}_5\mathbf{A}_6][\mathbf{A}_2\mathbf{A}_4\mathbf{A}_5\mathbf{A}_6] \\ &+[\mathbf{A}_1\mathbf{A}_4\mathbf{A}_5\mathbf{A}_6][\mathbf{A}_2\mathbf{A}_3\mathbf{A}_5\mathbf{A}_6] = 0, \end{aligned} \tag{2}$$

where (1) is with respect to 2D homogeneous vectors $\mathbf{a}_i, i = 1..5$, and (2) is with respect to 3D homogeneous vectors $\mathbf{A}_i, i = 1..6$. They are often used to simplify bracket computations, and to find the geometric meanings of a bracket equation [23]. For example: $[\mathbf{a}_1\mathbf{a}_2\mathbf{a}_5][\mathbf{a}_3\mathbf{a}_4\mathbf{a}_5] + [\mathbf{a}_1\mathbf{a}_4\mathbf{a}_5][\mathbf{a}_2\mathbf{a}_3\mathbf{a}_5]$ can be simplified as $[\mathbf{a}_1\mathbf{a}_3\mathbf{a}_5][\mathbf{a}_2\mathbf{a}_4\mathbf{a}_5]$ by (1), thus, $[\mathbf{a}_1\mathbf{a}_2\mathbf{a}_5][\mathbf{a}_3\mathbf{a}_4\mathbf{a}_5] + [\mathbf{a}_1\mathbf{a}_4\mathbf{a}_5][\mathbf{a}_2\mathbf{a}_3\mathbf{a}_5] = 0$ is equivalent to $[\mathbf{a}_1\mathbf{a}_3\mathbf{a}_5][\mathbf{a}_2\mathbf{a}_4\mathbf{a}_5] = 0$, which means at least that either $\mathbf{a}_1, \mathbf{a}_3, \mathbf{a}_5$ are collinear, or $\mathbf{a}_2, \mathbf{a}_4, \mathbf{a}_5$ are collinear.

Under a pinhole camera, a point $\mathbf{M}_i$ in space is projected to a point $\mathbf{m}_i$ in the image plane by:

$$x_i\mathbf{m}_i = \mathbf{K}(\mathbf{R}\ \mathbf{t})\mathbf{M}_i, \tag{3}$$

where $\mathbf{K}$ is the 3×3 matrix of camera intrinsic parameters, and $\mathbf{R}, \mathbf{t}$ are a 3×3 rotation matrix and a translation 3-vector, x_i is a nonzero scalar, called the depth of $\mathbf{M}_i$. If x_i is zero, then $\mathbf{M}_i$ could not be projected to the image plane. If the optical center, denoted as $\mathbf{O}$, is not at infinity, its non-homogeneous coordinate are given by $\hat{\mathbf{O}} = -\mathbf{R}^\tau\mathbf{t}$. We assume that $\mathbf{O}$ is not at infinity throughout this paper.

Consider points $\mathbf{M}_i = (\hat{\mathbf{M}}_i^\tau, 1)^\tau$ not at infinity, $i = 1..6$, where $\hat{\mathbf{M}}_i^\tau$ are non-homogeneous coordinates, then by (3) we have:

$$x_i x_j x_k [\mathbf{m}_i, \mathbf{m}_j, \mathbf{m}_k]$$
$$= [x_i \mathbf{m}_i, x_j \mathbf{m}_j, x_k \mathbf{m}_k]$$
$$= \det(\mathbf{K})[\mathbf{R}\hat{\mathbf{M}}_i + \mathbf{t}, \mathbf{R}\hat{\mathbf{M}}_j + \mathbf{t}, \mathbf{R}\hat{\mathbf{M}}_k + \mathbf{t}]$$
$$= \det(\mathbf{K})[\hat{\mathbf{M}}_i + \mathbf{R}^\tau \mathbf{t}, \hat{\mathbf{M}}_j + \mathbf{R}^\tau \mathbf{t}, \hat{\mathbf{M}}_k + \mathbf{R}^\tau \mathbf{t}] \tag{4}$$
$$= \det(\mathbf{K})[\hat{\mathbf{M}}_i - \hat{\mathbf{O}}, \hat{\mathbf{M}}_j - \hat{\mathbf{O}}, \hat{\mathbf{M}}_k - \hat{\mathbf{O}}]$$
$$= \det(\mathbf{K}) \begin{bmatrix} \hat{\mathbf{M}}_i, & \hat{\mathbf{M}}_j, & \hat{\mathbf{M}}_k, & \hat{\mathbf{O}} \\ 1 & 1 & 1 & 1 \end{bmatrix}$$
$$= \det(\mathbf{K})[\mathbf{M}_i, \mathbf{M}_j, \mathbf{M}_k, \mathbf{O}].$$

Thus, $[\mathbf{m}_i, \mathbf{m}_j, \mathbf{m}_k] = 0$ if and only if $[\mathbf{M}_i, \mathbf{M}_j, \mathbf{M}_k, \mathbf{O}] = 0$. Namely, $\mathbf{m}_i, \mathbf{m}_j, \mathbf{m}_k$ are collinear if and only if $\mathbf{M}_i, \mathbf{M}_j, \mathbf{M}_k, \mathbf{O}$ are coplanar.

For the optical center and five spatial points, there is a Grassmann-Plücker relation of (2) as:

$$[\mathbf{M}_i, \mathbf{M}_j, \mathbf{M}_m, \mathbf{M}_n][\mathbf{M}_k, \mathbf{O}, \mathbf{M}_m, \mathbf{M}_n] - [\mathbf{M}_i, \mathbf{M}_k, \mathbf{M}_m, \mathbf{M}_n][\mathbf{M}_j, \mathbf{O}, \mathbf{M}_m, \mathbf{M}_n]$$
$$+ [\mathbf{M}_j, \mathbf{M}_k, \mathbf{M}_m, \mathbf{M}_n][\mathbf{M}_i, \mathbf{O}, \mathbf{M}_m, \mathbf{M}_n] = 0.$$

From this equation and from (4), we have:

$$x_k [\mathbf{M}_i, \mathbf{M}_j, \mathbf{M}_m, \mathbf{M}_n][\mathbf{m}_k, \mathbf{m}_m, \mathbf{m}_n] - x_j [\mathbf{M}_i, \mathbf{M}_k, \mathbf{M}_m, \mathbf{M}_n][\mathbf{m}_j, \mathbf{m}_m, \mathbf{m}_n] \tag{5}$$
$$+ x_i [\mathbf{M}_j, \mathbf{M}_k, \mathbf{M}_m, \mathbf{M}_n][\mathbf{m}_i, \mathbf{m}_m, \mathbf{m}_n] = 0.$$

This equation is free of the image and world coordinate system.

In the following sections, for the notational convenience, if no ambiguity can be aroused, $\mathbf{M}_i$, $i = 1..6$ will be simply denoted as $\mathbf{1}, \mathbf{2}, \mathbf{3}, \mathbf{4}, \mathbf{5}, \mathbf{6}$.

Theorem 1 in [24] is recalled here. There exists a unique proper quadric cone with $\mathbf{1}$ as the vertex and passing through $\mathbf{2}, \mathbf{3}, \mathbf{4}, \mathbf{5}, \mathbf{6}$ with no three collinear and no four coplanar. Any point $\mathbf{X}$ belongs to this quadric cone if and only if:

$$\frac{[\mathbf{1246}][\mathbf{1356}]}{[\mathbf{1236}][\mathbf{1456}]} = \frac{[\mathbf{124X}][\mathbf{135X}]}{[\mathbf{123X}][\mathbf{145X}]}.$$

The above representation is not unique since the one obtained by a permutation of $\mathbf{2}, \mathbf{3}, \mathbf{4}, \mathbf{5}, \mathbf{6}$ is also a representation of the quadric cone.

Theorem 2 in [24] is recalled here too. There exists a unique proper twisted cubic passing through $\mathbf{1}, \mathbf{2}, \mathbf{3}, \mathbf{4}, \mathbf{5}, \mathbf{6}$ with no three collinear and no four coplanar. Any point $\mathbf{X}$ belongs to this twisted cubic if and only if:

$$\begin{cases} \dfrac{[\mathbf{1246}][\mathbf{1356}]}{[\mathbf{1236}][\mathbf{1456}]} = \dfrac{[\mathbf{124X}][\mathbf{135X}]}{[\mathbf{123X}][\mathbf{145X}]}, \\[2mm] \dfrac{[\mathbf{1246}][\mathbf{2356}]}{[\mathbf{1236}][\mathbf{2456}]} = \dfrac{[\mathbf{124X}][\mathbf{235X}]}{[\mathbf{123X}][\mathbf{245X}]}, \\[2mm] \mathbf{X} \text{ is not on the line } \mathbf{12} \text{ except for } \mathbf{1}, \mathbf{2}. \end{cases}$$

The above representation is not unique since the one obtained by a permutation of $\mathbf{1}, \mathbf{2}, \mathbf{3}, \mathbf{4}, \mathbf{5}, \mathbf{6}$ is also a representation of the twisted cubic.

3 Invariance of Six Points for the Quadric Cone and Twisted Cubic Cases

This section further assumes no four of $1, 2, 3, 4, 5, 6$ are coplanar. Then the bracket on spatial points is nonzero.

Proposition 1. Let Q be the quadric cone with $\mathbf{1}$ as the vertex and passing through $\mathbf{2}, \mathbf{3}, \mathbf{4}, \mathbf{5}, \mathbf{6}$. Then the camera optical center $\mathbf{O}$ lies on Q if and only if:

$$\frac{[\mathbf{1246}][\mathbf{1356}]}{[\mathbf{1236}][\mathbf{1456}]} = \frac{[\mathbf{m_1 m_2 m_4}][\mathbf{m_1 m_3 m_5}]}{[\mathbf{m_1 m_2 m_3}][\mathbf{m_1 m_4 m_5}]}.$$

Each side is a cross ratio invariant. After a permutation of $\mathbf{2}, \mathbf{3}, \mathbf{4}, \mathbf{5}, \mathbf{6}$ and their corresponding images, the equation is still an invariant relation of $\mathbf{O}$ lying on the same quadric cone, so is not independent of the one without permutation.

Proposition 1 is obtained as follows: according to Theorem 1 in [24] as shown in Section 2, there exists a unique proper quadric cone through $1, 2, 3, 4, 5, 6$ and with $\mathbf{1}$ as the vertex, and its equation is

$$\frac{[\mathbf{1246}][\mathbf{1356}]}{[\mathbf{1236}][\mathbf{1456}]} = \frac{[\mathbf{124X}][\mathbf{135X}]}{[\mathbf{123X}][\mathbf{145X}]},$$

where $\mathbf{X}$ is the varying vector. If $\mathbf{O}$ lies on it, then $\mathbf{O}$ satisfies the above equation. Furthermore by (4), we have:

$$\frac{[\mathbf{1246}][\mathbf{1356}]}{[\mathbf{1236}][\mathbf{1456}]} = \frac{[\mathbf{124O}][\mathbf{135O}]}{[\mathbf{123O}][\mathbf{145O}]} = \frac{[\mathbf{m_1 m_2 m_4}][\mathbf{m_1 m_3 m_5}]}{[\mathbf{m_1 m_2 m_3}][\mathbf{m_1 m_4 m_5}]},$$

and vice versa. We can permute $\mathbf{2}, \mathbf{3}, \mathbf{4}, \mathbf{5}, \mathbf{6}$ and their corresponding images in the above equation to obtain different equations, but among these equations, only one is independent since there is only one independent representation for a quadric cone.

Remark: If $\mathbf{2}, \mathbf{3}, \mathbf{4}, \mathbf{5}, \mathbf{6}$ are coplanar, $\mathbf{1}$ is not coplanar with them, and no bracket containing $\mathbf{1}$ is zero, the equation of Proposition 1 still means that $\mathbf{O}$ lies on the same proper quadric cone.

Similarly, according to Theorem 2 in [24] and (4) shown in Section 2, we have:

Proposition 2. The camera optical center $\mathbf{O}$ lies on the proper twisted cubic passing through $1, 2, 3, 4, 5, 6$ if and only if

$$\begin{cases} \dfrac{[\mathbf{1246}][\mathbf{1356}]}{[\mathbf{1236}][\mathbf{1456}]} = \dfrac{[\mathbf{m_1 m_2 m_4}][\mathbf{m_1 m_3 m_5}]}{[\mathbf{m_1 m_2 m_3}][\mathbf{m_1 m_4 m_5}]}, \\[3mm] \dfrac{[\mathbf{1246}][\mathbf{2356}]}{[\mathbf{1236}][\mathbf{2456}]} = \dfrac{[\mathbf{m_1 m_2 m_4}][\mathbf{m_2 m_3 m_5}]}{[\mathbf{m_1 m_2 m_3}][\mathbf{m_2 m_4 m_5}]}. \end{cases}$$

Each side is a cross ratio invariant. After a permutation of $1, 2, 3, 4, 5, 6$ and the corresponding images, the resulting equation system is still the invariant relationship of $\mathbf{O}$ lying on the same twisted cubic, but is not independent of the original one.

Note that in Theorem 2 of [24], there is another condition for the representation of the twisted cubic such as: $\mathbf{X}$ does not lie on the line through $\mathbf{1}$ and $\mathbf{2}$. Here, this additional condition is unnecessary because if $\mathbf{O}$ is on the line through $\mathbf{1}$ and $\mathbf{2}$, then $\mathbf{m}_1 = \mathbf{m}_2$, which is contrary to our assumption.

4   A Unified and Complete Framework of Invariance for Six Points

If $\mathbf{1}, \mathbf{2}, \mathbf{3}, \mathbf{4}, \mathbf{5}, \mathbf{6}$ are all coplanar, or five of them are coplanar, the invariant relationship between them and their images is reduced to the 2D case, which can be established easily. So, such cases will not be discussed and we always assume no five spatial points are coplanar in the subsequent discussion.

With the above assumptions, at least two coefficients of x_i, x_j, x_k in (5) are nonzero. For example, if the coefficients of x_i and x_j are zero, i.e. $[\mathbf{m}_k, \mathbf{m}_m, \mathbf{m}_n][\mathbf{M}_i, \mathbf{M}_j, \mathbf{M}_m, \mathbf{M}_n] = 0$ and $[\mathbf{m}_j, \mathbf{m}_m, \mathbf{m}_n][\mathbf{M}_i, \mathbf{M}_k, \mathbf{M}_m, \mathbf{M}_n] = 0$, then since the brackets on image points are assumed nonzero, we obtain $[\mathbf{M}_i, \mathbf{M}_j, \mathbf{M}_m, \mathbf{M}_n] = 0$ and $[\mathbf{M}_i, \mathbf{M}_k, \mathbf{M}_m, \mathbf{M}_n] = 0$, which mean that $\mathbf{M}_i, \mathbf{M}_j$, $\mathbf{M}_k, \mathbf{M}_m, \mathbf{M}_n$ are coplanar or $\mathbf{M}_i, \mathbf{M}_m, \mathbf{M}_n$ are collinear. This is contrary to our assumptions. So, at least two coefficients of x_i, x_j, x_k in (5) are nonzero, i.e. (5) is not identical to zero.

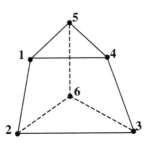

Fig. 1. For six spatial points, if no five of them are coplanar, and no three of them are collinear, then at most three brackets could be zero, and at least four brackets cannot be zero. Here only $[\mathbf{1234}] = 0$, $[\mathbf{1256}] = 0$, $[\mathbf{3456}] = 0$. And, $[\mathbf{1235}] \neq 0$, $[\mathbf{1236}] \neq 0$, $[\mathbf{1245}] \neq 0$, $[\mathbf{1246}] \neq 0$

On $\mathbf{1}, \mathbf{2}, \mathbf{3}, \mathbf{4}, \mathbf{5}, \mathbf{6}$, there are in total $C_6^4 = 15$ brackets as $S = \{[\mathbf{1234}], [\mathbf{1235}],$ $[\mathbf{1236}], [\mathbf{1245}], [\mathbf{1246}], [\mathbf{1256}], [\mathbf{1345}], [\mathbf{1346}], [\mathbf{1356}], [\mathbf{1456}], [\mathbf{2345}], [\mathbf{2346}],$ $[\mathbf{2356}], [\mathbf{2456}], [\mathbf{3456}]\}$. Under the assumption that no five spatial points are coplanar and no three spatial points are collinear, at most three brackets in S can be zero, or it is equivalent to that at least four brackets in S cannot be zero. For example, if $[\mathbf{1234}] = 0$, $[\mathbf{1256}] = 0$, $[\mathbf{3456}] = 0$, then any bracket in the set $S - \{[\mathbf{1234}], [\mathbf{1256}], [\mathbf{3456}]\}$ cannot be zero as shown in Fig. 1. Otherwise, five spatial points would be coplanar, or three spatial points would be collinear. So, we can always let $[\mathbf{1235}] \neq 0$, $[\mathbf{1236}] \neq 0$, $[\mathbf{1245}] \neq 0$, $[\mathbf{1246}] \neq 0$ by arranging the order of $\mathbf{1}, \mathbf{2}, \mathbf{3}, \mathbf{4}, \mathbf{5}, \mathbf{6}$.

The polynomial on the left side of (5) is denoted by $f_{ijk,mn}$, which means that the equation is linear with respect to x_i, x_j, x_k, and the brackets on image points share $\mathbf{m}_m$, $\mathbf{m}_n$, the brackets on spatial points share $\mathbf{M}_m$, $\mathbf{M}_n$. Let $\mathbf{P} = \mathbf{K}(\mathbf{R}, \mathbf{t})$. Among all the equations $f_{ijk,mn} = 0$, there are at most six independent ones. This is because from (3) there are in total 18 equations, and all possible $f_{ijk,mn} = 0$ are obtained by eliminating 12 parameters on $\mathbf{P}$ from these 18 equations. We list six $f_{ijk,mn} = 0$ as follows:

$$f_{123,45} = x_3[\mathbf{m}_3\mathbf{m}_4\mathbf{m}_5][1245] - x_2[\mathbf{m}_2\mathbf{m}_4\mathbf{m}_5][1345] + x_1[\mathbf{m}_1\mathbf{m}_4\mathbf{m}_5][2345] = 0,$$
(6)

$$f_{123,46} = x_3[\mathbf{m}_3\mathbf{m}_4\mathbf{m}_6][1246] - x_2[\mathbf{m}_2\mathbf{m}_4\mathbf{m}_6][1346] + x_1[\mathbf{m}_1\mathbf{m}_4\mathbf{m}_6][2346] = 0,$$
(7)

$$f_{124,35} = x_4[\mathbf{m}_3\mathbf{m}_4\mathbf{m}_5][1235] - x_2[\mathbf{m}_2\mathbf{m}_3\mathbf{m}_5][1345] + x_1[\mathbf{m}_1\mathbf{m}_3\mathbf{m}_5][2345] = 0,$$
(8)

$$f_{124,36} = x_4[\mathbf{m}_3\mathbf{m}_4\mathbf{m}_6][1236] - x_2[\mathbf{m}_2\mathbf{m}_3\mathbf{m}_6][1346] + x_1[\mathbf{m}_1\mathbf{m}_3\mathbf{m}_6][2346] = 0,$$
(9)

$$f_{125,36} = x_5[\mathbf{m}_3\mathbf{m}_5\mathbf{m}_6][1236] - x_2[\mathbf{m}_2\mathbf{m}_3\mathbf{m}_6][1356] + x_1[\mathbf{m}_1\mathbf{m}_3\mathbf{m}_6][2356] = 0,$$
(10)

$$f_{126,35} = x_6[\mathbf{m}_3\mathbf{m}_5\mathbf{m}_6][1235] - x_2[\mathbf{m}_2\mathbf{m}_3\mathbf{m}_5][1356] + x_1[\mathbf{m}_1\mathbf{m}_3\mathbf{m}_5][2356] = 0.$$
(11)

These equations are multi-linear with respect to x_i, brackets on image points, and brackets on spatial points.

The invariant relationship between spatial and image points is not related to the depths, so we are to eliminate x_i from the above equations in the following.

Since $[\mathbf{m}_3\mathbf{m}_4\mathbf{m}_5][1245] \neq 0$, we can solve x_3 from (6), and then substitute it into (7). After that, we obtain:

$$x_1([\mathbf{m}_1\mathbf{m}_4\mathbf{m}_5][\mathbf{m}_3\mathbf{m}_4\mathbf{m}_6][2345][1246] - [\mathbf{m}_3\mathbf{m}_4\mathbf{m}_5][\mathbf{m}_1\mathbf{m}_4\mathbf{m}_6][1245][2346])$$
$$= x_2([\mathbf{m}_2\mathbf{m}_4\mathbf{m}_5][\mathbf{m}_3\mathbf{m}_4\mathbf{m}_6][1345][1246] - [\mathbf{m}_3\mathbf{m}_4\mathbf{m}_5][\mathbf{m}_2\mathbf{m}_4\mathbf{m}_6][1245][1346]).$$
(12)

Similarly, we solve x_4 from (8), then substitute it into (9), and obtain:

$$x_1([\mathbf{m}_1\mathbf{m}_3\mathbf{m}_5][\mathbf{m}_3\mathbf{m}_4\mathbf{m}_6][2345][1236] - [\mathbf{m}_3\mathbf{m}_4\mathbf{m}_5][\mathbf{m}_1\mathbf{m}_3\mathbf{m}_6][1235][2346])$$
$$= x_2([\mathbf{m}_2\mathbf{m}_3\mathbf{m}_5][\mathbf{m}_3\mathbf{m}_4\mathbf{m}_6][1345][1236] - [\mathbf{m}_3\mathbf{m}_4\mathbf{m}_5][\mathbf{m}_2\mathbf{m}_3\mathbf{m}_6][1235][1346]).$$
(13)

Then (6), (7), (8), (9), (10), (11) are equivalent to (6), (12), (8), (13), (10), (11). Let the coefficients of x_1 and x_2 in (12) be c_1 and c_2, and the ones in (13) be e_1 and e_2. Since $x_1 \neq 0$ and $x_2 \neq 0$, we have: $c_1 = 0$ if and only if $c_2 = 0$, also $e_1 = 0$ if and only if $e_2 = 0$. Thus, there are only three cases: (i) $c_1 \neq 0$ and $e_1 \neq 0$; (ii) $c_1 = 0$ but $e_1 \neq 0$; Or $e_1 = 0$ but $c_1 \neq 0$; (iii) $c_1 = 0$ and $e_1 = 0$. We will discuss each case in the following.

Case (i). $c_1 \neq 0$ and $e_1 \neq 0$ (The general case already developed in the literature that is non-degenerate for camera parameter estimation. See the last paragraph of this section).

Since $c_1 \neq 0$ and $e_1 \neq 0$ (also $c_2 \neq 0, e_2 \neq 0$), solving out x_1/x_2 from (12) and (13) gives $x_1/x_2 = c_2/c_1 = e_2/e_1$. So we obtain $c_1e_2 - c_2e_1 = 0$. Simplify $c_1e_2 - c_2e_1 = 0$ by some Grassmann-Plücker relations of (1) on image points, we get its equivalent form as:

$$
\begin{aligned}
&[\mathbf{m_3m_4m_5}][\mathbf{m_1m_2m_6}][\mathbf{1235}][\mathbf{1245}][\mathbf{1346}][\mathbf{2346}]\\
+&[\mathbf{m_1m_2m_5}][\mathbf{m_3m_4m_6}][\mathbf{1345}][\mathbf{2345}][\mathbf{1236}][\mathbf{1246}]\\
+&[\mathbf{m_2m_3m_5}][\mathbf{m_1m_4m_6}][\mathbf{1245}][\mathbf{1345}][\mathbf{1236}][\mathbf{2346}]\\
+&[\mathbf{m_1m_4m_5}][\mathbf{m_2m_3m_6}][\mathbf{1235}][\mathbf{2345}][\mathbf{1246}][\mathbf{1346}]\\
-&[\mathbf{m_2m_4m_5}][\mathbf{m_1m_3m_6}][\mathbf{1235}][\mathbf{1345}][\mathbf{1246}][\mathbf{2346}]\\
-&[\mathbf{m_1m_3m_5}][\mathbf{m_2m_4m_6}][\mathbf{1245}][\mathbf{2345}][\mathbf{1236}][\mathbf{1346}] = 0.
\end{aligned}
\tag{14}
$$

This equation can be completely analyzed from the following three aspects:
(ia). $[\mathbf{1346}][\mathbf{2346}] \neq 0$ (Main case);
(ib). $[\mathbf{1346}] = 0$ (Four-point-coplanar case);
(ic). $[\mathbf{2346}] = 0$ (Four-point-coplanar case).

Let us have a close look on them respectively.

(ia). If $[\mathbf{1346}][\mathbf{2346}] \neq 0$, dividing each term by the first term in (14) gives:

$$
\begin{aligned}
&\frac{[\mathbf{m_1m_2m_5}][\mathbf{m_3m_4m_6}]}{[\mathbf{m_3m_4m_5}][\mathbf{m_1m_2m_6}]}I_1 - \frac{[\mathbf{m_1m_3m_5}][\mathbf{m_2m_4m_6}]}{[\mathbf{m_3m_4m_5}][\mathbf{m_1m_2m_6}]}I_2\\
+&\frac{[\mathbf{m_1m_4m_5}][\mathbf{m_2m_3m_6}]}{[\mathbf{m_3m_4m_5}][\mathbf{m_1m_2m_6}]}I_3 + \frac{[\mathbf{m_2m_3m_5}][\mathbf{m_1m_4m_6}]}{[\mathbf{m_3m_4m_5}][\mathbf{m_1m_2m_6}]}I_4\\
-&\frac{[\mathbf{m_2m_4m_5}][\mathbf{m_1m_3m_6}]}{[\mathbf{m_3m_4m_5}][\mathbf{m_1m_2m_6}]}I_5 + 1 = 0,
\end{aligned}
\tag{15}
$$

where

$$
I_1 = \frac{[\mathbf{1345}][\mathbf{2345}][\mathbf{1236}][\mathbf{1246}]}{[\mathbf{1235}][\mathbf{1245}][\mathbf{1346}][\mathbf{2346}]}, \quad I_2 = \frac{[\mathbf{2345}][\mathbf{1236}]}{[\mathbf{1235}][\mathbf{2346}]},
$$
$$
I_3 = \frac{[\mathbf{2345}][\mathbf{1246}]}{[\mathbf{1245}][\mathbf{2346}]}, \quad I_4 = \frac{[\mathbf{1345}][\mathbf{1236}]}{[\mathbf{1235}][\mathbf{1346}]}, \quad I_5 = \frac{[\mathbf{1345}][\mathbf{1246}]}{[\mathbf{1245}][\mathbf{1346}]}.
$$

It is easy to see that $I_1 = I_3I_4 = I_2I_5$. I_2, I_3, I_4, I_5 are all cross ratios on the six spatial points, and the coefficients of $I_i, i = 1..5$ are all cross ratios on the six image points. (15) is just the constraint pointed out by Quan [15, 16], and revisited by Carlsson [4, 5], Bayro-Crrochano and Banarer [2]. The following cases (ib) and (ic) were studied by Carlsson [4, 5], Roh and Kweon [17].

(ib). If $[\mathbf{1346}] = 0$, then $[\mathbf{1345}] \neq 0$ and $[\mathbf{2346}] \neq 0$ (otherwise five spatial points are coplanar, three spatial points are collinear). Let $[\mathbf{1346}] = 0$ in (14), and then divide each term by the last term in the result, we have:

$$
\begin{aligned}
&\frac{[\mathbf{m_1m_2m_5}][\mathbf{m_3m_4m_6}][\mathbf{2345}][\mathbf{1236}]}{[\mathbf{m_2m_4m_5}][\mathbf{m_1m_3m_6}][\mathbf{1235}][\mathbf{2346}]}\\
+&\frac{[\mathbf{m_2m_3m_5}][\mathbf{m_1m_4m_6}][\mathbf{1245}][\mathbf{1236}]}{[\mathbf{m_2m_4m_5}][\mathbf{m_1m_3m_6}][\mathbf{1235}][\mathbf{1246}]} - 1 = 0.
\end{aligned}
\tag{16}
$$

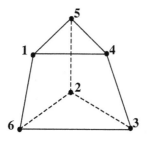

Fig. 2. $[\mathbf{1346}] = 0$ and $[\mathbf{2345}] = 0$

This is an equation on the invariants of spatial points and image points. Moreover if $[\mathbf{2345}] = 0$ (see Fig. 2), (16) becomes:

$$\frac{[\mathbf{1245}][\mathbf{1236}]}{[\mathbf{1235}][\mathbf{1246}]} = \frac{[\mathbf{m_2 m_4 m_5}][\mathbf{m_1 m_3 m_6}]}{[\mathbf{m_2 m_3 m_5}][\mathbf{m_1 m_4 m_6}]}. \tag{17}$$

Each side is a cross ratio invariant. In Fig. 2, if $[\mathbf{1256}] = 0$ additionally, then by the Grassman-Plücker relation of (2) as $[\mathbf{1234}][\mathbf{1256}] - [\mathbf{1235}][\mathbf{1246}] + [\mathbf{1236}][\mathbf{1245}] = 0$, there is $[\mathbf{1245}][\mathbf{1236}] = [\mathbf{1235}][\mathbf{1246}]$, (17) becomes:

$$[\mathbf{m_2 m_3 m_5}][\mathbf{m_1 m_4 m_6}] - [\mathbf{m_2 m_4 m_5}][\mathbf{m_1 m_3 m_6}] = 0.$$

Its geometric meaning is that the three lines $\mathbf{m_1 m_6}$, $\mathbf{m_2 m_5}$ and $\mathbf{m_3 m_4}$ intersect at a point, which means that the three spatial lines $\mathbf{16}$, $\mathbf{25}$, $\mathbf{34}$ are concurrent. Namely, at the time, the invariant relationship identically holds.

(ic). If $[\mathbf{2346}] = 0$, the discussion is similar to the above case **(ib)**.

Case (ii). $c_1 = 0$ but $e_1 \neq 0$; or $e_1 = 0$ but $c_1 \neq 0$ (The quadric cone case that is non-degenerate for camera parameter estimation).
If $c_1 = 0$ but $e_1 \neq 0$, then:

$$c_1 = [\mathbf{m_1 m_4 m_5}][\mathbf{m_3 m_4 m_6}][\mathbf{2345}][\mathbf{1246}] - [\mathbf{m_3 m_4 m_5}][\mathbf{m_1 m_4 m_6}][\mathbf{1245}][\mathbf{2346}]$$
$$= 0,$$
$$c_2 = [\mathbf{m_2 m_4 m_5}][\mathbf{m_3 m_4 m_6}][\mathbf{1345}][\mathbf{1246}] - [\mathbf{m_3 m_4 m_5}][\mathbf{m_2 m_4 m_6}][\mathbf{1245}][\mathbf{1346}]$$
$$= 0.$$

From the above equations, we can conclude that $[\mathbf{2345}][\mathbf{2346}][\mathbf{1345}][\mathbf{1346}] \neq 0$ (otherwise either five spatial points are coplanar or three spatial points are collinear). So, by division, the above equations are changed into:

$$\frac{[\mathbf{2345}][\mathbf{1246}]}{[\mathbf{1245}][\mathbf{2346}]} = \frac{[\mathbf{m_3 m_4 m_5}][\mathbf{m_1 m_4 m_6}]}{[\mathbf{m_1 m_4 m_5}][\mathbf{m_3 m_4 m_6}]},$$

$$\frac{[\mathbf{1345}][\mathbf{1246}]}{[\mathbf{1245}][\mathbf{1346}]} = \frac{[\mathbf{m_3 m_4 m_5}][\mathbf{m_2 m_4 m_6}]}{[\mathbf{m_2 m_4 m_5}][\mathbf{m_3 m_4 m_6}]}. \tag{18}$$

The geometric meaning of these equations is:

a) If no four of the spatial points are coplanar, by Proposition 1, (18) means that $1, 2, 3, 4, 5, 6, O$ lie on a proper quadric cone with 4 as the vertex, and we know that only one of the two equations in (18) is independent.

b) If there are four points to be coplanar, then the brackets possibly to be zero are only in $S_1 = \{[\mathbf{1234}], [\mathbf{1456}], [\mathbf{2456}], [\mathbf{3456}], [\mathbf{1256}], [\mathbf{1356}], [\mathbf{2356}]\}$.

If one of the first four brackets in S_1 is zero, then there will be three image points to be collinear, which is contrary to our assumption. For example, if $[\mathbf{1234}] = 0$, then by a Grassmann-Plücker relation, we have $[\mathbf{2345}][\mathbf{1246}] = [\mathbf{1245}][\mathbf{2346}]$. So, the first one of (18) is changed into $[\mathbf{m_3m_4m_5}][\mathbf{m_1m_4m_6}] - [\mathbf{m_1m_4m_5}][\mathbf{m_3m_4m_6}] = 0$, which means that $[\mathbf{m_1m_3m_4}] = 0$ or $[\mathbf{m_4m_5m_6}] = 0$ by a Grassmann-Plücker relation.

If one of the last three brackets in S_1 is zero, i.e. if one of $\{[\mathbf{1256}], [\mathbf{1356}], [\mathbf{2356}]\}$ is zero, (18) still means that $1, 2, 3, 4, 5, 6, O$ lie on the proper quadric cone with 4 as the vertex because no bracket in $\{[\mathbf{1256}], [\mathbf{1356}], [\mathbf{2356}]\}$ contains the vertex 4.

If two of the last three brackets in S_1 are zero, i.e. if two of $\{[\mathbf{1256}], [\mathbf{1356}], [\mathbf{2356}]\}$ are zero, then three spatial points will be collinear or five spatial points will be coplanar, which is contrary to our assumption.

Thus from above analyses of *a)* and *b)*, we know that the case $c_1 = 0$ and $e_1 \neq 0$ mean that $1, 2, 3, 4, 5, 6, O$ lie on a proper quadric cone. At the same time, only one of $\{[\mathbf{1256}], [\mathbf{1356}], [\mathbf{2356}]\}$ could be zero.

If $e_1 = 0$ but $c_1 \neq 0$, the invariant relationships from $e_1 = 0$, $e_2 = 0$ are:

$$\frac{[\mathbf{2345}][\mathbf{1236}]}{[\mathbf{1235}][\mathbf{2346}]} = \frac{[\mathbf{m_3m_4m_5}][\mathbf{m_1m_3m_6}]}{[\mathbf{m_1m_3m_5}][\mathbf{m_3m_4m_6}]},$$

$$\frac{[\mathbf{1345}][\mathbf{1236}]}{[\mathbf{1235}][\mathbf{1346}]} = \frac{[\mathbf{m_3m_4m_5}][\mathbf{m_2m_3m_6}]}{[\mathbf{m_2m_3m_5}][\mathbf{m_3m_4m_6}]}. \tag{19}$$

Similarly, the geometric meaning is that $1, 2, 3, 4, 5, 6, O$ lie on a proper quadric cone with 3 as the vertex, and only one of them is independent. At the same time, only one of $\{[\mathbf{1256}], [\mathbf{1456}], [\mathbf{2456}]\}$ could be zero.

Case (iii). $c_1 = 0$ and $e_1 = 0$ (The twisted cubic case that is degenerate for camera parameter estimation).

By $c_1 = 0$ and $e_1 = 0$, there are:

$$\begin{cases} \dfrac{[\mathbf{2345}][\mathbf{1246}]}{[\mathbf{1245}][\mathbf{2346}]} = \dfrac{[\mathbf{m_3m_4m_5}][\mathbf{m_1m_4m_6}]}{[\mathbf{m_1m_4m_5}][\mathbf{m_3m_4m_6}]}, \\[4mm] \dfrac{[\mathbf{2345}][\mathbf{1236}]}{[\mathbf{1235}][\mathbf{2346}]} = \dfrac{[\mathbf{m_3m_4m_5}][\mathbf{m_1m_3m_6}]}{[\mathbf{m_1m_3m_5}][\mathbf{m_3m_4m_6}]}. \end{cases} \tag{20}$$

By Proposition 2, the geometric meaning of the above equation system is that $1, 2, 3, 4, 5, 6, O$ lie on a proper twisted cubic if no four of the spatial points are coplanar. If there are four coplanar spatial points, by the analysis in Case (ii), we know that if $c_1 = 0$, only one of $\{[1256], [1356], [2356]\}$ could be zero. If $e_1 = 0$, only one of $\{[1256], [1456], [2456]\}$ could be zero. Thus, if $c_1 = 0$ and $e_1 = 0$, only $[1256]$ could possibly be zero. If $[1256] = 0$, this implies that three image points are collinear, which is contrary to our assumption. This is because eliminating $\dfrac{[2345]}{[2346]}$ in the first equation of (20) by the second equation of (20) gives:

$$\frac{[1235][1246]}{[1245][1236]} = \frac{[\mathbf{m}_1\mathbf{m}_3\mathbf{m}_5][\mathbf{m}_1\mathbf{m}_4\mathbf{m}_6]}{[\mathbf{m}_1\mathbf{m}_4\mathbf{m}_5][\mathbf{m}_1\mathbf{m}_3\mathbf{m}_6]}. \tag{21}$$

By the Grassmann-Plücker relation $[1234][1256] - [1235][1246] + [1245][1236] = 0$ and $[1256] = 0$, we obtain $[1235][1246] = [1245][1236]$. Then, (21) is changed into: $[\mathbf{m}_1\mathbf{m}_3\mathbf{m}_5][\mathbf{m}_1\mathbf{m}_4\mathbf{m}_6] - [\mathbf{m}_1\mathbf{m}_4\mathbf{m}_5][\mathbf{m}_1\mathbf{m}_3\mathbf{m}_6] = 0$. Again by the Grassmann-Plücker relation $[\mathbf{m}_1\mathbf{m}_3\mathbf{m}_4][\mathbf{m}_1\mathbf{m}_5\mathbf{m}_6] - [\mathbf{m}_1\mathbf{m}_3\mathbf{m}_5][\mathbf{m}_1\mathbf{m}_4\mathbf{m}_6] + [\mathbf{m}_1\mathbf{m}_4\mathbf{m}_5][\mathbf{m}_1\mathbf{m}_3\mathbf{m}_6] = 0$, we obtain $[\mathbf{m}_1\mathbf{m}_3\mathbf{m}_4][\mathbf{m}_1\mathbf{m}_5\mathbf{m}_6] = 0$. This is a contradiction.

Here is a summary: See Table 1. If no five spatial points are coplanar, and no three image points are collinear, the invariant relationships between six spatial points and their images consist of three cases: (i) the non-degenerate case including the general main case and four-point-coplanar case; (ii) the non-degenerate case when the camera optical center and the six spatial points lie on a proper quadric cone; (iii) the degenerate case when the camera optical center and the six spatial points lie on a proper twisted cubic. The first case has already been studied in the previous literature. The latter two cases are novel. The three cases constitute a complete and unified framework of invariance for six points. This is not only because the above three cases are the complete classification from (12) and (13), but also because (6), (7), (8), (9), (10), (11), or (6), (12), (8), (13), (10), (11) include all of the independent results after eliminating $\mathbf{P} = \mathbf{K}(\mathbf{R}, \mathbf{t})$ from those 18 equations $x_i\mathbf{m}_i = \mathbf{PM}_i, i = 1..6$. The equations (6), (12), (8), (13), (10), (11) are just six independent ones since $x_i, i = 2..6$ can be solved out and a further additional invariant relationship can be obtained from these six equations for Case (i) and (ii), and since for Case (iii), $x_i, i = 3..6$ can be solved out and two further additional invariant relationships can be obtained from these six equations.

Case (i) and Case (ii) are non-degenerate for camera parameter estimation from six points. Case (iii) is degenerate [8, 25]. Because, for Case (i) and (ii), $y_i = x_i/x_1, i = 2..6$ can be determined uniquely by spatial and image points, and then $\mathbf{P}$ can be estimated uniquely. While, (12) and (13) are identically zero for Case (iii). And so $y_2 = x_2/x_1$ could not be found and $\mathbf{P}$ has one degree of freedom. For details, see [25].

5   Experiments

The established invariance can be applied to detect the relative position information between camera and spatial points without any explicit estimation on

Table 1. A complete framework for the invariance of six spatial points and their images under the condition that no five spatial points are coplanar, and no three image points are collinear

Classified condition	Invariant relationship
Case(i) $c_1 \neq 0$ and $e_1 \neq 0$, the general case, non-degenerate:	
The general main case;	(15)
The one four-point-coplanar case;	(16)
The two four-point-coplanar case	(17)
Case(ii) $c_1 = 0$ but $e_1 \neq 0$; or $e_1 = 0$ but $c_1 \neq 0$	
The quadric cone case, non-degenerate	(18) or (19)
Case(iii) $c_1 = 0$ and $e_1 = 0$	
The twisted cubic case, non-degenerate	(20)

Fig. 3. A real image of a calibration grid

the camera projective matrix or the optical center. Based on the invariant relationship for the twisted cubic case, we construct an adaptive weighed criterion function g to recognize the nontrivial critical configuration of 3D reconstruction. The detailed algorithm and experiments can be found in [25] (Here, the details are omitted due to the space limit). Some experimental results are shown below.

We generate seven pairs of spatial and image points such that $1, 2, 3, 4, 5, 6, O$ do not lie on a twisted cubic, and $1, 2, 3, 4, 5, 7, O$ do lie on a twisted cubic. The image size is not greater than 1000×1000 pixels. Gaussian noise with mean 0 and standard deviation ranging from 0 to 6 pixels is added to each image points. With 100 runs under each noise level, the averaged value of the criterion function $g(\mathbf{6})$ from $1, 2, 3, 4, 5, 6$, and the averaged value of the criterion function $g(\mathbf{7})$ from $1, 2, 3, 4, 5, 7$ are computed. The results from eight independent groups of data, denoted as $D_i, i = 1..8$, are shown in Table 2. As it can be seen, our invariance can distinguish robustly between the degenerate configuration (that camera and spatial points lie on a twisted cubic) and the non-degenerate configuration, and the criterion function g is quite stable against noise.

Table 2. The values of $g(\mathbf{6}), g(\mathbf{7})$ under different noise levels

Noise level (pixel)		0	1	2	4	6
D_1	$g(\mathbf{6})$	7.40	7.40	7.41	7.41	7.45
	$g(\mathbf{7})$	0.00	0.04	0.09	0.18	0.31
D_2	$g(\mathbf{6})$	3.69	3.66	3.73	4.00	5.02
	$g(\mathbf{7})$	0.00	0.13	0.26	0.70	1.66
D_3	$g(\mathbf{6})$	2.62	2.62	2.62	2.62	2.62
	$g(\mathbf{7})$	0.00	0.01	0.02	0.04	0.06
D_4	$g(\mathbf{6})$	3.25	3.25	3.25	3.25	3.26
	$g(\mathbf{7})$	0.00	0.02	0.05	0.10	1.13
D_5	$g(\mathbf{6})$	2.12	2.13	2.13	2.17	2.23
	$g(\mathbf{7})$	0.00	0.04	0.09	0.17	0.31
D_6	$g(\mathbf{6})$	3.20	3.20	3.21	3.22	3.52
	$g(\mathbf{7})$	0.00	0.06	0.15	0.32	0.55
D_7	$g(\mathbf{6})$	2.00	2.00	2.00	2.00	2.00
	$g(\mathbf{7})$	0.00	0.02	0.03	0.06	0.09
D_8	$g(\mathbf{6})$	1.90	1.90	1.90	1.90	1.90
	$g(\mathbf{7})$	0.00	0.01	0.03	1.05	0.08

An experiment based on real data is also performed. The real image we used from a grid is shown in Fig. 3 (1024×768 size). We extract 108 pairs of spatial and image points, then calibrate camera parameters by the DLT method [1], and denote the calibrated results of intrinsic parameter matrix and pose parameters as $\mathbf{K}_0, \mathbf{R}_0, \mathbf{t}_0$. Among these 108 pairs of spatial and image points, 126 groups of six pairs are combined randomly with no three image points collinear, no four spatial points coplanar. The group with the maximal value of g, denoted as G_{max}, and the group with the minimal value of g, denoted as G_{min} are chosen to calibrate the camera respectively. We denote the calibrated results as $\mathbf{K}_1, \mathbf{R}_1, \mathbf{t}_1$ and $\mathbf{K}_2, \mathbf{R}_2, \mathbf{t}_2$ respectively, and compare them with $\mathbf{K}_0, \mathbf{R}_0, \mathbf{t}_0$:

$$\mathbf{K}_1 - \mathbf{K}_0 = \begin{pmatrix} 52.10, & 0.50, & -22.11 \\ 0, & 50.87, & -68.41 \\ 0, & 0, & 0 \end{pmatrix}, \mathbf{K}_0 - \mathbf{K}_2 = \begin{pmatrix} 1069.41, & -29.68, & 92.98 \\ 0, & 1180.05, & -147.71 \\ 0, & 0, & 0 \end{pmatrix},$$

$$\mathbf{R}_1 - \mathbf{R}_0 = \begin{pmatrix} 0.0066 & 0.0082 & -0.0036 \\ 0.0192 & 0.0213 & -0.0139 \\ -0.0005 & 0.0182 & 0.0283 \end{pmatrix}, \mathbf{R}_2 - \mathbf{R}_0 = \begin{pmatrix} -1.4450 & 1.3725 & -0.0208 \\ 0.2443 & 0.3405 & 1.8911 \\ 0.1759 & -0.0208 & 0.4848 \end{pmatrix}.$$

$$\mathbf{t}_1 - \mathbf{t}_0 = \begin{pmatrix} 0.3248 \\ 1.0234 \\ 0.6420 \end{pmatrix}, \quad \mathbf{t}_2 - \mathbf{t}_0 = \begin{pmatrix} -0.2954 \\ -6.3999 \\ -48.7613 \end{pmatrix}.$$

These results show that the accuracy of calibrated results from G_{max} is much higher than that from G_{min} except for the first element of the translation. For the first element of the translation, the absolute error of t_{11} (0.3248) is greater than the absolute error of t_{21} (0.2954), but the difference between the two errors is not as large as that for the second or third element of the translation.

Namely, the calibrated result from six pairs of spatial and image points with smaller value of the criterion function g (i.e. spatial points and optical center are near to the twisted cubic degenerate configuration) is not better than that with larger value of the criterion function g (i.e. spatial points and optical center are far from the twisted cubic degenerate configuration). Extensive experiments in [25] show that the criterion function g or the established invariance can be faithfully trusted for camera parameter estimation.

6   Conclusion

With bracket computations, we establish a unified and complete framework for the invariance of six points. It consists of three cases, one case is already fully studied in the literature, and two cases are novel as: 1) the case when the camera optical center and the spatial points lie on a quadric cone; 2) the case when the camera optical center and the spatial points lie on a twisted cubic. The established invariance framework can be applied to effectively detect the relative position geometric information between the camera and the spatial points without any explicit estimation on camera projective matrix or optical center. Experiments on the twisted cubic case are performed to validate the invariance. We believe this framework will have further useful implications in 3D reconstruction. For example, when applying RANSAC during the process of determining the camera parameters, the critical groups of data can be filtered out by the invariance or the criterion function in the twisted cubic case.

Acknowledgments

This work was supported by the National Natural Science Foundation of China under grant No. 60475009 and 60303021.

References

1. Abdel-Aziz, Y.I., Karara, H.M.: Direct linear transformation from comparator coordinates into object space coordinates in close-range photogrammetry. *Proc. ASP/UI Symp. on Close Range Photogrammetry*, (1971) 1-18.
2. Bayro-Crrochano, E., Banarer, V.: A geometric approach for the theory and applications of 3D projective invariants. *Journal of Mathematical Imaging and Vision*, **16**(2) (2002) 131-154.

3. Bayro-Corrochano, E., Rosenhahn, B.: A geometric approach for the analysis and computation of the intrinsic camera parameters. *Pattern Recognition*, **35** (2002) 169-186.
4. Carlsson, S.: Symmetry in perspective. *ECCV*, (1998) 249-263.
5. Carlsson, S.: View variation and linear invariants in 2-D and 3-D. Tech. Rep. ISRN KTH/NA/P–95/22–SE, Dec. 1995.
6. Csurka, G., Faugeras, O.: Algebraic and geometric tools to compute projective and permutation invariants. *IEEE Trans. Pattern Analysis and Machine Intelligence*, **21**(1) (1999) 58-65.
7. Faugeras, O., Luong, Q.T.: *The geometry of multiple images.* The MIT Press, 2001.
8. Hartley, R., Zisserman, A.: *Multiple view geometry in computer vision.* Cambridge University Press, 2000.
9. Hartley., R.: Projective reconstruction and invariants from multiple images. *IEEE Trans. Pattern Analysis and Machine Intelligence*, **16**(10) (1994) 1036-1041.
10. Hestenes, D., Sobczyk, G.: *Clifford algebra to geometric calculus.* D. Reidel, Dordrecht, Boston, 1984.
11. Lasenby, J., Bayro-Corrochano, E., Lasenby, A., Sommer. G.: A new methodology for computing invariants in computer vision. *Proc. 13th Int. Conf. on Pattern Recognition*, (1996) 393-397.
12. Li, H., Hestenes, D., Rockwood, A.: *Generalized homogeneous coordinates for computational geometry.* In: Sommer, G., Editor, Geometric computing with clifford algebra, Springer, Berline, pp. 27-52, 2001.
13. Mundy, L., Zisserman, A., Editors.: *Geometric invariant in computer vision.* The MIT Press, Cambridge, Mass., 1992.
14. Mundy, L., Zisserman, A., Forsyth, D., Editors.: *Applications of invariance in computer vision.* Lecture Notes in Computer Science, Vol. 825, Springer-Verlag, 1994.
15. Quan., L.: Invariants of 6 points from 3 uncalibrated images. *ECCV*, (1994) 459-470.
16. Quan., L.: Invariants of six points and projective reconstruction from three uncalibrated images. *IEEE Trans. Pattern Analysis and Machine Intelligence*, **17**(1) (1995) 34-46.
17. Roh, K.S., Kweon., I.S.: 3-D object recognition using a new invariant relationship by single-view. *Pattern Recognition*, **33** (2000) 741-754.
18. Schaffalitzky, F., Zisserman, A., Hartley, R., Torr., P.: A six point solution for structure and motion. *ECCV*, (2000) 632-648.
19. White, N.L., Editor.: *Invariant methods in discrete and computational geometry.* Proc. Curacao Conference, June, Kluwer Academic Publishers, 1994.
20. Sommer, G., Daniilidis, K., Pauli, J., Editors: *Computer analysis of images and patterns.* Lecture Notes in Computer Science, Vol. 1296, Springer-Verlag, 1997.
21. Sommer, G., Editor:, *Geometric computing with clifford algebra*, Springer, Berline, 2001.
22. Sturmfels, B.: *Algorithm in invariant theory.* Springer-Verlag, Wien New York, 1993.
23. Wu, Y.: *Bracket algebra, affine bracket algebra, and automated geometric theorem proving.* Ph. D. Dissertation, Institute of Systems Science, Chinese Academy of Sciences, Beijing, 2001.
24. Wu, Y., Hu, Z.: The invariant representations of a quadric cone and a twisted cubic. *IEEE Trans. on Pattern Recognition and Machine Inteligence*, **25**(10) (2003) 1329–1332.
25. Wu, Y., Hu, Z.: A unified and complete framework of invariance for six points and its application to recognize critical configuration for 3D reconstruction. Technical Report, RV-NLPR, Institute of Automation, Chinese Academy of Sciences, 2004.

An Introduction to Logical Animation

Jingzhong Zhang[1,2] and Chuanzhong Li[1]

[1] Institute of Educational Software, Guangzhou University,
510405 Guangzhou, China
{zjz101, czli1963}@yahoo.com.cn
[2] Chengdu Institute of Computer Applications,
Chinese Academy of Sciences, 610041 Chengdu, China
zjz101@hotmail.com

Abstract. In this paper the concept *logical animation* is described in a formal manner. It is abstracted from numerous popular animation applets made by dynamic geometry softwares. The concept is illustrated by a number of examples generated by the logical animation software *SuperSketchpad* (SSP), and some features and prospects related to applications of it are discussed.

Keywords: logical animation; dynamic geometry; educational software.

1 Introduction

A usual animation is a sequence of finite and discrete pictures. This kind of motion picture can be called *time-sequential animation.* In recent 20 years, another kind of animation appears, in the form of animation applets created by *dynamic geometry* softwares [1], [2], [3], [4], [5]. These applets are often used in education for teaching and studying mathematics, physics and other courses. Teachers and students like them, they like to construct the applets by themselves by working with dynamic geometry softwares.

In such an applet, a picture is generated by a set of rules, the user can drag one object in the picture to drive other objects by these rules. Such an animation is no longer restricted to geometric structures, as logical structures are also involved. So we call it *logical animation.*

The following is a simplest example: in a picture there are two line segments AB and CD, and their point of intersection P. If one drag with mouse any of the four endpoints, say point A, then line segment AB moves accordingly and so does the point of intersection P. If both A and B are on the same side of line CD, then point P vanishes.

Up to now, numerous logical animation applets have been created and employed in schools, with the accompanying instructions on how to create these applets with softwares. It is time to gather together the constructions of these applets and find their common properties. In this paper, we try to formulate a formal description of the concept logical animation and discuss its properties.

H. Li, P. J. Olver and G. Sommer (Eds.): IWMM-GIAE 2004, LNCS 3519, pp. 418–428, 2005.
© Springer-Verlag Berlin Heidelberg 2005

This paper is organized as follows: Section 2 gives the definition of logical animation, Section 3 shows some examples generated by our logical animation software *SuperSketchpad* (SSP), Section 4 analyzes some features and prospects of logical animation.

2   Logical Animation

Intuitively, an *object* means a pattern with some parameters to describe its property. In the picture, an object may be a point, a line segment, a circle, a curve, a block of text or a button, etc. The position, size, color and shape of an object depend on the values of the parameters. For example, a conic may be an ellipse, a hyperbola, or a parabola depending on different values of a parameter.

Let X be a set of objects and u be a set of parameters or expressions. A *construction* $A(X, u)$, is the command or function for creating a new object from X with u. Here A is the name for the class of the new object. The equality $Y = A(X, u)$ means that an object Y is created by executing the construction. When all parameters of object Y are fixed, we have a *state* of the object.

Definition 2.1. Let $X = \{X_1, X_2, \ldots, X_k\}$ be a set of objects. Let $X_0 = \emptyset$ be the empty set and let u be a set of parameters. If for $i \geq 1$, X_i is generated by construction $A_i(X_1, \ldots, X_{i-1}, u)$, then $X = A(X, u)$ is called an *ordered logical animation*. If at least one X_i is not generated in this way, then $X = A(X, u)$ is called a *non-ordered logical animation*.

Example 2.1. In our logical animation system SSP [5], there are over one hundred geometric constructions. The following basic ones will be used in this paper:

Point(a,b): a point with initial coordinates (a, b).
Segment(X,Y): a line segment with endpoints X and Y.
Midpoint(A,B): the midpoint of line segment AB.
IntersectionOfLine(L,M): the point of intersection of lines L and M.
VerticalLine(P,L): a line through point P, perpendicular to line L.
PointFlexRotate(A,B,r,a): a point P with PB=rAB and ∠PBA = a.
Square(A,B): a square with edge AB.
Circle(O,r): the circle with center O and radius r.
PointOnCurve(C,b): a point with parameter b on parametric curve C.
MeasureVariable(rand(0,1),3): a block of text showing the value of function rand(0,1), with 3 digits after the radix point. Here rand(0,1) generates a random number between 0 and 1.
Function(a*x^2+b*x+c,-6,6): curve $y = ax^2 + bx + c$ for $x \in [-6, 6]$.
Group(A,B, ... ,X): the set of objects A, B, ... , X.
Symmetric(X,L): the reflection of object X with respect to line L.
Translate(X,P,Q): the translation of object X along vector PQ.
DiscretePointCurve(A,B, ... ,X): Bezier curve through points A, B to X.
Animation(P): a button to make point P move or stop.

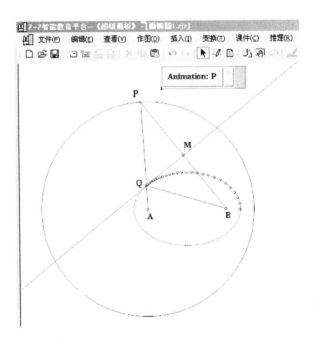

Fig. 1. A demo of Example 2.1

`AnimationVar(t,a,b)`: a button to make parameter `t` vary from `a` to `b`.
`Locus(A,B,C,...,X)`: the locus of point X when points A, B, C, ... move.
`Trace(X)`: the trajectory of object X when moving.

In the list, most parameters characterizing the properties of the objects are omitted, because their values are default. There are functions to list the parameters and change their values.

Now we construct a sequence of objects as follows (see Figure 1):

```
A=Point(0,0);
B=Point(0,a);
CR=Circle(A,r);
P=PointOnCircle(C,b);
PB=segment(P,B);
PA=segment(P,A);
M=Midpoint(P,B);
LN=VerticalLine(M,PB);
Q=Intersection(L,PA);
LC=Locus(P,Q);
TR1=Trace(Q);
TR2=Trace(LC);
```

It constitutes an ordered logical animation with 3 parameters a, r, b.

We explain below how to generate a sequence of pictures from an ordered or non-ordered logical animation.

Definition 2.2. Let $X = A(X, u) = \{X_1, X_2, \ldots, X_k\}$ be an ordered logical animation. Let parameters u take a series of values $u(i)$ for $i = 0, 1, \ldots, N$. Set

$$\begin{aligned}
X(i) &= \{X_1(i), X_2(i), \ldots, X_k(i)\}, && \text{for } i = 0, \ldots, N, \text{ where} \\
X_j(i) &= A_j(X_1(i), \ldots, X_{j-1}(i), u(i)), && \text{for } j = 1, \ldots, k.
\end{aligned} \tag{1}$$

The sequence $X(0), X(1), \ldots, X(N)$ is called a *demo* of the ordered logical animation $A(X, u)$. Here $X_j(i)$ is a state of X_j at the "parameter instant" $u(i)$, and $X(i)$ is the whole picture at the instant.

Ordered logical animations are convenient to construct by visualized operations. Up to now, most animation applets made by dynamic geometry are ordered logical animations.

Example 2.2. In Example 2.1, let a and r be constant, let b take a series of values so that point P moves on circle CR, then the locus of Q is an ellipse for a < r (Figure 1), and a hyperbola for a > r.

If we hide TR1, let b and r be constant, and let a take a series of values, we get another demo of this same logical animation, see Figure 2. The locus of Q will change and its trajectory is a series of conics varying from ellipses to hyperbolas.

Definition 2.3. Let $X = A(X, u) = \{X_1, X_2, \ldots, X_k\}$ be a non-ordered logical animation. Let parameters u take a series of values $u(i)$ for $i = 0, 1, \ldots, N$. Set

$$\begin{aligned}
X(i) &= \{X_1(i), X_2(i), \ldots, X_k(i)\}, && \text{for } i = 0, \ldots, N, \text{ where} \\
X_j(i) &= A_j(X_1(i^*), \ldots, X_k(i^*), u(i)), && \text{for } j = 1, \ldots, k.
\end{aligned} \tag{2}$$

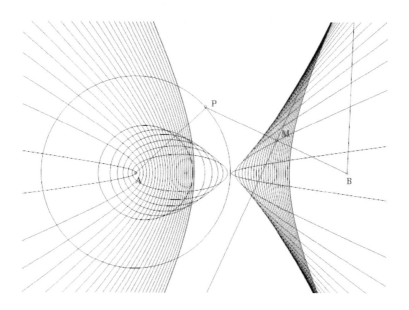

Fig. 2. Another demo of Example 2.1

In (2), we have $i^* = i$ in $X_h(i^*)$ if $i < h$ and $i^* = i-1$ otherwise. If $X_h(0)$ occurs then it takes some default initial value. The sequence $X(0), X(1), \ldots, X(N)$ is called a *demo* of the non-ordered logical animation $A(X, u)$.

Example 2.3. We construct a sequence of objects as follows:

```
A=Point(-5,-5);
B=Point(-5,5);
O=Point(0,0);
SQ=Square(A,B);
CR=Circle(O,5);
M1=MeasureVariable(rand(-5,5),12);
M2=MeasureVariable(rand(-5,5),12);
R=Point(M1,M2);
M3=MeasureVariable(sign(t)*(M3+1),0);
M4=MeasureVariable(sign(t)*(M4+sign(25-M1^2-M2^2)),0);
M5=MeasureVariable(4*M4/M3,4);
TR=Trice(R);
```

Here `sign(x)` equals 1 for $x > 0$ and 0 otherwise. Since object M3 appears in its own construction, so does M4, the above sequence is a non-ordered logical animation (Figure 3).

If the only parameter t is given a series of values $0 = t(0) < t(1) < \cdots < t(N) = 1$, then the varying pictures display an experiment of throwing bean R into square SQ randomly. At the initial value $t(0) = 0$, point $R(0)$ is at a random position. For $t(1) > 0$, both M1 and M2 are renewed, so is point R. By the construction, M3 is the number of beans in the square, and M4 is the number of beans in the disk. For larger number i, $M5(i)$ tends to π probabilistically.

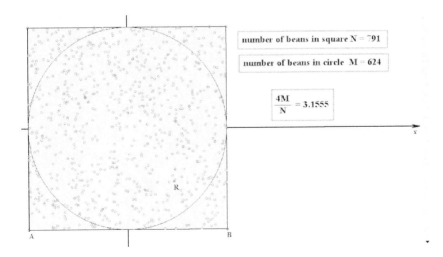

Fig. 3. Simulation of bean throwing

3   More Examples Generated by SSP

To disclose more features of logical animation, in this section we show more examples generated by SSP. In SSP, one can make construction by either visualized operations or by text commands.

Example 3.1. This is a simulation of the *Doppler phenomenon*, which shows the change in the observed frequency of an acoustic or electromagnetic wave due to some relative movement between the source and the observer (Figure 4).

Fig. 4. Doppler phenomenon

```
A=Point(t,0);
P(k)=Point(k,0); (k=1 to 15)
CR(k)=Circle(P(k),u*(t-k)*sign(t-k)); (k=1 to 15)
```

When parameter t takes a series of values increasing from 0 to 16, the source point A moves along the x-axis from $(0,0)$ to $(16,0)$. The waves represented by circles CR(k) enlarge gradually for $t > k$. The wave speed can be faster or slower than the source speed, depending on whether or not $u > 1$.

Example 3.2. Figure 5 shows a *planar tiling* with a great deal of variety. It is generated by the following constructions:

```
A=Point(x,0);
B=Point(y,0);
C=Point(-y,0);
K=Point(a,1);
Z=Point(0,0);
P(k)=Point(x(k),y(k)); (k=1 to 5)
CV(1)=DiscretePointCurve(B,P(1),P(2),C);
CV(2)=DiscretePointCurve(B,P(3),P(4),P(5),A);
CV(3)=Symmetric(CV(1),BC);
CV(4)=Symmetric(CV(2),AK);
CV(5)=Translate(CV(3),B,Z);
CV(6)=Translate(CV(4),A,C);
```

Planer Tiling

Drag points D E F G H to
change the shape of patterns

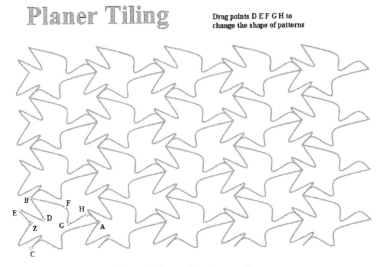

Fig. 5. Versatile planar tiling

```
CV=Group(CV(1),CV(2),CV(5),CV(6));
CU(1)=Translate(CV,Z,A);
CU(j+1)=Translate(CV(j),Z,A); (j=1 to 3)
CVS=Group(CV,CU(1),CU(2),CU(3),CU(4));
CVS1=Translate(CVS,C,B);
CVS2=Translate(CVS1,C,B);
CVS3=Translate(CVS2,C,B).
```

In Figure 5, curves CV(3) and CV(4) and point K are hidden, points P(1) to
P(5) are denoted by D to H respectively. If these points are dragged, or parameters
x and y are changed, then pattern CV and the whole picture change their shapes.

Example 3.3. Figure 6 shows a *fractal tree* created by about one thousand
objects. The fractal tree is generated by

```
A=Point(0,-3);
B=Point(0,0);
SG=Segment(A,B);
FT=mtree(A,B,r,s,x,y,8);
```

where the function mtree is defined recursively by

```
mtree(P,Q,r,s,a,b,k)
    {
  if (k==1) {tree(P,Q,r,s,a,b);}
    else {
        X=PointFlexRotate(P,Q,r,a);
        Y=PointFlexRotate(P,Q,s,b);
```

a fractal tree

Fig. 6. Fractal tree

```
M=Segment(Q,X);
N=Segment(Q,Y);
mtree(Q,X,r,s,a,b,k-1);
mtree(Q,Y,r,s,a,b,k-1);
}
```

and the pattern `tree` is constructed as follows:

```
tree(P,Q,r,s,a,b)
{X=PointFlexRotate(P,Q,r,a);
Y=PointFlexRotate(P,Q,s,b);
M=Segment(Q,X);
N=Segment(Q,Y);}
```

Line segment `AB` serves as the stem of the tree and is the first-level pattern. From one end of it grows two branches, which are the second-level pattern. In `mtree(A,B,r,s,x,y,1)`, two line segments `QX`, `QY` are constructed such that

$$QX = rPQ, \quad QY = sPQ, \quad \angle XQP = a, \quad \angle YQP = b. \tag{3}$$

The last parameter `k` in `mtree(A,B,r,s,x,y,k)` indicates the level of recursion.

If the four parameters `r,s,x,y` change periodically within a certain scope and at an assigned ratio of speed, then the fractal tree can sway, swing, spread, stretch, and shrink with various shapes and spectacles.

Example 3.4. Figure 7 shows a simulation of a dog chasing after a rabbit. The only constraint is that the chasing direction is always from the instant position of the dog to that of the rabbit.

This is a non-ordered animation. The construction is

```
X=MeasureVariable(r*cos(t),12);
Y=MeasureVariable(r*sin(t),12);
```

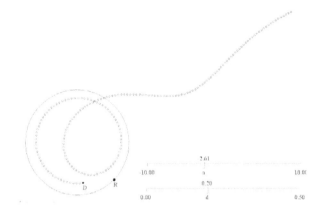

Fig. 7. Dog chasing after rabbit

```
R=Point(X,Y);
O=(0,0);
CR=Circle(O,r);
M=MeasureVariable(p*sign(a-t)+sign(t-a)*(M+d*(X-M)/K),12);
N=MeasureVariable(q*sign(a-t)+sign(t-a)*(N+d*(Y-N)/K),12);
D=Point(M,N);
K=MeasureLength(R,D,12);
TR=Trice(D);
```

Let point R be a rabbit running along a circle with radius r, let point D be a dog chasing after the rabbit. Let parameters r,p,q,a,d be constant, and let t take a series of values increasing from a − 1 to a + 100. At the beginning, the dog is at position (p, q). When t > a, the dog moves towards the rabbit with speed d relative to the rabbit.

Example 3.5. This is a simulation of Buffon's needle test (Figure 8).

```
LN(k)=Line(y=k); (k=0 to 8)
M0=MeasureVariable(rand(0,2*pi),12);
A=Point(a*cos(M0),a*sin(M0));
B=Point(0,0);
SG=Segment(B,A);
M1=MeasureVariable(rand(0,12),12);
M2=MeasureVariable(rand(0,8),12);
P=Point(M1,M2);
ND=Translate(SG,B,P);
TR=Trace(ND);
M3=MeasureVariable(sign(p-1)*(M3+1),0);
M4=MeasureVariable(sign(abs([M2+a*sin(M0)]-[M2])),0);
M5=MeasureVariable(sign(p-1)*(M5+M4),0);
M6= MeasureVariable(2*a*M3/M5,4);
```

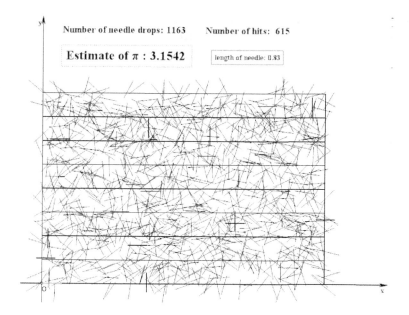

Fig. 8. Buffon's needle test

In the constructions, the LN(k)'s are a sequence of parallel lines with distance 1 between neighbors. Vector SG gives both the length and the direction of the needle. Point P gives the position of the tail of the needle, and ND the whole needle. [x] denotes the biggest integer not greater than x.

M4 equals 1 or 0 depending on whether or not the needle hits a line in {LN(k)}. M3 denotes the number of needle drops and M5 the number of hits. If parameter p takes a series of values increasing at a speed greater than 1, then M6 tends to π probabilistically for large values of M3.

4   Features and Prospects

From the previous examples we may draw some conclusions, which can explain why logical animation is a better choice for web-based education and training.

1. *The construction can be stated in plain language step by step.* The sequence of constructions for creating the objects may be used to analyze the behavior of different demos.
2. *Logical animation is easy to create.* One only need to design a basic pattern, a basic motion and the relationship between the basics and others.
3. *Logical animation costs less web resources.* By our software SSP, when the animation applets are transferred into html format, the file size of Examples 3.3 is 433k bytes, while for the other 6 examples, the sizes are 8k, 4k, 8k, 16k, 1k, 8k bytes respectively.

4. *Logical animation has nice interactivity.* It only needs the user to drag some point or adjust some parameter to control the pattern and style, which benefits the user to make detailed investigation as if in a scientific experiment.
5. *Logical animation can represent exact relations.* Since it can display the exact relations among the parameters and locations, logical animation is suitable for exploring the rules and laws intuitively.

Logical animation has a history no more than three decades. There remain many very important problems waiting for an answer. Here are some:

(1) How to describe the structure of a logical animation more intuitively? A graph or matrix may be used to represent the relations among the objects.

(2) Any continuous function can be approximated by polynomials. Can logical animations serve as a basis to approximate other animations?

(3) To meet different purposes in animation design, one needs to describe the behavior of the animation to be designed. How to make the description formally?

(4) To make logical animation easy to learn and practice, what programming language and platform should be designed? What standard may be proposed?

With the development of more powerful logical animation tools, the application of logical animation may be further extended to web-based education and training, telecommunication, commercial advertisement service, computer games and internet games.

References

1. S. Steketee and N. Jackiw, *The Geometer's Sketchpad* (GSP, software). Key Curriculum Press, 1987-2004. See http://www.keypress.com/sketchpad
2. J. Marie and F. Bellemain, *Cabri Geometry II* (software). Texas Instruments, 2001-2004. See http://education.ti.com/us/product/software/cabri
3. J. Richter-Gebert and U. Kortenkamp, *Cinderella* (software). Springer, Berlin, Heidelberg, 1998-2004. See http://www.cinderella.de/tiki-index.php
4. H. Fernandes and D. Bucknell, *Dr. Geo.* Ofset (part of the GNU project). See http://www.ofset.org/drgeo
5. C. Z. Li and J. Z. Zhang, *SuperSketchpad* (software). Beijing Normal UniversityBeijing 2004. See http://www.zplusz.org/en, http://ccmp.chiuchang.com.tw

Recent Methods for Reconstructing Surfaces from Multiple Images

Gang Zeng[1], Maxime Lhuillier[2], and Long Quan[1]

[1] Dep. of Computer Science, HKUST, Clear Water Bay, Kowloon, Hong Kong
{zenggang,quan}@cs.ust.hk
[2] LASMEA, UMR CNRS 6602, Université Blaise-Pascal, 63177 Aubière, France
Maxime.Lhuillier@lasmea.univ-bpclermont.fr

Abstract. Many objects can be mathematically represented as smooth surfaces with arbitrary topology, and smooth surface reconstruction from images could be cast into a variational problem. The main difficulties are the intrinsic ill-posedness of the reconstruction, image noise, efficiency and scalability. In this paper, we discuss the reconstruction approaches that use volumetric, graph-cut, and level-set optimization tools; and the objective functionals that use different image information, silhouette, photometry, and texture. Our discussion is accompanied by the implementations of these approaches on real examples.

1   Introduction

Surface reconstruction from multiple images is an old, fundamental yet difficult problem in computer vision, which has been extensively investigated over the past three decades. The aim is to create a 3D model of a scene using 2D images taken from arbitrary positions. An early application was robot navigation. Recent developments in camera calibration and multimedia computing have broadened interest in the problem, and its emphasis has shifted to generating more "appearance" views of the scene. Applications include virtual reality, movie making, entertainment and so on. Different from early applications, these require highly detailed surfaces that are constructed within a tolerable duration and limited memory. Such a task, however, is difficult to accomplish due to the intrinsic ill-posedness of the reconstruction, image noise, efficiency and scalability.

Traditional stereo vision has been based on image matching, using either intensity-based (direct) methods or feature-based methods. This class of reconstruction methods compute correspondences across images and then use triangulation to recover a 3D structure. They are effective with video sequences, and are especially successful when combined with graph-cut techniques to obtain optimal solutions. However, they rely on the assumptions that views must be close together and that correspondences must be maintained when spanning large changes in viewpoint. Moreover, occlusion is often ignored between different views.

H. Li, P. J. Olver and G. Sommer (Eds.): IWMM-GIAE 2004, LNCS 3519, pp. 429–447, 2005.

An alternative approach is the class of volumetric reconstruction methods. They divide the 3D scene space into a set of small units (voxels) and construct the scene by finding a subset of these units that are consistent with the input images. Instead of matching features in an image space, they perform a search in a scene space, and avoid several disadvantages of the first approach listed above. However, this leads to more computational cost especially when a high resolution is required. Moreover, the voxel suffers from noisy input and often lacks enough details for final surfaces.

A recent elegant addition is surface representation by a dynamic hyper-surface in 4D (3D space + 1D time). This representation is the basis of the level-set reconstruction methods, and has the advantage of being able to represent surface evolution with topological changes. In order to ensure convergence, they employ relatively high order derivatives in the dynamic surface evolution, and often produce over-smoothed results.

In this paper, we provide a general discussion of methods developed primarily for reconstructing natural, real world scenes using arbitrarily positioned off-the-shelf cameras. Most of these methods require accurate geometric information of the given cameras. We assume that the cameras have been fully calibrated, since camera calibration is itself a challenging problem with a large literature devoted to it [17]. The discussion is carried out according to visual cues of silhouette, photometry and texture. Therefore, the cues from single view, such as shading, focus, are not included, and medical imaging and active light methods are not covered. Meanwhile, we mainly focus on recent approaches, including volumetric methods, graph-cut methods and level-set methods. Several of our approaches are discussed in detail, and real scene examples demonstrate the quality of these reconstructions.

The remaining parts of the paper are organized as follows: Section 2 first describes visual cues of surface reconstruction from multiple views; Section 3 describes several volumetric methods, and two of the most recent approaches, *Silhouette Carving* and *Robust Carving for non-Lambertian Objects*, are focused on. In section 4 and 5, we discuss energy minimization methods, including graph-cut methods and level-set methods, and show details of two novel algorithms, *Surface Propagation* and *Surface Reconstruction by Integrating 3D and 2D Data of Multiple Views*. Finally, in section 6, we conclude the paper by providing a general analysis among different reconstruction methods and possible future research directions.

2   Visual Cues

2.1   Silhouette

Among different kinds of image cues, silhouette has been considered an effective psychophysical clue to shape understanding [1]. The 2D silhouette is obtained as a parallel or perspective projection of the object onto the image plane, which is a binary image, with the value at a point indicating whether or not the visual ray from the optical center through that image point intersects an object in the scene. The silhouette is usually obtained by the blue-screen technique, which is a computationally simple task. The algorithms for obtaining the silhouette are robust and perform well even with noisy images. Therefore, silhouette is a favorite cue for surface reconstruction.

In scene space, combined with camera calibration information, each point in the silhouette defines a ray that intersects the object at some unknown depth along this ray. The union of the rays for all points in the silhouette defines a generalized cone within which the 3D object must lie. The *visual hull* is then defined as the intersection of the generalized cones associated with a set of cameras, which is the maximal shape that gives the same silhouettes as the actual object for all views outside the convex hull of the object [25]. The visual hull has some interesting properties. First, it is a convex over-estimation of the object. Second, it depends on the number of views, the positions of the viewpoints and the complexity of the object. In practice, only a finite number of silhouettes are combined to reconstruct the scene, resulting in a rough shape approximation. Finally, it can be constructed efficiently, and thus the visual hull is often used as a first estimation for further refinements of surface reconstruction.

2.2   Photometry

Photo-consistency is another important image cue, which is especially useful when input images are grayscale or color rather than the binary images processed by shape-from-silhouette methods. The additional photometric information is used to improve the 3D reconstruction process. A set of images can be thought of as defining a set of constraints on the scene, in that a valid 3D scene model that is projected using the camera matrices associated with the input images must produce synthetic images that are the same as the corresponding real input images. The similarity measurement depends on the characteristics and the accuracy of both the geometric model and the reflectance model. It is important to notice here that photo-consistency does not guarantee a unique reconstruction, since there may be many 3D scenes that are consistent with a particular set of images.

In a simple case, when the reflectance model is assumed to be *Lambertian*, photo-consistency is defined to measure the similarity of its visible projections on the images. That is, a point on a surface is photo-consistent with a set of images if, for each image in which the point is visible, the radiance of that point is equal to the intensity of its projection pixel. Thus, photo-consistency can be defined in a number of ways as described in [37, 24]. The simplest is to threshold the variance of the intensities of the corresponding pixels. Experiments using other definitions of photo-consistency have also been conducted. This consistency can be defined as the standard deviation or, alternatively, the sum of the distances between all intensities and their expected value:

$$\sigma^2 = \frac{1}{K}\sum(I_i - \overline{I})^2 \quad \text{or,} \quad \sigma = \frac{1}{K}\sum|I_i - \overline{I}|$$

The *photo hull* is defined as the maximal shape that satisfies photo-consistency with the original images in [24]. Similar to the visual hull, the photo hull has some interesting properties. First, it is an over-estimation of the object. Second, it depends on the number of views, the positions of the viewpoints and the complexity of the object. In practice, the photo hull may contain several fault areas over intensity-homogeneous regions. Finally, to justify the photo-consistency criterion, a visibility test is necessary, which in turn is a difficult problem.

2.3   Texture

Texture correlation is the visual cue that is often used in the stereo matching problem to find correspondences between images of two cameras. It is based on a correlation of local windows or on matching of sparse features. Although texture correlation is based on the Lambertian assumption, the extended versions, e.g. Zero-mean Normalized Cross-Correlation, are quite tolerant to some non-Lambertian effects. This makes texture correlation quite useful when constructing textured surface with an unknown reflectance model. The correlation between the locations at $X_1 = (x_1, y_1)^T$ in the first image and at $X_2 = (x_2, y_2)^T$ in the second image is defined as:

$$ZNCC_{X_1, X_2} = \frac{\sum_i (I_1(X_1 + i) - \bar{I}_1)(I_2(X_2 + i) - \bar{I}_2)}{\sqrt{\sum_i (I_1(X_1 + i) - \bar{I}_1)^2 \sum_i (I_2(X_2 + i) - \bar{I}_2)^2}}$$

where $\bar{I}_1$ and $\bar{I}_2$ are the means of pixel luminance for the given windows centered at X_1 and X_2.

The biggest drawback concerning texture correlation is the need for an aggregation window of a certain size. The optimal size of this window highly depends on local variations of intensity and disparity. The window has to be both large, in order to decrease the influence of noise, and small, in order to avoid projective distortion effects. In practice, texture correlation is often combined with a certain kind of global optimization method, so that the errors caused by choosing window size are controlled.

3   Volumetric Methods

Several scene representations are proposed in different kinds of reconstruction methods, including 2D representations for view morphing [36], 3D volumetric representations [37] and 4D light fields [26]. Among these representations, octree [43] is space-efficient if the scene contains a large transparent or unseen region. Disparity space representation [6, 19] is easy to determine the visibility. Layer representation [2] divides the scene into a collection of parallel planar regions. Volumetric representations have developed quickly since their first introduction in the early 70's in the context of 3D medical imaging. The exponential growth of computational storage and processing during the last three decades have let these representations become more and more useful in computer vision.

The key point of classical two-view stereo vision is finding correspondences accurately and robustly. In the reconstruction process, these correspondences are converted to 3D points or lines in a world space. It has been put forward that correspondence algorithms have some difficulty in reasoning occlusion in image space [37]. Volumetric representations, on the other hand, are based on a world space formulation for which occlusion can be properly solved. Input images are considered as a set of constraints on the final surface, in that a valid model should produce the same images in the given camera positions. Thus, a hypothesized 3D scene model is tested by comparing the real and synthesized images. Moreover, volumetric representations also provide powerful flexibility to represent shapes of arbitrary topologies.

Volumetric representations assume that the scene lies in a big known bounding box and divide this box into small regions [37]. The most common approaches use regular cubes, called voxels, in Euclidean 3D space. The task of a volumetric reconstruction algorithm then, is to identify a set of voxels that accurately represent objects in the scene. For a scene without a transparent object, all voxels can be classified into one of two states, transparent or opaque. Transparent voxels represent empty space. Opaque voxels represent objects in the scene. Finally, surface voxels can be defined as opaque voxels which are adjacent to a transparent voxel. After reconstruction, a polygonal surface of the scene is constructed by the Marching Cubes algorithm [29]. Readers are referred to the survey [39].

3.1   Volumetric Visual Hulls

As mentioned in the above section, with camera calibration information, all silhouette pixels in one image define a generalized cone within which the object must lie. Then, the intersection of this set of generalized cones of all cameras is the visual hull [25, 30, 31]. Thus, the computation of the volumetric visual hull is quite straightforward. The determination about whether each voxel is transparent or not is made by projecting this voxel onto all images and testing whether it is contained by the silhouettes [43]. If all of its projection are foreground pixels, the voxel is opaque; otherwise, it is transparent. For example, the octree representation [40, 43, 14] starts with a low resolution voxel space. Each voxel is then projected onto the reference views. If all the pixels in a voxel's projection in any image are background pixels, the voxel must be transparent, so it is removed, or carved from the volume. If all the pixels in the voxel's projection in all images are foreground pixels, the voxel is opaque, so it remains in the volume. If in one of voxel's projection, there exist both silhouette pixels and background pixels. This voxel is divided into eight smaller voxels for further testing. The algorithm ends when high resolution is reached or all voxels have been classified.

The advantage of volumetric visual hull is its efficiency. It can be designed to finish the task in real time. Its applications includes virtual reality, real time human motion modeling, constructing light fields and lumigraphs [16], and building an initial coarse scene model [9]. Since the 90's, visual hulls have been computed from video streams originating from multiple cameras. There also exist more precise approaches [45, 41] that evaluate local curvature along occluding contours.

3.2   Silhouette Carving from Multiple Images of an Unknown Background

To compute the visual hull of the object, one needs to know the 2D silhouettes, which are often obtained by segmentation or the blue-screen technique. However, these silhouette extraction algorithms rely on a known background. Compared with them, our focus is given to a set of static images taken from different positions and the background is arbitrarily unknown. We describe, in this section, a novel method for visual hull computation (or silhouette extraction) for an unknown background. The details are in [49].

The algorithm is based on the *single-view consistency* criterion. We consider the continuity of the object in each image, as any geometric and radiometric continuity in 3D necessarily induces a continuity in 2D. In other words, we make a *Color-Continuity* assumption, as shown in Figure 1.

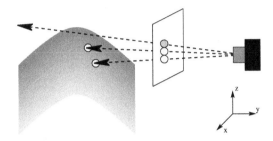

Fig. 1. *Color-Continuity Assumption*: Two adjacent 3D points on the object's surface with similar properties have similar projections on images, which often differ from those of the background

Based on the above color-continuity assumption, the projections of the object and the background are distinguishable. Thus, either foreground or background is a union of a certain set of regions. We have converted the color-continuity into geometric continuity, which guarantees that each reasonable part (region) fully belongs to either the foreground or the background. We call this as single-view consistency. Single-view consistency indicates that the silhouette comprises some region boundaries, which helps to locate the silhouette.

To achieve the goal, we use multiple calibrated images. Each input image is first segmented into regions. The algorithm is a space-carving-like algorithm, which is initialized with a bounding box of the object and gradually carves the inconsistent voxels. In each step, the algorithm first projects the current shape onto the images to form its silhouettes, these silhouettes then shrinks to some region boundaries to satisfy the single-view consistency criterion. The voxels that are projected outside the shrinked silhouettes are carved. The algorithm is iterated until all projections of the current shape are the union of certain regions. The correctness is guaranteed since the shape is always an over-estimation of the object during the process. An evolution example and results are in Figures 2-4.

Fig. 2. Silhouette estimations in the first 9 steps: Note that they are over-estimations. The new estimations are computed based on the previous estimations

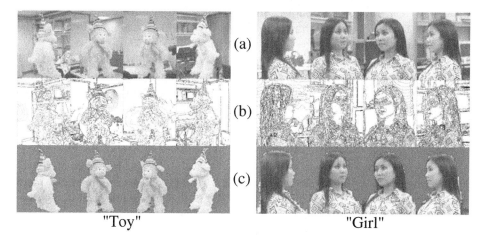

"Toy"   "Girl"

Fig. 3. Results of *Silhouette Carving* Algorithm: (a) Input images, (b) Segmentations, (c) Output silhouettes

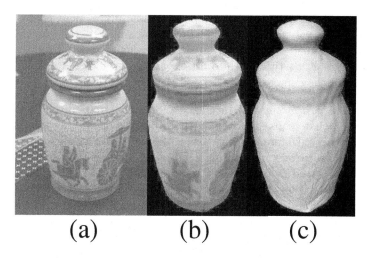

Fig. 4. Result of *Silhouette Carving* Algorithm: (a) One of input images, (b) Result with color, (c) Result without color

3.3   Volumetric Photo Hulls

Compared with silhouette, photo-consistency allows us to get a more accurate model using original intensity information. The volumetric photo hull is computed by gradually carving inconsistent voxels from an initial over-estimation. An important question is whether a voxel that is photo-consistent in the final model might become inconsistent in some step during the process. In fact, it does not happen if a suitable consistency measure is used. The measure must be monotonic [24]: If it finds a set of pixels to be inconsistent, it will find any superset of these pixels to be also inconsistent. Since the algorithm only carves opaque voxels, remaining opaque voxels become more and more visible as the algorithm runs. Thus, if the monotonic consistency measure finds a voxel to be inconsistent, this voxel will be inconsistent in the final model.

The remaining question is how to check the visibility, which is required for photo-consistency computation. As inconsistent voxels are carved, opaque voxels occlude each other from the input images in a complex and constantly changing pattern. Thus the visibility test has to be performed many times, resulting in a very slow process. Computing visibility is a difficult problem and several interesting variations have been developed.

Restricted Camera Placement. To simplify the visibility checking, Seitz and Dyer [37] imposed what they called the ordinal visibility constraint on the camera locations. It requires that all the voxels can be visited in a single scan in near-to-far order relative to every camera, like [8]. A simple example is that the cameras are placed on one side of the scene and voxels are visited in planes from near to far. In this way, the transparency of all voxels that might occlude a given voxel is determined before the given voxel is processed. It guarantees that the visibility of a voxel stops changing before computing the photo-consistency criterion. This results in an efficient Voxel Coloring algorithm [37]. Seitz and Dyer also showed that topological sorting of voxels is possible for any camera configuration in which the scene volume lies outside the convex hull of the cameras' optical centers. Thus, instead of a plane sweep at increasing distances from the cameras, the scene space is partitioned into an expanding front of layers at increasing distances from the cameras' convex hull.

Arbitrary Camera Placement. Kutulakos and Seitz proposed the Space Carving algorithm [24] that scans all voxels for photo-consistency by evaluating a plane of voxels at a time. Different from Voxel Coloring, it checks photo-consistency of each voxel in the current plane using a subset of cameras that are in front of the plane. Also, to speed up and limit the number of plane orientations swept, the planes that are parallel to the sides of the big bounding box are used. Compared with Space Carving, Generalized Voxel Coloring [10] computes exact visibility. It uses a data structure called item buffers (IBs) to record the closest surface voxel along the visual ray of the corresponding pixel. These IBs are computed by rendering all surface voxels, and the visibility of a voxel is found by the IBs of its projections. For efficiency, IBs are updated only after all inconsistent voxels of current surface are carved.

3.4   Robust Carving for Non-lambertian Objects

Most existing techniques discussed in the above section are based on the Lambertian assumption. However, this assumption is clearly insufficient for most common materials that have more complex behaviors like highlight, reflection, transparency or subsurface scattering. Handling all these effects in a unified way still appears unachievable, as some of them duplicate the characteristics of other objects. Here we focus on opaque and non-reflective objects and describe a general reconstruction for non-Lambertian objects without any specific constraint. The range of materials we consider is therefore broad: plastic, unpolished metal, general cloth, skin, and so on. The details of the algorithm are in [47].

We extend the carving criterion by considering a general specularity that introduces both reflectance models and robust statistics. We combine the Lambertian model and the Phong model to describe both view-independent and view-dependent effects: $I(\mathbf{v}) = \alpha + \sum \beta_i (\mathbf{r}_i \cdot \mathbf{v})^\sigma$, where $\mathbf{v}$ is the unit vector in the viewing direction and $\mathbf{r}_i$ is the unit vector in the mirror reflection (maximum reflection) direction of each light source, with α and β_i controlling the quantity of the Lambertian and non-Lambertian effects. σ indicates the shininess of the material. In practice, one light source is enough to model the reflection because of the only significant highlight. Thus, we approximate $\mathbf{r}$ by the viewing direction of the camera in which the point appears the brightest.

Like traditional photo-consistency, the criterion of *multi-view consistency* is defined as the mean deviation between the colors predicted by the reflectance model and the colors actually seen from the images:

$$MVC = \min_{\alpha,\beta_i,\sigma} \left\{ \sqrt{\frac{1}{k} \sum (I_j - I(\mathbf{v}_j))^2} \right\}$$

where k is the number of visible cameras, with I_j being the projection on the jth visible image and $\mathbf{v}_j$ being the corresponding viewing direction.

Finally, we simply ignore the worst matches to allow a significant but not major deviation from the Phong model. We handle image noise: we estimate the criterion through a process that starts with the above classical form and refines it by iteratively singling out samples whose distance to the current model is higher than the others. Thus, the multi-view consistency is defined as follows:

$$MVC_r = \min_{\alpha,\beta,\sigma} \left\{ \sqrt{\frac{1}{k-r} \sum w_j (I_j - I(\mathbf{v}_j))^2} \right\}$$

$$with \;\; w_j = 0, 1 \;\; and \;\; \sum w_j = k - r.$$

The algorithm includes two parts: We first use the multi-view and single-view consistency criteria to robustly reconstruct an overall shape of the object but lacks precision, which we call a *photo-visual hull*. We then use ZNCC to refine this shape into a finely detailed surface. This second step relies on the orientation information of the shape in the first step. We also introduce global and local optimizations to construct the final shape [47], and the results are shown in Figure 5 and 6.

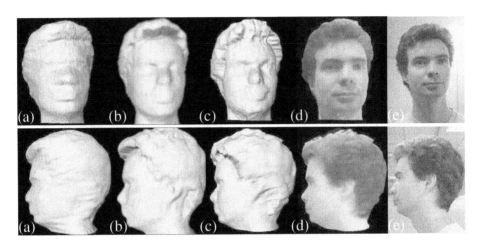

Fig. 5. Result of *Robust Carving* Algorithm: (a) Photo-visual hull, (b) Global optimization, (c-d) Local optimization, (e) Comparison with inputs

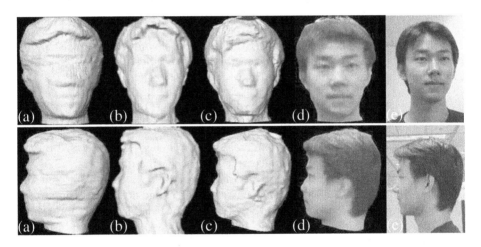

Fig. 6. Result of *Robust Carving* Algorithm: (a) Photo-visual hull, (b) Global optimization, (c-d) Local optimization, (e) Comparison with inputs

4   Graph Cut Methods

Surface reconstruction problems are often naturally expressed in terms of energy minimization. The energy minimization formalism has several advantages. It allows a clean specification of the problem to be solved, and naturally allows the use of soft constraints, such as spatial coherence. However, sometimes the energy is a non-convex function resulting in quite a heavy computational task. Recently, graph cut optimization methods have been developed, which are to construct a specialized graph so that the minimum cut on this graph minimizes the energy.

Suppose $\mathcal{G} = (\mathcal{V}, \mathcal{E})$ is a directed graph with nonnegative edge weights that has two special vertices called the source s and the sink t. A cut $C = S, T$ is a partition of the vertices in $\mathcal{V}$ into two disjoint sets S and T such that $s \in S$ and $t \in T$. The cost of the cut is the sum of costs of all edges that go from S to T:

$$c(S, T) = \sum_{u \in S, v \in T, (u,v) \in \mathcal{E}} c(u, v).$$

The minimum cut is the cut with the minimum cost, which is a classic problem that can be solved very efficiently by max flow algorithms [7] and [15].

Kolmogorov and Zabih [22] gave a characterization of the energy functions that can be minimized by graph cuts. Although their work are restricted to functions of binary variables, it is easily applicable to vision problems that involve large numbers of labels. They claim that, if E is a function of n binary variables from the class $\mathcal{F}^2$, i.e.,

$$E(x_1, x_2, \ldots, x_n) = \sum_i E^i(x_i) + \sum_{i<j} E^{i,j}(x_i, x_j),$$

it is graph-representable if and only if each term $E^{i,j}$ satisfies the inequality

$$E^{i,j}(0,0) + E^{i,j}(1,1) \le E^{i,j}(0,1) + E^{i,j}(1,0).$$

The authors also mentioned a general method to construct a graph to solve the problem. The readers are referred to [22] for the details.

The graph cut methods have been successfully used for a wide variety of vision problems, including image restoration, stereo and motion, image synthesis, image segmentation, voxel occupancy, multi-camera scene reconstruction, and medical imaging. Related to surface reconstruction, Kolmogorov and Zabih [23], Roy [35], and Ishikawa and Geiger [20] proposed direct discrete minimization formulations that are solved by graph cuts. These approaches achieve disparity maps with accurate contours but limited depth precision.

Boykov and Kolmogorov [3] proposed an extension to count for the geometric properties in graph cuts. They show a method for building a grid graph using data segmentation so that the cost of cuts is arbitrarily close to the length (area) of the corresponding contours (surface) for anisotropic Riemannian metric. Paris *et al* [34] also proposed a method to construct a graph to count for geometric information. This leads to an accurate global optimization. However their method only reconstructs the front part of the objects as they handle $z = f(x, y)$ surfaces.

Surface Reconstruction by Propagating 3D Stereo Data in Multiple 2D Images

Surface reconstruction is a natural extension of the point-based geometric methods, which robustly and accurately detect feature points to survive the geometry scrutiny of multiple images. Unfortunately, using 3D data from such a passive system in the same way as range scanner data is often insufficient for a direct surface reconstruction method. The main difficulties are that the 3D points are sparse, irregularly distributed and missing in large areas. On the other hand, most image-based surface reconstruction approaches equally consider all surface points. They ignore the *feature points* although these points can be precisely matched between multiple images and therefore lead to accurate 3D locations. These difficulties and limitations motivated our approach [48], which constructs surface representations from 3D stereo data [27], but using extra 2D image information that is still available from a passive system.

The idea is therefore to first construct small surface patches from some reliable 3D stereo points, then to progressively propagate these patches in their neighborhood from images into the whole surface. The problem reduces to searching for an optimal local surface patch going through a given set of 3D points from images. Since the patch is local, it is reasonable to parameterize it as a single valued height field $z = h(x, y)$. Then, we may look for a surface patch by minimizing a functional of type: $\iint_D c(h(x, y)) dx dy$, where $c(h(x, y))$ is a cost function accounting for the consistency of the surface point $(x, y, h(x, y))$ in multiple images, for instance, either photo-consistency or cross-correlation. We require that the optimized surface patch goes through the existing 3D points if they do exist in the specified neighborhood. We are therefore looking for an interpolating surface through the given 3D points. That is, to minimize

$$\iint_D c(h(x, y)) dx \, dy \quad \text{with} \quad h(x_i', y_i') = z_i',$$

where (x_i', y_i', z_i') is the local coordinates of a given stereo point. Finally, we follow a recent discrete graph-cut approach that reaches sharp results [34]. We simply add first derivative as smoothing terms

$$s(h(x, y)) = \left| \frac{\partial h}{\partial x}(x, y) \right| + \left| \frac{\partial h}{\partial y}(x, y) \right|$$

to minimize the following functional

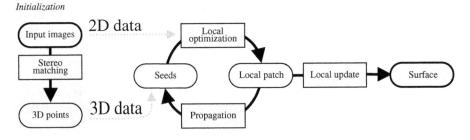

Fig. 7. Surface Propagation is centered on the loop that progressively extends the surface. Each iteration of the loop picks a seed (*i.e.* a 3D point), builds an optimal patch surrounding it, updates the surface and selects new seeds for further propagation. This process is initialized with 3D points computed by a stereoscopic method

$$\iint_{\mathcal{D}} (c(h(x,y)) + s(h(x,y)))\, dx\, dy \quad \text{with} \quad h(x_i', y_i') = z_i'.$$

This constrained optimization for a local surface patch could be nicely handled by a graph cut technique, and thus a detailed surface geometry is obtained by keeping the best 3D and 2D information. A summary of the algorithm is given in Figure 7, and samples reconstructions are shown in Figure 8.

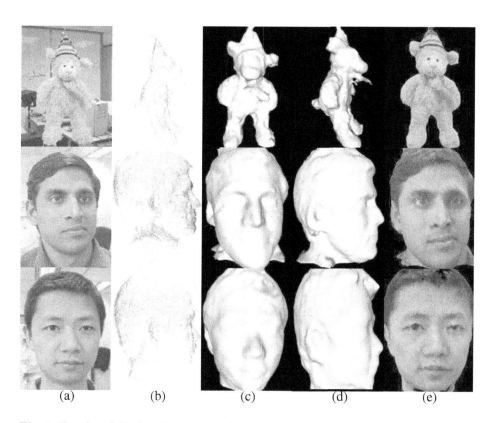

    (a)   (b)   (c)   (d)   (e)

Fig. 8. Results of *Surface Propagation* Algorithm: Each row shows the results for one example: (a) One of the input images, (b) Reconstructed 3D points, (c,d) Surface shape at two different viewpoints, (e) Surface shape with color

5   Level Set Methods

The level set method [33] is another approach to solve the energy minimization problem : the surface is represented by an implicit function, and its evolution toward the expected solution is driven by a steepest-descent defined by a partial differential equation (PDE). The major distinction between this approach and graph cuts is that the problem formulation of the former is continuous and implicit while that of the latter is discrete and parametric. The continuous energy minimization approach is known to lead to smooth surfaces.

The level set method is introduced by Osher and Sethian [33] for capturing moving interfaces (the object surface in our context). The time-evolving surface $\Gamma(t)$ is implicitly represented at time t by a level-set function $\phi(\mathbf{x}, t)$, such that $\phi(\Gamma(t), t) = 0$ and $\mathbf{x} \in R^3$. The value of this function is positive when $\mathbf{x}$ is outside the surface and negative inside. Differentiating the above equation with respect to t, leads to the Euler PDE : $\phi_t + \frac{d\Gamma(t)}{dt} \cdot \nabla\phi = 0$, or

$$\phi_t = -v_{\mathbf{n}}|\nabla\phi|$$

where $v_{\mathbf{n}}$ is the normal velocity of $\Gamma(t)$, to be defined below. Topological changes, accuracy, and stability of the PDE discretization are handled by using the proper numerical schemes developed by Osher and Sethian [33].

The surface Γ is defined by a weighted minimal surface formulation : we seek Γ minimizing the functional

$$p(\Gamma) = \int\int_\Gamma w ds$$

where w is positive (a weight) and ds is the infinitesimal surface element along surface Γ. The solutions of the minimization are given by PDEs : the Euler-Lagrange equation designated by $\nabla p = 0$. The normal velocity is chosen by

$$v_{\mathbf{n}} = -\nabla p.\mathbf{n}$$

where $\mathbf{n}$ is the normal of the evolving surface. Assuming that $w = w(\mathbf{x})$, the resulting expression of the Euler PDE is

$$\phi_t = \nabla w.\nabla\phi + w||\nabla\phi||div\frac{\nabla\phi}{||\nabla\phi||}.$$

This minimal surface formulation was introduced by Caselles and al [4,5] and Kichenassamy and al [21] for 3D segmentation. They define $w = g(||\nabla I||)$ where g is a decreasing function of the image gradient modulus $||\nabla I||$.

Faugeras and Keriven [12] developed a surface reconstruction from multiple images by minimizing the functional $\int\int w ds$ using a weighting function w that measures the consistency of the reconstructed objects reprojected onto 2D images. This measure is taken to be a function of the correlation functions $\rho(\mathbf{x}, \mathbf{n})$ between pairs of 2D images. The correlation function is dependent not only on the position $\mathbf{x}$ of the object surface, but also its orientation $\mathbf{n}$.

In a different context for surface reconstruction from dense and regular sets of scanned 3D point data, Zhao et al. [46] proposed to minimize the functional $\int\int w ds$ using a weighting function w to be the distance function of any surface point $\mathbf{x}$ to the set of 3D data points. Given a set of data points $\mathcal{P}$ and $d(\mathbf{x}, \mathcal{P})$ the Euclidean distance of the point $\mathbf{x}$ to $\mathcal{P}$, the weighting function is simply $w(\mathbf{x}) = d^p(\mathbf{x}, \mathcal{P})$, where p is a scaling factor.

Surface Reconstruction by Integrating 3D and 2D Data of Multiple Views

In this section, we describe a variational approach for surface reconstruction [28] that combines several types of the information, such as 3D stereo points, 2D image and silhouettes. The general methodology that we follow is inspired by all the works referenced in the previous paragraph.

It is important to emphasize that the set of 3D points considered here is not from scanned data like [18, 42, 11, 13, 32, 46, 44], but is derived from the given set of images by the first step of our surface reconstruction method using stereo and bundle-adjustment methods [27]. The "passive" 3D points are difficult to reconstruct by a surface as they are sparser and irregularly distributed and have more missing parts than the "active" data. The goal is to improve the insufficiency of 3D stereo data by using original 2D image information. Also a pure 2D approach like [12, 37, 24] is not satisfactory because of its local minima. We introduce two new intrinsic functionals $\int \int w ds$ which take into account both 3D data points and 2D original image informations, unlike previous works that consider either only 2D image information [12] or only scanned 3D data [46].

The first functional is defined by a weight w, which is a combination of 3D points and the visual hull defined implicitly by the silhouettes in the images :

$$w(\mathbf{x}) = min(d(\mathbf{x}, \mathcal{P}), \epsilon + d(\mathbf{x}, \mathcal{S})),$$

where d is the 3D Euclidean distance function; $\mathcal{P}$ is the set of 3D stereo points; $\mathcal{S}$ is the surface of the intersections of the cones defined by the silhouettes (the visual hull); and ϵ is a small constant favoring 3D points over the visual hull in the neighborhood of 3D points.

If the silhouettes are not available, a second functional is defined by a weight w, which is a combination of 3D points and photo-consistency/correlation function :

$$w(\mathbf{x}) = d(\mathbf{x}, \mathcal{P}) + \lambda e(\mathbf{x}, \mathbf{n}, I)$$

where $e(\mathbf{x}, \mathbf{n}, I)$ is a consistency measure of the reconstructed object in the original 2D image space. This second method assume that background and foreground colors are different.

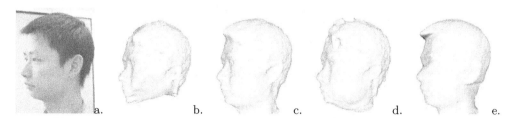

a.   b.   c.   d.   e.

Fig. 9. From left to right : (a) One of the original images; Surfaces computed from (b) 3D points only, (c) 3D points and silhouettes, (d) 3D points and photo-consistency, and (e) silhouettes only

Both methods are initialized by the surface evolution obtained with $w = d(x, \mathcal{P})$, and the surface will not be modified in the immediate neighborhood of $\mathcal{P}$. We also propose a "bounded regularization" method to accelerate the calculations thanks to the stability analysis in [28]. Reconstructed surfaces and experiments summary are shown in Figures 9 and 10.

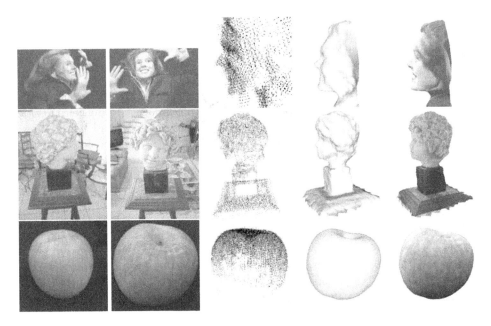

Fig. 10. Each row illustrates one example of reconstruction by showing two frames of the sequence, the reconstructed quasi-dense 3D points [27], Gouraud-shaded surface geometry, and the textured-mapped surface geometry. The Lady and Bust examples are reconstructed from 3D points only. The Apple example is reconstructed from 3D points and silhouettes

6   Conclusion

In this paper, we have discussed several recent approaches of surface reconstructions. They are mainly developed for reconstructing natural, real world scene using arbitrarily positioned off-the-shelf cameras. The discussion has been carried out according to visual cues of silhouette, photometry and texture. Meanwhile, we mainly focus on volumetric methods, graph-cut methods and level-set methods.

We have provided detailed description of the Silhouette Carving algorithm, which is based on the single-view consistency criterion to construct the visual hull without silhouettes. It is a volumetric method that gradually carves inconsistent voxels by projecting them onto the image and checking the continuity of the image regions. Although it only constructs a rough over-estimation, it is computationally efficient, and can be used as the initial shape for further refinements.

The Robust Carving algorithm aims at constructing surface geometries for non-Lambertian objects. It extends the traditional carving criterion by considering a general specularity that introduces both reflectance models and robust statistics. It also combines the Lambertian model and the Phong model to describe both view-independent and view-dependent effects. This results in a multi-view consistency criterion, which has been proved to be quite robust for real scene reconstructions.

Finally, we describe two energy minimization methods that combine both 3D stereo points and 2D image to obtain a more accurate reconstruction. This is motivated by the fact that only robust and accurate feature points that survived the geometry scrutiny of multiple images are reconstructed in space. The shape surface to be reconstructed should be constrained to go through these points. The density insufficiency and the inevitable holes in the stereo data should be filled in by using information from multiple images. These two methods are quite different from each other, since they are based on different formulations. The Surface Propagation algorithm employs a graph cut method to construct local surface patches, then use these patches to propagate the reliable stereo points into the whole surface, resulting in a sharp reconstruction. The Surface Integration algorithm combines multiple data within the level set framework, which produces a more smooth surface and is robust to handle the errors of the inputs.

Future work may be carried out to combine more additional image information, such as the shading and focus, and to exploit more accurate reflection models. An more efficient approach is also needed for a fast reconstruction of detailed surface geometry.

Acknowledgments

This project is supported by the Hong Kong RGC grant HKUST 6182/04E and 6188/02E. We thank Dayton Taylor for providing the Lady sequence. Also thanks to all people for allowing us to use their head images.

References

1. J. Aloimonos, Visual Shape Computation. *Proceedings of the IEEE*, **76(8)**, 1988, pp.899-916.
2. S. Baker, R. Szeliski and P. Anandan. A Layered Approach to Stereo Reconstruction. *Computer Vision and Pattern Recognition*, 1998, pp.434-441.
3. Y. Boykov and V. Kolmogorov. Computing Geodesics and Minimal Surfaces via Graph Cuts. *Int. Conf. on Computer Vision*, 2003, pp.26-33.
4. V. Caselles, R. Kimmel, and G. Sapiro. Geodesic active contours. *Int. Journal of Computer Vision*, **22**, 1997, pp.61-79.

5. V. Caselles, R. Kimmel, G. Sapiro, and C. Sbert. Minimal surfaces based object segmentation. *Trans. on Pattern Analysis and Machine Intelligence*, **19**, 1997, pp.394-398.

6. Q. Chen and G. Medioni. A Volumetric Stereo Matching Method: Application to Image-Based Modeling. *Computer Vision and Pattern Recognition*, 1999, pp.29-34.

7. B. V. Cherkassky and A. V. Goldberg. On implementing the push-relabel method for the maximum flow problem. *Algorithmica*, **19**, 1997, pp.390-410.

8. R.T. Collins. A Space-Sweep Approach to True Multi-Image Matching. *Computer Vision and Pattern Recognition*, 1996, pp.358-363.

9. G. Cross and A. Zisserman. Surface reconstruction from multiple views using apparent contours and surface texture. *Advanced Research Workshop on Confluence of Computer Vision and Computer Graphics*, 2000, pp.25-47.

10. W.B. Culbertson, T. Malzbender, and G. Slabaugh. Generalized voxel coloring. Volume 1883 of *LNCS*, 2000, pp.100-114.

11. B. Curless and M. Levoy. A volumetric method for building complex models from range images. *SIGGRAPH*, 1996, pp.303-312.

12. O. Faugeras and R. Keriven. Complete dense stereovision using level set methods. *European Conf. on Computer Vision*, 1998, pp.379-393.

13. P. Fua. From multiple stereo views to multiple 3d surfaces. *Int. Journal of Computer Vision*, **24**, 1997, pp.19-35.

14. B. García and P. Brunet. 3D Reconstruction with Projective Octrees and Epipolar Geometry. *Int. Conf. on Computer Vision*, 1998, pp.1067-1072.

15. A. Goldberg and S. Rao. Length functions for flow computations. Technical Report 97-055, *NEC Research Institute*, 1997.

16. S.J. Gortler, R. Grzeszczuk, R. Szeliski and M. Cohen. The Lumigraph. *SIGGRAPH*, 1996, pp.43-54.

17. R.I. Hartley and A. Zisserman. Multiple View Geometry in Computer Vision. *CU Press*, 2000.

18. H. Hoppe, T. Derose, T. Duchamp, J. McDonalt, and W. Stuetzle. Surface reconstruction from unorganized points. *Computer Graphics*, **26**, 1992, pp.71-77.

19. S.S. Intille and A.F. Bobick. Disparity-Space Images and Large Occlusion Stereo. *European Conf. on Computer Vision*, 1994, pp.179-186.

20. H. Ishikawa and D. Geiger. Occlusions, discontinuities, and epipolar lines in stereo. *European Conf. on Computer Vision*, 1998, pp.232-248.

21. S. Kichenassamy, A. Kumar, P. Olver, A. Tannenbaum, and A. Yezzi. Gradient flows and geometric active contour models. *Int. Conf. on Computer Vision*, 1995, pp.810-815.

22. V. Kolmogorov and R. Zabih. What Energy Functions can be Minimized via Graph Cuts? *European Conf. on Computer Vision*, 2002, pp.65-81.

23. V. Kolmogorov and R. Zabih. Multi-camera scene reconstruction via graph cuts. *European Conf. on Computer Vision*, 2002, pp.82-96.

24. K.N. Kutulakos and S.M. Seitz. A theory of shape by space carving. *Int. Conf. on Computer Vision*, 1999, pp.307-314.

25. A. Laurentini. The visual hull concept for silhouette-based image understanding. *Trans. on Pattern Analysis and Machine Intelligence*, **16(2)**, 1994, pp.150-162.

26. M. Levoy and P. Hanrahan, Light Field Rendering. *SIGGRAPH*, 1996, pp.31-42.

27. M. Lhuillier and L. Quan. Quasi-dense reconstruction from image sequence. *European Conf. on Computer Vision*, 2002, pp.125-139.

28. M. Lhuillier and L. Quan. Surface Reconstruction by Integrating 3D and 2D Data of Multiple Views. *Int. Conf. on Computer Vision*, 2003, pp.1313-1320.

29. W.E. Lorensen and H.E. Cline. Marching Cubes: A High Resolution 3D Surface Construction Algorithm. *SIGGRAPH*, 1987, pp.163-169.
30. W. Matusik, C. Buehler, R. Raskar, L. McMillan, , and S. Gortler. Image-based visual hulls. *SIGGRAPH*, 2000, pp.369-374.
31. W. Matusik, C. Buehler, R. Raskar, and L. McMillan. Polyhedral visual hulls for real-time rendering. *Eurographics Workshop on Rendering Techniques*, 2001, pp.115-126.
32. P.J. Narayanan, P.W. Rander and T. Kanade. Constructing Virtual Worlds using Dense Stereo. *European Conf. on Computer Vision*, 1998, pp.3-10.
33. S. Osher and J.A. Sethian. Fronts propagating with curvature-dependent speed: Algorithms based on hamilton-jacobi formulations. *Journal of Computational Physics*, **79(1)**, 1988, pp.12-49.
34. S. Paris and F. Sillion and L. Quan. A Surface Reconstruction Method Using Global Graph Cut Optimization. *Asian Conf. on Computer Vision*, 2004.
35. S. Roy. Stereo without epipolar lines : A maximum-flow formulation. *Int. Journal of Computer Vision*, **34**, 1999, pp.147-162.
36. S.M. Seitz and C.R. Dyer. View Morphing. *SIGGRAPH*, 1996, pp.21-30.
37. S.M. Seitz and C.R. Dyer. Photorealistic scene reconstruction by voxel coloring. *Computer Vision and Pattern Recognition*, 1997, pp.1067-1073.
38. J.A. Sethian. Level-set methods and fast marching methods. *CU Press*, 1999.
39. G. Slabaugh, B. Culbertson, T. Malzbender, and R. Schafer. A survey of methods for volumetric scene reconstruction from photographs. *VolumeGraphics 01*, 2001, pp.81-100.
40. S.K. Srivastava and N. Ahuja. Octree generation from object silhouettes in perspective views. *Computer Vision, Graphics and Image Processing*, **49**, 1990, pp.68-84.
41. R. Szeliski and R. Weiss. Robust shape recovery from occluding contours using a linear smoother. Technical Report DEC-CRL-93-7, *Digital Equipment Corporation, Cambridge Research Lab*, 1993.
42. R. Szeliski, D. Tonnesen, and D. Terzopoulos. Modelling surfaces of arbitrary topology with dynamic particles. *Computer Vision and Pattern Recognition*, 1993, pp.82-87.
43. R. Szeliski. Rapid octree construction from image sequences. *Computer Vision, Graphics and Image Processing*, **58**, 1993, pp.23-32.
44. C.K. Tang and G. Medioni. Curvature-augmented tensor voting for shape inference from noisy 3d data. *Trans. on Pattern Analysis and Machine Intelligence*, **24**, 2002, pp.858–864.
45. R. Vaillant and O. D. Faugeras. Using extremal boundaries for 3-D object modeling. *Trans. on Pattern Analysis and Machine Intelligence*, **14(2)**, 1992, pp.157-173.
46. H.K. Zhao, S. Osher, B. Merriman, and M. Kang. Implicit and non-parametric shape reconstruction from unorganized data using a variational level set method. *Computer Vision and Image Understanding*, **80**, 2000, pp.295-319.
47. G. Zeng, S. Paris and L. Quan. Robust Carving for Non-Lambertian Objects. *Int. Conf. on Pattern Recognition*, 2004, pp.119-122.
48. G. Zeng, S. Paris, L. Quan and M. Lhuillier. Surface Reconstruction by Propagating 3D Stereo Data in Multiple 2D Images. *European Conf. on Computer Vision*, 2004, pp.163-174.
49. G. Zeng and L. Quan. Silhouette extraction from multiple images of an unknown background. *Asian Conf. on Computer Vision*, 2004.

Author Index

Lecture Notes in Computer Science

For information about Vols. 1–3442

please contact your bookseller or Springer